# Physical Chemistry in Action

Physical Chemistry in Action presents volumes which outline essential physicochemical principles and techniques needed for areas of interdisciplinary research. The scope and coverage includes all areas of research permeated by physical chemistry: organic and inorganic chemistry; biophysics, biochemistry and the life sciences; the pharmaceutical sciences; crystallography; materials sciences; and many more. This series is aimed at students, researchers and academics who require a fundamental knowledge of physical chemistry for working in their particular research field. The series publishes edited volumes, authored monographs, and upper-level textbooks and encourages contributions from field experts working in all of the various disciplines.

Mineo Hashizume • Yoshiro Imura
Editors

# Colloids and Interfacial Dynamics

## Basics to Application

Springer

*Editors*
Mineo Hashizume
Department of Industrial Chemistry
Tokyo University of Science
Tokyo, Japan

Yoshiro Imura
Department of Industrial Chemistry
Tokyo University of Science
Tokyo, Japan

ISSN 2197-4349 ISSN 2197-4357 (electronic)
Physical Chemistry in Action
ISBN 978-981-96-6727-7 ISBN 978-981-96-6728-4 (eBook)
https://doi.org/10.1007/978-981-96-6728-4



This Springer imprint is published by the registered company Springer Nature Singapore Pte Ltd.
The registered company address is: 152 Beach Road, #21-01/04 Gateway East, Singapore 189721, Singapore

# Preface

When thinking about all the matter (or things) in this world, we can consider the surfaces and interfaces they come into contact with according to various sizes—molecules, nanoparticles, molecular assemblies, thin films, and bulk solids. If these materials behave dynamically, there should also be dynamic behavior on their surfaces and interfaces. While there are systems where the behavior of surfaces and interfaces greatly contributes to the whole material, there are also systems that do not affect the characteristics of the bulk and show surface/interface selective behavior. In other words, the contribution of surfaces and interfaces greatly depends on the size of the material and how its structure is formed. Moreover, understanding dynamic behavior is important not only for a single material but also in the dimension of colloids as the existence form of multiple materials.

This book was planned by the Division of Colloid and Interface Science, Research Institute for Science and Technology (RIST), Tokyo University of Science (TUS). There have been many researchers engaged in research related to colloid and interface science at TUS, and the division has been active for over 40 years. One of the recent main activities is focusing on research activities with interface dynamics as the main axis. For example, when thinking about the dynamics of structure–function correlation in the development of stimulus-responsive materials constructed from molecular assemblies, simply looking at the correspondence between input (stimulus) and output (function expression) makes their relationship a black box, and we cannot understand the details. We believe it is important to understand the linkage of "dynamics of molecules" → "dynamics of molecular assemblies" → "dynamics of the function of molecular assemblies" in response to stimulus input, and we are working on research aimed at understanding this. Through understanding such linkages, it is expected that design guidelines for the development of functional materials with new ideas will be found. This book was planned with the purpose of widely disseminating the importance of such perspectives to the world.

The features of this book are listed below.

- Introduces dynamics of various systems from surfactant assemblies to catalyst surfaces

- Shows commonalities of various systems in terms of dynamics of colloids and interfaces
- Provides a new perspective on nanomaterial development

This book is written by researchers who are leading research related to the dynamics of colloids and interfaces, not limited to members of the Division of Colloid and Interface Science. Research groups of the division members are responsible for half of the chapters in this book. Various systems such as molecular assemblies, thin films, biointerfaces, interface spectroscopy, nanoparticles and their assemblies, photocatalytic interfaces, polymer interfaces/colloids, surfactant assemblies, solid interfaces, etc., are discussed centered on each researcher's area of expertise (not limited to the field of chemistry). We hope that readers will not only deepen their understanding of the details of topics close to their own specialties but also comprehensively overview and understand topics different from their specialties, leading to new discoveries about the coordination of dynamics in various dimensions that were not visible before and leading to the development of new functional materials and understanding of biological or biomimetic systems.

Finally, we would like to express our gratitude to Dr. Shin'ichi Koizumi and Ms. Taeko Sato of Springer for their great efforts in publishing this book. We also thank RIST and TUS for their financial support for the publication of this book.

Tokyo, Japan Mineo Hashizume
Tokyo, Japan Yoshiro Imura

# Contents


1 **Dynamics of Photo-Responsive Systems Consisting of Molecular Self-Assemblies** . . . . . 1
Masaaki Akamatsu, Kenichi Sakai, and Hideki Sakai

2 **Dynamics of Molecules Loaded on Polysaccharide Composite Films** . . . . . 13
Takuya Sagawa and Mineo Hashizume

3 **Spatiotemporal Dynamic Interfaces in Biomaterials** . . . . . 27
Hidenori Otsuka

4 **Ultrafast Vibrational Dynamics of Water at Interfaces** . . . . . 59
Satoshi Nihonyanagi

5 **What Is the Role of Hydration Water in the Self-Assembly of Soft Materials Such as Lipids, Surfactants, Polymers, and Proteins?** . . . . . 75
Mafumi Hishida

6 **Dynamic Self-Assembly of Nanospheres Toward Functional Surfaces** . . . . . 97
Kai Liu and Junhui He

7 **Formation of Noble Metal Nanowires and Their Catalytic and Optical Properties** . . . . . 127
Yoshiro Imura and Clara Imura

8 **Photoexcited Carrier Dynamics and Photocatalytic Activity on Titanium Dioxide Surfaces** . . . . . 141
Kenichi Ozawa, Ryu Yukawa, Keita Hiromori, and Nobuo Nakajima

9 **Carrier Transfer from Photocatalyst to Cocatalyst** . . . . . 171
Ke-Hsuan Wang, Masaaki Yoshida, and Chechia Hu

**10 Dynamics of Polymer Chains Via Interfacial Viscoelasticity Analysis** ........ 189
Yoshihisa Fujii, Kenji Yamaoka, Shintaro Yasuda, and Akira Uchiyama

**11 Design and Applications of Stimuli-Responsive Polymer Interfaces** ........ 211
Takashi Miyata

**12 Roles of Cholesterol in Controlling Characteristics of Catanionic Vesicles** ........ 247
Filip Mravec, Chi-cheng Chiu, and Chien-Hsiang Chang

**13 Stability and Structural Analysis of Foams Formed by Surfactants Using Small-Angle Neutron Scattering** ........ 267
Shiho Yada and Tomokazu Yoshimura

**14 Surfactant-Induced Elution of Plasticizers from Medical Devices** ........ 277
Takehisa Hanawa

**15 Solubility Dynamics and Hydrophobic Interactions of Acidic Analgesics and Basic Anesthetics** ........ 293
Satoru Goto, Yuya Kurosawa, Chihiro Tsunoda, Kanji Hasegawa, Tomohiro Tsuchida, Kazushi Komatsu, and Hiroshi Terada

**16 Antioxidative Mechanism of Radical Scavenger and Local Anesthetics** ........ 313
Tomohiro Tsuchida and Satoru Goto

**17 In Situ Observations of Friction Interfaces** ........ 335
Kaisei Sato, Hikaru Okubo, Shouhei Kawada, Seiya Watanabe, and Shinya Sasaki

**18 Friction Dynamics on Human-Skin Surfaces** ........ 349
Yoshimune Nonomura

**Index** ........ 365

# Chapter 1
# Dynamics of Photo-Responsive Systems Consisting of Molecular Self-Assemblies

**Masaaki Akamatsu, Kenichi Sakai, and Hideki Sakai**

**Abstract** Controlling the morphology of the molecular assemblies formed by surfactants via photoirradiation enables the controlled release of incorporated substances, which can be applied to delivery systems for drugs and active ingredients. Micelles formed by azobenzene-based cationic surfactants exhibited changes in the solubilization capability of model perfumes upon photoirradiation. This controlled release was derived from changes in micelle formation by photoirradiation using small-angle neutron scattering (SANS) analysis. Conventional photo-responsive surfactants and molecular assemblies have slow response rates, making it difficult to control their functions at desired times. The recombination reaction was accelerated using the confined inner space of the micelles formed by the surfactants. In situ SANS measurements revealed rapid morphological changes in the micelles formed by amphiphilic lophine dimers upon ultraviolet (UV) light irradiation. Moreover, rapid controlled release of a model drug was achieved by UV light irradiation using photo-responsive micelles. By focusing on the dynamics of photo-responsive systems, the rapid system can be realized for the controlled release of drugs at the desired time, thereby developing an efficient and precise delivery system for active ingredients (such as drugs, fragrances, and nutrients).

**Keywords** Surfactant · Self-assembly · Stimuli-responsiveness · Photochromism

---

M. Akamatsu (✉)
Research Institute for Sustainable Humanosphere, Kyoto University, Kyoto, Japan
e-mail: akamatsu.masaaki.6t@kyoto-u.ac.jp

K. Sakai · H. Sakai
Department of Pure and Applied Chemistry, Tokyo University of Science, Tokyo, Japan
e-mail: hisakai@rs.tus.ac.jp



M. Hashizume, Y. Imura (eds.), *Colloids and Interfacial Dynamics*, Physical Chemistry in Action, https://doi.org/10.1007/978-981-96-6728-4_1

## 1.1 Introduction

Biological functions are dynamically regulated in response to chemical and physical stimuli. For example, objects can be seen, because the photoisomerization of retina in rod cells upon light stimulation is immediately recognized by the brain as visual information [1]. Photosynthesis [2] and the germination of plant seeds under certain environmental conditions (such as temperature, light, and moisture) [3] are examples of excellent stimuli-responsive molecular systems. If molecules and their molecular assemblies exhibit stimulus-responsive properties, it is possible to control their desired physical properties at any given time. However, to fully understand the function of a stimulus-responsive molecular assembly and to design its function, it is necessary to understand not only the final amount of change but also the process or dynamics.

Introducing a stimuli-responsive moiety into an amphiphilic molecule enables the control of the structure of molecular assemblies through the application of external stimuli [4]. This stimuli-responsive amphiphile or surfactant can control the interfacial properties (such as surface/interfacial tension, wettability, and dispersibility) at the required times. In addition, the active ingredients (such as drugs, fragrances, and nutrients) incorporated into the molecular assemblies can be released by external stimuli at the desired time (Fig. 1.1a). Light, electricity, magnetic fields, and changes in pH and temperature have been used as the external stimuli. Among these, "light" is clean, has excellent wavelength selectivity and temporal/spatial resolution, and is an excellent external stimulus for dynamically controlling the physical properties of materials [5, 6].

Photochromic molecules that undergo a reversible structural change upon light irradiation and change their color tone and other photo- and electronic properties are often used as photo-responsive moieties. Examples include photoisomerization of azobenzene (*trans–cis*), stilbene (*trans–cis*, dimerization), diarylethene (open ring–closed ring), and spiropyran (spiropyran-type–merocyanine-type) (Fig. 1.1b) [5, 6].

Shinkai et al. designed a surfactant with an azobenzene moiety as the photo-responsive moiety and reported that the micellar structure changed upon light irradiation [7, 8]. Hatton et al. found that the balance of the hydrophilic and hydrophobic groups (chain length) of azobenzene-modified nonionic surfactants significantly affected the surface tension of aqueous solutions upon photoirradiation [9]. The structural change in the micelles was used to control the thermal conductivity of the solution by photoirradiation and was found to be applicable in efficient heat-exchange systems. Thus, by combining photo-responsive molecules with various characteristics, in addition to the interfacial properties and molecular assembly ability of surfactants, photo-responsive surfactants and molecular assemblies with desired functions can be developed. In this chapter, we introduce our recent research, focusing on the dynamics of changes in the physical properties of photo-responsive molecular assemblies.

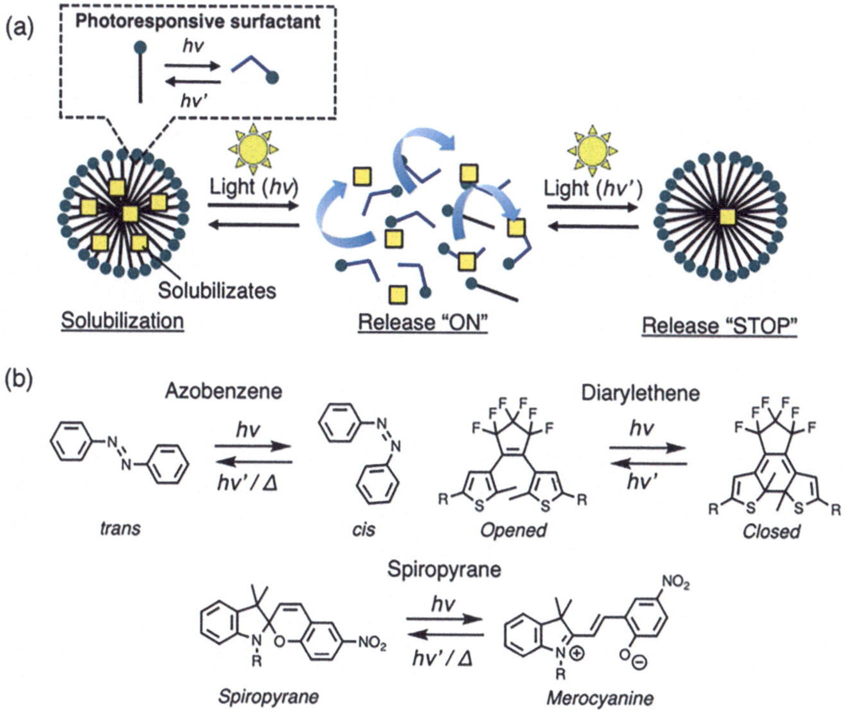


**Fig. 1.1** (**a**) Controlled release of solubilizates with the micelles formed by photo-responsive surfactants as an example of stimuli-responsive system. (**b**) Photochromism of azobenzene, diarylethene, and spiropyrane as typical photochromic compounds

## 1.2 Azobenzene-Based Cationic Surfactant

A cationic surfactant containing an azobenzene moiety, as shown in Fig. 1.2a, was used as the photo-responsive surfactant (4-butylazobenzene-4′-(oxyethyl)trimethylammonium bromide: AZTMA). AZTMA undergoes *trans–cis* isomerization under ultraviolet (UV) or visible light irradiation in aqueous solutions [10]. The critical micelle concentration (cmc) of AZTMA was 2.7 mM for the *trans* form and 8.2 mM for the *cis* form, indicating a decrease in the micelle formation ability due to photoisomerization. This can be attributed to the change in the morphology and polarity of the azobenzene moiety upon photoisomerization. These results reveal that the micelle formation ability could be controlled by irradiation with UV and visible light. Furthermore, we succeeded in controlling the solubilization of model flavors such as ethylbenzene using photoirradiation [11]. This depends on the micelle formation ability, that is, the concentration and morphology of the micelles, and was analyzed using small-angle neutron scattering (SANS) to verify the details, including the time variation of the micelle structure. SANS analysis provides information on the structure of molecular assemblies (shape, size, number, and interactions) within several tens to hundreds of nanometers in size.

SANS measurements of 5 mM AZTMA in $D_2O$ are shown in Fig. 1.2b. Model analysis of the scattering profile showed that the micelle shape was ellipsoidal, and its long and short radii ($r_a$ and $r_b$) were ~ 39 and ~ 21 Å, respectively [11]. Next, when the solution was irradiated with UV light to promote photoisomerization from the *trans* to the *cis* form, the scattering intensity decreased nearly to the background level (Fig. 1.2c), confirming the deformation of the micelles. Subsequent isomerization to *trans* by visible light irradiation restored the profile to its initial state. From these results, it was confirmed that the collapse and formation of micelles were reversible upon photoirradiation in 5 mM AZTMA solution, a concentration near the cmc level.

The same analysis was performed for 10 mM AZTMA in $D_2O$, a concentration that exceeded the cmc of the *trans* and *cis* forms. The obtained profiles were fitted as ellipsoids with sizes comparable to those of the 5 mM system. By contrast, UV irradiation decreases its long radius and volume fraction corresponding to the number of micelles (Fig. 1.2d). Although the volume fraction decreased as in the 5 mM system, it did not reach zero, and the micelles did not completely deform. This is because the concentration of 10 mM *cis*-AZTMA was higher than the cmc of *cis*-AZTMA (8.2 mM) obtained by UV irradiation. Subsequent visible light irradiation restored these values to their pre-irradiation values.

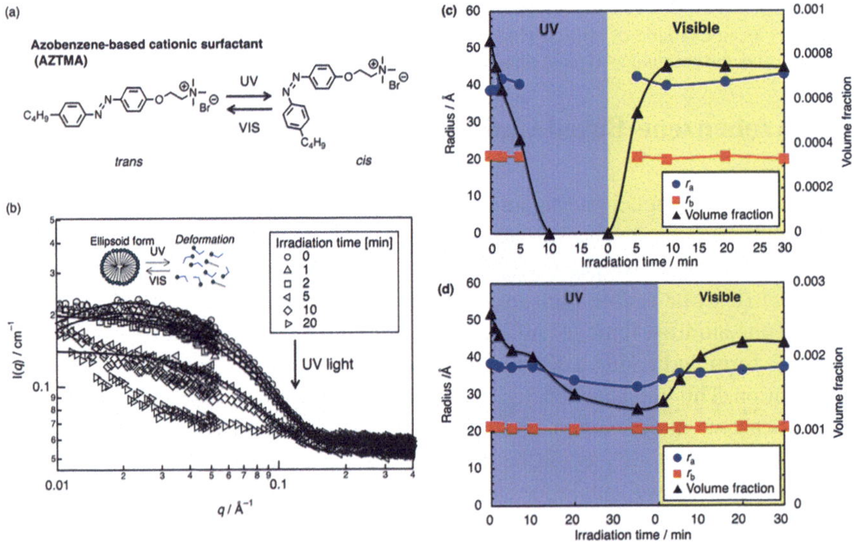


**Fig. 1.2** (**a**) Chemical structure and photochromism of an azobenzene-based cationic surfactant (AZTMA). (**b**) Variations in SANS profiles of 5 mM AZTMA in $D_2O$ upon UV light irradiation. Variation in SANS fit values for 5 and 10 mM (**c**, **d**). Reprinted with permission from ref. 11. Copyright 2015 American Chemical Society

From the above results, it was found that the previous examples of photo-controlled solubilization of ethylbenzene originated from changes in the number density of the micelles, as indicated by the volume fraction. In addition, it was also found that the micelle formation or morphology of azobenzene-based cationic surfactant, AZTMA, was controlled on the order of "minutes" upon photoirradiation.

## 1.3 Acceleration of Photochromism of Lophine Dimer Using the Inside of the Micelle

Several studies have been conducted to control the molecular assembly structures using photo-responsive surfactants. However, in the cases reported to date, including our results above, light irradiation on the order of minutes to hours is required to change the structure of molecular assemblies, and it is difficult to express the desired function at any given moment. For example, if this limitation can be overcome, the required amount of drug can be released into the affected area at the desired time, and an efficient, precise, and on-demand drug delivery system can be realized.

To achieve this, it is necessary to control the structure of the molecular assemblies at high speeds. Therefore, we focused on photochromic molecules called lophine dimers (Fig. 1.3). Upon UV irradiation, this molecule dissociates into two lophyl radicals, and these active radical species thermally recombine. However, the

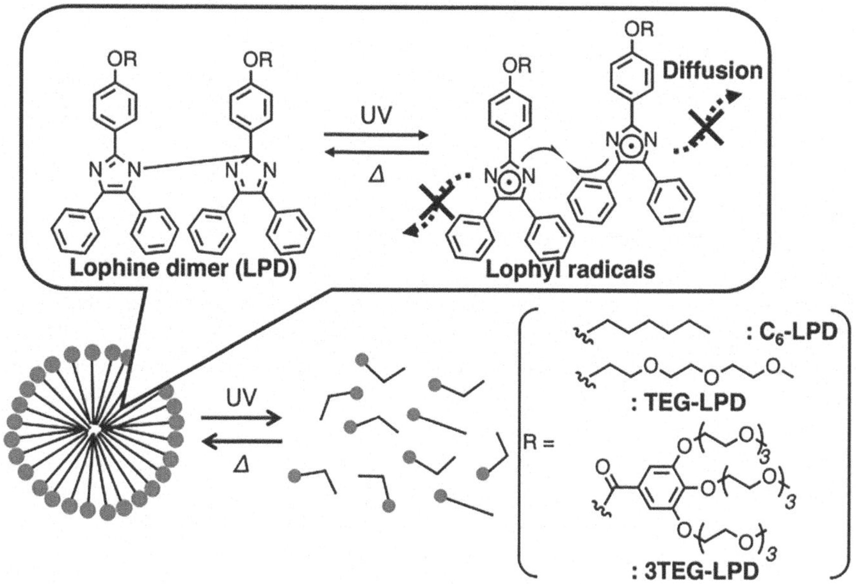


**Fig. 1.3** Photochromism in the micelles and chemical structures of lophine dimer derivatives

recombination reaction is extremely slow because lophyl radicals diffuse freely in solution. Abe et al. reported that the cross-linking of the lophine skeletons by a covalent bond to suppress the diffusion of lophyl radicals accelerated the recombination reaction from the millimeter to the femtosecond scale [12]. Based on this finding, the author previously used the confined space inside the micelles to suppress the diffusion of lophyl radicals and improve the speed of the recombination reaction (Fig. 1.3).

The solubilization of alkylated lophine dimer derivatives ($C_6$-LPD) in aqueous micellar solutions of cetyltrimethylammonium bromide (CTAB) enhanced the rate of recombination of lophyl radicals by more than 100-fold compared to organic solvents, where the radical species can diffuse freely [13] (Fig. 1.3). Furthermore, when a lophine dimer derivative containing a triethylene glycol (TEG) group (TEG-LPD) as a nonionic hydrophilic group was solubilized in CTAB micelles to increase its solubilization capability into the micelles, the recombination rate was further enhanced [14]. This indicated that the recombination reaction was successfully accelerated inside the micelles.

To further confirm that the confined space inside the micelles inhibited the diffusion of lophyl radicals, we performed the same evaluation using aqueous micelle solutions of different sizes prepared using cationic surfactants with alkyl chains of various lengths ($C_{12}$TAB, $C_{14}$TAB, and $C_{16}$TAB). SANS measurements (0.5 mM TEG-LPD/50 mM surfactant in $D_2O$) were performed using the SANS instrument TAIKAN (BL-15) at the MLF of J-PARC in Japan to clarify the morphology and size of the micelles. The SANS data fitted well using a uniform long ellipsoid model interacting via a screened Coulomb potential [15]. As the alkyl chain length increased, the long radius ($r_a$) and short radius ($r_b$) increased, meaning the increase in the volume of the individual micelles ($V_m$).

The apparent recombination rate ($k'$) of the lophyl radicals in these micelles was then examined by UV/Vis absorption spectroscopy, and $k'$ was plotted against the micelle volume $V_m$ obtained by SANS analysis (Fig. 1.4). The results showed that the recombination reaction was enhanced in micelles with smaller volumes. To observe a clear trend toward this assumption, the free volume per lophyl radical in the micelle ($V_f$) was calculated according to $V_m$. This value is consistent with this trend, indicating that the $C_{12}$TAB micelles have the smallest volume required for the diffusion of lophyl radicals. The kinetic parameters were also evaluated, and it was understood that smaller micelles have a higher randomness of lophyl radicals, which promotes the reaction probability. From these results, it was found that the confined space inside the micelles and limited diffusion of lophyl radicals promoted the photochromic reaction. These results indicate that the internal volume of the micelles as a recombination reaction space can be adjusted by choosing the surfactant.

Based on these results, we attempted to construct micelles formed solely from lophine dimers to increase the concentration of lophine dimers inside the micelles and reduce the diffusion distance of the lophyl radical. Therefore, an amphiphilic lophine dimer with multiple TEG groups (3TEG-LPD) was designed and synthesized [14] (Fig. 1.3). From surface tension measurements of the synthesized 3TEG-LPD solution, cmc and $\gamma_{cmc}$ were determined to be 0.80 μM and 46.5 mN/m,

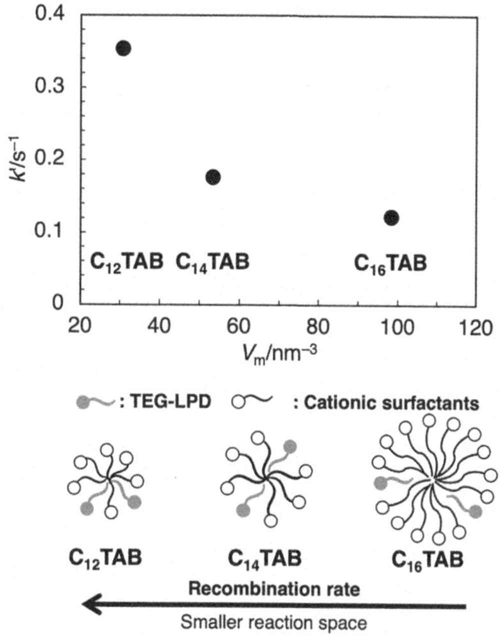


**Fig. 1.4** Apparent reaction rates of recombination of the lophyl radicals ($k'$) versus the volume of the micelle with different alkyl chain length of cationic surfactants and schematic of micellar structures formed between TEG-LPD and cationic surfactants. Reprinted with permission from ref. 15. Copyright 2022 Elsevier

respectively. Dynamic light scattering measurements suggested that TEG-LPDs formed micelles with diameters of approximately 7.4 nm. Similarly, in aqueous 3TEG-LPD micelle solutions, the recombination of lophyl radicals produced by UV irradiation was approximately 800 times more accelerated than that in organic solvents.

These results indicated that the use of the confined space inside the micelles facilitated the recombination of lophyl radicals produced by photoirradiation (Fig. 1.3). The recombination rate was dependent on the concentration of the lophine dimer inside the micelle. This indicates that the distance between active radical species affects the collision frequency. Thus, the aggregation state inside the molecular assembly is a suitable reaction field for the high-speed photoisomerization of the lophine dimer in order to achieve high speed.

Next, we investigated the rapid photoisomerization of 3TEG-LPD to control its interfacial properties. A 5 mM solution of 3TEG-LPD was irradiated with UV light, and the surface tension decreased in seconds [14]. When light irradiation was stopped, the surface tension recovered to its original value in a similar manner within seconds. This suggests that the lophyl radicals produced by light irradiation pack tightly at the air/water interface and stabilize it. In addition, the proximity of these radical species to the Gibbs membrane and molecular assemblies at the air/water interface promotes rapid recombination reactions.

In conclusion, we succeeded in accelerating the recombination reaction of lophyl radicals and in the high-speed control of surface tension with photoirradiation by utilizing the confined space inside molecular assemblies.

## 1.4 Rapid Control of the Morphology of Micelles Formed by Amphiphilic Lophine Dimer with Photoirradiation

Morphological changes and its dynamics of micelles formed by the amphiphilic LPD derivative (3TEG-LPD) upon UV light irradiation were investigated. However, conventional laboratory-scale analytical techniques for colloidal particles (e.g., light scattering and laser diffraction) cannot provide accurate information on the structure of molecular assemblies because of the time resolution and overlapping optics for photochromic reactions and analyses. To overcome this problem, we developed an in situ SANS system (TAIKAN, MLF, J-PARC, Japan) equipped with a lamp and a UV–visible absorption spectrophotometer to monitor the photochromic reaction and morphology of the photo-responsive molecular assemblies [16]. Figure 1.5a shows the SANS profile of 3TEG-LPD in $D_2O$. The experimental data were well-fitted using a uniform ellipsoid model with a long radius ($r_a$) of ~47 Å and a short radius ($r_b$) of ~28 Å [17]. After UV irradiation, the long radius ($r_a$) increased

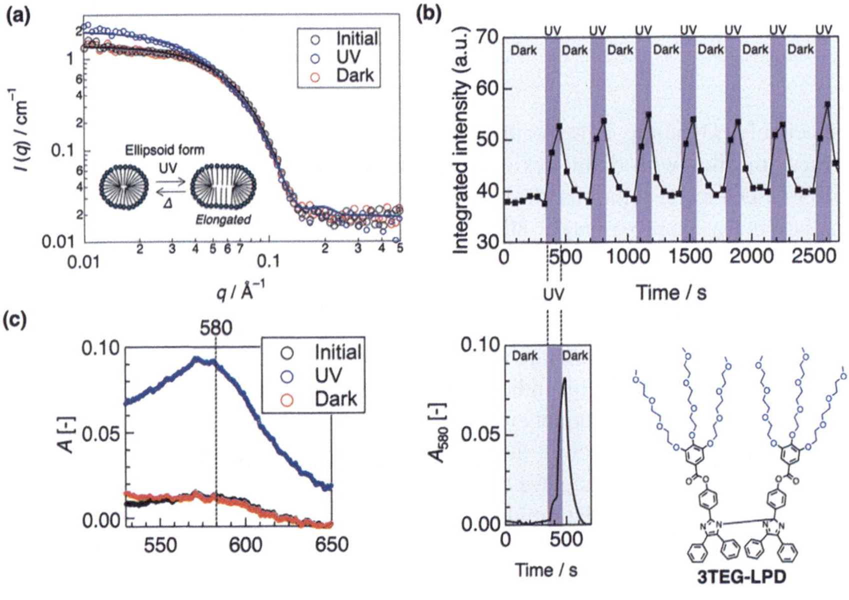


**Fig. 1.5** In situ SANS profiles and UV/Vis absorption data. (**a**) SANS profiles (actual data and fitted curves) of 10 mM 3TEG-LPD in $D_2O$ before and after 2 min UV light irradiation and 4 min idle time in the dark. (**b**) Integrated scattering intensity of the SANS profiles of 3TEG-LPD in the *q*-range of 0.01–0.05 $Å^{-1}$ during multiple cycles of irradiation and idle periods in the dark. (**c**) UV/Vis absorption spectra before and after UV irradiation and idle storage in the dark, and the temporal changes in absorption at 580 nm during the cycle using an in situ UV/Vis absorption spectrometer. Reprinted with permission from ref. 17. Copyright 2021 Springer Nature

to ~70 Å, which is consistent with the SANS profile of the micelle. This indicated that UV light irradiation elongated the long radius of the micelles owing to a photo-induced morphological change in the molecular structure between the dimer and monomer.

To see the time course of the morphological changes in the SANS profile, the integrated scattering intensity in the $q$ region of 0.01–0.05 $Å^{-1}$ was plotted versus time for a cycle of 2 min UV irradiation and 4 min dark rest, as shown in Fig. 1.5b. The results confirmed that the elongation of the elliptical micelles was completed after 60 s of irradiation and reversed after 60 s of dark incubation. UV/Vis absorption spectra were collected simultaneously during the SANS measurements. As shown in Fig. 1.5c, a new absorption band originating from the lophyl radical is observed at 580 nm during UV irradiation. As shown in Fig. 1.5b, c, a new absorption band originating from the lophyl radical is observed at 580 nm during UV irradiation. The change in the 580 nm absorption band during the irradiation cycle revealed that the morphological changes of the LPD micelles and photochromism occurred without a time lag within the analytical time resolution (Figs. 1.5b, c). Based on these results, the developed in situ SANS system was successfully used to observe the rapid morphological changes in LPD micelles.

## 1.5 Rapid Control of Solubilization/Release Using Micelles Formed by Amphiphilic Lophine Dimer with Photoirradiation

We investigated the controlled release of solubilized materials using these fast, photo-responsive 3TEG-LPD micelles. As a model drug, fluorescent calcein was solubilized in an aqueous solution of the 3TEG-LPD micelles. The fluorescence spectra showed a calcein-derived peak at 545 nm, which shifted to longer wavelengths and decreased in intensity upon UV irradiation [17] (Fig. 1.6a). This indicated that calcein was released from the micelles upon photoirradiation, resulting in aggregation and precipitation.

The fluorescence peak intensity plotted against time shows that the fluorescence change was completed earlier than 60 s after UV irradiation (Fig. 1.6b). This indicates that the change in solubilizing capacity was sufficient for the release of calcein before the structural change of the UV-irradiated micelles reached a steady state. Similarly, calcein-release experiments were conducted using an amphiphilic lophine dimer with 12 TEG groups (6TEG-LPD), which did not undergo micellar structural changes upon photoirradiation. No changes in fluorescence were observed. This confirms that the fluorescence change in the 3TEG-LPD micelle system is not caused by the photodegradation of calcein but by light-irradiation-induced changes in the micelle structure. In conclusion, we found that 3TEG-LPD micelles can rapidly release solubilized drugs.

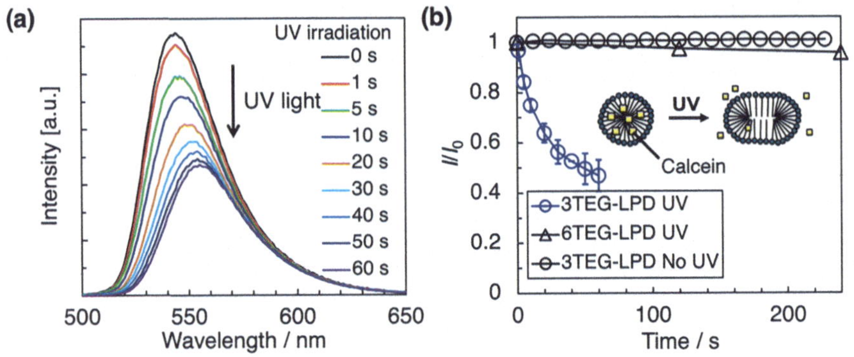


**Fig. 1.6** (**a**) Variations in fluorescence spectra of 1.0 mM calcein/5.0 mM 3TEG-LPD aqueous solution upon UV light irradiation. (**b**)The normalized transient changes in fluorescence intensity at the peaks for 1.0 mM calcein/5.0 mM 3TEG-LPD and 6TEG-LPD aqueous solutions during UV irradiation. Reprinted with permission from ref. 17. Copyright 2021 Springer Nature

## 1.6 Conclusion

In this chapter, we present our recent research on the construction of photo-responsive molecular assemblies and the dynamics of the photoirradiation-induced changes in their physical properties. We found that the solubilization of a micellar aqueous solution of an azobenzene-modified cationic surfactant (AZTMA) in ethylbenzene changed reversibly upon photoirradiation. Furthermore, small-angle neutron scattering (SANS) analysis revealed that the change in micelle size and volume fraction with photoirradiation occurred on the order of minutes, indicating that the change in solubilization was due to a change in micelle number density.

To enhance the photo-response speed, we constructed on-demand photo-responsive molecular assemblies using lophine dimer derivatives. The recombination of lophyl radicals produced from lophine dimers upon UV light irradiation was promoted in the confined spaces inside the micelles. This was because the free diffusion of lophyl radicals was suppressed in the confined space, and the frequency of the reaction was enhanced. Next, we used this finding to synthesize and evaluate amphiphilic lophine dimers. In situ SANS measurements showed that the micelles formed by the amphiphilic lophine dimer (3TEG-LPD) underwent a morphological change on the order of seconds upon photoirradiation. Rapid photoirradiation of the model drug was observed. In conclusion, we successfully constructed fast photo-responsive molecular assemblies by utilizing the inner space of micelles formed by surfactants. By taking advantage of its rapid photo-response, this system is expected to be applied to on-demand delivery systems of drugs and active ingredients, as well as actuators and heat-exchange materials. However, in this system, different hierarchical phenomena, such as photochemical reactions, changes in micellar morphology, and solubilization ability, can occur upon light irradiation. Understanding the dynamics of each hierarchy and their relationships enables us to control for the different dimensions of hierarchy, time, and space.

**Acknowledgments** The authors thank Prof. Gregory G Warr, Dr. Paul A. FitzGerald (University of Sydney), and Dr. Hiroki Iwase (CROSS) for the collaborative work on neutron scattering measurements. This research was partially supported by a JSPS KAKENHI Grant-in-Aid for Early-Career Scientists (Grant Number JP20K15248).

## References

1. Kukura P, McCamant DW, Yoon S, Wandschneider DB, Mathies RA (2005) Structural observation of the primary isomerization in vision with femtosecond-stimulated raman. Science 310:1006–1009
2. Allen JF (2002) Photosynthesis of ATP—electrons, proton pumps, rotors, and poise. Cell 110:273–276
3. Schütz W, Milberg P, Lamont B (2002) Germination requirements and seedling responses to water availability and soil type in four eucalypt species. Acta Oecol 23:23–30
4. Brown P, Butts CP, Eastoe J (2013) Stimuli-responsive surfactants. Soft Matter 9:2365–2374
5. Chen S, Costil R, Leung FK, Feringa BL (2021) Self-assembly of photoresponsive molecular amphiphiles in aqueous media. Angew Chem Int Ed 60:11604–11627
6. Akamatsu M, Sakai K, Sakai H (2023) Dynamic control of interfacial properties and self-assembly with photoirradiation. Chem Lett 52:573–581
7. Shinkai S, Matsuo K, Sato M, Sone T, Manabe O (1981) Photocontrol of micellar catalysis. Tetrahedron Lett 22:1409–1412
8. Shinkai S, Matsuo K, Harada A, Manabe O (1982) Photocontrol of micellar catalyses. J Chem Soc Perkin Trans 2(10):1261–1265
9. Shang T, Smith KA, Hatton TA (2003) Photoresponsive surfactants exhibiting unusually large, reversible surface tension changes under varying illumination conditions. Langmuir 19:10764–10773
10. Orihara Y, Matsumura A, Saito Y, Ogawa N, Saji T, Yamaguchi A, Sakai H, Abe M (2001) Reversible release control of an oily substance using photoresponsive micelles. Langmuir 17:6072–6076
11. Akamatsu M, FitzGerald PA, Shiina M, Misono T, Tsuchiya K, Sakai K, Abe M, Warr G, Sakai H (2015) Micelles structure in a photoresponsive surfactant with and without solubilized ethylbenzene from small-angle neutron scattering. J Phys Chem B 119:5904–5910
12. Kishimoto Y, Abe J (2009) A fast photochromic molecule that colors only under UV light. J Am Chem Soc 131:4227–4229
13. Akamatsu M, Suzuki T, Tsuchiya K, Masaki H, Sakai K, Sakai H (2018) Accelerated recombination of lophyl radicals solubilized in micelles. Chem Lett 47:113–115
14. Akamatsu M, Kobayashi K, Sakai K, Sakai H (2019) Accelerated recombination of lophyl radicals and control of the surface tension with amphiphilic lophine dimers. Chem Commun 55:9769–9772
15. Akamatsu M, Morita T, Kobayashi K, Iwase H, Sakai K, Sakai H (2022) Controlled recombination rate of lophyl radicals in cationic surfactant micelles. Colloids Surf A Physicochem Eng Asp 638:128319
16. Iwase H, Akamatsu M, Inamura Y, Sakaguchi Y, Morikawa T, Kasai S, Oh-uchi K, Kobayashi K, Sakai H (2023) New measurement system based on small-angle neutron scattering for structural analysis of light-responsive materials. J Appl Crystallogr 56:110–115
17. Akamatsu M, Kobayashi K, Iwase H, Sakaguchi Y, Tanaka R, Sakai K, Sakai H (2021) Rapid controlled release by photo-irradiation using morphological changes in micelles formed by amphiphilic lophine dimers. Sci Rep 11:10754

# Chapter 2
# Dynamics of Molecules Loaded on Polysaccharide Composite Films

**Takuya Sagawa and Mineo Hashizume**

**Abstract** Natural polysaccharides have biocompatibility, biodegradability, and nontoxicity, and therefore these can be used as raw materials for biomaterials. Recently, we have developed processes for the fabrication of polysaccharide composite films from polyion complexes composed of chondroitin sulfate C (CS) and chitosan (CHI), denoted as CS/CHI films. The films showed model drug loading and release ability and molecular permeability. On the other hand, the interactions between the molecules and polysaccharides in the films have not been thoroughly evaluated, which should affect the dynamics of these events. In this chapter, the dynamics of molecules in the polysaccharide composite films are described, focusing mainly on two cases. Firstly, the preparation of fluorescein (FL)-loaded polysaccharide composite films (CS/CHI-FL) and evaluation of the ionization and release behavior of FL in CS/CHI-FL were carried out to discuss the dynamics of the loaded molecules in polysaccharide composite films. Furthermore, to clarify the details of different molecular permeation behavior through molecularly imprinted films (IFs), IFs using methylene blue (MB) and a cationic porphyrin derivative (TMPyP) as template molecules (MB-IF, TMPyP-IF) were prepared, and the permeation behavior was investigated when the permeation molecules were changed. These results suggested that the polysaccharide composite films can be applied to pH-responsive materials and drug delivery system (DDS) carriers.

**Keywords** Polysaccharide · Polyion complex · Free-standing film · Molecular loading · Molecular permeation

---

T. Sagawa · M. Hashizume (✉)
Department of Industrial Chemistry, Faculty of Engineering, Tokyo University of Science, Tokyo, Japan
e-mail: mhashizu@ci.tus.ac.jp



M. Hashizume, Y. Imura (eds.), *Colloids and Interfacial Dynamics*, Physical Chemistry in Action, https://doi.org/10.1007/978-981-96-6728-4_2

## 2.1 Introduction

Biomaterials are materials that contact to human body or tissues directly or indirectly and are required to have high biocompatibility. Recently, biomaterials using stimuli-responsive polymers, which respond to external stimuli such as light, temperature, and solution pH, have been reported [1, 2]. Natural polymers have been used as the raw materials for these biomaterials. In particular, natural polysaccharides have received attention because of their high biocompatibility and biodegradability. Notably, stimuli-responsive drug carriers and cell scaffolds made of polysaccharides have been developed [3, 4].

Polysaccharides can be classified into water-insoluble and water-soluble polysaccharides. Water-insoluble polysaccharides include cellulose and chitin and are difficult to handle and mold into desired shapes on the microscale. On the other hand, water-soluble polysaccharides consist of nonionic polysaccharides, such as amylose and pullulan, and ionic polysaccharides. The representative ones in the latter are chitosan (CHI) and sodium chondroitin sulfate C (CS) (Fig. 2.1), which have the advantage of being easier to handle than water-insoluble polysaccharides [5]. Further, these water-soluble polysaccharides have cationic or anionic functional groups in their sugar units, and they behave as polyions (also called polyelectrolytes) under appropriate pH conditions. Therefore, they can electrostatically interact with inorganic salts, proteins, or oppositely charged polysaccharides [6–8], and also charged water-soluble polysaccharides. The resulting composites can be used as pH-responsive materials. However, materials composed of water-soluble polysaccharides do not have sufficient mechanical strength and durability for biomaterials. To give good mechanical properties and high durability, the formation of polyion complexes (PICs) has been investigated [9, 10]. When CHI, a cationic polysaccharide, and CS, an anionic polysaccharide, were mixed, PICs were formed through multivalent electrostatic interactions. PIC formation does not require cross-linking agents or chemical modifications, and therefore PICs maintain the intrinsic properties of the polysaccharide itself [10]. Thus, structural materials with various properties can be obtained by formation of PICs with water-soluble polysaccharides.

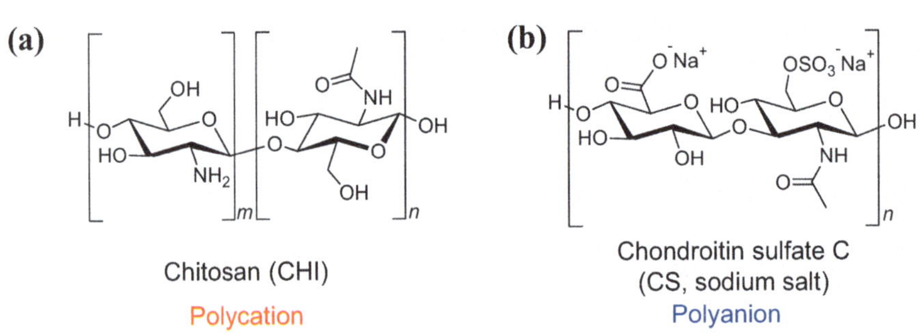


**Fig. 2.1** Chemical structures of (**a**) chitosan (CHI) and (**b**) chondroitin sulfate C sodium salt (CS)

Biocompatible films are often used as biomaterials such as cell scaffolds and drug carriers. Such films have been fabricated by solution casting and layer-by-layer (LbL) assembling techniques [11, 12]. However, the control of the thickness and density in the casting method is difficult, and the LbL assembling generally provides films with the thickness of a few tens of nanometers. Therefore, the fabrication of films on the order of micrometers takes a long time. On the other hand, our group has reported the fabrication of free-standing polysaccharide composite films from PICs consisting of CS and CHI (CS/CHI films) by hot press techniques (Fig. 2.2) [13–20]. The simple method can fabricate free-standing films with an easy handle. The CS/CHI films do not dissolve in water, but swell instead. The swelling property depends on the species of polysaccharide and the pH of the solution. Therefore, the polysaccharide composite films can be used as pH-responsive materials. We have also investigated the functions of these polysaccharide composite films. For example, the loading/release properties of model drug molecules such as methylene blue (MB) (Fig. 2.2) [15, 16]. We also evaluated the molecular permeability of the films, and the permeability depended on electrostatic interactions of the permeant with the polysaccharides in the film, unlike cellulose dialysis membranes [17]. These properties were correlated with the swelling ratio of the film and the pH of the solution. The fabrication of model drug-loaded polysaccharide composite films, their release behavior, and the molecular permeability of the films have been investigated in detail. However, the interactions between the loaded molecules and polysaccharides in the films have not been thoroughly evaluated, especially in terms of their dynamics. A sufficient understanding of this aspect of the films is expected to lead to a detailed understanding of molecular loading, release, and permeation, and to expand the range of applications of biomaterials.

In this chapter, we select and describe the following two cases to discuss the dynamics of the loaded molecules in polysaccharide composite films. The first one is the preparation of fluorescein (FL)-loaded polysaccharide composite films (CS/CHI-FL) and the evaluation of the ionization behavior of FL in CS/CHI-FL [18]. FL

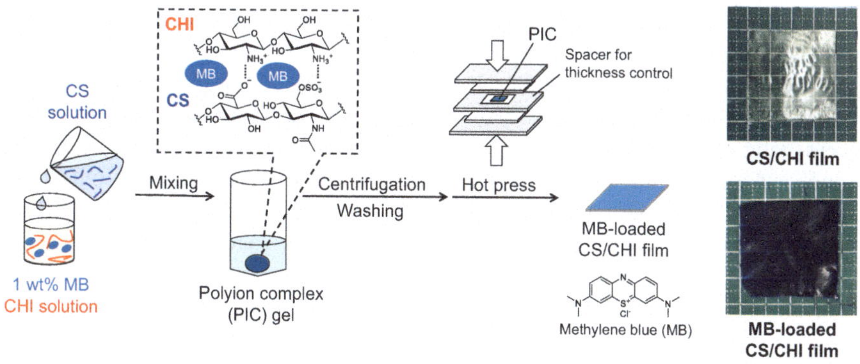


**Fig. 2.2** Schematic illustration for the fabrication of MB-loaded films and photographs of CS/CHI film and MB-loaded CS/CHI film

shows cationization or anionization of the functional groups contained in the molecule, which changes its spectroscopic properties. The ionization degree of FL depends on the pH of the solution. Therefore, the interaction between FL and polysaccharides in the film can be evaluated by the change in the color and fluorescence of the film using FL as the probe molecule. Furthermore, the molecularly imprinted film (IF), whose preparation is utilized the interaction between the loaded molecules and polysaccharides in the film [19, 20], is described as the second topic. The IFs show different molecular permeation behaviors from the film without molecular imprinting (non-IF). To clarify its details, polysaccharide composite films using MB and a cationic porphyrin derivative (TMPyP) as the template molecules (MB-IF, TMPyP-IF) were prepared, and the permeation behaviors of these films were evaluated using several different kinds of permeation molecules.

## 2.2 pH Responsiveness of Molecules Loaded on Polysaccharide Composite Films

CS/CHI films showed the loading and release abilities of model drug molecules, and the release rate of the molecules depended on the pH of the medium [15, 16]. When a molecule is incorporated in the film, the interaction between the molecule and polysaccharides in the film should occur, which should affect the function or property of the molecule compared to those in solutions. To clarify this point, fluorescein (FL)-loaded polysaccharide composite films were prepared. The pH responsiveness of FL was utilized to evaluate the interaction between FL and polysaccharides in the films [18]. It is known that the structure of FL, in other words, the degree of ionization, changes depending on the pH of the solution (Fig. 2.3a,b). The FL-loaded films (CS/CHI-FL films) were prepared in the same manner as CS/CHI films using hot press techniques, except that FL was added to the solution used for the preparation of polysaccharide PICs. The obtained films appeared pale yellow in color and emitted fluorescence upon irradiation with UV light at 365 nm, which reflected the spectroscopic properties of FL (Fig. 2.3c). When the films were immersed in phosphate buffer solutions at pH 2.0, 3.0 (actually they are not buffers), 5.8, 7.4, and 8.0, a portion of the FL molecules were released from the films. The FL release behavior changed depending on the pH of the solution. This was due to the change of the electrostatic interaction between polysaccharides (the ionization degree of carboxy groups in CS and amino groups of CHI changed in these pH ranges), the ionization structure of FL, and the interaction between FL and polysaccharides, under different pH conditions. Although the parts of FL molecules were released, more than half of the FL molecules still remained in the films even after the release experiment, and FL-derived fluorescence was observed in all films upon UV light irradiation (Fig. 2.3d). Based on the ionic ratio of FL in each pH solution, the film was expected to exhibit blue emission at pH 2.0, weak green fluorescence at pH 3.0 and 5.8, and strong green fluorescence at pH 7.4 and 8.0. However, in practice, blue

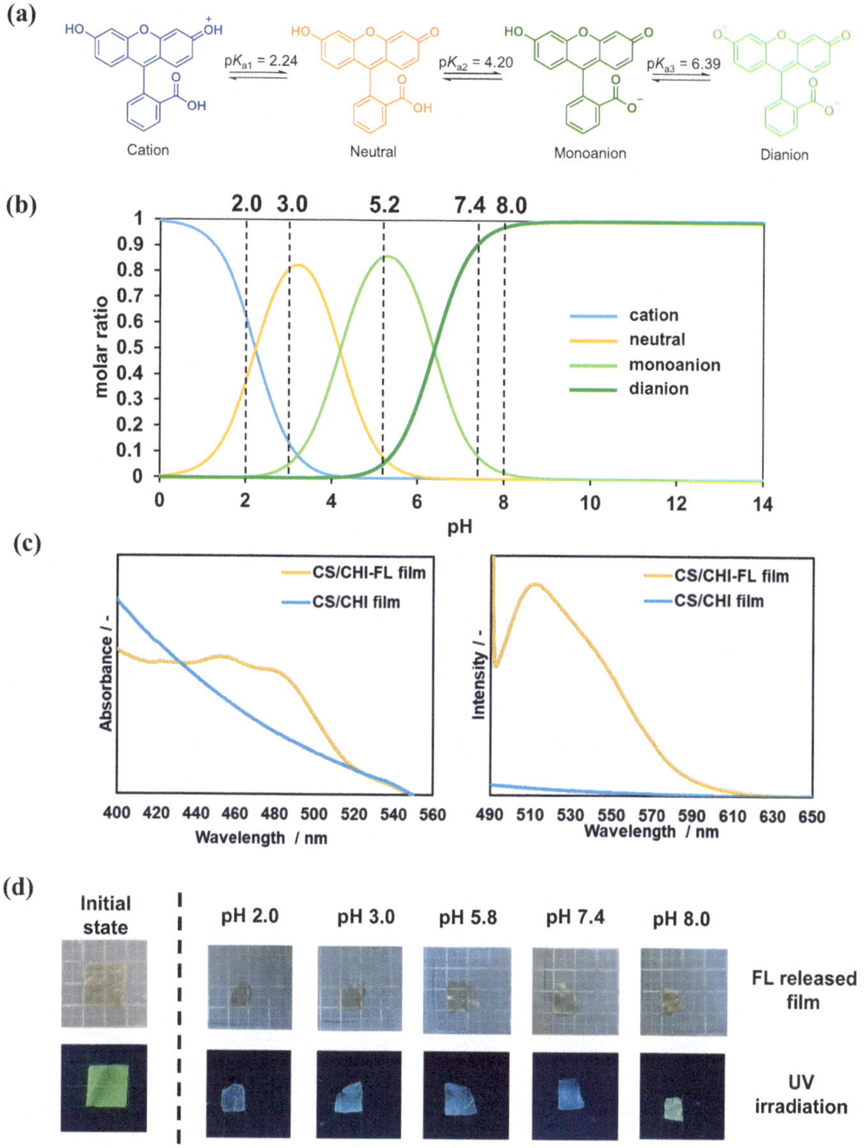


**Fig. 2.3** (**a**) Changes of FL structures and their p$K_a$, (**b**) molar ratio of each ionic structure of FL, (**c**) UV–vis (left) and fluorescence (right, excitation wavelength: 487 nm) spectra of CS/CHI-FL and CS/CHI films. (**d**) Macroscopic images of CS/CHI-FL films after release experiments (parts of FL molecules remained in the film). Reproduced from Ref. 18 with permission from the Royal Society of Chemistry

fluorescence was observed at pH 2.0, 3.0, 5.8, and 7.4, and green fluorescence at pH 8.0. Therefore, the p$K_a$ of FL in the film might change due to hydrogen bonds and electrostatic interactions with polysaccharides.

Furthermore, the structure of FL in the film was evaluated by measurement of the fluorescence spectra of the films after FL release experiments. It was confirmed that the main ionic structure of FL in the film at pH 2.0 was "cation." At pH 3.0, the neutral, zwitterionic structure (the isomer of the neutral structure in Fig. 2.3a) of FL was presented, at pH 5.8 and 7.4, the monoanion was presented, and at pH 8.0, the dianion was presented. From the above, the estimated structure of FL in each film is shown in Fig. 2.4. The ionized FL interacted with the polysaccharides in the films through electrostatic interactions and hydrogen bonding. Thus, at pH 2.0, 3.0, 5.8, and 8.0, the ionization was the same as that of FL in solution. Meanwhile, at pH 7.4, the monoanion and dianion structures existed in a ratio of 1:4 in solution, which is different from the structure of FL in the films, as estimated by the spectra of the films. When FL becomes the dianion structure, deprotonation of the hydroxy groups in FL is necessary. However, the deprotonation was suppressed by hydrogen bonds between the hydroxy groups of polysaccharides and FL. Accordingly, the loaded molecules in the films were ionized in the same way as in solutions, and the FL ionization by changing the pH of the solution was difficult (in other words, the dynamics of the pH-dependent ionization change was suppressed), due to the interactions between the loaded molecules and polysaccharides such as hydrogen bonds.

Moreover, to confirm the pH responsiveness of the FL in the films, the dried films were put under hydrogen chloride or ammonia atmosphere. The ionic state of the FL in the films changed in both cases (Fig. 2.5). These results indicate that the

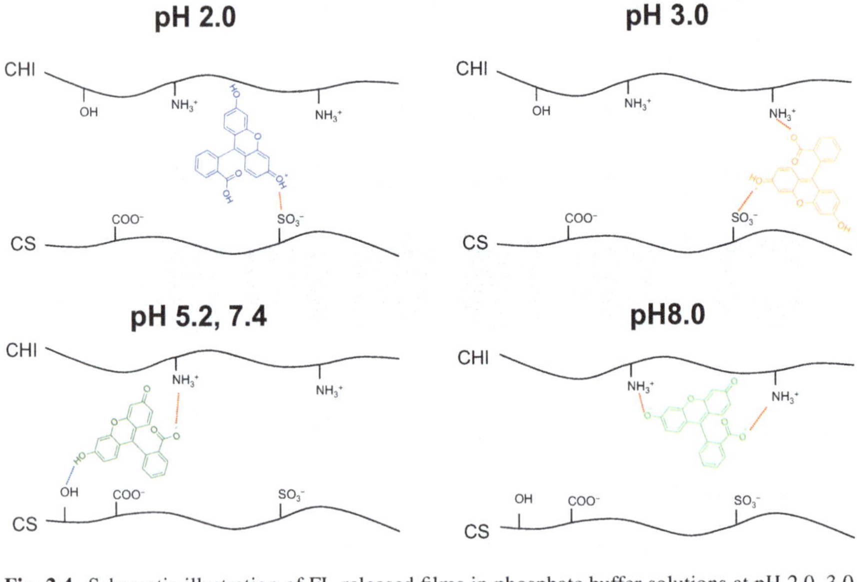


**Fig. 2.4** Schematic illustration of FL-released films in phosphate buffer solutions at pH 2.0, 3.0, 5.8, 7.4, and 8.0. Reproduced from Ref. 18 with permission from the Royal Society of Chemistry

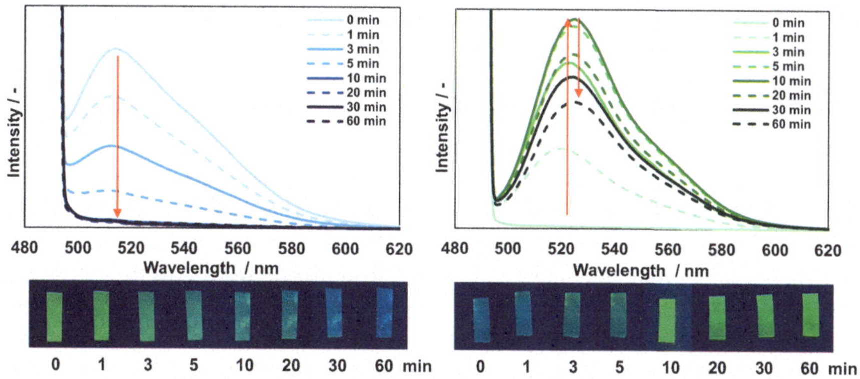


**Fig. 2.5** Fluorescence spectra and color change of the CS/CHI-FL films under UV light with exposure to HCl gas (left) and $NH_3$ gas (right). Reproduced from Ref. 18 with permission from the Royal Society of Chemistry

ionic state of the loaded molecules can be controlled within the film, and immobilization of the molecules in a specific ionic state can be achieved in the films. Furthermore, when these films load molecules, the molecules are three-dimensionally surrounded by CS and CHI, depending on the shape and charge state of the molecules, therefore it is possible to load a variety of molecules. These properties indicate that these films can be used as pH-responsive materials in biomedical applications such as carriers for drug delivery systems (DDS).

## 2.3 Molecular Permeability of Molecularly Imprinted Films

Molecular imprinting is an approach to design molecular recognition sites in various materials, which mimic natural recognition functions such as antibodies and receptors [19, 21, 22]. Molecularly imprinted polymers have been generally prepared using the target molecules or their derivatives as a template. On the other hand, the CS/CHI films developed by us showed molecular permeability [17], and it is expected that the molecular imprinting method could be used to control the molecular permeability of the films. When charged molecules are loaded on CS/CHI films as a template, the molecules are surrounded by the CS and CHI as described above, and especially, they are electrostatically interacted with the oppositely charged polysaccharides. The loaded molecules are then released from the films under an appropriate pH, and the pores of the same size as the molecule are generated in the films. Here, the opposite charges of the released molecules should remain near the pores. It has been reported that such pore formation by molecular imprinting actually has a significant effect on the molecular permeability of the film [22]. Herein, to clarify and utilize the interaction between the loaded molecules and polysaccharides in the films, the preparation of molecularly imprinted films using

MB and TMPyP (see below) (MB-IF and TMPyP-IF) and comparison of their molecular permeability with non-imprinted films (non-IF, CS/CHI films) using various permeants including the template molecules were conducted (Fig. 2.6) [20].

MB-IF and TMPyP-IF were prepared as follows: CS/CHI films loaded with monocationic MB or tetravalent cationic porphyrin derivative TMPyP (Fig. 2.6b,e) were respectively prepared, and these films were immersed in PBS (pH 7.4) for 60 min to release the MB or TMPyP from the films. The loading amounts of MB and TMPyP in the MB-loaded film and the TMPyP-loaded film (ca. $4 \times 4$ cm$^2$ each) were $4.16 \pm 0.31$ μmol and $5.43 \pm 0.33$ μmol, respectively (Fig. 2.7a,b), indicating that the numbers of loaded molecules were comparable. To obtain MB-IF and TMPyP-IF, phosphate-buffered saline (PBS, pH 7.4) was used for the fabrication process to maintain the pH during the molecular imprinting (in other words, the

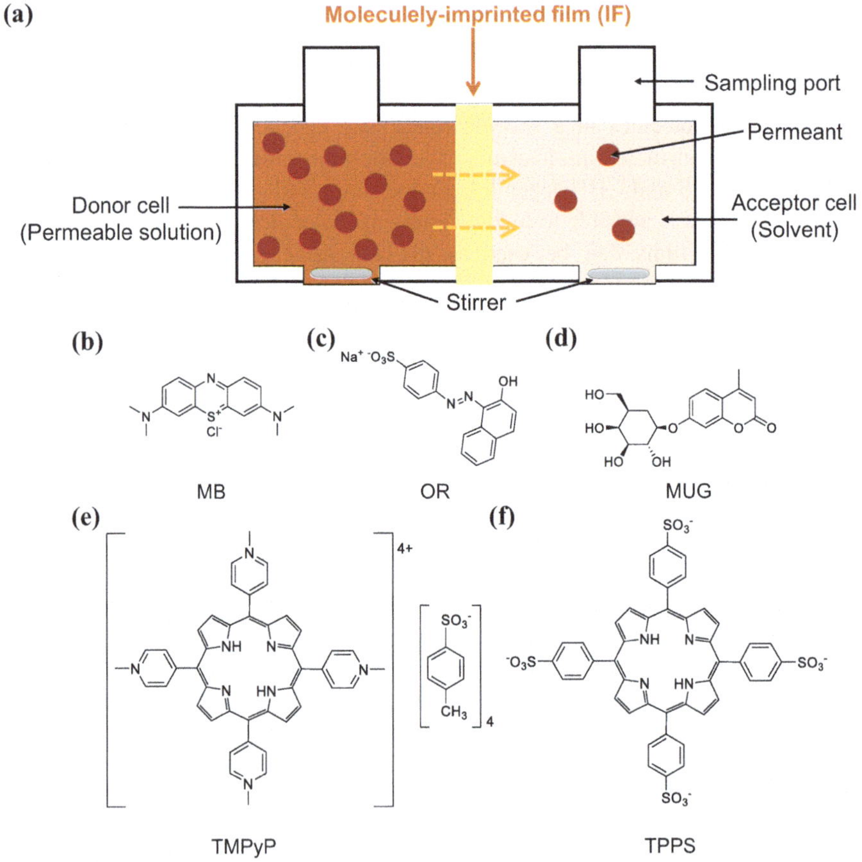


**Fig. 2.6** (**a**) Schematic illustration of the horizontal diffusion apparatus. Chemical structures of (**b**) methylene blue (MB), (**c**) orange II (OR), (**d**) 4-methylumbelliferyl-β-D-galactopyranoside (MUG), (**e**) α,β,γ,δ-tetrakis(1-methylpyridinium-4-yl)porphyrin *p*-toluenesulfonate (TMPyP), and (**f**) 5,10,15,20-tetrakis(4-sulfophenyl)porphyrin (TPPS, deprotonated structure in PBS is shown)

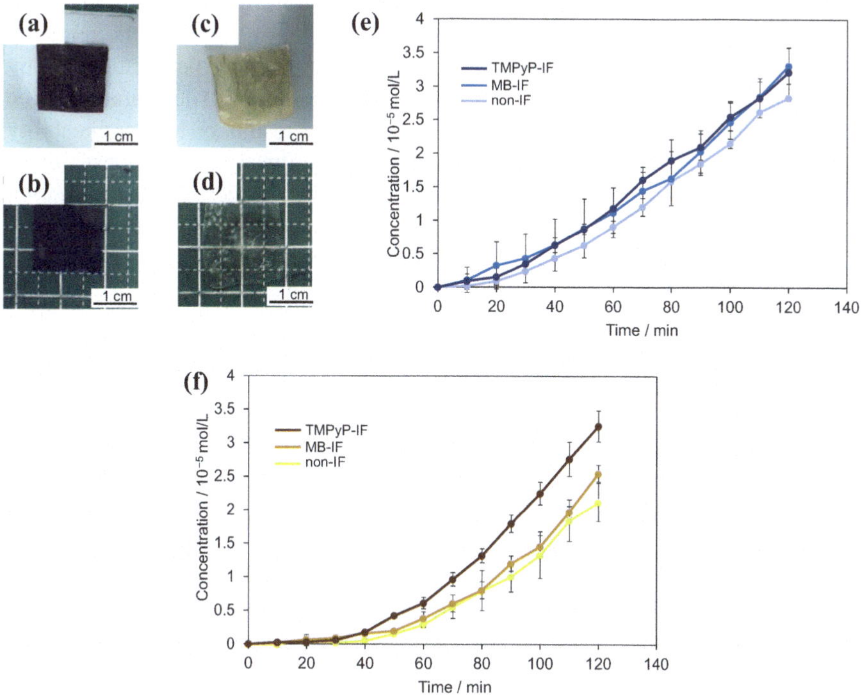


**Fig. 2.7** Photographs of (**a**) MB-loaded film, (**b**) TMPyP-loaded film, (**c**) MB-IF, and (**d**) TMPyP-IF. Permeation behavior of (**e**) MB and (**f**) TMPyP for the TMPyP-IF, MB-IF, and non-IF. Reproduced from Ref. 20, Sagawa, T. et al., Effects of template molecules on the molecular permeability of molecularly imprinted polysaccharide composite films, *Bull. Chem. Soc. Jpn.* **2024**, *97(12)*, uoae124, by permission of the Chemical Society of Japan

release of MB or TMPyP). After the release treatments, the color of both films was pale yellow (Fig. 2.7c,d), similar to non-IF, suggesting that almost all template molecules in MB- or TMPyP-loaded films were released. A TMPyP molecule has four cations, and therefore the release of TMPyP from polysaccharide composite films should be more difficult than that of MB due to multivalent electrostatic interactions between TMPyP and polysaccharides. The actual release ratio of MB from the film was 107.0 ± 2.9%, while the release ratio for TMPyP-loaded film was 88.6 ± 4.0%, according to the spectroscopic evaluations. The surface charges of the obtained IFs were evaluated. The order of zeta potentials of the films at pH 7.4 was as follows: TMPyP-IF (−60.3 ± 13.2 mV) < MB-IF (−28.6 ± 1.3 mV) < non-IF (−9.7 ± 1.4 mV). It indicated that the imprinting to polysaccharide composite films by cationic template molecules gave anionic charges on the film surface [19], and the anionic charges were larger when the cationic template with higher valence was used.

The permeation behavior of MB and TMPyP through MB-IF and TMPyP-IF was evaluated and compared with that of non-IF. Figure 2.6a shows an apparatus for the permeation experiments, and Fig. 2.7e, f show the permeation curves of MB and TMPyP for each film.

As for the MB permeation, it was confirmed that both IFs permeated more MB than non-IF (Fig. 2.7e). Due to the formation of the imprinted pores, these IFs are expected to exhibit two opposite effects: (1) acceleration of the permeation rate by permeants with a suitable size and shape to pass through, and (2) deceleration of the diffusion rate by selective capture of permeants. The results of MB permeability indicated that the effect of (1) is advantageous in this film from PICs consisting of CS and CHI. The time lags ($\theta$) were obtained from the intercept of the permeation curves and were used to evaluate the diffusion of permeants through each film. The $\theta$ for MB permeation was 21.5 ± 3.2, 17.6 ± 3.5, and 9.4 ± 2.9 min for non-IF, MB-IF, and TMPyP-IF, respectively. The shorter $\theta$ means faster molecular diffusion in the film, supporting the effect of the molecular imprinting. It should be noted that TMPyP-IF also exhibited the acceleration effect on MB permeation, even though MB was not the template.

On the other hand, regarding the TMPyP permeation, both IFs also permeated more TMPyP than non-IF, and TMPyP-IF exhibited a larger acceleration effect than MB-IF (Fig. 2.7f). The $\theta$ values for TMPyP permeation were 86.1 ± 18.9, 76.1 ± 3.3, and 47.8 ± 4.4 min for non-IF, MB-IF, and TMPyP-IF, respectively. Therefore, the order of ease of TMPyP diffusion in the films was as follows: TMPyP-IF > MB-IF > non-IF. Furthermore, when compared with respect to the permeants, the $\theta$ was shorter for MB than that for TMPyP, suggesting that the molecular diffusion in the films was affected by the size and valence of the permeant. Accordingly, molecular permeation through IFs depended on the template molecules and the permeants.

Next, the effects of permeant species on the permeation behavior through each IF were investigated in detail. In addition to MB and TMPyP, other permeants with different sizes and charges, OR (monoanionic), MUG (nonionic), and TPPS (tetravalent anionic porphyrin derivative) were also used (Figs. 2.6b–f and 2.8). When MB was used as the permeant, the order of the permeation amount was as follows: TMPyP-IF ≈ MB-IF > non-IF. On the other hand, the permeation amounts of OR were smaller than that of MB, and the order was the opposite: non-IF > MB-IF >

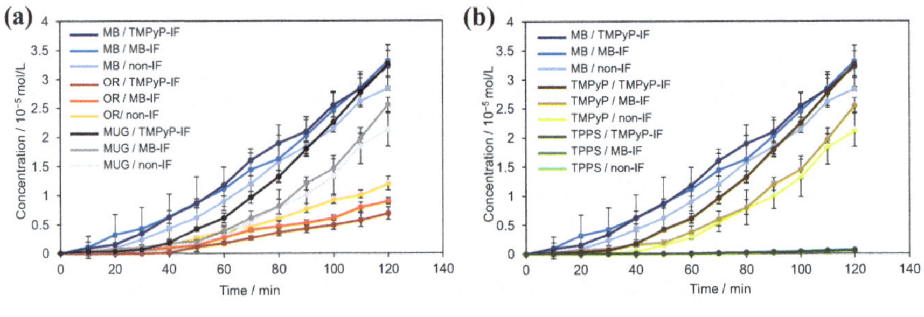


**Fig. 2.8** Permeation behavior of (**a**) MB, OR, and MUG, and (**b**) MB (same as (**a**)), TMPyP, and TPPS for the TMPyP-IF, MB-IF, and non-IF in PBS (pH 7.4). Reproduced from Ref. 20, Sagawa, T. et al., Effects of template molecules on the molecular permeability of molecularly imprinted polysaccharide composite films, *Bull. Chem. Soc. Jpn.* **2024**, *97(12)*, uoae124, by permission of the Chemical Society of Japan

TMPyP-IF (Fig. 2.8a). It suggested that the film surface was negatively charged due to the $COO^-$ groups of CS in the films.

In case of using MUG as the permeant, the order of permeation amount was the same as MB; TMPyP-IF > MB-IF > non-IF (Fig. 2.8a). MUG is a nonionic compound, and therefore the negative charges of the film surface should not affect the permeability. It was attributed that the difference in the molecular permeability of IFs was mainly due to the size of the imprinted pores, which depended on the size of the template molecule. The molecular size of TMPyP was larger than that of MB and MUG, and thus TMPyP-IF showed a larger permeation amount of MUG than MB-IF and non-IF.

When TMPyP and TPPS, tetravalent cationic and anionic molecules, were used as the permeant, the permeation amounts were lower than monocationic MB and monoanionic OR, respectively. It indicated that the charge valence also affected the permeability, in addition to the molecular size. In particular, the permeation amounts of TPPS for these films were few (Fig. 2.8b). Accordingly, the permeation behaviors through IFs were affected by both the surface charges of the films and the size of template molecules, from the focus on the film side. The difference in the molecular permeability for these cases also evaluated using kinetic parameters such as permeability coefficient and diffusion coefficient that obtained by analyzing the permeation profiles [20]. These results support that molecular imprinting can control the molecular permeability, in other words, the dynamics of molecular mobility, of the polysaccharide composite films.

The explanation of the mechanism for the permeation of the molecules through polysaccharide composite films is described (Fig. 2.9). In general, molecularly imprinted polymers are synthesized in the presence of a template molecule, and then the template molecule is removed from the polymer to form pores, which memorize the shape of the template molecule. Therefore, the polymers exhibit high selectivity and affinity for the template molecule [19–21]. As a result, the permeation of template molecules through a molecularly imprinted film should be accelerated or decelerated compared to non-imprinted films, as described above. On the other hand, the present polysaccharide composite films consist of PICs that are formed by electrostatic interactions between the amino groups of CHI and the carboxy and sulfo groups of CS, and thus the inner structure of the films should depend on the pH of the solution. When PIC gels with cationic template molecules are formed, the template molecules interact with the anionic functional groups in CS (Fig. 2.9a). The template molecule is then released, and a significant amount of the $NH_3^+$ groups deprotonate to $NH_2$ groups because the release is performed in PBS at pH 7.4. Therefore, the IF has molecular-sized pores and anionic groups, resulting in a negatively charged surface (Fig. 2.9b). The negative charge is larger for TMPyP, a tetravalent cation, than for MB, a monovalent cation. Therefore, the use of TMPyP as the template molecule for IF can provide more anionic groups to the film surface than that of MB. Such a difference in the formed pores should have different degrees of effect on the molecular permeability.

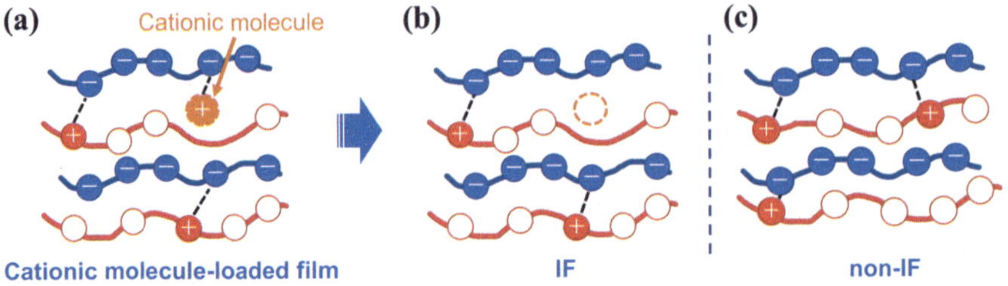


**Fig. 2.9** Presumable microscopic structures in template-loaded film (**a**), IF (**b**), and non-IF (**c**). Reproduced from Ref. 20, Sagawa, T. et al., Effects of template molecules on the molecular permeability of molecularly imprinted polysaccharide composite films, *Bull. Chem. Soc. Jpn.* **2024**, *97(12)*, uoae124, by permission of the Chemical Society of Japan

## 2.4 Conclusion

This chapter focuses on the dynamic behaviors of the molecules in the polysaccharide composite films and mainly describes two cases related to this topic. The pH responsiveness of FL in the films was different from that in solution. The ionization of FL by increasing solution pH was suppressed in the films. The molecular permeabilities of molecularly imprinted films were affected by the sizes and charges of the template and permeant molecules, and the former affected the film property. For example, MB permeation was accelerated for MB-IF and TMPyP-IF. These examples clearly show that the interactions, particularly electrostatic interactions, between the loaded molecules and polysaccharides in the films can endow significant effect on the function and property, and their dynamics, of the loaded molecules. This is because these polysaccharide composite films consist of polyion complexes of anionic and cationic polysaccharides, and it is different from the cases of other polymer film materials. Such a unique property of the polysaccharide composite films can make them useful functional film materials with dynamic properties like stimuli-responsive drug carriers or nanoreactors.

**Acknowledgments** This work was partly supported by JSPS KAKENHI Grant Number 19K05588 (M. H.).

## References

1. Xu X, Liu Y, Fu W, Yao M, Ding Z, Xuan J, Li D, Wang S, Xia Y, Cao M (2020) Poly(N-isopropylacrylamide)-based thermoresponsive composite hydrogels for biomedical applications. Polymers 12(3):580
2. Rizwan M, Yahya R, Hassan A, Yar M, Azzahari AD, Selvanathan V, Sonsudin F, Abouloula CN (2017) pH sensitive hydrogels in drug delivery: brief history, properties, swelling, and release mechanism, material selection and applications. Polymers 9(4):137
3. Fwu-Long M, Shin-Shing S, Yu-Bey W, Sung-Tao L, Jen-Yeu S, Rong-Nan H (2001) Fabrication and characterization of a sponge-like asymmetric chitosan membrane as a wound dressing. Biomaterials 22(2):165–173

4. Khan MUA, Razak SIA, Arjan WSA, Nazir S, Anand TJS, Mehboob H, Amin R (2021) Recent advances in biopolymeric composite materials for tissue engineering and regenerative medicines: a review. Molecules 26(3):619
5. Muzzarellia RAA, Greco F, Busilacchi A, Sollazzo V, Gigante A (2012) Chitosan, hyaluronan and chondroitin sulfate in tissue engineering for cartilage regeneration: a review. Carbohydr Polym 89(3):723–739
6. Vijayalekshmi V, Khastgir D (2017) Eco-friendly methanesulfonic acid and sodium salt of dodecylbenzene sulfonic acid doped cross-linked chitosan based green polymer electrolyte membranes for fuel cell applications. J Membr Sci 523(1):45–59
7. Water JJ, Schack MM, Velazquez-Campoy A, Maltesen MJ, van de Weert M, Jorgensen L (2014) Complex coacervates of hyaluronic acid and lysozyme: effect on protein structure and physical stability. Eur J Pharm Sci 88(2):325–331
8. Wu J, Wang X, Keum JK, Zhou H, Gelfer M, Avila-Orta C, Pan H, Chen W, Chiao S, Hsiao BS, Chu B (2007) Water soluble complexes of chitosan-g-MPEG and hyaluronic acid. J Biomed Mater Res Part B 80(4):800–812
9. Delair T (2011) Colloidal polyelectrolyte complexes of chitosan and dextran sulfate towards versatile nanocarriers of bioactive molecules. Eur J Pharm Sci 78(1):10–18
10. Li Z, Ramay HR, Hauch KD, Xiao D, Zhang M (2005) Chitosan–alginate hybrid scaffolds for bone tissue engineering. Biomaterials 26(18):3919–3928
11. Decher GF (1997) Nanoassemblies: toward layered polymeric multicomposites. Science 277(5330):1232–1237
12. Ariga K, Yamauchi Y, Rydzek G, Ji Q, Yonamine Y, Wu KC-W, Hill JP (2014) Layer-by-layer nanoarchitectonics: invention, innovation, and evolution. Chem Lett 43(1):36–68
13. Hashizume M, Kobayashi H, Ohashi M (2011) Preparation of free–standing films of natural polysaccharides using hot press technique and their surface functionalization with biomimetic apatite. Colloids Surf B: Biointerfaces 88(1):534–538
14. Hashizume M, Ohashi M, Kobayashi H, Tsuji Y, Iijima K (2015) Free-standing polysaccharide composite films: improved preparation and physical properties. Colloids Surf A Physicochem Eng 483(20):18–24
15. Hashizume M, Murata Y, Iijima K, Shibata T (2016) Drug loading and release behaviors of freestanding polysaccharide composite films. Polym J 48(4):545–554
16. Iijima K, Suzuki A, Hashizume M (2018) Preparation of swellable and highly operable composite films made of polysaccharides and supporting substrates. Kobunshi Ronbunshu 75(2):195–202
17. Iijima K, Kimura T, Sato R, Takahashi T, Hashizume M (2017) Kinetic analysis of molecular permeabilities of free-standing polysaccharide composite films. Macromol Chem Phys 218(3):1600391
18. Takagi K, Sagawa T, Hashizume M (2023) The pH responsiveness of fluorescein loaded in polysaccharide composite films. Soft Matter 19(46):8945–8953
19. Sagawa T, Oishi M, Yataka Y, Sato R, Iijima K, Hashizume M (2022) Control of the molecular permeability of polysaccharide composite films utilizing a molecular imprinting approach. Polym J 54(4):125033
20. Sagawa T, Yamamoto T, Hashizume M (2024) Effects of template molecules on the molecular permeability of molecularly-imprinted polysaccharide composite films. Bull Chem Soc Jpn 97(12):uoae124
21. BelBruno JJ (2019) Molecularly imprinted polymers. Chem Rev 119(1):94–119
22. Yoshikawa M, Tharpa K, Dima Ş-O (2016) Molecularly imprinted membranes: past, present, and future. Chem Rev 116(19):11500–11528

# Chapter 3
# Spatiotemporal Dynamic Interfaces in Biomaterials

**Hidenori Otsuka**

**Abstract** Surface engineering techniques for cellular micropatterning are emerging as important tools to clarify the effects of the microenvironment on cellular behavior, because cells usually integrate and respond the microscale environment, such as chemical and mechanical properties of the surrounding fluid and extracellular matrix, soluble protein factors, small signal molecules, and contacts with neighboring cells. Furthermore, recent progress in cellular micropatterning has contributed to the development of cell-based biosensors for the functional characterization and detection of drugs, pathogens, toxicants, and odorants. In this regard, the ability to control shape and spreading of attached cells and cell–cell contacts through the cell-adhesive patches with high precision is important. Commitment of stem cells to different specific lineages depends strongly on cell shape, implying that controlled microenvironments through engineered surfaces may not only be a valuable approach toward fundamental cell-biological studies but also of great importance for the design of cell culture substrates for tissue engineering. In particular, surface engineering techniques for cellular micropatterning as three-dimensional (3D) spheroids are focused on this review. To develop this kind of cellular microarray composed of a cell-resistant surface and cell attachment region, micropatterning a protein-repellent surface is important, because cellular adhesion and proliferation are regulated by protein adsorption. Thus, the focus of this review is on the surface engineering chemistries of biologically motivated micropatterning of two-dimensional non-fouling surfaces.

Another exciting approach to cell delivery for tissue engineering is the use of 3D polymer structures (i.e., hydrogels) that can be injected into the body. This approach enables the clinician to transplant the cell and polymer combination in a minimally

H. Otsuka (✉)
Department of Applied Chemistry, Faculty of Science, Tokyo University of Science, Tokyo, Japan
e-mail: h.otsuka@rs.tus.ac.jp; https://www.rs.kagu.tus.ac.jp/otlabo/index.html



M. Hashizume, Y. Imura (eds.), *Colloids and Interfacial Dynamics*, Physical Chemistry in Action, https://doi.org/10.1007/978-981-96-6728-4_3

invasive manner. Hydrogels have structural similarity to the biomacromolecular-based components in the body and are considered biocompatible. Tissue engineering is a most recent application of hydrogels, in which they are used as scaffolds to engineer new tissues. In this review, we discuss the critical design parameters of hydrogels to be used in tissue engineering. Currently used hydrogels with potential applications in tissue engineering are divided into their natural or synthetic origin. Limitations of widely used gels from natural polymers have motivated approaches to modify these polymers as well as to use various synthetic polymers. A wide range of synthetic polymers may potentially have suitable chemical and physical properties for these applications. In addition, incorporation of growth factors and the role of mechanical signals to enhance tissue development will be discussed as future directions.

**Keywords** Biointerface · 3D culture · Micropatterning · PEGylation · Cell array · Hydrogel scaffold

## 3.1 Introduction

In the fields of toxicology and drug testing, in vivo work has an advantage over in vitro work in that it takes into account the entire biological system in determining the time-dependent response to a chemical challenge. However, it is often not possible to do in vivo chemical toxicity studies with human subjects. Therefore, the development of a microscale cell culture analog system, an in vitro human surrogate, is another promising area using cell culture and microfabrication technologies. Recent progress in cell culture and microfabrication technologies has contributed to the development of cell-based biosensors for the functional characterization and detection of such drugs, pathogens, toxicants, odorants, and other chemicals [1–9].

On the other hand, cell culture in two dimensions has been routinely and diligently undertaken in thousands of laboratories worldwide for the past four decades. However, the culture of cells in two dimensions is arguably primitive and does not reproduce the anatomy or physiology of a tissue for informative or useful study. Creating a third dimension for cell culture is clearly more relevant, but requires a multidisciplinary approach and expertise. When entering the third dimension, researchers need to consider the design of scaffolds for supporting the organization of cells or the use of bioreactors for controlling nutrient and waste product exchange. As 3D culture systems become more mature and relevant to human and animal physiology, the ability to design and develop cocultures becomes possible as does the ability to integrate stem cells. The primary objectives for developing 3D cell culture systems vary widely and range from engineering tissues for clinical delivery to the development of models for drug screening. Cellular spheroids are simple three-dimensional models that can be generated from a wide range of cell types and form due to the tendency of adherent cells to aggregate. They are typically created from single culture or coculture techniques such as hanging drop, rotating culture, or concave plate methods. Spheroids do not require scaffolds and can readily be imaged by light, fluorescence, and confocal microscopy. Consequently, spheroids have seen a use in modeling solid tumor growth and metastasis studies and are also used in a multitude of therapeutic studies, for example, for high-throughput screening (HTS). The intention of this review is to provide a general overview of the common approaches and techniques for designing 3D culture models. In particular, surface engineering techniques for cellular micropatterning [3, 10–29] are focused as important tools to obtain spheroid micropattern.

To develop this kind of cellular microarray composed of a cell-resistant surface and cell attachment region, micropatterning a protein-repellent surface is important, because cellular adhesion and proliferation are regulated by protein adsorption. The different engineering approaches aiming at a precise control of cell adhesion and spreading, through chemically and spatially designed surfaces, are the main focus of this review. In particular, the importance of non-fouling surface chemistries is discussed.

Alternatively, hydrogel as a 3D cell scaffold is of great interest because of their inherent biochemical and biophysical properties, such as cellular adhesion, degradation, and viscoelasticity. Hydrogels can mimic features of many tissues, and there have been great advances in the tailoring of hydrogel properties (e.g., mechanics, degradation) for widespread biomedical applications. Cross-linking methods to form biopolymer networks are discussed in detail, including (a) covalent cross-linking, (b) dynamic covalent cross-linking, and (c) physical cross-linking. Finally, recent advances in the use of chemically modified biopolymer hydrogels for tissue scaffolds, therapeutic delivery, as well as the formation of interpenetrating network biopolymer hydrogels, are highlighted.

## 3.2 Cell-Patterning Techniques

Microfabrication techniques are used to generate patterns of cells on surfaces. This cellular patterning is a necessary component for cell-based biosensors, cell culture analogs, tissue engineering, and fundamental studies of cell biology. We have accomplished the dry etching (or plasma etching) technique applied to make cell micropatterning. The photolithographic technique is also highly developed and has been widely used for patterning cells. However, this technique has some disadvantages for certain biological applications, especially biocompatibility. A set of alternative techniques, such as microcontact printing, microfluidic patterning using microchannels, and laminar flow patterning, has been developed for use in biological applications. In this paper, three representative methods are reviewed.

### 3.2.1 *The Basis of Cellular Patterning: Non-fouling Surface Chemistries*

In any cellular patterning using surface modification, the ability to nonspecific interactions between the surface and the protein-containing media is crucial in order to generate unbiased experimental outcomes. Advances in surface chemistry have made possible the synthesis of so-called non-fouling surfaces that significantly reduce or eliminate the nonspecific adsorption of proteins and other biomolecules from biological fluids such as cell culture media. Several types of native molecules have the ability to reduce the adsorption of proteins at surfaces, for example, carbohydrates such as agarose and mannitol as well as albumin [30, 31]. Due to their limited efficiency and stability, a number of synthetic materials have been developed [32].

The use of polymer molecules as surface modifiers has been shown to be a very promising way of manipulating the interactions of surfaces with proteins [33]. The proper tuning of these interactions will hopefully lead us to the rational design of a modified surface that will improve biocompatibility. For example, the inclusion of tethered polymers on the surfaces of liposomes results in their increased longevity in the blood stream, hence, these aggregates can be used as drug delivery systems [34]. Furthermore, tethered polymers can be used for the specific targeting of liposomes to cells by chemically inserting a specific binding reagent to the free ends of the tethered polymers [35, 36].

In this review, we will discuss the recent advances in our understanding of the ability of tethered polymer chains to prevent protein adsorption. The present experimental status of these systems will be discussed together with theoretical studies that enable us to calculate the interactions of proteins with tethered polymer layers as well as protein adsorption on modified surfaces. First, we will discuss some of the general properties of tethered polymer layers.

The ability of tethered polymer layers to prevent protein adsorption is believed to be due to the steric repulsion associated with the perturbation that the protein molecules cause on the polymer layer [37] (see Fig. 3.1) [38]. In order to understand this repulsion, it is appropriate to understand the structure of the polymer layers. The general picture in the good solvent regime (i.e., when the polymer is soluble in the solvent) is that [39], for very low surface coverages of polymers, the tethered chains are isolated and their structure is not very different from that of the polymers in bulk. This is called the "mushroom" regime. As the surface coverage increases the chains start to repel each other, and they have to stretch out of the surface in order to avoid each other and thus reduce the polymer–polymer repulsions. The regime of highly stretched chains is generally called the "brush" regime [40, 41].

Detailed studies performed by computer simulations and molecular theories, in particular the single-chain mean-field (SCMF) theory, show that the treatment of tethered polymers in terms of only the two above-defined regimes is not appropriate for short to intermediate chain length polymers, which is from a few segments to a few hundred segments. Furthermore, the intermediate regime, that is, the

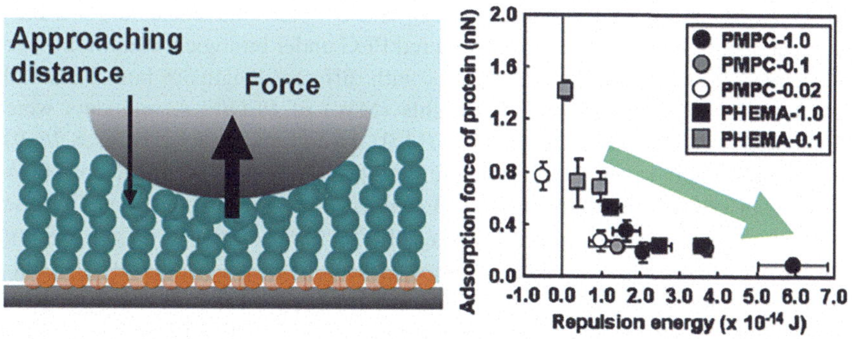


**Fig. 3.1** Hydrophilic poly(2-methacryloyloxyethyl phosphorylcholine) (PMPC) and poly(2-hydroxyethyl methacrylate) (PHEMA) brush layers with different thicknesses and graft densities were prepared to construct a model surface to elucidate protein–surface interactions. In particular, we focused on the steric repulsion of hydrophilic polymer layers as one of the surface properties that strongly influence protein adsorption and employed force-versus-distance (f–d) curve measurements obtained via atomic force microscopy to quantitatively evaluate the steric repulsion force, which is also referred to as the "elastic repulsion energy." Two types of hydrophilic polymer brush surfaces with grafted PMPC and PHEMA chains were prepared on initiator-immobilized silicon wafers using SI-ATRP using 3-(2-bromoisobutyryl)decyl dimethylchlorosilane (BrC10DMCS) as a surface-attachable initiator. Circles: PMPC brush surface from an initiator-immobilized surface with (black) 100% BrC10DMCS (PMPC-1.0), (gray) 10% BrC10DMCS (PMPC-0.1), and (white) 2% BrC10DMCS (PMPC-0.02). Squares: PHEMA brush surface from an initiator-immobilized surface with (black) 100% BrC10DMCS (PHEMA-1.0), (gray) 10% BrC10DMCS (PHEMA-0.1) [38]. (reprinted from ref. 38)

"mushroom" to "brush" transition region, is the one that corresponds to most experimental studies. This has been shown to be the case in experimental observations of tethered polyethylene oxide (PEO) (also called polyethylene glycol [PEG]), which is the most widely used polymer in biological-related applications of tethered polymer layers. This conclusion was reached by comparisons of experimental observations with the predictions of the SCMF theory on structural and thermodynamic properties.

The most widely used system is polyethylene glycol (PEG) with the monomeric repeat unit [-$CH_2$-$CH_2$-O-]. The factors governing protein resistance of a PEG-graft copolymer were recently investigated in detail, with quantitative information provided on the interfacial architecture of PEG chains and their influence on protein resistance [37, 42–58]. Many different PEG surface-immobilization strategies have been successfully applied. One example is the use of triblock copolymers PEG-PPO-PEG [59–65] that assemble spontaneously on hydrophobic surfaces through hydrophobic–hydrophobic interactions. This polymer class is also known as poloxamers or Pluronics and has been extensively studied [49, 66–69].

We have reported the novel approaches for the construction of functionalized polyethylene glycol (PEG) layer on surfaces using heterobifunctional PEG/Polylactide (PLA) block copolymers [70, 71]. One of the objectives in the study is to investigate the effect of the variation in PEG chain length on surface properties.

This includes assays on protein adsorption and cellular attachment to get a biochemical insight into the behavior of tethered PEG under biological conditions. For this purpose, numerous acetal-PEG/PLAs with different lengths of both PEG and PLA were synthesized. Molecular weights (MW) of PEG/PLA segments were abbreviated as follows: PEG/PLA(0.65/11.0, 1.8/7.0, 3.3/5.4, 5.0/4.6, 8.7/6.9) where the numbers in parentheses denote the MW of the PEG segments and PLA segments in kg/mol, respectively. The PEG-brushed layer was constructed on the silanized glass surface by the spin coating of 4% (w/v) solution of PLA solution/toluene, followed by the 2% (w/v) solution of acetal-PEG/PLA/toluene. The wettability of the surface covered with PEG/PLA block copolymers was estimated both in air and in water by contact angle measurement (Table 3.1) [70]. In water-in-air measurements, coating of PEG/PLA block copolymer onto a PLA surface increased its wettability with increasing PEG molecular weight, as indicated by a decrease in static contact angle. A similar trend was observed in air-in-water measurements. The contact angle of a water droplet in air decreased remarkably in the range between PEG/PLA(0.65/1.10) and PEG/PLA(5.0/4.6). The decrease became moderate in the region with higher PEG molecular weights. Since the top few angstroms can be sensed by a contact angle measurement, the relatively high contact angles on the surfaces containing the lower molecular weight PEG are most likely to be attributed to an incomplete coverage of the uppermost surface by PEG chains. The dynamic contact angle was then measured to estimate the dynamics of the uppermost surface. The coating of PEG/PLA block copolymers reduces both the advancing and the receding angles of the substrates, although the change depends on the PEG molecular weight, which is consistent with the result of static contact angle. The maximum hysteresis was observed for the substrate with a medium PEG chain

**Table 3.1** Molecular weights of PEG/PLA block copolymers and contact angle analysis [70] (reprinted from ref. 70)

| | | PEG | | | PLA | Static contact angle[b] | |
|---|---|---|---|---|---|---|---|
| Sample[a] | | $M_n$ | $M_w$ | $M_w/M_n$ | $M_n$ | Dry[c] | Wet[d] |
| 1 | PLA | – | – | – | 20,000 | 80.2 | 69.6 |
| 2 | PEG/PLA(0.65/11.5) | 650 | 720 | 1.10 | 11,470 | 78.9 | 51.4 |
| 3 | PEG/PLA(1.8/7.0) | 1880 | 1930 | 1.03 | 7020 | 75.6 | 48.7 |
| 4 | PEG/PLA(3.3/5.4) | 3340 | 3470 | 1.04 | 5410 | 59.7 | 40.2 |
| 5 | PEG/PLA(5.0/4.6) | 5050 | 5210 | 1.03 | 4640 | 47.2 | 37.7 |
| 6 | PEG/PLA(8.7/6.9) | 8730 | 8810 | 1.01 | 6940 | 47.9 | 32.6 |

[a]Molecular weights (MW) of PEG/PLA segments were abbreviated as follows: PEG/PLA(1.8/7.0, 5.0/4.6, 8.7/6.9) where the numbers in parentheses denote the MW of the PEG segments and PLA segments in kg/mol, respectively

[b]The values (degree) after hydration in PBS for 24 h

[c]The values in dry state were measured by a sessile droplet technique, where a water droplet was placed on the film surface

[d]The values in wet state were measured by a captive bubble technique, where a sample film was immersed in water and a small air bubble was placed on the film from the bottom using a curved needle

length such as PEG/PLA(3.3/5.4). Hysteresis in the dynamic contact angle may be caused by the hydration of PEG segments. In the dry state, the PEG chain should assume a conformation flat to the surface experienced by the advancing contact line. Upon hydration, however, the PEG chain should extend from the surface due to the hydration of PEG chains. As a result, the receding contact line experiences a more hydrophilic surface than the advancing contact line. It is likely that this is the origin of the hysteresis observed on these surfaces.

Protein adsorption on PEG/PLA surfaces was then estimated by using bovine serum albumin (BSA) as a model protein. On PLA surfaces, BSA adsorbed significantly, while on PEG-coated surfaces, BSA adsorption is clearly decreased mainly due to the steric stabilization by PEG chains. Moreover, minimum adsorption was obtained at a medium PEG chain length, that is, PEG/PLA(3.3/5.4). Note that this surface revealed the maximum hysteresis in dynamic contact angle measurement. Protein adsorption may be related to the hysteresis observed in the dynamic contact angle, which is likely to depend on particular surface properties such as the density and mobility of tethered PEG chains on the surface.

Another widely used PEG-chemistry-based approach relies on oligo-EG or PEG-modified alkanethiolate self-assembled monolayers (SAMs) [3, 5, 72–82]. Although these PEG-modified SAMs show substantial reduction of protein adsorption, they still adsorb significant amounts of serum proteins. Moreover, they tend to oxidize under ambient conditions, restricting their use to short-term cell culture studies [83]. Bearinger et al. proposed an attractive alternative for modification of gold surfaces [84] based on PPS–PEG diblock or PEG–PPS–PEG triblock copolymers with poly(propylene sulfide) (PPS) as the central block (that binds to gold surfaces) and PEG-grafted chains. Other polymer architectures such as gels and polymeric SAMs have also been successfully utilized to link PEG chains to surfaces. For example, Healy and coworkers developed a gel-like interpenetrating polymeric network (IPN) of poly(acrylamide) and poly(ethylene glycol) [P(AAm-co-EG)] [85, 86], while Toner et al. used a poly(ethylene glycol) diacrylate (PEG-DA) hydrogel [26].

The choice of protein/cell resistant chemistry is often dictated by the type of substrate material to be used. Some chemistries are highly versatile and can be applied to different surfaces, while others require specific substrates as in the case of the gold-thiol system. However, each approach has its specific strengths and weaknesses. Covalently bound molecules have higher binding strength than physisorbed adlayers; however, non-covalent immobilization offers many attractive ways to modify surfaces. It should be pointed out that cell-patterning investigations frequently use serum-free or serum-depleted cell culture media. Some groups completely exclude serum from the media, while others initially plate the cells in serum-free conditions and later add adequate amounts of serum to keep cells alive. These measures are sometimes necessary in order to prevent the cells to attach to the background, which would result in poor cell pattern quality. We note that such protocols have been frequently used in publications reporting the use of $EG_3$- or $EG_6$-modified alkanethiols. This is not so surprising, since it is known that such coatings are not highly resistant to the adsorption of proteins.

Notably, our previous studies [87, 88] have presented the long-term stability of cell patterning; PEG chain density high enough to inhibit outgrowth of endothelial cells from the cell-adhering region in the horizontal direction could be obtained only by employing formation of a short filler layer of PEG (2 kDa) in the preconstructed longer PEG-brushed layer (5 kDa), which prevented nonspecific protein adsorption almost completely. In this way, a completely micropatterned array of endothelial cells with long-term viability was obtained. This clearly indicated the importance of a short, underbrushed PEG layer in minimizing nonspecific protein adsorption for long-term maintenance of the active cell pattern.

### *3.2.2 2D Cell Patterning*

Heterotypic cell interaction between parenchymal cells and nonparenchymal neighbors has been reported to modulate cell growth, migration, and/or differentiation. In both the developing and adult liver, cell–cell interactions are imperative for coordinated organ function. In vitro, cocultivation of hepatocytes and nonparenchymal cells has been used to preserve and modulate the hepatocyte phenotype. Toner et al. summarized frontier studies in this area as well as previously reported advances in microfabrication that have allowed for more precise control over cell–cell interactions through "cellular patterning" or "micropatterning" [13].

Recent advances in microfabrication techniques allowed novel studies of the role of the heterotypic interface and the ratio of cell populations as independent variables, resulting in some new insights on the complex modes of cell–cell communication in these cocultures. Future improvements in microfabrication technology will allow these approaches to be tailored to explore specific hypotheses on a broad range of materials, with a variety of biomolecules and an achievable spatial resolution in the submicron range. Furthermore, some preliminary design criteria were determined for the use of cocultures in bioreactors for clinical support of several tissue failures. In addition to their utility in bioreactors for clinical support, cellular cocultures will have applications in fundamental studies of cell communication, organogenesis, and physiology as well as in development of functional tissue constructs for medical applications. The following is a summary of typical cell micropatterning methods.

#### 3.2.2.1 Photolithography

Photolithography, developed for fabricating semiconductor devices, is a process for transferring geometric features of a photomask to a substrate by UV irradiation (Fig. 3.2a) [89]. The substrate is coated with a light-sensitive polymer called photoresist and irradiated through a photomask. Depending on its type, the photoresist becomes either soluble or insoluble by this irradiation. After soluble parts are removed by the developing process, a patterned photoresist remains. Then, a

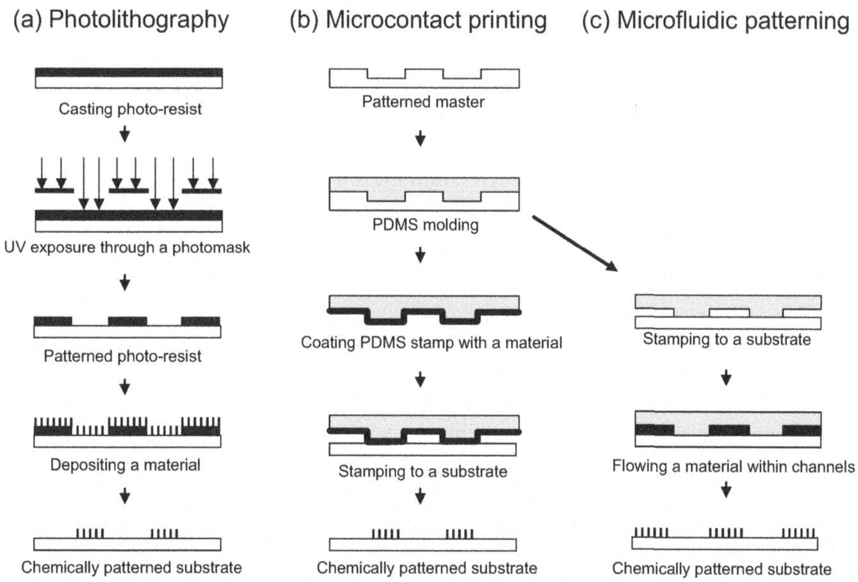


**Fig. 3.2** Schematic illustration of surface micropatterning technique. (**a**) Photolithography (**b**) Microcontact printing (**c**) Microfluidic patterning. Each procedure is discussed in the text [89]. (reprinted from ref. 89)

material of interest is deposited onto the exposed regions of the substrate. Development of the remaining photoresist yields a micropattern of the material on the substrate.

#### 3.2.2.2 Soft Lithography

Soft lithography is based on a soft elastomeric stamp with patterned relief [90]. The benefit of soft lithography is that the photolithographic process is only necessary for the fabrication of a master, and multiple copies of elastomeric stamps can be prepared by replica molding against the master. Because of this, the use of a clean-room environment is minimized; hence, it is more convenient and cheaper than standard photolithography. There are two major soft-lithographic methods for surface patterning: microcontact printing [91] and microfluidic patterning [92].

#### 3.2.2.3 Microcontact Printing

Microcontact printing is a process of transferring materials from an elastomeric stamp having a patterned relief to a substrate at the area contacted by the stamp (Fig. 3.2b). By casting a solution of poly(dimethylsiloxane) (PDMS) prepolymer onto a photolithographically fabricated patterned master, a PDMS stamp is replica molded. This stamp is immersed in a solution of the material of interest, and then stamped to a substrate via conformal contact to the surface in order to transfer the material to the substrate. Alkanethiols are often microprinted onto a gold surface to prepare cell micropatterns [93]. In this case, the microprinted substrates are back-filled with PEG-thiol followed by immersion in a fibronectin solution. This process yields selective adsorption of fibronectin onto alkanethiol-stamped regions, which allows the selective attachment of cells onto the stamped regions.

#### 3.2.2.4 Microfluidic Patterning

Microfluidic patterning is a process related to microcontact printing (Fig. 3.2c). Instead of stamping a PDMS mold having a relief of the master, one having a microfluidic network is stamped onto a substrate [92, 94]. By injecting a particular solution through the microchannels, the substrate is exposed to the flow, resulting in patterning of the material. Although this method is less frequently used than microcontact printing, it is more useful to micropattern multiple components on a surface, because single microcontact printing divides the surface into only two regions unless a multilevel stamp is used [95]. Furthermore, microfluidic patterning allows a directed delivery of cells onto a surface, as well as a selective deposition of materials, as described above [96, 97].

#### 3.2.2.5 Dry Etching (Plasma Etching)

Folch et al. [14] describe a new technique for micropatterning of cells that is based on the use of oxygen plasma as a patterning tool. The technique consists of (1) homogeneously grafting a glass substrate with a protein-repellent interpenetrating polymeric network (IPN) of poly(acrylamide) and poly(ethylene glycol) [P(AAm-co-EG)] prepared with commercially available reagents and (2) selectively removing this coating using oxygen plasma. They use elastomeric stencils (i.e., self-sealing membranes with through-holes, Fig. 3.3a) and microchannels (Fig. 3.3b) as removable masks for the selective oxygen plasma etch of the IPN areas that are not protected by the mask [98]. The stencil or microchannels are peeled off to reveal cell-adhesive regions separated by the nonadhesive coating. Their method offers a convenient way of patterning a robust protein-repelling material and allows for independently controlling the chemistry of the regions reserved for cell attachment.

### 3.2.3 *Patterning of Spheroids as 3D-Microorganized Cells*

Dry etching refers to the removal of material, typically a masked pattern of semiconductor material, by exposing the material to a bombardment of ions (usually a plasma of nitrogen, chlorine, and boron trichloride) that dislodge portions of the material from the exposed surface. Unlike with many (but not all, see isotropic

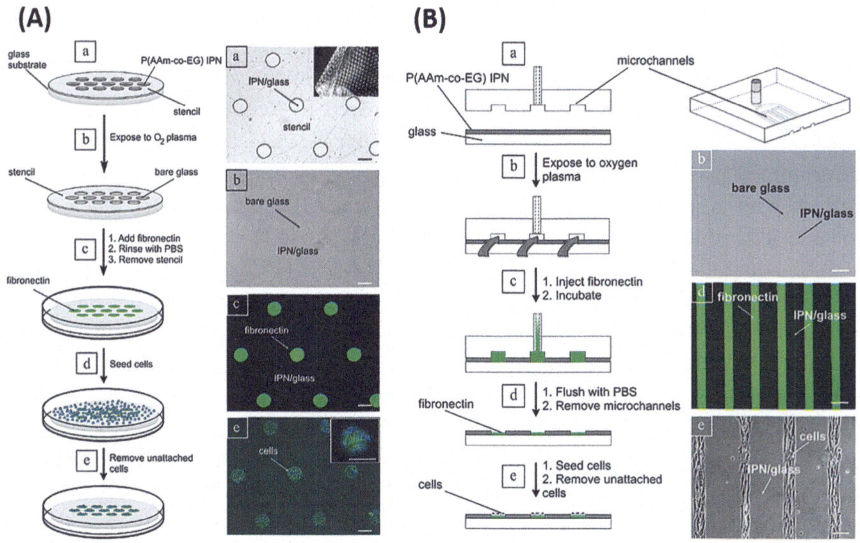


**Fig. 3.3** (**A**) Schematic illustration (left column) and demonstration (right column) of the procedure for cellular micropatterning based on the use of PDMS stencils. (a) The stencil is placed onto a glass substrate homogeneously grafted with the IPN. A picture of the stencil before (inset) and after application onto the surface is shown on the right. (b) The stencil serves as a mask for selective etching of the IPN in an oxygen plasma. The etching leaves islands of IPN-free glass surrounded by IPN (as shown in the corresponding phase-contrast image on the right). (c) Fluorescently tagged fibronectin (Fn) is adsorbed onto the plasma-etched islands of glass, and the stencil is removed. (d) The fibronectin/IPN-patterned substrate is incubated with a cell suspension. (e) Unattached cells are removed by exchanging the medium. The inset shows a fluorescence image of C2C12 myoblasts (blue) attached to the fibronectin islands (green). Scale bars are 100 μm. (**B**) Schematic illustration (left column) and demonstration (right column) of the procedure for cellular micropatterning based on the use of PDMS microchannels. (a) The PDMS mold is placed onto a glass substrate homogeneously grafted with the IPN. (b) The IPN is selectively removed after exposure of the surface to oxygen plasma, which leaves lines of IPN-free glass surrounded by IPN (see phase-contrast image on the right). (c and d) Fluorescently tagged fibronectin is injected into the microchannel network. After flushing with PBS, the microchannels are removed, leaving lines of fibronectin surrounded by IPN (see fluorescence image of the adsorbed fibronectin on the right). (e) The fibronectin/IPN-patterned substrate is incubated with a cell suspension, and unattached cells are removed by exchanging the medium. Shown on the right is a phase-contrast image of C2C12 myoblasts attached to the lines of fibronectin. Scale bars are 100 μm [98] (reprinted from ref. 98)

etching) of the wet chemical etchants used in wet etching, the dry etching process typically etches directionally or anisotropically. By using this dry etching technique, a two-dimensional microarray of ten thousand (100 × 100) hepatocyte heterospheroids, underlaid with endothelial cells, was successfully constructed with a 100-μm spacing in an active area of 20 × 20 mm on microfabricated glass substrates that were coated with polyethylene glycol (PEG) brushes [29]. Cocultivation of hepatocytes with endothelial cells was essential to stabilize hepatocyte viability and liver-specific functions, allowing us to obtain hepatocyte spheroids with a diameter of 100 μm, functioning as a miniaturized liver to secret albumin for at least 3 weeks.

Micropatterned PEGylated substrates with two-dimensional arrays of plasma-etched circular domains ($\phi$100-μm) were prepared by sequential spin coating of polylactide (PLA) and α-lactosyl-polyethylene glycol (PEG)/PLA block copolymer on silanized glass slide dishes, followed by plasma etching through a metal mask pattern with circular holes (Fig. 3.4a) [29]. Round, 100-μm diameter holes separated by 100-μm (edge-to-edge distance) spacing was used to mask a $N_2 + H_2$ plasma etch, forming the patterned α-lactosyl (or methoxy)-PEG/PLA surface. Bovine aortic endothelial cells (BAECs) at passage 13 were then seeded onto the patterned surfaces with $\phi$100-μm circular domains that were edge-to-edge spaced in $l$100-μm intervals (Fig. 3.4a), and cultured at 37 °C for 24 h in a 10% fetal bovine serum medium. Obviously, BAECs adhered only onto the circular domains exposing a glass substrate (Fig. 3.4b). Preferentially adsorbed extracellular matrix (ECM) proteins, including fibronectin, vitronectin, and laminin on the glass circular domains, may promote the adhesion of anchorage-dependent BAECs. Rat primary hepatocytes, suspended in a culture medium, were then applied to the patterned dishes with cultured endothelial cells selectively located in the circular domains. Interestingly, rat primary hepatocytes formed spheroids within 24 h only on the circular regions of existing endothelial cells, generating a 2D-arrayed structure of the hepatocyte spheroids (Fig. 3.4c). In contrast, on the same patterned α-lactosyl-PEG/PLA surface without pre-adhered BAECs, hepatocytes attached to and spread on both the PEG layer and the glass regions, without spheroid formation (Fig. 3.4d). These results demonstrate the significant role of BAEC as a feeder layer for the formation of hepatocyte spheroids. The cell viability of the obtained spheroid was assessed with a LIVE/DEAD Viability/Cytotoxicity assay kit. The result suggested that cell viability was well retained in the spheroid structure, interacting with the underlaid BAEC layer.

Hepatocyte spheroids in contact with BAECs were characterized by an immunohistochemical double staining method. In situ fluorescent staining was done with an anti-rat albumin antibody for cellular albumin synthesis, a characteristic phenotype of hepatocytes. Rhodamine-conjugated phalloidin was used for F-actin. Figure 3.5 demonstrates a three-dimensional view of a multicellular spheroid of hepatocytes underlaid with BAECs as a feeder cell [29]. This was reconstructed from a stack of two-dimensional image volumes. It should be noted that spheroids significantly express a stable level of liver-specific functions (albumin secretion) even after 3 weeks, showing intense green fluorescence, compared to the usual cell monolayers. Note that these albumin secretions, cytoskeleton as well as cell–cell junction, are maintained intact in the spheroids, presumably due to the heterotypic cell interaction through the hepatocyte–BAEC contact [99–102].

TBB and CBB (tissue and cell-based biosensors), based on the spheroid formation presented here, offer the promise of responding to environmental perturbations such as toxicants, pathogens, or other agents in a physiologically relevant manner. There is overwhelming evidence that in vitro three-dimensional tumor cell cultures more accurately reflect the complex in vivo microenvironment than simple

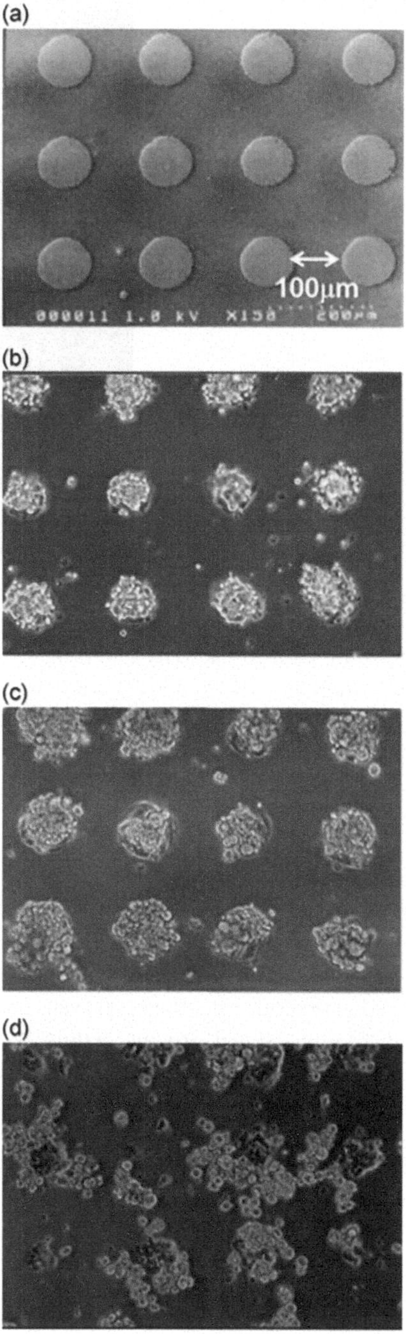


**Fig. 3.4** Patterned 3D coculture of hepatocyte spheroids and endothelial cells (BAECs). (**a**) Micropatterned α-lactosyl-PEG/PLA coated dish with $\phi$100-μm circular domains spaced in $l$100-μm intervals. (**b**) Patterned culture of BAECs on substrate (**a**) for 24 h at 37 °C. (**c**) Organized pattern of hepatocytes spheroids underlaid with BAECs. (**d**) Hepatocytes directly seeded on substrate (**a**) without pre-adhered BAECs [29] (reprinted from ref. 29)

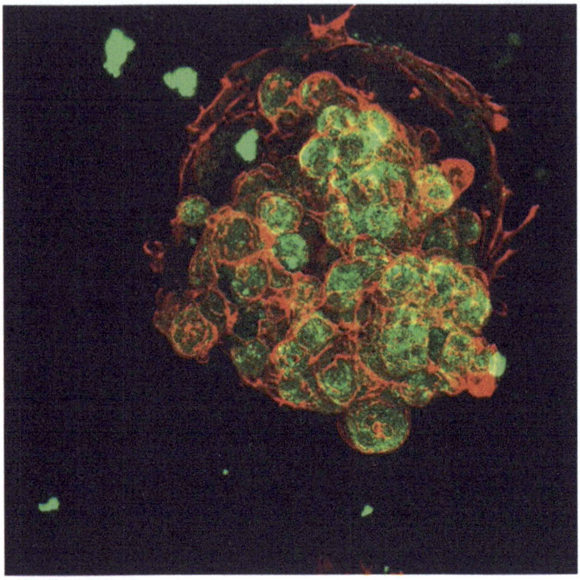

**Fig. 3.5** Three-dimensional view of spheroids underlaid with endothelial cells as a feeder layer [29] (reprinted from ref. 29)

two-dimensional cell monolayers, not least with respect to gene expression profiles, signaling pathway activity and drug sensitivity. However, most currently available three-dimensional techniques are time-consuming and/or lack reproducibility; thus standardized and rapid protocols are urgently needed. Eccles et al. have established and validated a suite of highly reproducible tumor microplate three-dimensional functional assays to enhance the biological relevance of early preclinical cancer studies (Fig. 3.6) [103]. To deploy three-dimensional cultures in cancer drug discovery and target validation, they established a standardized microplate method with the following desirable characteristics: (a) 96-well suspension culture; (b) a single spheroid/well, centered for ease of optical imaging; (c) high reproducibility; (d) simple harvesting for further analysis; (e) fully automated imaging and quantitative analysis. They used ultra-low attachment (ULA) 96-well round-bottomed plates that do not require coating to prevent cell adhesion. Tumor cell suspensions formed a three-dimensional structure within 24–48 h. The well shape promotes the formation of single, centrally located spheroids of reproducible size (Fig. 3.6a). Optimal seeding densities were established such that tumor spheroids for every cell line tested (regardless of their proliferative potential and cell cycle time) fell within a size range of 300–500 μm in diameter on day 4. For rapid, routine imaging and analysis of tumor spheroids, they utilized a Celigo imaging cytometer (Nexcelom Bioscience LLC, Lawrence, MA 01843, USA) (Fig. 3.6b), which is a benchtop in situ cellular analysis system providing high-quality, full or partial images of wells using bright field or fluorescent illumination. Images were acquired and analyzed using the Colony Counting Embryoid Body (EB) application that identifies and counts individual three-dimensional structures (EBs or spheroids) and small clusters of cells. With centrally located spheroids and by selecting 1/16 field of view/

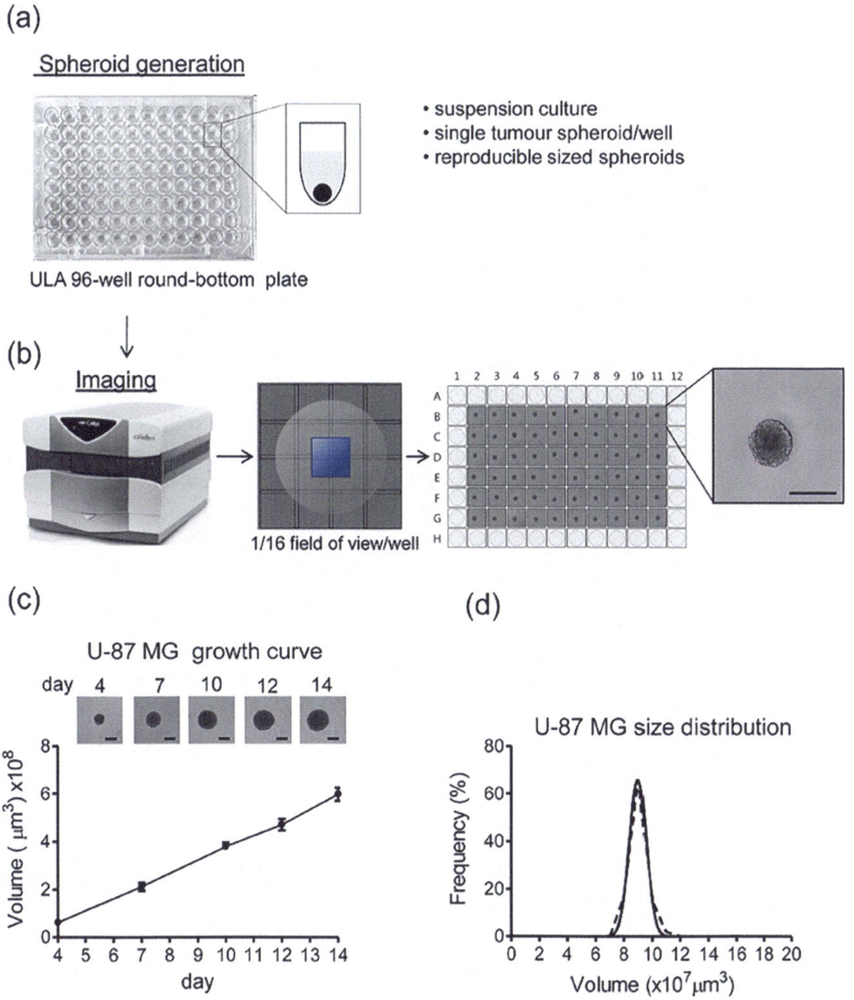


**Fig. 3.6** High-throughput (HT) standardized method for generation, imaging, and analysis of tumor spheroids. (**a**) Ultra-low attachment (ULA) 96-well round-bottomed plates were used to generate suspension cultures of reproducibly sized, single spheroids in each well. (**b**) Fully automated imaging and analysis were carried out on a Celigo cytometer using the Colony Counting Embryoid Body application with 1/16 field of view/well. A screen print shows reproducibly sized U-87 MG spheroids across the plate and details of the segmentation analysis around the spheroid. Scale bar: 500 μm. (**c**) U-87 MG tumor spheroid growth curve. Values are means ± SD ($n = 6$). Representative images obtained on Celigo for each timepoint are shown. Scale bar 500 μm. (**d**) Frequency plots of four-day U-87 MG spheroid volume ($\mu m^3 \times 10^7$) across a plate containing 60 spheroids [103] (reprinted from ref. 103)

well (Fig. 3.6b), the Celigo cytometer scans a ULA 96-well plate in 8 minutes, allowing fast, automated multiparametric analysis including measurements of spheroid diameter, perimeter, and area. Images are acquired at intervals, and growth curves are rapidly and easily generated, showing linear volume increases up to 14 days following cell seeding. An example using U-87 MG glioblastoma is shown in Fig. 3.6c. Small standard deviations demonstrate the reproducibility and robustness of the method. They believe these assays will increase the translational predictive value of in vitro drug evaluation studies and reduce the need for in vivo studies by more effective triaging of compounds.

This miniaturized artificial tissue-like array has the ability to rapidly evaluate drugs and environmental perturbations for potential risk to health and to make predictions for the effects of exposure. Furthermore, the patterned array of cell-organized structures such as spheroids on surfaces may have a particular importance for constructing tissue-engineered liver by seeding spheroids into three-dimensional scaffolds. This is also a useful tool to obtain insights into the mechanism of cell–cell interaction, a central research topic in cell biology.

## 3.3 Hanging Drop

Takayama et al. describe a 384-well format hanging drop culture plate that makes spheroid formation, culture, and subsequent drug testing on the obtained 3D cellular constructs as straightforward to perform and adapt to existing high-throughput screening (HTS) instruments as conventional 2D cultures [104]. Using this platform, they show that drugs with different modes of action produce distinct responses in the physiological 3D cell spheroids compared to conventional 2D cell monolayers. Specifically, the anticancer drug 5-fluorouracil (5-FU) that inhibits cellular proliferation has higher antiproliferative effects on 2D cultures, whereas the hypoxia-triggered cytotoxin tirapazamine (TPZ) that causes DNA damage is more effective against 3D cultures. The multiplexed 3D hanging drop culture and testing plate provides an efficient way to obtain biological insights that are often lost in 2D platforms.

## 3.4 Cell Assembly for Tissue Engineering

Artificial microtissues can also be fabricated by inducing the aggregation of one or more cell types [105]. These microtissues may be beneficial for applications such as pancreatic, liver, vascular, and cardiac tissue engineering, as well as drug discovery. Layering of cells has been used to engineer myocardial tissues by assembling multiple sheets of cardiomyocytes [106] or to engineer blood vessels by fabricating cylindrically rolled sheets of endothelial cells [107]. Although such approaches may be suitable for some tissue engineering applications, they lack the complexity associated with the architecture of more complex organs. Microscale approaches may provide a solution to this challenge as templates to generate microtissues in a reproducible manner. For example, by using a combination of microcontact printing and micromachining, hepatocyte spheroids have been formed [108]. More recently, nonadhesive PEG microwells have been used as templates for the formation of aggregates of various cell types, including ES cells [109]. This approach aims to overcome the disadvantages associated with the hanging drop and suspension culture methods [105] by providing control over the size, shape, and other features of the cellular assembly in a scalable manner. The controlled formation of embryoid bodies may also be important in generating more homogeneous cultures that are capable of directing the differentiation of ES cells. In addition, template-based assembly of cells could be used to organize multiple cell types into specific geometries relative to each other within these aggregated tissue sections. It is envisioned that with the integration of such technologies with biomaterials such as photocrosslinkable gels and microfluidics, more complex tissue sections for therapeutic applications can be fabricated.

For tissue engineering applications, the cells are subsequently incorporated into three-dimensional polymer scaffolds that act as analogs to the natural extracellular matrices found in tissues. These scaffolds deliver the cells to the desired site in the patient's body, provide a space for new tissue formation, and potentially control the structure and function of the engineered tissue [110, 111]. A critical element in virtually all tissue engineering approaches is the polymer scaffold. The polymer potentially mimics many roles of extracellular matrices found in tissues. Extracellular matrices, comprised of various amino acids and sugar-based macromolecules, bring cells together and control the tissue structure, regulate the function of the cells, and allow the diffusion of nutrients, metabolites, and growth factors [112]. Various types of polymers have been three-dimensionally formulated to be utilized as suitable cell scaffolds [113]. Among them, hydrogels have found numerous applications in tissue engineering, because they have structural similarity to the macromolecular-based components in the body and are considered biocompatible [114]. Hydrogels, currently used or with potential applications in tissue engineering, are divided into two categories, according to their natural or synthetic origin. Hydrogels from natural polymers have been widely used for tissue engineering approaches and are expected to have better interaction with cells, along with increased cell proliferation and differentiation. However, limitations of gels from natural polymers have motivated approaches to modify these polymers as well as to use various synthetic polymers. A wide range of synthetic polymers may potentially have suitable chemical and physical properties for these applications.

Here, we will discuss typical natural polymers (divided into polysaccharides and polypeptides) and synthetic polymers, which have been widely used for hydrogel formation, and then their properties and applications (Fig. 3.7) [115].

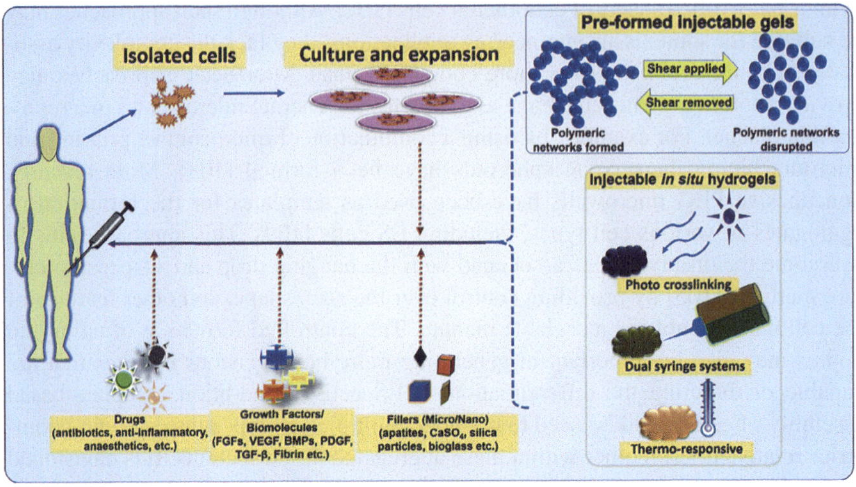


**Fig. 3.7** Schematic outlining preformed injectable hydrogels along with different in situ forming hydrogels and the possibilities of using them as carriers for cell, drug, growth factor/biomolecules, and filler [115] (reprinted from ref. 115)

## 3.5 Hydrogels as Three-Dimensional Cell Scaffolds

### *3.5.1 Synthetic Polymers*

Specific applications in tissue regeneration require certain modifications to the polymer structure. This reengineering is quite difficult with natural polymers; hence, synthetic polymers, such as aliphatic polyesters including poly(glycolic acid) (PGA), poly(lactic acid) (PLA), their copolymers (PLGA), and polycaprolactone (PCL), are widely used in tissue engineering. PCL is a relatively cheap and elastic synthetic polymer that has been approved by the United States Food and Drug Administration (FDA). PLGA is another FDA-approved synthetic biopolymer used in tissue engineering. The major advantages of PLGA include biocompatibility and controlled degradation.

### *3.5.2 Biopolymers*

Tissue engineering employs biopolymer-based biocompatible and biodegradable natural materials that provide a suitable bioactive environment and necessary mechanical support to promote the growth of new tissue in defect sites. Due to their superior biocompatibility and minute negative immunological influence, natural biopolymers, including polysaccharides and polypeptides such as collagen, chitosan, alginate, silk, fucoidan, elastin, gelatin, and hyaluronic acid, are extensively used in regenerative medicine.

### 3.5.3 Polysaccharides

Polysaccharides are substances composed of a number of monosaccharide molecules joined by glycosidic bonds and are composed of repeating units of monosaccharides or disaccharides, providing important structural and biological functionality in living organisms (Fig. 3.8) [116]. Polysaccharides are obtained as biosynthetic products by living organisms and are often isolated from renewable resources such as plants and microorganisms. Hydrogels with a wide range of functions have been developed by cross-linking chemically modified polysaccharides.

**Polysaccharides**

| Name | Hyaluronic Acid (HA) | Chondroitin Sulfate (CS) | Heparin Sulfate (HS) | Dextran |
|---|---|---|---|---|
| Chemical structure | 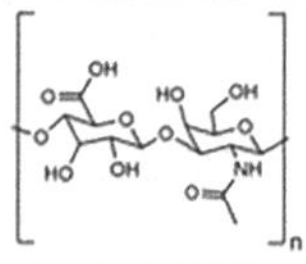 | 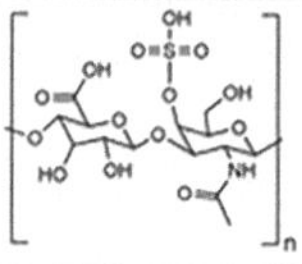 | 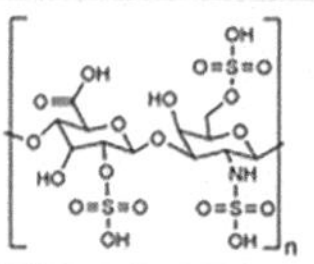 | 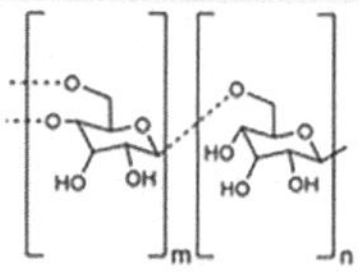 |
| Type | Linear GAG | Linear GAG | Linear GAG | Branched |
| Native function | Major component of ECM in human connective tissue | Major component of ECM in human connective tissue | Stored in mast cells in humans and secreted into vasculature upon injury | Component of some bacterial ECM |
| Source for biomedical applications | Bacterial culture; rooster comb[19] | Animal sources (e.g., bovine trachea, porcine nasal septa)[29] | Animal sources (e.g., porcine intestines)[34] | Bacterial culture[39] |
| Name | Alginate | Cellulose | Chitin | Chitosan |
| Chemical structure | 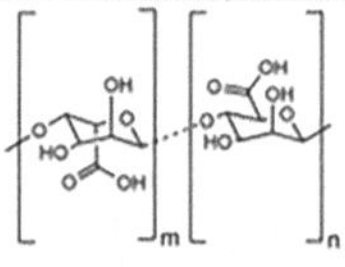 | 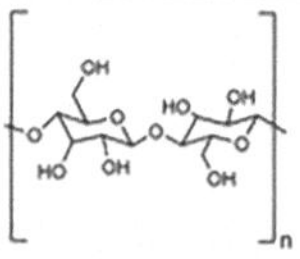 | 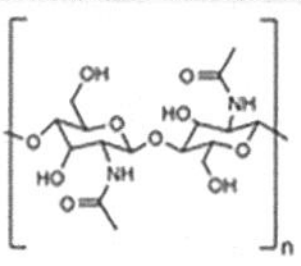 | 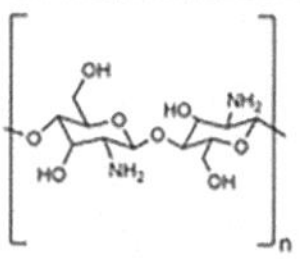 |
| Type | Linear | Linear | Linear | Linear |
| Native function | Structural support in seaweed | Structural support in plant cell walls; component in some bacterial ECM | Structural support in insect and crustacean exoskeletons | n/a |
| Source for biomedical applications | Brown algae[42] | Bacterial culture[49] | Shrimp and crab shells from food industry waste products[56] | Deacetylated derivative of chitin[57] |

**Fig. 3.8** Polysaccharide-based biopolymers. Representative chemical structures, function, and sources of various polysaccharide biopolymers that have been modified to form biomedical hydrogels [116] (reprinted from ref. 116)

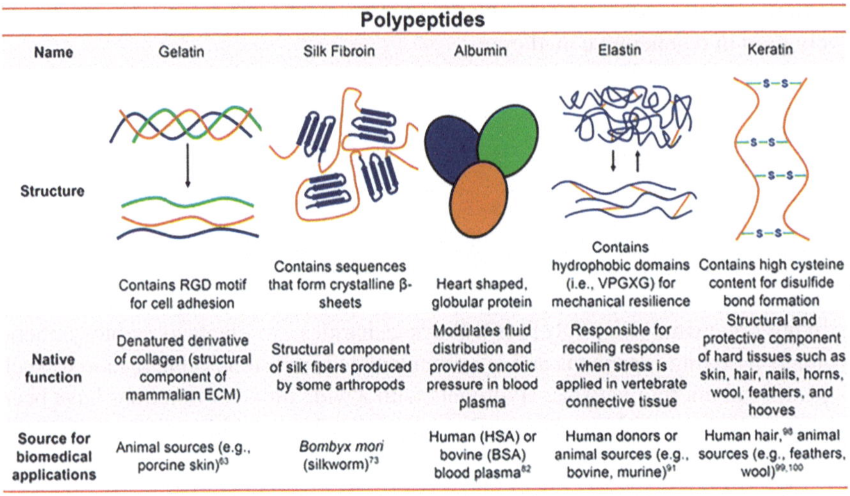


**Fig. 3.9** Polypeptide-based biopolymers. Representative structures, function, and sources of various polypeptide biopolymers that have been modified to form biomedical hydrogels [116] (reprinted from ref. 116)

### *3.5.4 Polypeptide*

A polypeptide is a biopolymer consisting of linear repeating chains of amino acids; if it consists of more than 50 amino acids, it is defined as a protein (Fig. 3.9) [116]. Hydrogels made from polypeptides have many cell interaction sites and can mimic the function of the extracellular matrix, making them suitable for biomedical applications. Polypeptides can be isolated from animal or plant sources or synthetically biomanufactured using peptide synthesizers or recombinant protein production as a function of total host cell biomass synthetic capacity.

### *3.5.5 Methodologies of Cross-Linking*

Gelation is caused by cross-linking of polymers to form a polymer network with an internal solvent. There are two main types of cross-linking: chemical cross-linking and physical cross-linking. Chemical cross-linking is formed by covalent bonding of functional groups within the biopolymer or synthetic polymer and is generally strong and stable. In contrast, physical cross-linking is formed by non-covalent bonding, such as hydrogen bonding, between polar polymer chains. Although the mechanical strength may be low, gels with self-healing and shear-thinning properties are often obtained due to the reversibility of the physical interaction. These typical methods are described below.

### *3.5.6 Chemical Cross-Linking*

In general, covalent bonds are stable and enable the synthesis of hydrogels with high mechanical strength. Among covalent bonds, the introduction of dynamic covalent bonds, which are reversible reactions, makes it possible to give gels self-healing properties while maintaining high mechanical strength. In addition, the gelation and degradation behavior may be controlled in a stimulus-responsive manner. Because of these advantages, dynamic covalent gels have been considered for many applications, such as scaffold materials in tissue engineering and drug delivery systems. In this section, we will discuss the covalent click chemistry, active-ester reaction and dynamic covalent Schiff bases, as representative examples.

#### 3.5.6.1 Click Chemistry

Click chemistry was proposed by K. B. Sharpless as a versatile, substrate-specific reaction to proceed in high yield and without by-products. The advantages of click chemistry are that it is simple to operate, and the substrates and products are stable to water and oxygen and proceed under a wide variety of reaction conditions. Because of its biocompatibility, reliability, and bioorthogonal specificity, it is widely used in biomedical applications such as biomaterials engineering and drug discovery. Typical click chemistry used for hydrogel cross-linking includes thiol-ene Michael addition reaction, thiol-ene radical addition reaction, azide-alkyne reaction, and tetrazine-norbornene cycloaddition.

#### 3.5.6.2 Active-Ester Reaction

A more, mild cross-linking to form hydrogels involves the application of active esters of carboxylic acids. To obtain these molecules, carboxylic acids contained in the polymers are converted into esters of N-hydroxy succinimide (NHS), which yields an ester that is readily attacked by free amine groups and subsequently converted into stable amide linkages. One drawback of this cross-linking is that two different components have to be mixed thoroughly shortly before use to obtain homogeneous gels.

#### 3.5.6.3 Schiff Bases

A Schiff base is a bond formed by the condensation reaction of a carbonyl group with a primary amine and is represented by the general formula $R_1R_2C{=}NR_3$. Only water is produced as a by-product. The reversible reaction of Schiff base formation proceeds under mild conditions and is pH-dependent. In the biomaterials field, hydrogels are formed by mixing aldehyde-functionalized and amine-functionalized polymers. It is also known that the neighboring atoms of primary amines used for Schiff base formation affect the stability of cross-linking. A Schiff base with a nitrogen atom next to the primary amine is called a hydrazone bond, which can form a more stable hydrogel than a normal Schiff base.

### 3.5.7 *Physical Cross-Linking*

Physical cross-links are formed by physical interactions between biomacromolecules, such as hydrogen bonding, hydrophobic interactions, and electrostatic attraction. Although mechanical strength may be lower than gels cross-linked by chemical cross-linking, physical cross-linking can be used to enhance cell migration and diffusion of cells. In addition, physical interactions are reversible and can impart self-healing properties to hydrogels. Physical cross-linking is usually formed by one-pot mixing of two or more components without the need for an external catalyst or initiator.

### 3.5.8 *Effect of Gel Properties on Cell Function*

Gel properties such as pore size, stiffness, degradation rate, adhesion, and viscoelasticity have been shown to influence cell function in vitro and in vivo. In particular, the pore size, decomposition rate, and viscoelasticity of the gel are discussed as follows.

### 3.5.9 *Pore Size of Gel*

Hydrogels have a porous structure with many pores, which enables cell migration and transport of nutrients. The pore size is influenced by the polymer composition that constitutes the hydrogel and the cross-linking method. If the hydrogel absorbs more water due to the degree of hydrophilicity, the mesh size will correspondingly increase. Furthermore, recent studies on the microstructure of scaffolds have shown that the optimal pore size for cell function is cell type specific and influences cell shape, differentiation, and extracellular matrix secretion. Several studies have estimated whether the average pore size of the scaffold affects the initial adhesion of mesenchymal stem cells (MSCs) and later cell differentiation. Hydrogel scaffolds with larger average pore size (300 nm) have been reported to stimulate higher cell proliferation, cartilage gene expression, cartilage matrix deposition, and to have higher volumetric modulus after in vitro culture than gels with smaller pore size (94 and 130 nm). The advantage of scaffolds with large pore size may be attributed to improved initial cell attachment and proliferation and later cell migration to form tissue-like cell spheroids, as well as to their effect on morphology. Gels with large pore size may have supported the formation of spheroid-friendly rounded cell morphology and promoted chondrogenic differentiation.

### 3.5.10 Degradation of Hydrogels

The use of hydrogels in vivo in patients requires their substitution with extracellular matrix via degradation, and several strategies have been developed for this purpose. To be removed from the body, the hydrogel must be degraded to a water-soluble polymer small enough to pass through the renal barrier; in the case of PEG derivatives, the upper size limit has been reported to be approximately 40,000 Da. Hydrolytically cleavable bonds such as ester bonds or substances that are degraded by enzymes are used. In natural polymers, the hydrogel can also be removed by physiological metabolism of the monomer, for example, by digestion. Important factors in these degradations include the degree of swelling depending on network structures and the toxicity of the released degradation products leading to immune rejection. With regard to the rate of scaffold degradation of aliphatic polyesters including PLA and PGA, the degradation products of biodegradable polymers have been reported largely nontoxic, but the production of localized acidic by-products, which depend on the rate of scaffold degradation, can have adverse effects on cells. Hak-Joon Sung et al. used fast-degrading polymer (poly(lactic-co-glycolic acid)) and slow-degrading polymer (polycaprolactam) to differentiate degradation rates and produce anti-porous scaffolds of uniform thickness. The effect of degradation was tested in vitro, and cell viability was found to be inversely proportional to degradation rate. Rapid degradation of the scaffold, possibly due to acidification of the local environment caused by polymer degradation, has been reported to adversely affect cell viability and migration. The results of these studies indicate the importance of the influence of the physical properties of hydrogels on cellular function in tissue regeneration.

### 3.5.11 Influence of Viscoelasticity

The mechanical properties of hydrogels, such as viscosity and elasticity, are important properties to withstand mechanical forces and also influence cell differentiation during stem cell culture. The mechanical stability of hydrogels is affected by several parameters, including polymer content, cross-linking density, and cross-linking method. Increasing these parameters increases gel strength, but there are concerns that a significant decrease in mesh size may prevent the growth of cells necessary for tissue formation. It has been reported that the stiffness of the scaffold material, regulated by these factors, affects stem cell differentiation. Depending on the stiffness of the scaffold material, differentiation lineages are determined as follows:

neurons in brain tissue stiffness (0.1–1 kPa), myoblastic neurons in muscle tissue stiffness (8–17 kPa), and osteoblastic neurons in new bone stiffness (25–40 kPa). Therefore, it is important to prepare the appropriate viscoelasticity and stiffness for the target tissue.

### *3.5.12 Injectable IPN Gels as New and Minimally Invasive Applications*

Interpenetrating polymer network (IPN) gels, which are defined as a mixture of multiple interconnected cross-linked polymer network structures, have attracted much attention in recent years. IPN gels are expected not only to improve mechanical strength, but also to enhance biocompatibility by mimicking the extracellular matrix compared to single-network gels, and are expected to be applied to tissue regeneration. In addition, by using multiple materials and designing gels that take advantage of their characteristics, gels with numerous functions have been developed, such as mechanical strength, affinity to cells, self-healing properties, and chondrogenic differentiation properties. Normally, IPN gels were synthesized through multiple steps: the first network was formed, and then chemical reagents or external stimuli were added to form the second network. However, those additives and stimuli are likely to be toxic in in situ gels that are injected into the body. Therefore, a method was devised to synthesize IPN gels in a one-pot, injectable, and time-dependent manner by utilizing two different network-forming driving forces. An exciting alternative approach to cell delivery for tissue engineering is the use of polymers (i.e., hydrogels) that can be injected into the body. This approach enables the clinician to transplant the cell and polymer combination in a minimally invasive manner. Among hydrogels, injectable gels, which form in situ when administered in vivo, are attracting attention as carriers for drug delivery and scaffolds for tissue engineering. Gels containing cells, therapeutic drugs, growth factors, and other substances can be easily prepared by simply mixing them with a gel precursor polymer solution and injecting them. Ishikawa et al. moderately chemical cross-linked the amino groups of chitosan (CH) with both terminal succinimidated poly(ethylene glycol) (NHS-PEG-NHS) gels, followed by the addition of the self-aggregating peptide chain RADA16 to form IPNs, which showed high utility and safety in terms of mechanical strength and chondrogenic differentiation properties [117]. This structure is formed by the physical cross-linking of the self-assembling peptide chain RADA16 immediately after the three-component mixture of RADA16, CH, and NHS-PEG-NHS, followed by the chemical cross-linking of chitosan and polyethylene glycol. This one-pot and in situ method of synthesizing IPN gels allows for a wide variety of material combinations and is expected to be a widely applicable technology not only in tissue regeneration but also in drug delivery, diagnosis, etc.

## 3.6 Conclusion

A number of techniques utilized for cell patterning have been separated into two categories: those where the cells are passively patterned by random seeding on surfaces modified with cytophilic and cytophobic regions and those where the cells are actively deposited on the surfaces via optical or electrical forces or even directly printed. Although this paper only reviewed the former category, we believe that there is great potential for the future in combining specific techniques with the aim of solving a number of problems inherent to long-term culturing of active cell arrays.

Microfabrication using dry etching and photolithography is rapidly expanding, and "soft lithography" is being increasingly employed for biological applications [118]. Many microfluidic devices have also been developed, and these systems are rapidly being applied in the biomedical and pharmaceutical industries [119]. Particular importance to form multicellular aggregates as spheroids was suggested for long-term drug screening and tissue engineering. Typically, cirrhosis refers to irreversible liver damage where healthy tissue is replaced by scar tissue, resulting in impaired liver function. There is no cure, and current treatments only prevent further liver damage; thus, novel therapeutic options are urgently needed. We make a proposition for a new approach that enables the formation of self-assembled 3D spheroids of mesenchymal stem cells (MSCs), and human hepatocytes via injectable hydrogels that can be injected into the body. This approach enables the clinician to transplant the cell and polymer combination in a minimally invasive manner. Compared with the surgeon to make incisions (cuts) sufficiently large to enable placement of the polymer/cell constructs, the injectable hydrogel-based MSC spheroids offer a notable benefit in tissue-structure formation and paracrine factor secretion.

**Acknowledgments** Financial support for this work was partly provided by Special Coordination Funds for Promoting Science and Technology and also supported by Research Promotion Bureau under contract Nos. 15-99 and 15-396, both from the Ministry of Education, Culture, Sports, Science, and Technology (MEXT), Japan. Part of this work was also supported financially by a Grant-in-Aid for Research on Health Sciences for Drug Innovation (KHD1027), the Ministry of Health, Labor, and Welfare of Japan, and The New Energy and Industrial Technology Development Organization (NEDO).

## References

1. Whitesides GM, Ostuni E, Takayama S, Jiang X, Ingber DE (2001) Soft lithography in biology and biochemistry. Annu Rev Biomed Eng 3:335–373. https://doi.org/10.1146/annurev.bioeng.3.1.335
2. Jeon NL, Baskaran H, Detringer SKW, Whitesides GM, Water LV, Toner M (2002) Neutrophil chemotaxis in linear and complex gradients of interleukin-8 formed in a microfabricated device. Nat Biotechnol 20:826–830. https://doi.org/10.1038/nbt712

3. Britland S, Clark P, Connolly P, Moores G (1992) Micropatterned substratum adhesiveness: a model for morphogenetic cues controlling cell behavior. Exp Cell Res 198:124–129. https://doi.org/10.1016/0014-4827(92)90157-4
4. Healy KE, Lom B, Hockberger PE (1994) Spatial distribution of mammalian cells dictated by material surface chemistry. Biotechnol Bioeng 43:792–800. https://doi.org/10.1002/bit.260430814
5. Hickman JJ, Bhatia SK, Quong JN, Shoen P, Stenger DA, Pike CJ et al (1994) Rational pattern design for in-vitro cellular networks using surface photochemistry. J Vac Sci Technol A 12(3):607–616. https://doi.org/10.1116/1.581256
6. Bekos EJ, Ranien JP, Aebischer P, Gardella JA, Bnght FV (1995) Structural changes of bovine serum albumin upon adsorption to modified fluoropolymer substrates used for neural cell attachment studies. Langmuir 11:984–989. https://doi.org/10.1021/la00003a048
7. Singhvi R, Kumar A, Lopez GP, Stehanopoulos GN, Wan DIC et al (1994) Engineering cell shape and function. Science 264:696–698. https://doi.org/10.1126/science.8171320
8. Spargo BJ, Testoff MA, Nielsen TB, Stenger DA, Hickman JJ, Rudolph AA (1994) Spatially controlled adhesion, spreading, and differentiation of endothelial cells on self-assembled molecular monolayers. Proc Natl Acad Sci USA 91:11070–11074. https://doi.org/10.1073/pnas.91.23.11070
9. Stenger DA, Georger JH, Dulcey CS, Hickman JJ, Rudolph AS, Nielsen TB, McCort SM, Calvert JM (1992) Coplanar molecular assemblies of amino- and perfluorinated alkylsilanes: characterization and geometric definition of mammalian cell adhesion and growth. J Am Chem Soc 114:8435–8442. https://doi.org/10.1021/ja00048a013
10. van den Berg A, Lammerink TSJ (1998) Micro total analysis systems: microfluidic aspects, integration concept and applications. In: Manz A, Becker H (eds) Microsystem technology in chemistry and life science. Topics in current chemistry, vol 194. Springer, Berlin, Heidelberg, pp 21–49. https://doi.org/10.1007/3-540-69544-3_2
11. Manz A, Graber N, Widmer HM (1990) Miniaturized total chemical analysis systems: a novel concept for chemical sensing. Sens Actuat B Chem 1:244–248. https://doi.org/10.1016/0925-4005(90)80209-I
12. Dittrich PS, Manz A (2006) Lab-on-a-chip: microfluidics in drug discovery. Nat Rev Drug Discov 5:210–218. https://doi.org/10.1038/nrd1985
13. Bhatia SN, Balis UJ, Yarmush ML, Toner M (1999) Effect of cell-cell interactions in preservation of cellular phenotype: cocultivation of hepatocytes and nonparenchymal cells. FASEB J 13:1883–1900. https://doi.org/10.1096/fasebj.13.14.1883
14. Folch A, Toner M (2000) Microengineering of cellular interactions. Annu Rev Biomed Eng 2:227–256. https://doi.org/10.1146/annurev.bioeng.2.1.227
15. Tsang VL, Bhatia SN (2004) Three-dimensional tissue engineering. Adv Drug Deliv Rev 56:1635–1647. https://doi.org/10.1016/j.addr.2004.05.001
16. Chen CS, Mrksich M, Huang S, Whitesides GM, Ingber DE (1997) Geometric control of cell life and death. Science 276:1425–1428. https://doi.org/10.1126/science.276.5317.1425
17. Zamir E, Katz BZ, Aota KM, Yamada KM, Geiger B, Kam Z (1999) Molecular diversity of cell-matrix adhesions. J Cell Sci 112:1655. https://doi.org/10.1242/jcs.112.11.1655
18. Geiger B, Bershadsky R, Pankov R, Yamada KM (2001) Transmembrane crosstalk between the extracellular matrix and the cytoskeleton. Nat Rev Mol Cell Biol 2:793–805. https://doi.org/10.1038/35099066
19. Chou L, Firth JD, Uitto VJ, Brunette DM (1995) Substratum surface topography alters cell shape and regulates fibronectin mRNA level, mRNA stability, secretion and assembly in human fibroblasts. J Cell Sci 108:1563–1573. https://doi.org/10.1242/jcs.108.4.1563
20. Clark P, Connolly P, Curtis ASG, Dow JAT, Wilkinson CDW (1990) Topographical control of cell behaviour: II. Multiple grooved substrata. Development 108:635–644. https://doi.org/10.1242/dev.108.4.635

21. Stenger DA, Gross GW, Keefer EW, Shaffer KM, Andreadis JD, Ma W, Pancrazio JJ (2001) Detection of physiologically active compounds using cell-based biosensors. Trends Biotechnol 19:304–309. https://doi.org/10.1016/S0167-7799(01)01690-0
22. Kononen J, Bubendorf L, Kallioniemi A, Barlund M, Schraml P, Leighton S, Torhorst J, Mihatsch MJ, Sauter G, Kallioniemi OP (1998) Tissue microarrays for high-throughput molecular profiling of tumor specimens. Nat Med 4:844–847. https://doi.org/10.1038/nm0798-844
23. Ziauddin J, Sabatini DM (2001) Microarrays of cells expressing defined cDNAs. Nature 411:107–110. https://doi.org/10.1038/35075114
24. Michalopoulos GK, DeFrances MC (1997) Liver regeneration. Science 276(5309):60–66. https://doi.org/10.1126/science.276.5309.60
25. Anderson DG, Levenberg S, Langer R (2004) Nanoliter-scale synthesis of arrayed biomaterials and application to human embryonic stem cells. Nat Biotechnol 22:863–866. https://doi.org/10.1038/nbt981
26. Revzin A, Tompkins RG, Toner M (2003) Surface engineering with poly(ethylene glycol) photolithography to create high-density cell arrays on glass. Langmuir 19:9855–9862. https://doi.org/10.1021/la035129b
27. Thielecke H, Mack A, Robitzki A (2001) A multicellular spheroid-based sensor for anti-cancer therapeutics. Biosens Bioelectron 16:261–269. https://doi.org/10.1016/S0956-5663(01)00140-3
28. Mack AR, Thielecke H, Robitzki AA (2002) 3D-biohybrid systems: applications in drug screening. Trends Biotechnol 20:56–61. https://doi.org/10.1016/S0167-7799(01)01880-7
29. Otsuka H, Hirano A, Nagasaki Y, Okano T, Horiike Y, Kataoka K (2004) Two-dimensional multiarray formation of hepatoytes sphroids on a microfabricated PEG-brush surface. Chembiochem 5:850–855. https://doi.org/10.1002/cbic.200300822
30. Luk YY, Kato M, Mrksich M (2000) Self-assembled monolayers of alkanethiolates presenting mannitol groups are inert to protein adsorption and cell attachment. Langmuir 16:9604–9608. https://doi.org/10.1021/la0004653
31. Nelson CM, Raghavan S, Tan JL, Chen CS (2003) Degradation of micropatterned surfaces by cell-dependent and -independent processes. Langmuir 19:1493–1499. https://doi.org/10.1021/la026178b
32. Ratner BD, Bryant SJ (2004) Biomaterials: where we have been and where we are going. Ann Rev Biomed Eng 6:41–75. https://doi.org/10.1146/annurev.bioeng.6.040803.140027
33. Elbert D, Hubbell J (1996) Surfaee treatments of polymers for biocompatibility. Annu Rev Mater Sci 26:365–394
34. Lasic DD, Papahadjopoulos D (1996) Liposomes and biopolymers in drug and gene delivery. Curr Opin Solid State Mater Sci 1:392–400
35. Lee RJ, Low PS (1994) Delivery of liposomes into cultured kb cells via folate receptor-mediated endocytosis. J Biol Chem 269:3198
36. Lee RJ, Low PS (1995) Folate-mediated tumor cell targeting of liposome-entrapped doxorubicin in vitro. Biochim Biophys Acta 1233:134
37. Harris JM (ed) (1992) Poly(ethylene glycol) chemistry. Biotechnical and biomedical applications. Plenum Press, New York, NY, USA. https://doi.org/10.1007/978-1-4899-0703-5
38. Inoue Y, Nakanishi T, Ishihara K (2013) Elastic repulsion from polymer brush layers exhibiting high protein repellency. Langmuir 29:10752–10758. https://doi.org/10.1021/la4021492
39. Szleifer I (1996) Statistical thermodynamics of polymers near surfaces. Curr Opin Colloid Interface Sci 1:416–423
40. Milner ST (1991) Polymer brushes. Science 251:905-914.12
41. Halperin A (1994) On polymer brushes and blobology: an introduction. In: Rubin Y, Bruisma R (eds) Soft order in physical systems. Prenum Press, New York, London, pp 33–56
42. Lee J, Martic P, Tan J (1989) Protein adsorption on pluronic copolymer-coated polystyrene particles. J Colloid Interface Sci 131:252–266. https://doi.org/10.1016/0021-9797(89)90166-5

43. Lee J, Kopecek J, Andrade J (1989) Protein-resistant surfaces prepared by PEO-containing block copolymer surfactants. J Biomed Mater Res 23:351–368. https://doi.org/10.1002/jbm.820230306
44. Tan J, Martic P (1990) Protein adsorption and conformational change on small polymer particles. J Colloid Interface Sci 136:415–431. https://doi.org/10.1016/0021-9797(90)90389-6
45. Lshihara K, Ziats N, Tierney B, Nakabayashi N, Anderson J (1991) Protein adsorption from human plasma is reduced on phospholipid polymers. J Biomed Mater Res 25:1397–1407. https://doi.org/10.1002/jbm.820251107
46. Desai N, Hubbell J (1991) Biological responses to polyethylene oxide modified polyethylene terephthalate surfaces. J Biomed Mater Res 25:829–843. https://doi.org/10.1002/jbm.820250704
47. Gombotz W, Guanghui W, Horbett T, Hoffman (1991) A. Protein adsorption to poly(ethylene oxide) surfaces. J Biomed Mater Res 25: 1547–1562. https://doi.org/10.1002/jbm.820251211
48. Ham DK, Jeong S, Kim Y, Min B, Cho H (1991) Negative cilia concept for thromboresistance: synergistic effect of PEO and sulfonate groups grafted onto polyurethanes. J Biomed Mater Res 25:561–575. https://doi.org/10.1002/jbm.820250502
49. Amiji M, Park K (1992) Prevention of protein adsorption and platelet adhesion on surfaces by PEO/PPO/PEO triblock copolymers. Biomaterials 13:682–692. https://doi.org/10.1016/0142-9612(92)90128-B
50. Bergström K, Holmberg K, Safranj A, Hoffman AS, Edgell MJ, Kozlowski A, Hovanes BA, Harris JM (1992) Reduction of fibrinogen adsorption on PEG-coated polystyrene surfaces. J Biomed Mater Res 26:779–790. https://doi.org/10.1002/jbm.820260607
51. Osterberg E, Bergström K, Holmberg K, Riggs JA, Alstine JMV, Schuman TP, Burns NL, Harris JM (1993) Comparison of polysaccharide and poly(ethylene glycol) coatings for reduction of protein adsorption on polystyrene surfaces. Colloids Surf A Physicochem Eng Aspects 77:159–169. https://doi.org/10.1016/0927-7757(93)80112-R
52. Fujimoto K, Lnoue H, Lkada Y (1993) Protein adsorption and platelet adhesion onto polyurethane grafted with methoxy-poly(ethylene glycol) methacrylate by plasma technique. J Biomed Mater Res 27:1559–1567. https://doi.org/10.1002/jbm.820271213
53. Llanos GR, Sefton MV (1993) Immobilization of poly(ethylene glycol) onto a poly(vinyl alcohol) hydrogel: 2. Evaluation of thrombogenicity. J Biomed Mater Res 27:1383–1391. https://doi.org/10.1002/jbm.820271105
54. Lin YS, Hlady V, Gölander CG (1994) The surface density gradient of grafted poly(ethylene glycol): preparation, characterization and protein adsorption. Colloids Surf B: Biointerfaces 3:49. https://doi.org/10.1016/0927-7765(93)01114-7
55. McPherson TB, Lee SJ, Park K (1995) Analysis of the prevention of protein adsorption by steric repulsion theory. In: Horbett TA, Brash JL (eds) Proteins at interfaces II. American Chemical Society, Washington, DC, USA, pp 395–404. https://doi.org/10.1021/bk-1995-0602.ch028
56. Pasche S, Vörös J, Griesser HJ, Spencer ND, Textor M (2005) Effects of ionic strength and surface charge on protein adsorption at PEGylated surfaces. J Phys Chem B 109(37):17545–17552. https://doi.org/10.1021/jp050431+
57. Pasche S, Paul SMD, Vörös J, Spencer ND, Textor M (2003) Poly(l-lysine)-graft-poly(ethylene glycol) assembled monolayers on niobium oxide surfaces: a quantitative study of the influence of polymer interfacial architecture on resistance to protein adsorption by ToF-SIMS and in situ OWLS. Langmuir 19:9216–9225. https://doi.org/10.1021/la034111y
58. Wagner MS, Pasche S, Castner DG, Textor M (2004) Characterization of poly(l-lysine)-graft-poly(ethylene glycol) assembled monolayers on niobium pentoxide substrates using time-of-flight secondary ion mass spectrometry and multivariate analysis. Anal Chem 76:1483–1492. https://doi.org/10.1021/ac034873y
59. Booth C, Attwood D (2000) Effects of block architecture and composition on the association properties of poly(oxyalkylene) copolymers in aqueous solution. Macromol Rapid Commun 21:501–527. https://doi.org/10.1002/1521-3927(20000601)21:9<501::AID-MARC501>3.0.CO;2-R

60. Lopes JR, Loh W (1998) Investigation of self-assembly and micelle polarity for a wide range of ethylene oxide–propylene oxide–ethylene oxide block copolymers in water. Langmuir 14:750–756. https://doi.org/10.1021/la9709655
61. Alexandridis P, Holzwarthf JF, Hatton TA (1994) Micellization of poly(ethylene oxide)-poly(propylene oxide)-poly(ethylene oxide) triblock copolymers in aqueous solutions: thermodynamics of copolymer association. Macromolecules 27:2414–2425. https://doi.org/10.1021/ma00087a009
62. Kozlov MY, Melik-Nubarov NS, Batrakova EV, Kabanov AV (2000) Relationship between pluronic block copolymer structure, critical micellization concentration and partitioning coefficients of low molecular mass solutes. Macromolecules 33:3305–3313. https://doi.org/10.1021/ma991634x
63. Wanka G, Hoffmann H, Ulbricht W (1994) Phase diagrams and aggregation behavior of poly(oxyethylene)-poly(oxypropylene)-poly(oxyethylene) triblock copolymers in aqueous solutions. Macromolecules 27(15):4145–4159. https://doi.org/10.1021/ma00093a016
64. Alexandridis P, Hatton TA (1995) Poly(ethylene oxide)-poly(propylene oxide) -poly(ethylene oxide) block copolymer surfactants in aqueous solutions and at interfaces: thermodynamics, structure, dynamics, and modeling. Colloids Surf A: Physicochem Eng Aspects 96:146. https://doi.org/10.1016/0927-7757(94)03028-X
65. Thissen H, Hayes JP, Kingshott P, Johnson G, Harvey EC, Griesser HJ (2002) Nanometer thickness laser ablation for spatial control of cell attachment. Smart Mater Struct 11(5):792–799. https://doi.org/10.1088/0964-1726/11/5/326
66. Li JT, Caldwell KD (1996) Plasma protein interactions with Pluronic™-treated colloids. Colloids Surf B-Biointerface 7:9–22. https://doi.org/10.1016/0927-7765(96)01269-6
67. Neff JA, Tresco PA, Caldwell KD (1999) Surface modification for controlled studies of cell–ligand interactions. Biomaterials 20:2377–2393. https://doi.org/10.1016/S0142-9612(99)00166-0
68. Amiji MM, Park KJ (1994) Analysis on the surface adsorption of PEO/PPO/PEO triblock copolymers by radiolabelling and fluorescence techniques. Appl Polym Sci 52:539–544. https://doi.org/10.1002/app.1994.070520409
69. Liu VA, Jastromb WE, Bhatia SN (2002) Engineering protein and cell adhesivity using PEO-terminated triblock polymers. J Biomed Mater Res 60:126–134. https://doi.org/10.1002/jbm.10005
70. Otsuka H, Nagasaki Y, Kataoka K (2000) Surface characterization of functionalized polylactide through the coating with heterobifunctional poly(ethylene glycol)/Polylactide block copolymers. Biomacromolecules 1:39–48. https://doi.org/10.1021/bm990005s
71. Otsuka H, Akiyama Y, Nagasaki Y, Kataoka K (2001) Quantitative and reversible lectin-induced association of gold nanoparticles modified with α-Lactosyl-ω-mercapto-poly(ethylene glycol). J Am Chem Soc 123:8226–8230. https://doi.org/10.1021/ja010437m
72. Nuzzo RG, Allara DL (1983) Adsorption of bifunctional organic disulfides on gold surfaces. J Am Chem Soc 105:4481–4483. https://doi.org/10.1021/ja00351a063
73. Ostuni E, Chapman RG, Liang MN, Meluleni G, Pier G, Ingber DE, Whitesides GM (2001a) Self-assembled monolayers that resist the adsorption of proteins and the adhesion of bacterial and mammalian cells. Langmuir 17:6336–6343. https://doi.org/10.1021/la010552a
74. Ostuni E, Chapman RG, Holmlin RE, Takayama S, Whitesides GM (2001b) A survey of structure–property relationships of surfaces that resist the adsorption of protein. Langmuir 17:5605–5620. https://doi.org/10.1021/la010384m
75. Schwendel D, Dahint R, Herrwerth S, Schloerholz M, Eck W, Grunze M (2001) Temperature dependence of the protein resistance of poly- and oligo(ethylene glycol)-terminated alkanethiolate monolayers. Langmuir 17:5717–5720. https://doi.org/10.1021/la010576h
76. Kumar A, Biebuyck HA, Whitesides GM (1994) Patterning self-assembled monolayers: applications in materials science. Langmuir 10:1498–1511. https://doi.org/10.1021/la00017a030

77. Mrksich M, Whitesides GM (1995) Patterning self-assembled monolayers using microcontact printing: a new technology for biosensors? Trends Biotechnol 13:228–235. https://doi.org/10.1016/S0167-7799(00)88950-7
78. Xia Y, Whitesides GM (1995) Use of controlled reactive spreading of liquid alkanethiol on the surface of gold to modify the size of features produced by microcontact printing. J Am Chem Soc 117(32):74–75. https://doi.org/10.1021/ja00116a033
79. Jackman RJ, Wilbur JL, Whitesides GM (1995) Fabrication of submicrometer features on curved substrates by microcontact printing. Science 269:664–666. https://doi.org/10.1126/science.7624795
80. Dulcey CS, Georger JH, Krauthamer V, Stener DA, Fare TL, Calvert JM (1991) Deep UV photochemistry of chemisorbed monolayers: patterned coplanar molecular assemblies. Science 252:551–554. https://doi.org/10.1126/science.2020853
81. Kleinfeld D, Kahler KH, Hockber PE (1988) Controlled outgrowth of dissociated neurons on patterned substrates. J Neurosci 8:4098–4120. https://doi.org/10.1523/JNEUROSCI.08-11-04098.1988
82. Frisbie CD, Wollma EW, Wrighton MS (1995) High lateral resolution imaging by secondary ion mass spectrometry of photopatterned self-assembled monolayers containing aryl azide. Langmuir 11:2563–2571. https://doi.org/10.1021/la00007a040
83. Schoenfisch MH, Pemberton JE (1998) Air stability of alkanethiol self-assembled monolayers on silver and gold surfaces. J Am Chem Soc 120:4502–4513. https://doi.org/10.1021/ja974301t
84. Bearinger JP, Terrettaz S, Michel R, Tirelli N, Vogel H, Textor M, Hubbell JA (2003) Chemisorbed poly(propylene sulphide)-based copolymers resist biomolecular interactions. Nat Mater 2:259–264. https://doi.org/10.1038/nmat851
85. Barber TA, Golledge SL, Castner DG, Healy KE (2003) Peptide-modified p(AAm-co-EG/AAc) IPNs grafted to bulk titanium modulate osteoblast behavior in vitro. J Biomed Mater Res Part A 64A:38–47. https://doi.org/10.1002/jbm.a.10321
86. Bearinger JP, Castner DG, Healy KE (1998) Biomolecular modification of p(AAm-co-EG/AA) IPNs supports osteoblast adhesion and phenotypic expression. J Biomater Sci Polym Ed 9:629–652. https://doi.org/10.1163/156856298X00064
87. Satomi T, Nagasaki Y, Kobayashi H, Otsuka H, Kataoka K (2007) Density control of poly(ethylene glycol) layer to regulate cellular attachment. Langmuir 23:6609–6703. https://doi.org/10.1021/la0624384
88. Satomi T, Nagasaki Y, Kobayashi H, Tateishi T, Otsuka H (2007) Physicochemical characterization of densely packed poly(ethylene glycol) layer for minimizing nonspecific protein adsorption. J Nanosci Nanotechnol 7(7):2394–2399. https://doi.org/10.1166/jnn.2007.695
89. Nakanishi J, Takarada T, Yamaguchi K, Maeda M (2008) Recent advances in cell micropatterning techniques for bioanalytical and biomedical sciences. Anal Sci 24:67–72. https://doi.org/10.2116/analsci.24.67
90. Xia YN, Whitesides GM (1998) Soft lithography. Annu Rev Mater Sci 28:153–184. https://doi.org/10.1146/annurev.matsci.28.1.153
91. Kumar A, Whitesides GM (1993) Features of gold having micrometer to centimeter dimensions can be formed through a combination of stamping with an elastomeric stamp and an alkanethiol ink followed by chemical etching. Appl Phys Lett 63:2002–2004. https://doi.org/10.1063/1.110628
92. Delamarche E, Bernard A, Schmid H, Michel B, Biebuyck H (1997) Patterned delivery of immunoglobulins to surfaces using microfluidic networks. Science 276:779–781. https://doi.org/10.1126/science.276.5313.779
93. Mrksich M, Dike LE, Tien J, Ingber DE, Whitesides GM (1997) Using microcontact printing to pattern the attachment of mammalian cells to self-assembled monolayers of alkanethiolates on transparent films of gold and silver. Exp Cell Res 235:305–313. https://doi.org/10.1006/excr.1997.3668

94. Delamarche E, Bernard A, Schmid H, Bietsch A, Michel B, Biebuyck H (1998) Microfluidic networks for chemical patterning of substrates: design and application to bioassays. J Am Chem Soc 120(3):500–508. https://doi.org/10.1021/ja973071f
95. Tien J, Nelson CM, Chen CS (2002) Fabrication of aligned microstructures with a single elastomeric stamp. Proc Natl Acad Sci USA 99(4):1758–1762. https://doi.org/10.1073/pnas.042493399
96. Chiu DT, Jeon NL, Huang S, Kane RS, Wargo CJ, Choi IS, Ingber DE, Whitesides GM (2000) Patterned deposition of cells and proteins onto surfaces by using three-dimensional microfluidic systems. Proc Natl Acad Sci USA 97(6):2408–2413. https://doi.org/10.1073/pnas.040562297
97. Takayama S, Ostuni E, LeDuc P, Naruse K, Ingber DE, Whitesides GM (2003) Selective chemical treatment of cellular microdomains using multiple laminar streams. Chem Biol 10(2):123–130. https://doi.org/10.1016/S1074-5521(03)00019-X
98. Tourovskaia A, Barber T, Wickes BT, Hirdes D, Grin B, Castner DG, Healy KE, Folch A (2013) Micropatterns of chemisorbed cell adhesion-repellent films using oxygen plasma etching and elastomeric masks. Langmuir 19:4754–4764. https://doi.org/10.1021/la0267948
99. Matsumoto K, Yoshitomi H, Rossant J, Zaret KS (2001) Liver organogenesis promoted by endothelial cells prior to vascular function. Science 294:559–563. https://doi.org/10.1126/science.1063889
100. Zaret KS (2002) Regulatory phases of early liver development: paradigms of organogenesis. Nat Rev Genet 3:499–512. https://doi.org/10.1038/nrg837
101. LeCouter J, Moritz DR, Li B, Phillips GL, Liang H, Gerber HP, Hillan KJ, Ferrara N (2003) Angiogenesis-independent endothelial protection of liver: role of VEGFR-1. Science 299:890–893. https://doi.org/10.1126/science.1079562
102. Ferrara N, Gerber HP, LeCouter J (2003) The biology of VEGF and its receptors. Nat Med 9:669–676. https://doi.org/10.1038/nm0603-669
103. Vinci M, Gowan S, Boxall F, Patterson L, Zimmermann M, Court W, Lomas C, Mendiola M, Hardisson D, Eccles SA (2012) Advances in establishment and analysis of three-dimensional tumor spheroid-based functional assays for target validation and drug evaluation. BMC Biol 10:29. https://doi.org/10.1186/1741-7007-10-29
104. Tung YC, Hsiao AY, Allen SG, Torisawa Y, Ho M, Takayama S (2011) High-throughput 3D spheroid culture and drug testing using a 384 hanging drop array. Analyst 136:473–478. https://doi.org/10.1039/C0AN00609B
105. Kelm JM, Fussenegger M (2004) Microscale tissue engineering using gravity-enforced cell assembly. Trends Biotechnol 22:195–202. https://doi.org/10.1016/j.tibtech.2004.02.002
106. Shimizu T, Yamato M, Kikuchi A, Okano T (2003) Cell sheet engineering for myocardial tissue reconstruction. Biomaterials 24:2309–2316. https://doi.org/10.1016/S0142-9612(03)00110-8
107. L'Heureux N, Paquet S, Labbe R, Germain L, Auger FA (1998) A completely biological tissue-engineered human blood vessel. FASEB J 12:47–56. https://doi.org/10.1096/fsb2fasebj.12.1.47
108. Fukuda J, Mizumoto H, Nakazawa K, Kajiwara T, Funatsu K (2004) Hepatocyte organoid culture in elliptic hollow fibers to develop a hybrid artificial liver. Int J Artif Organs 27:1091–1099. https://doi.org/10.1177/039139880402701213
109. Khademhosseini A, Yeh J, Jon S, Eng G, Suh KY, Burdick JA, Langer R (2004) Molded polyethylene glycol microstructures for capturing cells within microfluidic channels. Lab Chip 4:425–430. https://doi.org/10.1039/B404842C
110. Putnam AJ, Mooney DJ (1996) Tissue engineering using synthetic extracellular matrices. Nat Med 2:824–826. https://doi.org/10.1038/nm0796-824
111. Marler JJ, Upton J, Langer R, Vacanti JP (1998) Transplantation of cells in matrices for tissue regeneration. Adv Drug Deliv Rev 33:165–182. https://doi.org/10.1016/s0169-409x(98)00025-8
112. Alberts B, Bray D, Lewis J, Raff M, Roberts K, J.D. (1994) Watson in molecular biology of the cell. Garland Publishing, New York, p 971

113. Kim BS, Mooney DJ (1998) Development of biocompatible synthetic extracellular matrices for tissue engineering. Trends Biotechnol 16:224–230. https://doi.org/10.1016/s0167-7799(98)01191-3
114. Jhon MS, Andrade JD (1973) Water and hydrogels. J Biomed Mater Res 7:509–522. https://doi.org/10.1002/jbm.820070604
115. Sivashanmugam A, Kumar RA, Priya MV, Nair SV, Jayakumar R (2015) An overview of injectable polymeric hydrogels for tissue engineering. Eur Polym J 72:543–565. https://doi.org/10.1016/j.eurpolymj.2015.05.014
116. Muir VG, Burdick JA (2021) Chemically modified biopolymers for the formation of biomedical hydrogels. Chem Rev 121:10908–10949. https://doi.org/10.1021/acs.chemrev.0c00923
117. Ishikawa S, Iijima K, Matsukuma D, Asawa Y, Hoshi K, Osawa S, Otsuka H (2020) Interpenetrating polymer network hydrogels via a one-pot and in situ gelation system based on peptide self-assembly and orthogonal cross-linking for tissue regeneration. Chem Mater 32:2353–2364. https://doi.org/10.1021/acs.chemmater.9b04725
118. Jensen KF (1999) Microchemical systems: status, challenges, and opportunities. AICHE J 45:2051–2054. https://doi.org/10.1002/aic.690451003
119. Stone HA, Kim S (2001) Microfluidics: basic issues, applications, and challenges. AICHE J 47:1250–1254. https://doi.org/10.1002/aic.690470602

# Chapter 4
# Ultrafast Vibrational Dynamics of Water at Interfaces

**Satoshi Nihonyanagi**

**Abstract** Water at interfaces plays essential roles in life and industry. Despite its importance, our understanding of water at interfaces is still limited mainly because of the technical difficulty in studying water at interfaces selectively with ultrafast time resolution. This difficulty has been overcome by developing a novel nonlinear spectroscopy, named time-resolved heterodyne-detected vibrational sum frequency generation (TR-HD-VSFG) spectroscopy. TR-HD-VSFG enabled us to study the ultrafast dynamics of water at interfaces without interference from the vast amount of bulk water. In this chapter, I will review our recent works that revealed remarkable new insights into the interfacial water dynamics. In particular, the vibrational dynamics of water at the air/water interface were investigated from the viewpoint of the vibrational relaxation time ($T_1$). It was revealed that the $T_1$ of water at the air/water interface is almost independent of the excitation frequency and is consistent with bulk value upon hydrogen bond OH stretch excitation. On the other hand, $T_1$ of the free OH, which exists only at the interface, is unique to the interface. Because $T_1$ of the free OH does not change with the isotopic dilution, vibrational relaxation of the free OH is considered to proceed with diffusive rotation of the free OH. Moreover, the vibrational dynamics of water at charged monolayer interfaces have been investigated with various lipid headgroups from the viewpoint of spectral diffusion dynamics. It was revealed that the spectral diffusion dynamics of water at the lipid interfaces depend on the charge as well as the hydrogen bonding ability of the headgroup, reflecting the differences in the hydrogen bond fluctuation of water at the interfaces.

**Keywords** Water · Dynamics · Hydrogen bond · Interfaces

---

S. Nihonyanagi (✉)
Molecular Spectroscopy Lab, RIKEN, Wako, Japan
e-mail: nsatoshi@riken.jp


M. Hashizume, Y. Imura (eds.), *Colloids and Interfacial Dynamics*, Physical Chemistry in Action, https://doi.org/10.1007/978-981-96-6728-4_4

## 4.1 Introduction

Water is ubiquitous on Earth and is the most important liquid for life, the environment, sciences, and so on. Despite the ubiquitousness of water, its physical properties are anomalous in various aspects. The anomalies of water stem from its high ability to form strong, three-dimensionally networked hydrogen bonds. Because water hydrogen bonding is highly dynamic and its microscopic structure changes with an ultrafast (sub-picosecond) time scale, understanding water requires elucidation of the ultrafast dynamics of hydrogen bonds. The ultrafast hydrogen bond dynamics of bulk water have been intensively investigated using ultrafast time-resolved vibrational spectroscopies, such as transient infrared (IR)/Raman and two-dimensional IR spectroscopies [1].

When water is in contact with another immiscible phase, an interface is formed. Water interfaces exist everywhere on Earth and play essential roles in various phenomena. For example, cells consist of membrane bilayers, and fundamental biochemical events, such as transports and enzymatic reactions occur at the membrane/water interface. In environmental science, many important chemical reactions occur at the seawater and aerosol surfaces. Electrochemical and photovoltaic water-splitting reactions occur at material/water interfaces. All these heterogeneous chemical processes are mediated by water at the interfaces. Therefore, understanding the water dynamics at interfaces is essential for various scientific fields.

Interface selectivity is required to elucidate the structure of water interfaces in molecular details, because the vast amount of bulk water signal overwhelms the tiny interface signal in conventional spectroscopy. Vibrational sum frequency generation (VSFG) is a unique tool that enables us to measure vibrational spectra of interfaces. Our group has developed heterodyne-detected vibrational sum frequency generation (HD-VSFG) spectroscopy, a phase-sensitive extension of VSFG, which is crucial to making proper interpretations of the interface spectra [2]. Furthermore, elucidation of the dynamic properties of molecules at interfaces requires not only interface selectivity but also ultrafast time resolution. To achieve this demanding requirement, our group developed Time-Resolved Heterodyne-Detected Vibrational Sum Frequency Generation (TR-HD-VSFG) spectroscopy [3, 4]. As explained in the next section, TR-HD-VSFG is a combination of HD-VSFG and the pump–probe scheme which enables us to measure the femtosecond time-resolved vibrational spectrum of interfaces. Using this technique, we have revealed remarkable new insights into the interfacial water dynamics, which was completely veiled until the last decade.

The principles and applications of conventional VSFG spectroscopy have been thoroughly reviewed in the past three decades [5–8]. Several reviews are available also for HD-VSFG studies of interfacial water structure [9, 10] and TR-HD-VSFG studies of interfacial water dynamics [11]. The principles and instrumentation of HD-VSFG and TR-HD-VSFG are well-described in these review articles. In this chapter, I will first briefly summarize the methodology of TR-HD-VSFG spectroscopy. For readers who are not familiar with molecular spectroscopy, intuitive simple explanations are given, rather than precise, mathematical descriptions. Then, I will describe the water dynamics at interfaces revealed in recent TR-HD-VSFG studies. In the early days, there were some incorrect conclusions due to the inaccuracy of measurements and limited experimental results. Because these early discussions were well-summarized in the former reviews, in this chapter, I will focus on the new insights and current understanding of the structures and dynamics of water at interfaces.

## 4.2 Methodology

In this section, I briefly introduce the basics of TR-HD-VSFG spectroscopy which is used for studying the vibrational dynamics of interfaces. As TR-HD-VSFG is a combination of HD-VSFG spectroscopy and the pump–probe technique, I first describe the essence of HD-VSFG spectroscopy and then the principle of the pump–probe technique, followed by TR-HD-VSFG spectroscopy.

When strong laser pulses with distinct frequencies are focused on an interface, a new light with a frequency that is the sum of the two input lights is generated as shown in Fig. 4.1a. This is called sum frequency generation (SFG) and is one of the second-order nonlinear optical processes. Because second-order nonlinear optical processes are forbidden in the bulk of a medium with inversion symmetry, the signal SFG light is generated only at the interface, where inversion symmetry is necessarily broken. Therefore, SFG can selectively probe an interface without interference from vast bulk molecules. This interface selectivity makes SFG a unique tool for investigating the microscopic properties of interfacial molecules. In particular, vibrational sum frequency generation (VSFG) is widely used for investigating the structure and orientation of molecules at various interfaces. In VSFG spectroscopy, a tunable or broadband infrared (IR) light and a narrowband frequency-fixed visible (or near-infrared) light are focused on the sample interface, and the intensity of the emitted sum frequency light is measured. Because the SFG signal is enhanced when the incident IR frequency is in resonance with a vibrational state of the molecule at the interface, it can be used to measure the vibrational spectra of an interface. However, there are critical drawbacks to conventional VSFG spectroscopy: because the light intensity is the modulus square of the corresponding electric field, the intensity of the SFG signal is proportional to the modulus square of the nonlinear optical susceptibility ($|\chi^{(2)}|^2$). Thus, conventional VSFG intensity measurement provides no phase information, resulting in a loss of sign information and spectral distortion, which makes the interpretation of VSFG spectra very difficult.

The drawbacks of conventional VSFG spectroscopy can be solved by a phase-resolved extension of VSFG, heterodyne-detected (HD-) VSFG spectroscopy developed in RIKEN [2, 12] and other places [13, 14]. HD-VSFG spectroscopy enables us to measure the amplitude and phase of second-order susceptibility ($\chi^{(2)}$) spectra (Fig. 4.1b). In HD-VSFG, infrared and visible pulses are first focused onto a thin crystal of Y-cut quartz, which generates SFG light used as a local oscillator (LO).

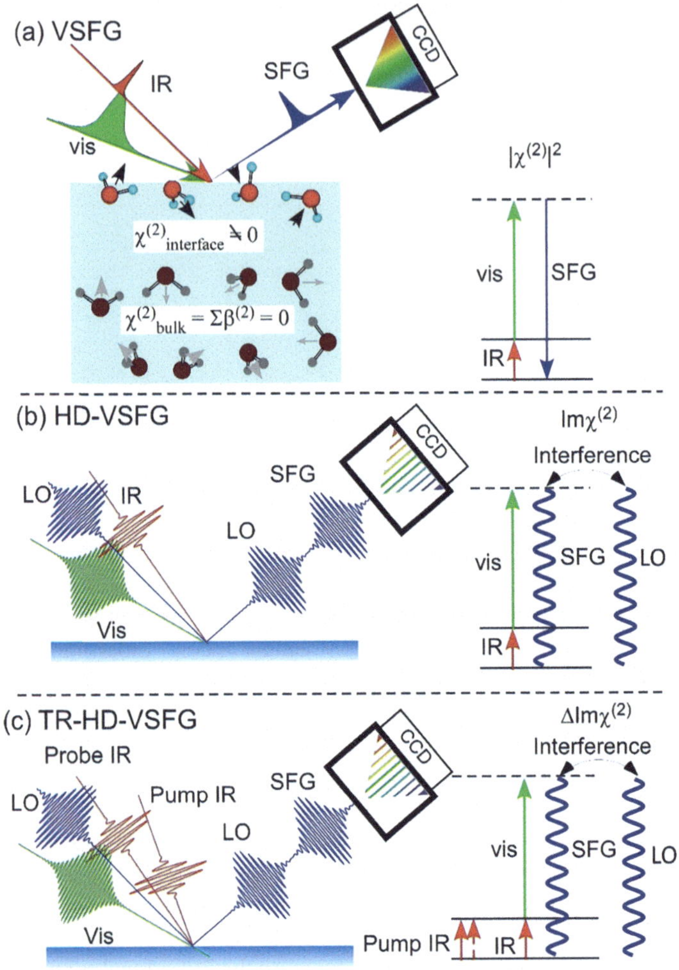


**Fig. 4.1** Conceptual sketches for (**a**) VSFG, (**b**) HD-VSFG, and (**c**) TR-HD-VSFG spectroscopy. The energy diagram of the VSFG process is depicted on the right-hand side of (**a**)

The transmitted LO, IR, and visible pulses are refocused on the sample interface and generate SFG signals of the sample interface. These two SFG lights generated from the quartz and sample interfaces interfere in the polychromator, and the frequency-domain interference fringe spectrum is measured with a CCD. The recorded interference fringe spectrum is inversely Fourier-transformed and filtered in the time domain to extract the heterodyned signal. The filtered interference spectrum is Fourier-transformed back to the frequency domain and normalized by the filtered fringe spectrum obtained from a z-cut quartz surface at the sample position, providing a complex $\chi^{(2)}$ spectrum. The imaginary part of $\chi^{(2)}$ ($\mathrm{Im}\chi^{(2)}$) spectrum is an interface analog of the absorption spectrum which is given basically by $\mathrm{Im}\chi^{(1)}$ ($\chi^{(1)}$:

the linear optical susceptibility). Therefore, $\mathrm{Im}\chi^{(2)}$ spectrum is free from spectral distortion and can be easily interpreted as we do for the bulk absorption/Raman spectra. Nevertheless, it is important to note that $\mathrm{Im}\chi^{(2)}$ is not proportional to the density of molecules but to the density times orientation factor R(<cosθ>) (<cosθ>: ensemble average of the cosine of the angle between the transition dipole moment and the surface normal), reflecting the symmetry-sensitive nature of the second-order process. As <cosθ> flips its sign with an up/down orientation, the sign of $\mathrm{Im}\chi^{(2)}$ includes the information about the up/down orientation. Because of the distortion-free interface spectrum and the sign information, HD-VSFG spectroscopy is now considered a very powerful tool for studying molecular structures of interfaces, particularly of liquid interfaces which cannot be brought into an ultrahigh vacuum.

Time-resolved vibrational spectra of an interface can be obtained by time-resolved (TR-) HD-VSFG which combines HD-VSFG spectroscopy as an interface-selective probe and another IR pulse as a pump light using the principle of the pump–probe technique as depicted in Fig. 4.1c. In this technique, a pump-induced change in an $\mathrm{Im}\chi^{(2)}$ spectrum ($\Delta\mathrm{Im}\chi^{(2)}$) is directly obtained as in the case of transient IR absorption measurements in which a pump-induced change in absorption ($\Delta\mathrm{Im}\chi^{(1)}$) spectra are measured. Thus, $\Delta\mathrm{Im}\chi^{(2)}$ can be regarded as an interfacial analog of transient absorption, and it can be easily analyzed. It is important to note that heterodyne detection is necessary for time-resolved VSFG measurements. This can be readily understood by considering the pump-induced change in the intensity spectrum:

$$\Delta I_{\mathrm{SFG}} = \left|\chi^{(2)}_{\mathrm{steady}} + \Delta\chi^{(2)}\right|^2 - \left|\chi^{(2)}_{\mathrm{steady}}\right|^2 = \chi^{(2)}_{\mathrm{steady}}\Delta\chi^{(2)*} + \chi^{(2)}_{\mathrm{steady}}{}^{*}\Delta\chi^{(2)} + \left|\Delta\chi^{(2)}\right|^2$$

where * denotes the complex conjugate. The resultant quantity is complicated and not even proportional to the $\Delta\chi^{(2)}$. Because of this complexity, the analyses and interpretation of time-resolved VSFG intensity spectra are challenging and have misled a lot of incorrect conclusions in the past decades, although several groups have tried time-resolved VSFG intensity measurements [15]. Thus, the linearity of spectra obtained by heterodyne detection is essential for time-resolved VSFG spectroscopy.

The principle of the pump–probe technique is illustrated in Fig. 4.2. When the probe pulse arrives at the sample before the pump pulse arrives, only the ground state of the sample molecules is populated, so the probe measures the steady-state spectrum (Fig. 4.2a). When the pump pulse arrives at the sample (defined as $T = 0$), the pump pulse vibrationally excites the molecules of interest (Fig. 4.2b, creating the population in the vibrationally excited state ($v = 1$). Then the subsequent probe pulse monitors the pump-induced changes in the vibrational spectra (Δ) at a certain delay time between the pump and probe pulses ($T$) (Fig. 4.2c). By analyzing the plot of Δ vs. $T$, one can obtain time constants such as the vibrational relaxation time ($T_1$) (Fig. 4.2d). Alternatively, by analyzing the spectral features in the Δ spectra, one can discuss how the excess vibrational energy is redistributed in time, that is, how the spectral diffusion dynamics proceeds, as explained below.

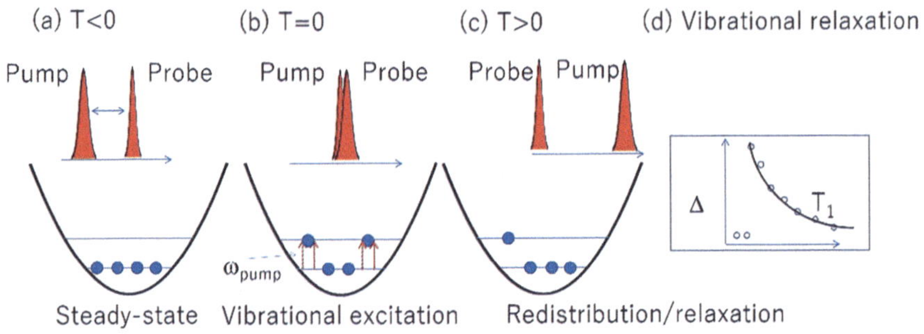


**Fig. 4.2** Schematic drawings of the pump–probe technique. Pulse arrivals and populations of vibrational states at (**a**) T<0, (**b**) T=0, (**c**) T>0. (**d**) a time trace of the pump-induced changes

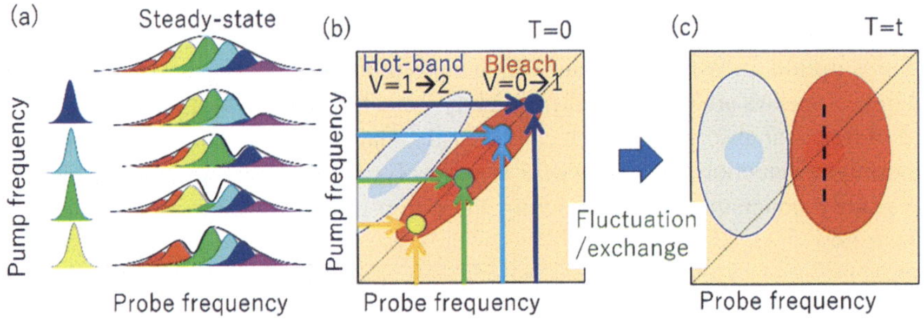


**Fig. 4.3** Schematic drawings of the principle of 2D spectrum. (**a**) Hole-burnt spectra with various pump frequencies. Contour plots of (**a**) at (**b**) T=0 and (**c**) T=t

The principle of 2D spectroscopy is illustrated in Fig. 4.3. Let us consider a broad OH band that consists of several narrow sub-ensembles (represented by different colors in Fig. 4.3a). The steady-state spectrum provides information only about the sum of these sub-ensembles. When this broad OH band is excited with a narrow band pump pulse with various center frequencies, only the sub-ensemble having the same frequency as the pump pulse would be excited. Therefore, a narrow bleach band due to the depletion of the ground-state transition ($v = 0 \rightarrow 1$) is observed at the excitation frequency right after the excitation. This is the so-called "hole-burning." By combining the hole-burnt spectra with a series of excitation frequencies, a two-dimensional (2D-) spectrum can be obtained as depicted in Fig. 4.3b. At T = t, the bleach lobe elongates completely toward the diagonal reflecting the consistency between the excitation frequency and the frequency of the depleted sub-ensemble. In addition to the bleach lobe, another lobe would be observed due to the increase of the excited-state transition ($v = 1 \rightarrow 2$). This hot band lobe is usually observed in the lower frequency side of the bleach lobe due to the anharmonicity of the OH band (i.e., the energy of $v = 1 \rightarrow 2$ transition is lower than that of $v = 0 \rightarrow 1$ transition). After a certain delay time, because of fluctuation and exchange of hydrogen bonds, this narrow bleach band becomes broader and the hole-burnt feature is lost, resulting in the vertically elongated lobe, as shown in Fig. 4.3c. In this case, the response no longer depends on pump frequency. The process of losing the pump-frequency dependence is called "spectral diffusion" and is an important quantity to differentiate the ultrafast hydrogen bonding of water. By taking the center line slope (CLS) or nodal line slope of the lobes as a function of delay time, one can quantify the spectral diffusion dynamics.

In our TR-HD-VSFG spectrometer, Ti: Sapphire regenerative amplifier produces 80 femtoseconds pulses with 5 mJ of energy at 1 kHz [16]. The probe IR is generated by a commercial optical parametric amplifier (TOPAS prime, Light Conversion) and difference frequency generation (DFG). The visible light is a part of the amplifier output centered at 795 nm which is narrowed by a bandpass filter down to 25 $cm^{-1}$. The bandwidth of the visible light practically determines the frequency resolution of the probe spectra. A pump IR pulse is generated by taking the difference frequency of a part of the amplifier output and frequency-doubled idler light from TOPAS prime. The bandwidth of the pump IR is relatively narrow (~150 $cm^{-1}$), and the center frequency can be scanned in the OH stretch region to obtain the 2D spectrum. Because of this scanning method, the time resolution of the instrument is limited to about 200 fs. Typical pulse energies of Vis, probe IR, and pump IR pulses are all around 10 μJ. The polarization of SFG, Vis, probe IR, and pump IR pulses are s-, s-, p-, and p-polarized, respectively.

## 4.3 Vibrational Relaxation of Water at the Air/ Water Interface

The steady-state $\text{Im}\chi^{(2)}$ spectrum of the air/water interface exhibits two features, a negative broadband around 3450 $cm^{-1}$ and a positive relatively sharp peak at 3700 $cm^{-1}$ [17]. The frequency of the former band is almost the same as the bulk water spectrum and hence is assigned to hydrogen-bonded OH of water near the interface. The negative sign of the band in the $\text{Im}\chi^{(2)}$ spectrum indicates that the hydrogen-bonded OH of interfacial water points its hydrogen toward the bulk water phase. The high-frequency positive band is attributed to the non-hydrogen-bonded OH of the outermost water molecules sticking out of the air. This OH band is often called "free OH." Because the free OH exists only at the interface, elucidating the properties of the free OH is key to understanding the overall physics of water at the interface. The vibrational dynamics of water at the air/water interface have been investigated using TR-HD-VSFG in terms of vibrational relaxation time ($T_1$).

As shown in Fig. 4.4a, when the free OH is selectively excited, $\Delta\text{Im}\chi^{(2)}$ spectrum of the air/$H_2O$ interface shows negative and positive features at 0 ps [18]. The negative signal at ~3700 $cm^{-1}$ is readily assignable to the bleach of the free OH because the pump pulse excited only the free OH. The bleach band should be accompanied by an increase in the excited-state transition in the lower frequency side (hot band).

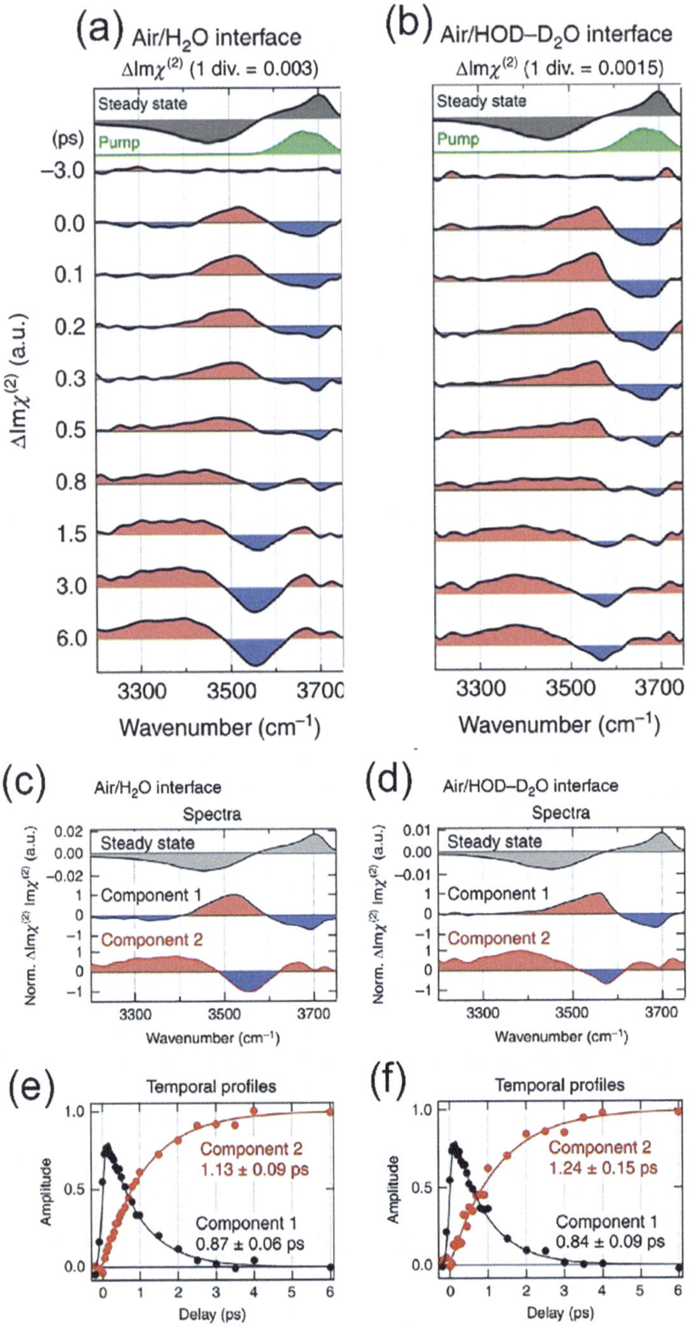


**Fig. 4.4** (**a**) $\Delta\mathrm{Im}\chi^{(2)}$ spectra of the air/$H_2O$ interface in the OH stretch region at various delay times indicated on the left axis. The corresponding steady-state $\mathrm{Im}\chi^{(2)}$ spectrum is shown on the top. (**c**) The spectral components obtained by SVD analysis of (**a**). (**e**) The temporal profiles of the spectral components obtained by SVD analysis of (**a**). (**b**), (**d**), and (**f**) are the same as (**a**), (**c**), and (**e**) but of the isotopically diluted water ($H_2O$/HOD/$D_2O$ = 1/2/1) interfaces. Reprinted from Ref. [18] under the Creative Commons Attribution 4.0 International license

The positive signal at ~3500 cm$^{-1}$ is thus assigned to the hot band of the free OH band. As delay time increases, the excited free OH spectrum decreases, and finally, strong positive and negative features appear in the hydrogen-bonded OH region. These signals reflect a blue shift of the hydrogen-bonded OH due to the temperature increase caused by the thermalization process. The observed transient spectra were decomposed using a mathematical analysis called singular value decomposition (SVD) into two spectral components: Component 1 is associated with the excited free OH spectrum which exponentially decays with $T_1$ = 0.87 ps. Component 2 is the thermalized spectrum which rises with $T_{th}$ = 1.1 ps (Fig. 4.4c, e).

In a prior homodyne TR-VSFG study, the vibrational dynamics of the free OH have been discussed by hypothesizing two relaxation pathways of the excited free OH [20]: First, it relaxes through energy transfer from a free OH to an HB-OH or to a bend overtone in the same water molecule. Second, it relaxes through the rotation of the interfacial water molecules. In this case, an excited free OH goes out of the detection angle, and an unexcited free OH comes into the detection angle. Because the energy transfer is highly frequency-dependent, it is strongly affected when the counterpart H of a free OH is deuterated. In contrast, the rotation of a free OH is not influenced by the isotope substitution of the counterpart. Thus, an isotopic dilution experiment is the key to distinguishing these two mechanisms.

Surprisingly, the $T_1$ of the free OH of a 50% $H_2O/D_2O$ mixture is almost identical to that of the $H_2O$ interface, showing no isotopic dilution effect in the free OH relaxation (Fig. 4.4b, d, f). The absence of the isotopic dilution effect on the $T_1$ of the free OH suggests that the excited free OH relaxes solely by the reorientation mechanism. This is rationalized as frequencies of the free OH and hydrogen-bonded OH are not overlapped and hence resonant energy transfer is negligible. Thus it was concluded that the free OH vibration is energetically decoupled from the surrounding HB-OHs and decays with the diffusive rotation mechanism.

The vibrational relaxation dynamics of the free OD stretch were also investigated by TR-HD-VSFG experiments using $D_2O$ and isotopically diluted heavy water ($D_2O$ in $H_2O$) [21]. It was concluded that the $T_1$ of the free OD is about 1.8 ps, independent of the isotopic dilution. It means that the excited free OD also relaxes through the rotation but the free OD rotation is 2 times slower than the free OH rotation.

The relaxation times $T_1$ of hydrogen-bonded (HB) OH as well as the free OH of water at the air/water interface have been examined with various excitation frequencies by monitoring the excited-state transition (hot band) as depicted in Fig. 4.5b [19]. The $T_1$ with different excitation frequencies determined from the decay of the hot band [19] is consistent with that determined from the recovery of the ground-state bleach (GSB) [18, 22]. The results (Fig. 4.5c) clearly show that the $T_1$ of the OH stretch vibration of HB OH of water at the air/water interface is almost the same as those found for the bulk water, exhibiting no significant pump frequency dependence. It was considered that because the spectral diffusion of the HB OH is ultrafast (<100 fs), the memory of the excited frequency is rapidly swept out and hence it does not influence the $T_1$ of the HB OH. In addition, intramolecular vibrational coupling possibly contributes to the frequency-independent $T_1$ of the HB OH by

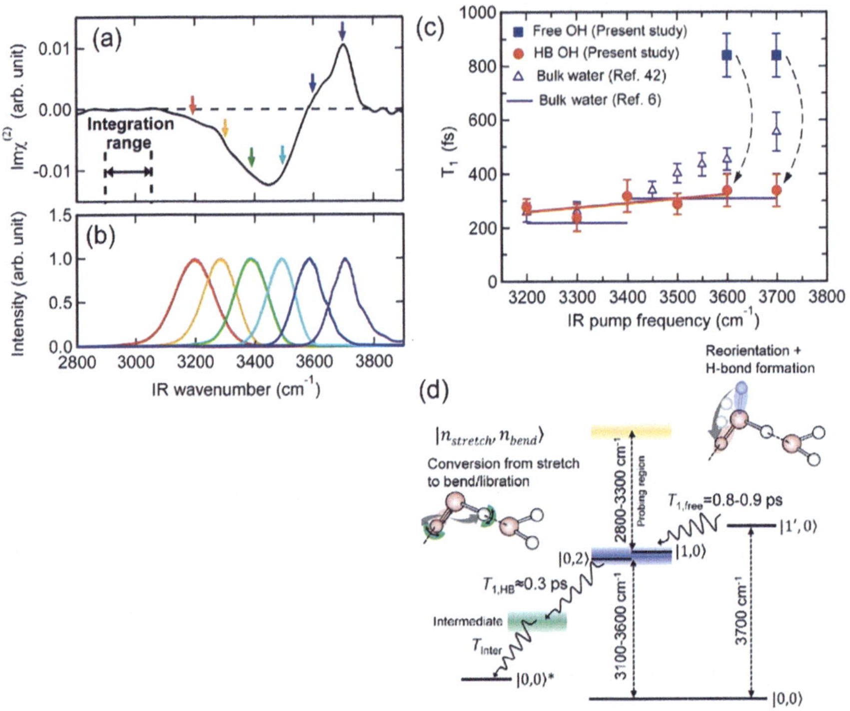


**Fig. 4.5** (**a**) The steady-state $Im\chi^{(2)}$ spectrum of the air/water interface in the OH stretch region. (**b**) Pump-IR spectra measured as SFG intensity spectra from quartz. (**c**) The vibrational relaxation ($T_1$) times of water OH stretch with various pump frequencies. (**d**) Schematic drawing of the relaxation pathway of vibrational energy of water at the interface. Reprinted from Ref. [19] under the Creative Commons Attribution 4.0 International license

broadening the homogeneous bandwidth. Experiments with isotopically diluted water will soon clarify the role of vibrational couplings on the frequency-independence of the $T_1$.

In addition, the delayed rise of the hot band signal was observed upon excitation of the free OH. This is consistent with the scenario in which the excited free OH relaxes via the rotation with $T_1$ ~ 0.9 ps as derived from the recovery of the GSB. The hot band measurement revealed that after the rotation of the excited free OH, it forms a hydrogen bond immediately and then relaxes as an HB OH with a time constant of ~0.3 ps.

## 4.4 Spectral Diffusion Dynamics of Water at Charged Lipid Monolayer/Water Interfaces

Lipid monolayer/water interfaces are good model systems to study interactions between interfacial water and functional groups of lipids at interfaces. We have studied prototypical positively and negatively charged lipid interfaces [23]. Figure 4.6a shows the steady-state Im$\chi^{(2)}$ spectra of the isotopically diluted water (HOD-$D_2O$) at a sodium dipalmitoyl-sn-glycero-3-phosphorylglycerol (DPPG) and 1,2-dipalmitoyl-3-trimethyl-ammonium propane chloride (DPTAP) interfaces. DPPG has a phosphorylglycerol headgroup providing a negative charge and several

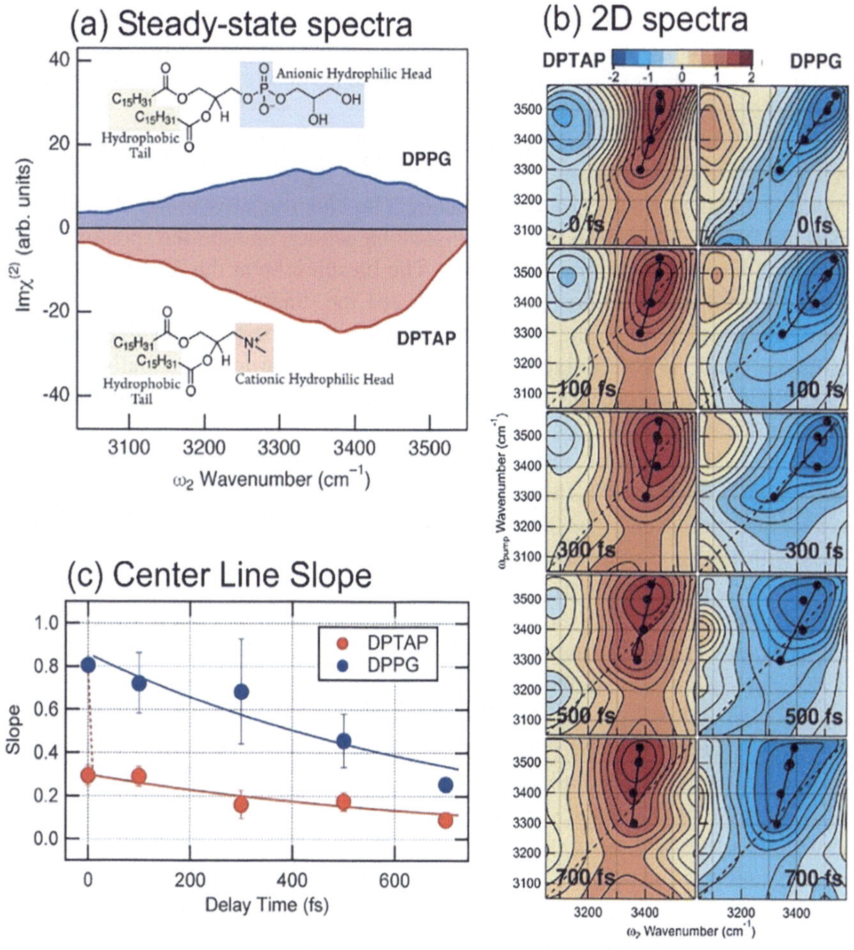


**Fig. 4.6** (**a**) Steady-state HD-VSFG spectra, (**b**) 2D-HD-VSFG spectra, and (**c**) the center line slope decay of the DPPG/water and the DPTAP/water interfaces. Reprinted from Ref. [23] under the Creative Commons Attribution 4.0 International license

hydrogen bonding sites, while DPTAP has a positive charge but no hydrogen bonding sites. Here, HOD-$D_2O$ was used to eliminate the vibrational coupling which makes the $H_2O$ spectrum very complicated [24, 25]. The Im$\chi^{(2)}$ spectrum of the DPPG interface shows a broad OH band with a positive sign, indicating the H-up orientation of water. The H-up orientation is attributed to the electrostatic interaction between interfacial water and the negative charges at the DPPG monolayer interface. It is expected that interfacial water forms H-bonds with the DPPG headgroup. In contrast, a negative OH band was observed at the positively charged DPTAP interface, indicating the H-down orientation. For the DPTAP interface, hydrogen bonding is not expected, because the headgroup of DPTAP has no hydrogen bonding site. Nevertheless, apart from the opposite sign, both interfaces show one broadband centered around 3400 $cm^{-1}$, which is similar to the bulk HOD-$D_2O$ absorption spectrum. Although intensity and band shape are somewhat different, it is difficult to identify the character of the interfacial water molecules at these two lipid interfaces having very different chemical structures, based on the time-averaged steady-state spectra.

Figure 4.6b shows the 2D HD-VSFG spectra of the DPPG/ HOD-$D_2O$ and the DPTAP/HOD-$D_2O$ interfaces. The horizontal axis represents probe frequency, and the vertical axis represents pump frequency. The blue (negative) lobe of the DPPG interface can be assigned to the ground-state bleaching, and the red (positive) lobe is the excited-state transition (hot band). The bleach lobe at the DPPG/HOD-$D_2O$ interface is elongated almost perfectly toward the diagonal, which indicates the strong inhomogeneity of the broad OH band which persists longer than the time resolution (~200 fs). As delay time increases, the bleach lobe gradually becomes vertical with a time constant of 700 fs as revealed by the center line slope (CLS) decay shown in Fig. 4.6c.

For the DPTAP interface, the 2D spectrum at $t = 0$ ps differs from that observed for the DPPG interface. The bleach lobe (shown by red) is elongated toward a more vertical direction even at $t = 0$, although the subsequent spectral diffusion also proceeds with the time constant of 700 fs. Summarizing the experimental data, the DPPG interface shows a high initial CLS value and decay with 700 fs. On the other hand, the DPTAP interface shows a small initial CLS value and decay with a similar time constant. The difference most clearly appears in the initial CLS value.

A former bulk water study reported that bulk HOD-$D_2O$ shows an ultrafast spectral diffusion within 50 fs due to the fast fluctuation of hydrogen bonds, followed by the sub-ps component due to the hydrogen bond rearrangement [26]. Considering the bimodal decay of the bulk HOD-$D_2O$, it is considered that there is a very fast decay component also in the spectral diffusion of the DPTAP interface. Because the given time resolution is around 200 fs, a sub-50 fs decay component cannot be detected and appears only as a small initial CLS value. Therefore, the spectral diffusion dynamics at the DPTAP interface is considered to be similar to that in the bulk. On the other hand, the spectral diffusion of the DPPG interface differs from that of the bulk water and does not exhibit fast fluctuation. The absence of fast hydrogen bond fluctuation at the DPPG interface is attributed to the H-bonds of interfacial water with the phosphorylglycerol headgroup. Because the DPPG headgroup provides multiple H-bonding sites, it can immobilize the interfacial water molecules and suppress the fluctuation of interfacial water. On the other hand, the DPTAP headgroup provides no hydrogen bonding sites, and water molecules at the DPTAP interface are in the H-down orientation. Thus, the interfacial water at the DPTAP interface H-bonds only with other water molecules in the deeper water phase, and hence the interfacial water can fluctuate as it does in the bulk.

Although the H-bonds can explain the difference between the DPPG and DPTAP interfaces quite well, it is also possible that up/down orientation may explain the difference to some extent. This possibility was examined by measuring the octadecylamine (ODA) interface [27]. ODA provides a negative charge as in the case of DPTAP but also has hydrogen bonding sites. As shown in Fig. 4.7, the CLS plot obtained from the ODA interface shows an initial CLS value higher than that of the DPTAP interface. This indicates that hydrogen bonding plays a role even when the interfacial water molecules are in the H-down orientation. At the DPPG interface, the fluctuation of interfacial water is suppressed most effectively, because the interfacial water molecules, which are H-up-orientated, hydrogen bond with the large phosphorylglycerol headgroup. At the ODA interface, the interfacial water molecules are in the H-down orientation and accept hydrogen bonds from the ODA headgroup, which partially suppresses the fluctuation of the interfacial water. At the

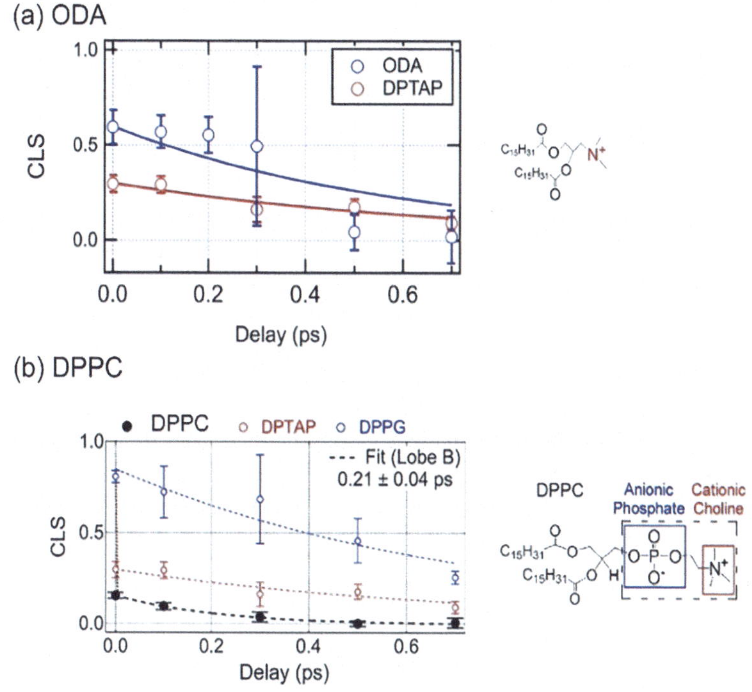


**Fig. 4.7** The center line slope decay of (**a**) the ODA/water and (**b**) the DPPC/water interfaces. These figures (**a**) and (**b**) are adapted with permission from Refs [27] and [28], respectively

DPTAP interface, the interfacial water molecules are H-down-oriented and receive no H-bond from the monolayer headgroup, resulting in bulk-like fast fluctuation.

Another interesting question is how the coexistence of phosphate and choline moieties influences the ultrafast fluctuation dynamics of water at the interface. This was examined by observing the DPPC interface, where the DPPC headgroup is regarded as a combination of the DPPG (phosphate) and DPTAP (choline) headgroups. Surprisingly, however, the initial CLS value for the DPPC interface is not intermediate between the DPPG and DPTAP interfaces but is even smaller than that observed at the DPTAP interface [28]. This means that the hydrogen bond fluctuation dynamics of water at the DPPC interface is not a simple sum or an average of those in the hydration shells of phosphate and choline but is more cooperative. It suggests that the hydrogen bond network of the DPPG interface is larger than that of the phosphate of the DPPC interface and that ultrafast fluctuation of water in the hydration shell of the choline perturbs the hydration shell of the phosphate group.

## 4.5 Conclusion

In the last decade, time-resolved, phase-sensitive, interface-selective spectroscopy (named TR-HD-VSFG and 2D HD-VSFG) has been established, and it revealed remarkable new physical insights into the ultrafast dynamics of water at aqueous interfaces. The vibrational dynamics of water at the air/water interface has been studied from the viewpoint of the vibrational relaxation time ($T_1$) of the OH stretches. $T_1$ of water at the air/water interface is categorized into two. One is the free OH moiety, which converts to the HB OH via diffusive rotation in the time scale of 0.9 ps. The other is the HB OH stretches, which relax with the time scale of 0.3–0.4 ps, almost independent of the excitation frequency. The frequency-independent $T_1$ time of the HB OH is considered due to fast spectral diffusion at the air/water interface. However, the spectral diffusion dynamics of HB OH at the air/water interface has not been elucidated, because it is too fast for the present experiment. On the other hand, the vibrational dynamics of water at the monolayer/water interfaces have been studied from the viewpoint of spectral diffusion dynamics which can be slowed down by interacting with monolayer headgroups. In particular, the initial value of the center line slope obtained from the 2D spectrum at 0.0 ps reflects the hydrogen bond fluctuation of the interface, which is sensitive to the chemical structure of the headgroup of the monolayer. Ultrafast fluctuation is a new yardstick for identifying the nature of interfacial water, which enables us to deepen our understanding.

Contrary to the remarkable progress in understanding the hydrogen bond dynamics of water at the air/water and air/monolayer/water interfaces, TR-HD-VSFG spectroscopy has not even been applied to solid/liquid interfaces, despite its importance. In fact, solid/liquid interfaces are the main target of interface-related research fields, because novel functionality can be developed by controlling interfacial properties by designing and applying a variety of functional groups at solid/liquid interfaces. Certainly, solid/liquid interfaces will be the main target of the studies in the ultrafast dynamics of water at interfaces, in the next decade.

**Acknowledgments** The author acknowledges former and current members of Molecular Spectroscopy Lab, RIKEN. This work was partly supported by JSPS KAKENHI Grant Numbers 23H00292, Japan.

## References

1. Bakker HJ, Skinner JL (2010) Vibrational spectroscopy as a probe of structure and dynamics in liquid water. Chem Rev 110:1498–1517
2. Nihonyanagi S, Yamaguchi S, Tahara T (2009) Direct evidence for orientational flip-flop of water molecules at charged interfaces: a heterodyne-detected vibrational sum frequency generation study. J Chem Phys 130:204704
3. Nihonyanagi S, Singh PC, Yamaguchi S, Tahara T (2012) Ultrafast vibrational dynamics of a charged aqueous Interface by femtosecond time-resolved heterodyne-detected vibrational sum frequency generation. Bull Chem Soc Jpn 85:758–760
4. Singh PC, Nihonyanagi S, Yamaguchi S, Tahara T (2012) Ultrafast vibrational dynamics of water at a charged interface revealed by two-dimensional heterodyne-detected vibrational sum frequency generation. J Chem Phys 137:094706–094706
5. Shen YR (1996) A few selected applications of surface nonlinear optical spectroscopy. Proc Natl Acad Sci USA 93:12104–12111
6. Richmond GL (2002) Molecular bonding and interactions at aqueous surfaces as probed by vibrational sum frequency spectroscopy. Chem Rev 102:2693–2724
7. Gopalakrishnan S, Liu D, Allen HC, Kuo M, Shultz MJ (2006) Vibrational spectroscopic studies of aqueous interfaces: salts, acids, bases, and nanodrops. Chem Rev 106:1155–1175
8. Shen YR, Ostroverkhov V (2006) Sum-frequency vibrational spectroscopy on water interfaces: polar orientation of water molecules at interfaces. Chem Rev 106:1140–1154
9. Nihonyanagi S, Mondal JA, Yamaguchi S, Tahara T (2013) Structure and dynamics of interfacial water studied by heterodyne-detected vibrational sum-frequency generation. Ann Rev Phys Chem 64:579–603
10. Shen YR (2013) Phase-sensitive sum-frequency spectroscopy. Ann Rev Phys Chem 64:129–150
11. Nihonyanagi S, Yamaguchi S, Tahara T (2017) Ultrafast dynamics at water interfaces studied by vibrational sum frequency generation spectroscopy. Chem Rev 117:10665–10693
12. Yamaguchi S, Tahara T (2008) Heterodyne-detected electronic sum frequency generation: "up" versus "down" alignment of interfacial molecules. J Chem Phys 129:101102
13. Ji N, Ostroverkhov V, Tian CS, Shen YR (2008) Characterization of vibrational resonances of water-vapor interfaces by phase-sensitive sum-frequency spectroscopy. Phys Rev Lett 100:096102
14. Stiopkin IV, Weeraman C, Pieniazek PA, Shalhout FY, Skinner JL, Benderskii AV (2011) Hydrogen bonding at the water surface revealed by isotopic dilution spectroscopy. Nature 474:192–195
15. Zhang Z, Piatkowski L, Bakker HJ, Bonn M (2011) Ultrafast vibrational energy transfer at the water/air interface revealed by two-dimensional surface vibrational spectroscopy. Nat Chem 3:888–893
16. Inoue K, Nihonyanagi S, Singh PC, Yamaguchi S, Tahara T (2015) 2D heterodyne-detected sum frequency generation study on the ultrafast vibrational dynamics of H2O and HOD water at charged interfaces. J Chem Phys 142:212431
17. Nihonyanagi S, Kusaka R, Inoue K-I, Aniruddha A, Yamaguchi S, Tahara T (2015) Accurate determination of complex $\chi(2)$ spectrum of the air/water interface. J Chem Phys 143:124707

18. Inoue K, Ahmed M, Nihonyanagi S, Tahara T (2020) Reorientation-induced relaxation of free OH at the air/water interface revealed by ultrafast heterodyne-detected nonlinear spectroscopy. Nat Commun 11:5344
19. Sung W, Inoue KI, Nihonyanagi S, Tahara T (2024) Unified picture of vibrational relaxation of OH stretch at the air/water interface. Nat Commun 15:1258
20. Hsieh C-S, Campen RK, Okuno M, Backus EHG, Nagata Y, Bonn M (2013) Mechanism of vibrational energy dissipation of free OH groups at the air–water interface. Proc Natl Acad Sci USA 110:18780–18785
21. Ahmed M, Nihonyanagi S, Tahara T (2022) Ultrafast vibrational dynamics of the free OD at the air/water interface: negligible isotopic dilution effect but large isotope substitution effect. J Chem Phys 156:224701
22. Inoue K-I, Ishiyama T, Nihonyanagi S, Yamaguchi S, Morita A, Tahara T (2016) Efficient spectral diffusion at the air/water Interface revealed by femtosecond time-resolved heterodyne-detected vibrational sum frequency generation spectroscopy. J Phys Chem Lett 7:1811–1815
23. Singh PC, Inoue K, Nihonyanagi S, Yamaguchi S, Tahara T (2016) Femtosecond hydrogen bond dynamics of bulk-like and bound water at positively and negatively charged lipid interfaces revealed by 2D HD-VSFG spectroscopy. Angew Chem Int Ed 55:10621–10625
24. Nihonyanagi S, Yamaguchi S, Tahara T (2010) Water hydrogen bond structure near highly charged interfaces is not like ice. J Am Chem Soc 132:6867–6869
25. Mondal JA, Nihonyanagi S, Yamaguchi S, Tahara T (2010) Structure and orientation of water at charged lipid monolayer/water interfaces probed by heterodyne-detected vibrational sum frequency generation spectroscopy. J Am Chem Soc 132:10656–10657
26. Roberts ST, Ramasesha K, Tokmakoff A (2009) Structural rearrangements in water viewed through two-dimensional infrared spectroscopy. Acc Chem Res 42:1239–1249
27. Inoue K-I, Ahmed M, Nihonyanagi S, Tahara T (2019) Effect of hydrogen-bond on ultrafast spectral diffusion dynamics of water at charged monolayer interfaces. J Chem Phys 150:054705
28. Inoue K-I, Singh PC, Nihonyanagi S, Yamaguchi S, Tahara T (2017) Cooperative hydrogen-bond dynamics at a Zwitterionic lipid/water Interface revealed by 2D HD-VSFG spectroscopy. J Phys Chem Lett 8:5160–5165

# Chapter 5
# What Is the Role of Hydration Water in the Self-Assembly of Soft Materials Such as Lipids, Surfactants, Polymers, and Proteins?

**Mafumi Hishida**

**Abstract** Many soft materials and biomolecules such as lipids, surfactants, polymers, and proteins self-assemble in water and form various aggregate structures. It is not fully understood whether the surrounding water plays an active role in the formation of these aggregate structures or is merely a background solvent. On the other hand, an attempt has recently been made to measure the dynamics of water molecules around soft materials using a technique called terahertz spectroscopy and to accurately determine the hydration state from the dynamical changes. This technique allows the observation of water that is weakly affected by the solute whose dynamics is slightly slowed down. As a result, it has been found that "weakly affected water"spread over long distances on molecular surfaces is deeply involved in the formation of aggregate structures and phase behavior of soft materials and biomolecules.

**Keywords** Soft material · Biomolecule · Water dynamics · Hydration · Terahertz spectroscopy · Self-assembly · Weakly affected water

## 5.1 Introduction: The Role of Water in the Self-Assembly of Soft Materials and Biomolecules

Many soft materials and biomolecules, such as amphiphilic surfactants, phospholipids, water-soluble polymers, and proteins, self-assemble in water to form various structures. Since the self-assembled structure greatly affects the macroscopic

M. Hishida (✉)
Department of Chemistry, Faculty of Science, Tokyo University of Science, Tokyo, Japan
e-mail: hishida@rs.tus.ac.jp



M. Hashizume, Y. Imura (eds.), *Colloids and Interfacial Dynamics*, Physical Chemistry in Action, https://doi.org/10.1007/978-981-96-6728-4_5

properties of these solutions, it is very important to control the self-assembled structure when making industrial products such as detergents, cosmetics, and paints. In addition, it is known that the self-assembled structures of biomolecules are closely related to the emergence of biological phenomena. Therefore, the mechanism of self-assembly of such soft materials and biomolecules in water and methods to control it have been studied for a long time.

Self-assembly is mainly governed by weak interactions (e.g., van der Waals interactions) between molecules and self-assembled structures of soft materials and biomolecules [1]. Therefore, the investigation of the mechanism of self-assembly has focused on elucidating the relationship among various interactions between soft materials and biomolecules. In most of these studies, water as a solvent has been assumed to be a homogeneous background, and its effect on self-assembly has been assumed to be almost negligible. On the other hand, when a soluble molecule dissolves in water, the water molecules are thought to interact with the polar groups of the solute molecules, and the solutes change water's dynamics [1, 2]. Water with this altered dynamics is generally referred to as hydration water or bound water. Hydration water is thought to exist on the surface of soft materials dispersed in water. In hydration water, water molecules are expected to have different physical properties from those of bulk water (water that has not undergone the hydration effect, sometimes called free water), because their dynamics have changed. In other words, water is never homogeneous in the presence of solutes, and the state of water can be involved in the self-assembly of soft materials.

However, it is still unclear how (or whether) the presence of hydration water is involved in the self-assembly of soft materials and biomolecules [3, 4]. In fact, there is still no unified view on the extent of the presence of hydration water. This chapter describes the results of the measurement of the hydration state of soft materials using terahertz spectroscopy, [5, 6] a spectroscopic technique that has been developed in recent years, which revealed that the strong relation of hydration state of molecules to their self-assembled structures.

## 5.2 Observation of Hydration Water

The definition of hydration water is ambiguous. Differential scanning calorimetry (DSC) and other thermal measurement methods are often used to observe hydration water in soft materials [7–11]. By this method, water in a sample solution is frozen, and the amount of water that has not frozen below 0 °C is estimated from the amount of latent heat of water when it thaws, and this is defined as the hydration number. For this reason, it is also called unfreezable (or non-freezing) water. For example, on phospholipid membranes, which are the basic structures of biological membranes, it is known that there is about one layer of unfreezable water on the membrane surface [8, 12]. As can be seen from this, unfreezable water is considered to be water that undergoes a strong binding effect from the solute surface.

From a viewpoint in the molecular scale, hydration water is often defined as water molecules bound by interaction with hydrophilic groups of solutes. Their mobility is altered by the interaction. Water molecules are in translational and rotational diffusion dynamics on a timescale from sub-picoseconds ($10^{-13}$ s) to ten picoseconds ($10^{-11}$ s) [13–15]. For example, the relaxation time of collective rotational dynamics is about 8 ps (= $8 \times 10^{-12}$ s) at room temperature. Focusing on these dynamics of water molecules, hydration water has been studied using NMR and neutron quasi-elastic scattering. Both of these techniques classically have observed molecular dynamics in the time region slower than picoseconds ($10^{-9}$–$10^{-11}$ s at the fastest) [16, 17]. Therefore, these techniques only observe water whose dynamics is greatly slowed down by hydration. In the case of phospholipid bilayers, the number of hydration water molecules per phospholipid molecule has been determined to be about 5–6 molecules [8]. This corresponds to the presence of only one layer of hydration water on the surface of the hydrophilic group, indicating that these techniques have observed only water that interacts directly with the hydrophilic group and is strongly bound to it.

However, it has been pointed out that since water forms hydrogen bond networks between water molecules, the hydration effect of the solute can extend farther through the hydrogen bond network, and weakly affected hydration water exists outside of the strongly bound water layer (Fig. 5.1) [2, 18]. If the hydration effect thus extends over a long distance, there may be a large amount of water that is more weakly affected by hydration. The strongly bound layer is sometimes referred to as the first hydration shell, and the weakly affected layer as the second hydration shell [1]. It has also been suggested that in the weakly affected hydration layers, the hydration effect by solute molecules may not only constrain the dynamics of water but may also accelerate its dynamics probably because of the breaking of hydrogen bond network [1, 19]. However, the actual nature of the weakly affected water is still not well understood. If weakly affected hydration layer exists in a long-range, it is possible that this water may play a significant role in the structural formation of various soft matter and biomolecules because the length-scale of the interaction between aggregated soft materials [1] (e.g., van der Waals interaction) corresponds to several layers of water molecules. In order to observe such weakly affected water, it is necessary to observe the dynamics of water in a time domain equivalent to that of bulk water dynamics. This is where terahertz spectroscopy comes in.

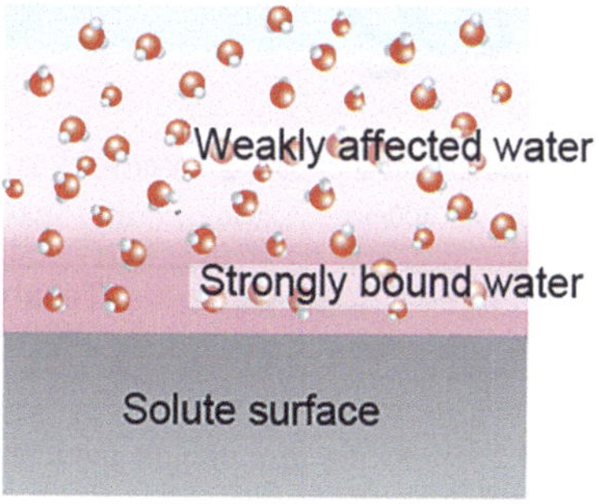


**Fig. 5.1** Hierarchical structure of hydration water

## 5.3 Terahertz Time-Domain Spectroscopy (THz-TDS)

### *5.3.1 Terahertz Spectroscopy on Water and Aqueous Solution*

Terahertz (THz) spectroscopy is a spectroscopy using light with a terahertz (THz = $10^{12}$ Hz) frequency [5, 6]. 1 THz light has a wavenumber of 33.3 $cm^{-1}$, a wavelength of 0.3 mm, and an energy of 4.13 meV. This frequency range is intermediate between infrared light and radio waves, such as millimeter waves (Fig. 5.2). Terahertz light was difficult to generate and detect for long, and has been the only unexplored region of electromagnetic waves, but recent developments in optical technology have led to the development of various generation and detection methods.

If terahertz light is used for spectroscopic measurements, it is possible to measure molecular dynamics in the time region of $10^{-12}$ s, which is the reciprocal of the frequency of terahertz light. In particular, several modes of water molecular dynamics are known to exist in this region [13, 20, 21]. It is known that there are five components of water absorption in the terahertz frequency region (generally referring to the region from about 0.1 to 10 THz). From the lower frequencies, there are a relaxation mode around 20 GHz (0.02 THz), which is attributed to the collective rotational relaxation of water (slow relaxation), a relaxation mode at 0.4 THz that is thought to be caused by the dynamics of isolated water molecules from hydrogen bond network (fast relaxation), a bending vibration between water molecules with a peak at 1 THz (but with very low intensity), a stretching vibration mode between water molecules with a peak at 6 THz, and a liberation mode between water molecules with a peak at 15 THz. In the frequency region below 3 THz, the effect of the hem of slow relaxation is more dominant than the other components, since THz light absorption is quite effective by the slow relaxation.

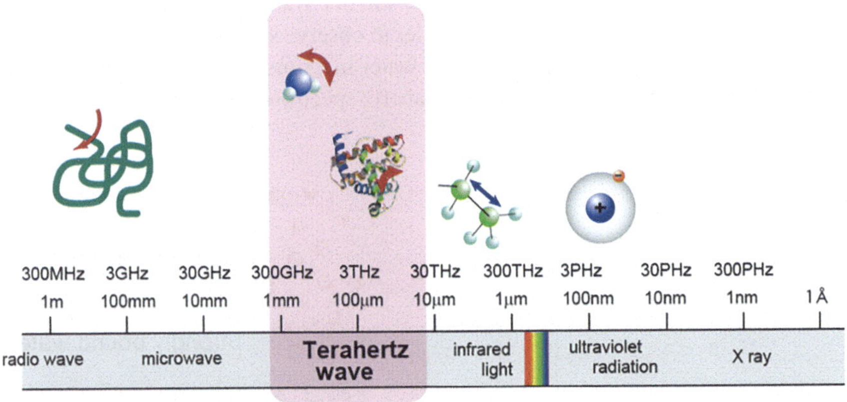


**Fig. 5.2** Frequency map of light and molecular dynamics

Consider the case where a solute dissolves in water and some water becomes hydration water (Fig. 5.3). First, we consider the case where water is bound by the solute and water dynamics is slowed down (Fig. 5.3a) [6, 22–25]. In particular, the hydration effect is clearly observed on the "slow relaxation". Because only the hem of the slow relaxation is observed in the terahertz region, the spectrum of the slow relaxation of hydration water is no longer observed in the terahertz region when the dynamics slows down and the peak frequency becomes lower by the hydration effect. Therefore, the observed spectrum is only that derived from the remaining bulk water, and the spectrum of slow relaxation of bulk water decreases in proportion to the amount of hydration water. Therefore, the amount of hydration water can be estimated from the decrease in slow relaxation derived from bulk water. On the other hand, the slow relaxation can become more mobile than bulk water due to solute-induced hydration effects (Fig. 5.3b) [1, 19, 26, 27]. In such cases, the slow relaxation of hydration water shifts to a higher frequency region than bulk water, that is, the terahertz region. Then, the absorption of light in the terahertz region becomes larger than that of pure water. Therefore, the degree of increase can be used to evaluate the degree of faster slow relaxation. The water with faster rotational relaxation dynamics than bulk water is considered to be equivalent to what has been called "hypermobile water." [28].

In the GHz region, where the slow relaxation of hydration water exists, various dynamics of solute itself also exist and mode of hydration water is difficult to separate from them. In contrast, in the THz region, there is little solute absorption and water dynamics is dominant, so only water dynamics is observed and accurate hydration information can be obtained. Modes other than slow relaxation also change slightly with hydration [29]. Since our measured frequency region is at 0.3–3 THz, fast relaxation is also important in addition to slow relaxation. Since the fast relaxation is a transient dynamics of an isolated water molecule due to the breaking or recombination of a hydrogen bond network, it can be assumed that the fast relaxation is expected to remain in the hydration water and only the intensity can change. The intensity of the fast relaxation increases when the hydrogen bonds are broken, and the number of isolated water molecules increases. In this case, some of the slow relaxation of the bulk water will become fast relaxation, and the amount of hydration water will correspondingly decrease slightly.

To evaluate the change in each mode of water dynamics from the obtained THz spectra and to determine the amount of hydration water, a functional fitting is performed on the spectra obtained from the experiment using the following Eq. (5.1) [21, 29, 30].

$$\tilde{\varepsilon}(\omega) = c\left(\frac{\Delta\varepsilon_{\text{slow}}}{1+i\omega\tau_{\text{slow}}} + \frac{\Delta\varepsilon_{\text{fast}}}{1+i\omega\tau_{\text{fast}}} + \frac{A_{\text{s}}}{\omega_{\text{s}}^{2}-\omega^{2}+i\omega\gamma_{\text{s}}}\right). \tag{5.1}$$

The first term in parentheses represents the slow relaxation, the second term fast relaxation, and the third term intermolecular stretching vibration of water. Since bending vibration is very weak, it is ignored usually. $c$ represents the volumetric

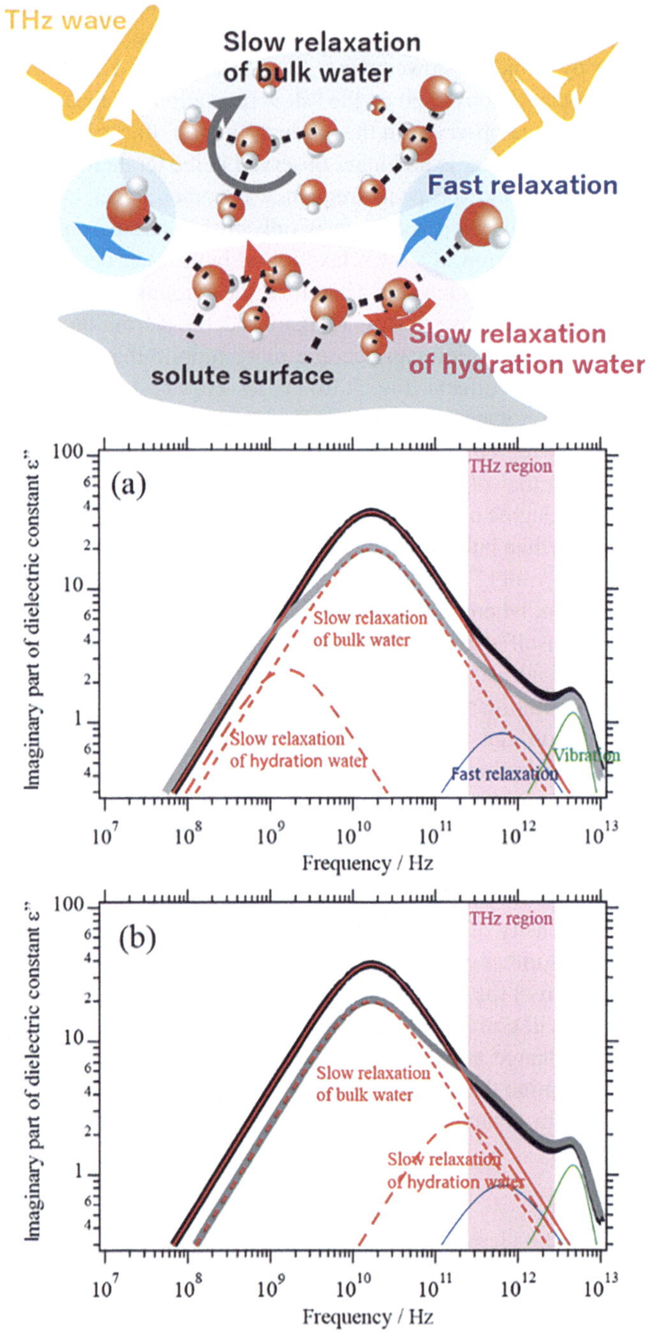


**Fig. 5.3** THz spectroscopy of water and solution. (**a**) and (**b**) are the imaginary part of dielectric constant in GHz to THz region. (**a**) In the case that a solute bound water. (**b**) In the case that the slow relaxation dynamics becomes more mobile due to hydration

fraction of total water in the system, which represents the volumetric effect (the fraction of water per unit volume becomes less than that of pure water due to the presence of solutes, regardless of the presence of hydration water). In THz spectroscopy measurement, since the imaginary part of the dielectric constant generally is obtained with higher accuracy than the real part, fitting is often performed only for the imaginary part. From this fitting, the intensity of slow and fast relaxation $\Delta\varepsilon_{\text{slow}}$ and $\Delta\varepsilon_{\text{fast}}$ are obtained for pure water and solution, and then the amount of hydration per molecule (or per gram) of solute is determined using the following Eq. (5.2) [29].

$$n = N\left(\frac{\dfrac{\Delta\varepsilon_{\text{slow}}^{\text{water}} - \Delta\varepsilon_{\text{slow}}^{\text{solution}}}{g_{\text{slow}}} - \dfrac{\Delta\varepsilon_{\text{fast}}^{\text{solution}} - \Delta\varepsilon_{\text{fast}}^{\text{water}}}{g_{\text{fast}}}}{\dfrac{\Delta\varepsilon_{\text{slow}}^{\text{water}}}{g_{\text{slow}}} + \dfrac{\Delta\varepsilon_{\text{fast}}^{\text{water}}}{g_{\text{fast}}}}\right) \tag{5.2}$$

$g_{\text{slow}}$ and $g_{\text{fast}}$ are the Kirkwood's correlation factors for the slow and fast relaxation, respectively, which were set to 2.9 and 1.0, respectively [29]. The fraction of hydration water to all the water in the system is calculated in parentheses. If $N$ is the total weight of water per unit weight of solute in the solution, $n$ becomes the weight of hydration water per unit weight of the solute, whereas if $N$ is the number of water molecules per solute molecule, $n$ is the number of hydration water molecules per solute molecule. If slow relaxation becomes faster due to the solute hydration effect, it should be observed in the THz region. In such case, the spectrum should be fitted by superposing four modes including the slow relaxation of hydration water. However, it is difficult to accurately evaluate the degree of acceleration in the THz region, which likely depends on the solute. Therefore, fitting and calculation of hydration volume are performed using Eqs. (5.1) and (5.2) even in the case that hydration water becomes faster, which enable us to use the same model to evaluate the slow relaxation regardless of the binding or acceleration of water relaxation dynamics. In this case, the slow relaxation of bulk water is treated as if it were increased over pure water, and the hydration volume is calculated as a negative value [27]. This has no physical meaning, but it is a good way to evaluate the magnitude of the effect of faster slow relaxation.

### *5.3.2 THz Time-Domain Spectroscopy (THz-TDS)*

Spectroscopy in THz frequency region has been conducted with various methods. Time-domain spectroscopy (TDS) is one of the important methods that has been actively studied recently [5, 6]. Time-domain spectroscopy is, as the name implies, the observation of light not in the frequency domain, but in the time domain, that is, the observation of the time waveform of the electromagnetic field. By generating and detecting light with as short a pulse width as possible, the frequency domain is broadened when the light is Fourier transformed, enabling spectroscopy over a wide bandwidth. Another feature of time-domain spectroscopy is that the real and imaginary parts of the dielectric permittivity can be simultaneously obtained without using the Kramers–Kronig transform by directly Fourier transform of the time waveform.

Figure 5.4 shows a schematics of the typical terahertz time-domain spectrometer (THz-TDS). The terahertz light is generated and detected using a photoconductive antenna designed by microfabrication technology. Terahertz light is generated by irradiating a femtosecond ultrafast pulsed laser in the near-infrared region on the photoconductive antenna. The detection is done by synchronizing the near-infrared pulsed light, which is guided by a bifurcation of the laser used for generation, with the electric field of the terahertz light at the detector position. By delaying the arrival of the near-infrared pulsed light, the time variation (time waveform) of the terahertz optical field can be obtained. First, the waveform of the terahertz light is obtained in the absence of a sample, and then the sample is allowed to pass through. The extent to which the arrival time of the terahertz light is delayed and the intensity of the terahertz light is reduced can be used to determine the dielectric properties of the sample.

The most common sample cell for THz spectroscopy is a transmission type. However, when dealing with aqueous solutions, it is difficult to obtain precise spectrum because the terahertz light is greatly absorbed by the water, resulting in extremely poor transmission. In such cases, attenuated total reflection (ATR) cells are used [20]. In this method, terahertz light is guided through a prism made of Si crystal (or other materials) and totally reflected at the surface on which the sample is placed. In this case, only the evanescent field interacts with the sample. Therefore, extreme attenuation of terahertz light intensity is avoided, and accurate measurements of the absorption can be made. The authors perform the experiments using a terahertz time-domain spectroscopy equipment of their own design, but recently, commercial devices are also available.

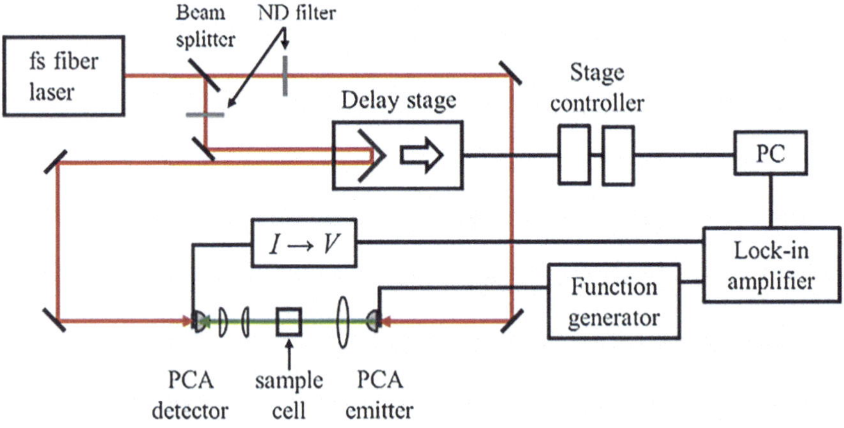


**Fig. 5.4** Typical system of terahertz time-domain spectroscopy (THz-TDS)

## 5.4 Hydration State of Phospholipid Bilayers and Its Effect on Membrane Structural Change

The cell membrane, which separates cells from the outside solution, and the biomembrane, which constitutes the intracellular organelle, are mainly composed of a phospholipid bilayer as their basic structure [31]. When phospholipids are dispersed in water, they spontaneously form a bilayer structure, and in many cases, the bilayer forms a lamellar structure with a water layer in between.

To understand how phospholipid bilayers form the lamellar structure, it is necessary to understand the interactions between the bilayers [1, 32]. It has been considered that four main interactions act between phospholipid bilayers, and the balance of these interactions determines the stability of the lamellar structure and the distance between the membranes. The repulsive interaction is the van der Waals interaction. The other three are all repulsive interactions. The first is electrostatic repulsion, the second is steric repulsion (called Helfrich repulsion), which occurs when membranes fluctuate and collide with each other, and the last is short-range repulsion. The origin of short-range repulsion is still a matter of debate, but the most likely candidate is the protruding motion of phospholipid molecules from the membrane surface. In the past, the presence of hydration water was also a candidate for the short-range repulsion, because a strongly bound hydration water of about 5–6 molecules (about one layer on the surface) was well suited to the decay length of the short-range repulsion [8]. However, since similar short-range repulsion has been observed in solvents other than water, [1] it is now believed that hydration is not the origin of the short-range repulsion.

The heterogeneity of water, that is, the presence of hydration water, has not been fully considered when considering the interactions between membranes. When a lamellar structure is formed with bilayers of neutral phospholipids, the thickness of the water layer between the bilayers is only about 25 Å (about 10 layers of water) at most. Since the water layer thickness is such a molecular scale, the structure of the water network between the membranes cannot be ignored in the interaction between the membranes. Especially when hydration water exists over a longer distance than previously thought, the effect should be more pronounced.

How much hydration water is actually present when such lamellar structures are formed? Using terahertz spectroscopy, we observed the hydration state of 1,2-dimy ristoyl-*sn*-glycero-3-phosphocholine (DMPC), a common neutral phospholipid [33]. Figure 5.5a shows the imaginary part of the dielectric constant in the terahertz frequency region for phospholipid and water mixture. In this frequency region, there was almost no absorption by the phospholipids themselves, so this spectrum is water-derived. As the concentration of phospholipids is increased, the dielectric constant decreases. This means that the absorption of terahertz light by bulk water is decreasing because of the volumetric effect and the existence of hydration water.

While considering the volumetric effect calculated from the lipid concentration, the relaxation intensities of slow and fast relaxation were obtained by functional fitting with Eq. (5.1), and the hydration volume was obtained from Eq. (5.2) as

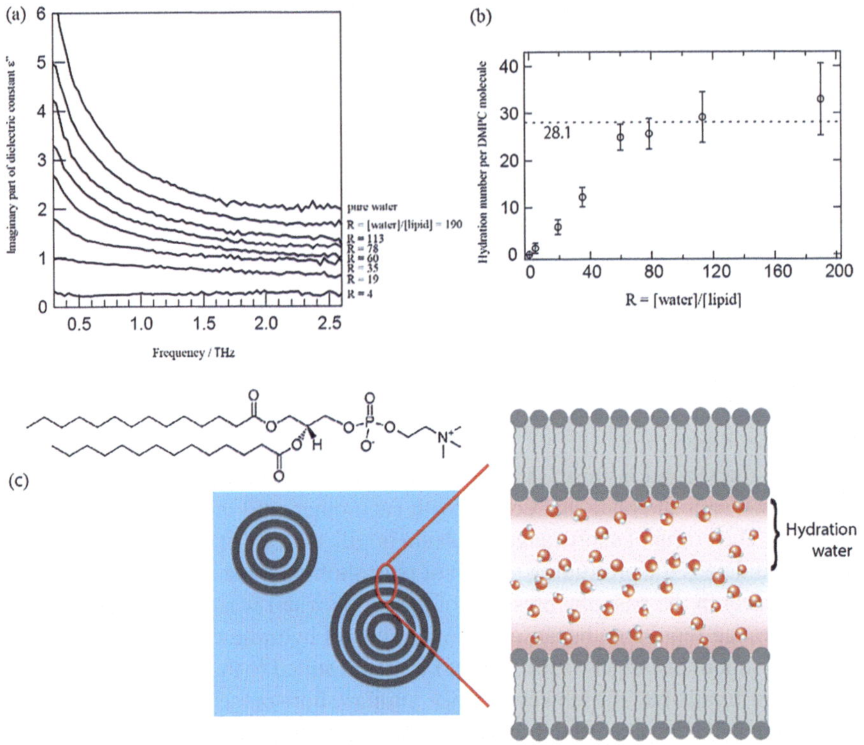


Fig. 5.5 THz spectroscopy result of DMPC membranes [33]. (**a**) THz spectra for different concentration of DMPC. (**b**) Hydration number per DMPC. (**c**) Schematic image of hydration water in lamellar phases of DMPC. (The graph is reprinted with permission from [33]. Copyright 2011 American Physical Society)

shown in Fig. 5.5b (Although different equations are used in the cited references [33], the model is essentially the same and the results become similar.) It can be seen that the hydration number increases as the amount of water is increased, but is almost constant from a certain concentration ($R$ = [water]/[lipid] = 30–40). On the other hand, when the total amount of water contained between the bilayers was estimated by small-angle X-ray diffraction, it was found that the amount of water between the membranes increases with the ratio of water when the lipid concentration is high, but that no more water is incorporated between the membranes when R exceeds 40. In other words, since the amount of water between the membranes no longer changes above $R = 40$, it is not surprising that the amount of hydration water remained constant. The consistency between two completely independent experiments confirms that the derivation of the hydration number by terahertz spectroscopy is valid. For $R > 40$, the thickness of the water layer was about 25 Å. The average number of hydration at $R > 40$ was about 28 per molecule of phospholipid. This corresponds to 4–5 layers of water (about 10 Å), which means that in the case of water sandwiched between two membranes with thickness of 25 Å, most of the water between the membranes can be defined as hydrated water, since it is affected from both sides of the phospholipid membranes (Fig. 5.5c). From past studies using NMR, neutron quasi-elastic scattering in $10^{-9}$–$10^{-11}$ s, and thermal measurements, it has been known that there was one layer of strongly bound water [8]. Therefore, the fact that four to five layers of hydration were estimated in the present results means that the hydration obtained by terahertz spectroscopy includes a weakly affected hydration water. If the heterogeneity of water is found to extend over such a long distance, it is possible that hydration water may be involved in interactions between membranes. Recently, neutron quasi-elastic scattering was performed in a faster time region than previously performed, and water mobility was observed at a similar time region to that observed in terahertz spectroscopy [34, 35]. The results confirmed the existence of water with bound dynamics in the vicinity of phospholipids (DMPC) over several layers.

Since hydration water is produced by interaction with hydrophilic groups, the hydration number may also vary depending on the chemical structure of the hydrophilic groups. Fig. 5.6 shows terahertz spectroscopy result with two phospholipid solutions (1-palmitoyl-2-oleoyl-*sn*-glycero-3-phosphocholine (POPC; PC lipid) and1-palmitoyl-2-oleoyl-*sn*-glycero-3-phosphoethanolamine (POPE; PE lipid)) [26]. Although all other conditions (two hydrophilic groups, concentration, temperature, and phase) are the same except for the structure of the hydrophilic group, a large difference can be seen in the THz spectra (Fig. 5.6a). This indicates that the hydration water state is largely different between the two lipids. The imaginary part of the dielectric constant is smaller for POPC solution than for pure water, as in the DMPC solution (also PC lipid) mentioned above. This means that in PC lipid solution, a lot of water is bound. On the other hand, when the spectrum of the PE lipid solution was normalized by concentration, the imaginary part of the dielectric constant was larger than that of pure water. This suggests that the collective relaxation dynamics of water becomes more mobile due to hydration. Since PE lipid is the lipid molecule in which all three methyl groups of the choline group in PC lipid are replaced with hydrogen atoms, it appears to be more hydrophilic at first glance than PC lipid. Therefore, it is very interesting that water binding is stronger for PC lipid.

Molecular dynamics simulations were performed to investigate the degree of water binding and hydrogen bonding structure on the surface of these hydrophilic groups [36]. It was found that water in the first layer around the ammonium group of PE lipid was better attracted than PC lipid, while hydrogen bonds between water molecules in the first and second layers were less than those in pure water around PE lipid. In other words, the attraction of the water in the first layer to the PE lipid is thought to have caused the breaking of the hydrogen bonds between the water molecules (Fig. 5.6b). It is assumed that the collective reorientation relaxation was accelerated by the breaking of hydrogen bonds between water, which made the water more mobile. For the case of PC lipid, the binding of the water in the first layer was not so strong compared to PE lipid because of three methyl groups at the ammonium group. On the other hand, it was known that there were more hydrogen bonds between water molecules than bulk water in the first and second hydration layers. Therefore, the dynamics of water molecules were slowed down due to

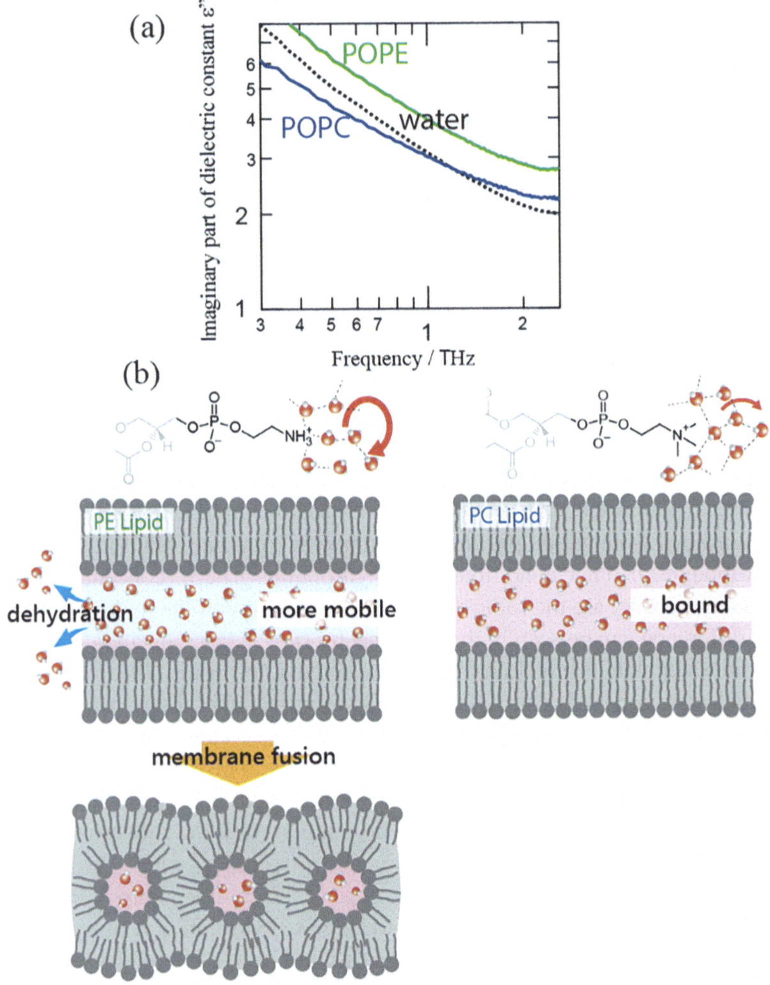


**Fig. 5.6** (**a**) THz spectroscopy result of PC and PE lipid solutions. (**b**) Schematic image of hydration state of PC and PE lipids, and membrane fusion of PE lipid [26]

binding each other through the hydrogen bonding between them. This mechanism is considered to be similar to the formation of ice-berg structure by so-called hydrophobic hydration [37, 38]. The same tendency between the hydrophilic group dependence of water dynamics observed by terahertz spectroscopy and that in the second hydration layer observed by molecular dynamics simulations confirms that the hydration water observed by terahertz spectroscopy still contains weakly affected water in the second layer on the surface.

Using grazing incidence small-angle X-ray scattering (GISAXS), the phase transition behavior of the POPC and POPE phospholipid films were observed as decreasing the ambient humidity [26]. The phase transition from the lamellar phase to the inverted hexagonal phase, a non-bilayer structure, was observed for the PE lipid, and deep decreasing of humidity is needed for the transition of PC lipid. It is thought that water evaporation was more difficult in POPC due to the presence of more bound hydration water, suggesting that the difference in the hydration state of the phospholipid molecules affected even the phase transition behavior (Fig. 5.6b). The lamellar to inverted-hexagonal phase transition of phospholipids with PE lipid was considered to be an important process in the membrane fusion of biomembranes, [39] and it is also known that the membrane fusion does not occur without PE lipid [40]. Thus, it is assumed that the difference in hydration state of the two lipids can affect the ease of this phase transition occurs, that is, membrane fusion in vivo.

## 5.5 Involvement of Hydration in Structural Transitions of Surfactant

Here, I introduce how the behavior of the surrounding hydration water changes when surfactants undergo a phase transition that alters the meso-scale aggregation structure. $C_{12}E_5$ is a type of polyoxyethylene alkyl ether nonionic surfactant. When the concentration is 45 wt%, $C_{12}E_5$ shows the phase transition from the hexagonal phase at low temperatures, to the micellar phase at intermediate temperatures, and to the lamellar phase at high temperatures. The phase transition temperatures are about 25 and 50 °C, respectively. Terahertz spectroscopy was performed on a 45 wt% aqueous solution of $C_{12}E_5$ while changing the temperature [41]. Fig. 5.7 shows the temperature dependence of hydration number per surfactant molecule. Jumps at 20–25 and 50 °C in the hydration number are seen with increasing temperature, which coincide well with the two-phase transition points of the surfactant. This means that when the aggregative structure of the surfactant changes with the phase transition, the surrounding water also exhibits transition-like behavior. It is not yet clear whether water plays an active role in the phase transition or whether the state of water simply changes along with the phase transition. In any case, it can be said that the surrounding hydration water is not just a homogeneous background for the phase transition of surfactants.

## 5.6 The Role of Hydration Water in Protein Structural Stability

Many proteins express their functions by having a defined folded conformation in water, but the mechanism by which the conformation is stabilized is still not fully understood. Attempts have long been made to understand this mechanism from a

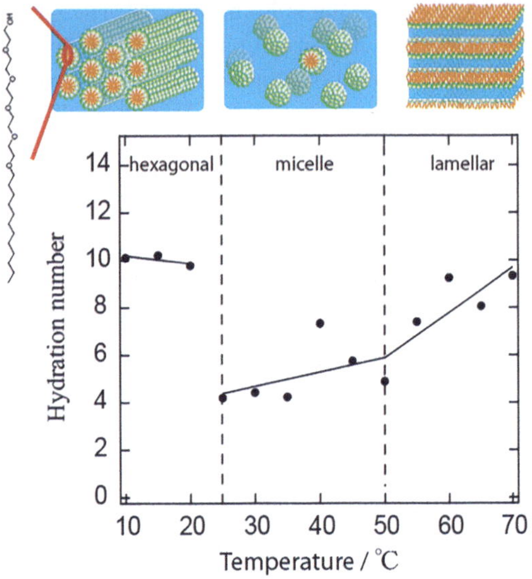


**Fig. 5.7** Temperature dependence of hydration number of $C_{12}E_5$. The graph is reprinted with permission from [41] (Copyright 2012 IOP Publishing Ltd)

thermodynamic viewpoint [7, 42, 43]. As a result, it is known that not only the intramolecular interactions and entropy of the protein itself but also the interactions and entropy related to the surrounding water are significantly involved in the free energy balance between the folded state (native state) and the unfolded state (denatured state) of the protein. In particular, recent detailed energetic analyses have shown that the denatured state is greatly stabilized by the interaction between water and proteins, which results in a decrease in the entropy of water and a decrease in the enthalpy due to the interaction between water and proteins [43]. In other words, water is structured with proteins in the denatured state, and the structuring of the water network plays a role in lowering the free energy of the denatured state (making it easier to denature than in the absence of water). However, these understandings are only at the level of macroscopic thermodynamics, and it is not fully understood how the water surrounding the protein (hydration water) changes upon denaturation at the microscopic level, and what physical properties of water are involved in stabilization.

Using BSA as a model protein, we investigated the change in hydration state upon thermal denaturation using terahertz spectroscopy [44]. In the native state (below 60 °C), there were about three water layers of bound hydration water on the surface, which decreased with increasing temperature. This may be due to the increased thermal motion of the water. Furthermore, the amount of hydration water increased with thermal denaturation at about 60 °C (Fig. 5.8). On the other hand, using thermal measurements, the amount of unfreezable water was also estimated from the freezing and thawing behavior of water when the protein solution was cooled. Since high concentrations of BSA aggregate once denatured and do not exhibit refolding upon cooling, the amount of unfreezable water in the denatured state can be determined by cooling the protein solution after heating it to the denatured state. The amount of unfreezable water against the temperature to which the sample was heated is shown by the red dots in Fig. 5.8. It can be seen that the amount of unfreezable water decreases with thermal denaturation. Unfreezable water is considered to be water that is "strongly bound" to the protein. In fact, a smaller amount of water than the amount of hydration obtained by terahertz spectroscopy has been quantified as unfreezable water content.

Combining the results of terahertz spectroscopy and thermal measurements, it is found that "strongly bound water" decreases and "weakly affected water" increases with denaturation. Furthermore, the OH stretching vibration of water observed by infrared spectroscopy indicates that the hydrogen-bonded OH increases with denaturation (Fig. 5.8). This suggests that in the denatured state, water molecules bind to each other through the increased hydrogen bonding between water molecules, which causes an increase in "weakly affected water." This is thought to be a similar trend as in the case of PC lipids (POPC and DMPC).

As mentioned above, past thermodynamic studies have suggested that water is more structured in the denatured state than in the native state and that this difference has a significant effect on the free energy of the system, and our results suggest that

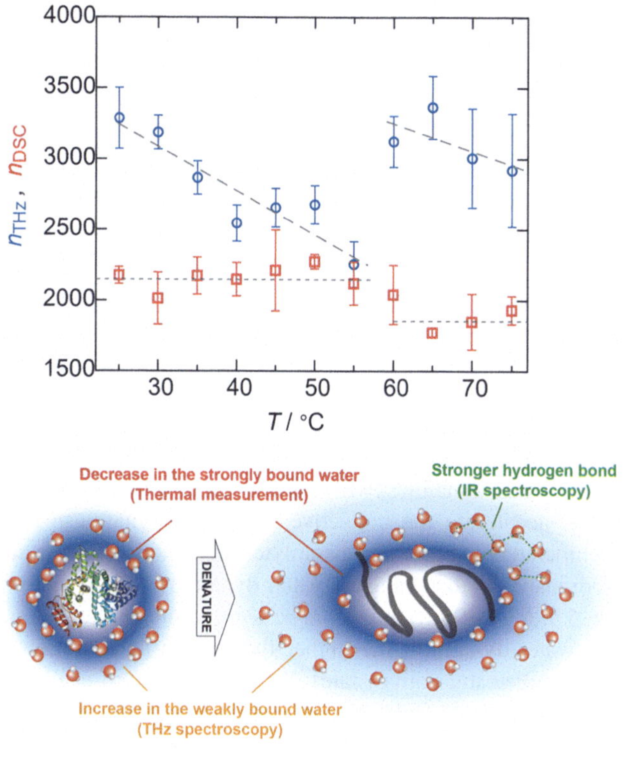


**Fig. 5.8** Hydration number of BSA protein. $n_{THz}$ is the result of THz spectroscopy, while $n_{DSC}$ is the number of unfreezable water. Reprinted from [44] under the Creative Commons Attribution 4.0 International license

this structured water is the "weakly affected water." Furthermore, its origin is due to increased hydrogen bonding between water molecules rather than stronger bonding to proteins. When proteins denature, they expose their internal hydrophobic regions to water to some extent. The "weakly affected water" observed in the denatured state is expected to be the so-called hydrophobic hydration water that is produced around the hydrophobic chains. It has long been said that hydrophobic hydration is due to the structuring of water by hydrogen bonds between water (ice-berg structure [37, 38]), and there is a large possibility that this is observed in terahertz spectroscopy. In other words, there is a large possibility that hydrophobic hydration makes a significant contribution to the free energy balance of the protein (the free energy of the denatured state is greatly reduced due to hydrophobic hydration).

## 5.7 Effect of Organic Molecules (Osmolyte) on Protein Stability

It is also known that the stability of protein folded structure is altered by the addition of various small organic molecules (osmolytes) such as urea, sugars, or TMAO, making the folded structure more or less stable [45–47]. For example, urea is known to decrease the denaturation temperature, while sugars and TMAO increase it. However, the mechanism of the stability change in a unified manner is not fully understood, and there have been two main theories. One is related to the interaction between proteins and osmolyte molecules: osmolytes with greater denaturing effects interact more strongly with proteins, while osmolytes with greater stabilizing effects are excluded from proteins (osmophobic effect) and proteins are stabilized by the presence of water surrounding the proteins [48]. Another theory is related to hydration. That is, osmolyte molecules can alter the hydrogen-bond network structure and/or dynamics of water in the hydration water layer of proteins, thereby affecting protein stability [45, 49]. However, it has not been known which is more essential.

The effects of 15 different osmolytes on water were investigated by terahertz spectroscopy using protein-free osmolyte solutions [27]. It was found that osmolytes that increase the protein denaturation temperature, such as TMAO and sugars, bind the collective water dynamics, whereas urea and guanidine HCl, known denaturing osmolytes, accelerate the collective relaxation dynamics. The change in denaturation temperature when these osmolytes were added was examined using RNase A and compared to the change in water dynamics studied by terahertz spectroscopy in Fig. 5.9. A good correlation can be seen, meaning that the more effective osmolytes bind water dynamics, the higher the protein denaturation temperature (the more they stabilize the native state). Whereas the more effective osmolytes increase water mobility (faster collective relaxation movement) than bulk water, the more the protein is destabilized and the denaturation temperature is lowered. There have been few studies comparing the effects of many osmolytes in such a unified manner, but these good correlations in the present results strongly suggest that the effects of osmolytes on proteins can be mainly indirect, through changes in the state of water. Of course, in some cases, direct interactions can act, but even in such a case, the effects are possibly combined with water-mediated effects.

Furthermore, when the origin of the change in water mobility was investigated by osmolyte using MD simulations, [50] the hydrogen bonding probability between water molecules increased around osmolytes that bind the rotational dynamics of water, whereas the hydrogen bonding probability between water decreased around the osmolyte that makes the rotational dynamics of water more mobile. This tendency is similar to the results of PC and PE lipids, described above. For the former, the increased hydrogen bonding between the water molecules makes them more tightly bound to each other and less mobile, while for the latter, the opposite is thought to be true; the decrease in hydrogen bonding compared to that in pure water makes them more mobile. In particular, this difference in mobility is produced significantly in water from the second hydration layer from the osmolyte surface. It remains to be unknown whether water mobility or hydrogen bonding structure essentially determines protein stability. It is possible that changes in water mobility also change the protein's own mobility and/or hydration state (especially the amount and/or mobility of "weakly affected water"), which in turn changes the stability of the folded structure, but this is still speculative and awaits further study.

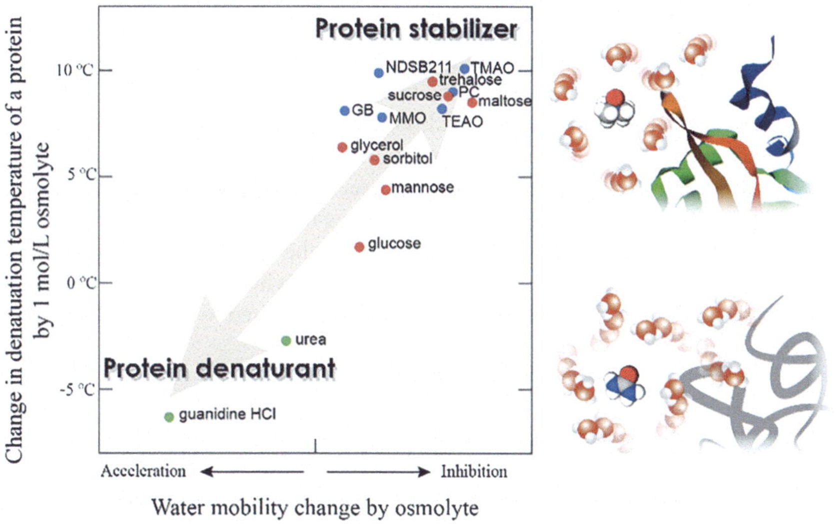


**Fig. 5.9** Relation between water mobility change by osmolyte and their effect on protein denaturation temperature. (Reprinted with permission from [48]. Copyright 2022 American Chemical Society)

## 5.8 Thermoresponsive Property of Polymers and Their Hydration States

Water hydration is also deeply involved in the thermoresponsive property of synthetic polymers, not only of proteins. Certain water-soluble polymers are known to dehydrate and aggregate with increasing temperature, and this aggregation temperature is called the cloud point [24]. The phase behavior is mentioned as LCST (lower critical solution temperature) behavior. However, the mechanism of how this cloud point temperature is determined and what kind of polymers have a cloud point are not fully understood. The relationship between structure, hydration, and cloud point in water for polymers was systematically investigated by changing the ratio of hydrophilic (polyethylene glycol [51]) and hydrophobic (alkyl chain) side chains [52]. These amphiphilic polymers form micellar structure in water. Electron density analysis of the micelles by small-angle X-ray scattering measurement revealed that the electron density in the hydrophilic region in the micelles increases with increasing hydrophobic chain fraction. This indicates that the molecular packing of the hydrophilic side chains becomes denser in the micelles. This can be explained by the fact that the curvature of the micelle decreases as the hydrophobic chains increase due to the volume balance between hydrophilic and hydrophobic groups (Critical Packing Parameter [1]).

Terahertz spectroscopy on these micelles revealed that the amount of hydration water decreased as the hydrophobic chain fraction increased [52]. In other words, the hydration amount seems to have decreased as the hydrophilic groups become more densely packed, making them less able to hydrate. The cloud point temperature also decreased with an increase in the fraction of hydrophobic chains. This can be interpreted as dehydration at the cloud point temperature occurring more easily due to the decrease in the amount of hydration. In other words, the relationships between the structure and physical properties are successfully interpreted in a systematic manner: the molecular structure determines the micelle structure, which in turn determines hydration, which in turn determines the cloud point temperature. It is particularly important to note that hydration water, which determines the cloud point, includes weakly affected water, which can be measured by terahertz spectroscopy (Fig. 5.10).

## 5.9 Conclusion: Importance of Weakly Affected Water on the Properties of Soft Materials and Biomolecules

This chapter describes the results of hydration state analysis of soft matter and biomolecules mainly using terahertz spectroscopy and the role of water in the formation of self-assembled structures revealed by this analysis. Unlike other methods previously used in hydration state analysis, terahertz spectroscopy enables us to quantify the amount of hydrated water, including weakly affected water, and to observe the acceleration of water by hydration, which is called negative hydration.

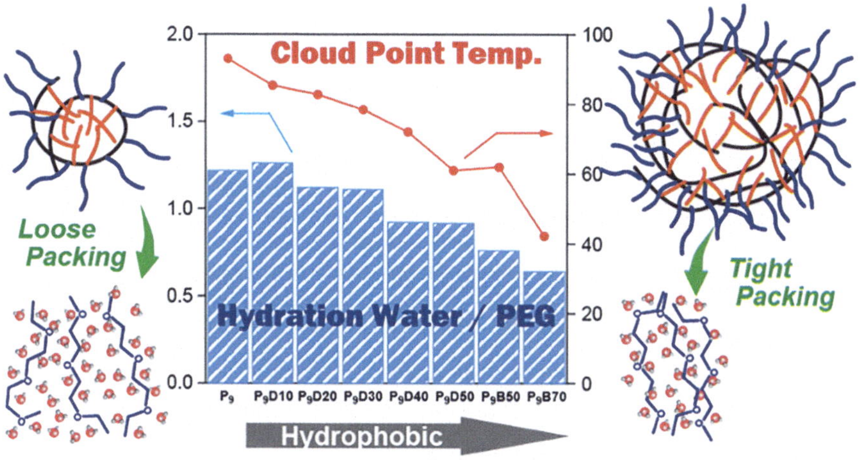


**Fig. 5.10** Relation between micellar structure, hydration state, and cloud point temperature of block copolymers. Reprinted from [52] under the Creative Commons Attribution 4.0 International license

Using this method, it has become clear that differences in the hydration states in long-range are closely linked to structural formation and phase transitions in various systems, such as structural phase transitions in phospholipid membranes and surfactants, protein denaturation and the effect of small molecules on it, and thermoresponsive property of polymers. Molecular dynamics simulations and neutron quasi-elastic scattering have also confirmed these results, increasing the importance of this measurement method. Water has often been treated as a mere background in the self-assembly of soft matter and biomolecules, but in fact it is actively involved in self-assembly mechanism. This strongly suggests the possibility that water is deeply involved in biological phenomena even in living organisms.

At present, however, the relationship between the hydration water observed by other methods and the water state observed by this method remains unknown, and it will be important to clarify the essence of the phenomenon of hydration by comparing the results of many methods in the future. It is still unclear what physical properties of the solute surface essentially determine the hydration behavior. A wide range of research is needed, from comprehensive measurements to theoretical considerations.

**Acknowledgments** Each work was the collaboration with Prof. K. Tanaka (Kyoto Univ.), Profs. K. Saito and Y. Yamamura (Univ. Tsukuba), Profs. M. Tanaka, T. Anada, and Y. Higuchi (Kyushu Univ.), Prof. T. Terashima (Kyoto Univ.), Prof. H. Seto (KEK), and Dr. T. Tominaga (CROSS). These works were partly supported by JSPS KAKENHI Grant Numbers JP19H05717, JP24740289, by the Sumitomo Foundation (No. 090930), and by the Murata Science Foundation (H23-54). GISAXS experiments were performed at National Synchrotron Radiation Research Center, Taiwan (Proposal No. 2011-1-150-5), and SAXS experiments were performed at Photon Factory, High Energy Accelerator Research Organization, Japan (Proposal Nos. 2009G602, 2009G217, 2011G516, 2011G550, and 2021G586).

## References


1. Israelachvili JN (2011) Intermolecular and surface forces. Elsevier
2. Frank HS, Wen W-Y (1957) Ion-solvent interaction. Structural aspects of ion-solvent interaction in aqueous solutions: a suggested picture of water structure. Discuss Faraday Soc 24:133–140
3. Ball P (2008) Water as an active constituent in cell biology. Chem Rev 108:74–108
4. Dargaville BL, Hutmacher DW (2022) Water as the often neglected medium at the interface between materials and biology. Nat Commun 13:4222
5. Neu J, Schmuttenmaer CA (2018) Tutorial: an introduction to terahertz time domain spectroscopy (THz-TDS). J Appl Phys 124:231101
6. Penkov NV (2023) Terahertz spectroscopy as a method for investigation of hydration shells of biomolecules. Biophys Rev 15:833–849
7. Makhatadze GI, Privalov PL (1994) Hydration effects in protein unfolding. Biophys Chem 51:291–309
8. Milhaud J (2004) New insights into water–phospholipid model membrane interactions. Biochim Biophys Acta Biomembr 1663:19–51
9. Nishida K, Anada T, Tanaka M (2022) Roles of interfacial water states on advanced biomedical material design. Adv Drug Deliv Rev 186:114310
10. Yoshida H, Hatakeyama T, Hatakeyama H (1992) Characterization of water in polysaccharide hydrogels by DSC. J Therm Anal 40:483–489
11. Hatakeyama T, Tanaka M, Hatakeyama H (2010) Thermal properties of freezing bound water restrained by polysaccharides. Aust J Biol Sci 21:1865–1875
12. Bach D, Miller IR (1998) Hydration of phospholipid bilayers in the presence and absence of cholesterol. Biochim Biophys Acta Biomembr 1368:216–224
13. Shiraga K, Tanaka K, Arikawa T et al (2018) Reconsideration of the relaxational and vibrational line shapes of liquid water based on ultrabroadband dielectric spectroscopy. Phys Chem Chem Phys 20:26200–26209
14. Volkov AA, Chuchupal SV (2022) Dielectric spectra of liquid water: Ultrabroadband modeling and interpretation. J Mol Liq 365:120044
15. Buchner R, Barthel J, Stauber J (1999) The dielectric relaxation of water between 0°C and 35°C. Chem Phys Lett 306:57–63
16. König S, Sackmann E, Richter D et al (1994) Molecular dynamics of water in oriented DPPC multilayers studied by quasielastic neutron scattering and deuterium-nuclear magnetic resonance relaxation. J Chem Phys 100:3307–3316
17. Swenson J, Kargl F, Berntsen P, Svanberg C (2008) Solvent and lipid dynamics of hydrated lipid bilayers by incoherent quasielastic neutron scattering. J Chem Phys 129:045101
18. Ebbinghaus S, Kim SJ, Heyden M, Yu X, Heugen U, Gruebele M, Leitner DM, Havenith M (2007) An extended dynamical hydration shell around proteins. Proc Natl Acad Sci 104:20749–20752
19. Engel G, Hertz HG (1968) On the negative hydration. A nuclear magnetic relaxation study. Ber Bunsenges Phys Chem 72:808–834
20. Nagai M, Yada H, Arikawa T, Tanaka K (2006) Terahertz time-domain attenuated total reflection spectroscopy in water and biological solution. Int J Infrared Millimeter Waves 27:505–515
21. Rønne C, Keiding SR (2002) Low frequency spectroscopy of liquid water using THz-time domain spectroscopy. J Mol Liq 101:199–218
22. Arikawa T, Nagai M, Tanaka K (2008) Characterizing hydration state in solution using terahertz time-domain attenuated total reflection spectroscopy. Chem Phys Lett 457:12–17
23. Shiraga K, Ogawa Y, Kondo N (2016) Hydrogen bond network of water around protein investigated with terahertz and infrared spectroscopy. Biophys J 111:2629–2641
24. Ono Y, Shikata T (2006) Hydration and dynamic behavior of poly( *N* -isopropylacrylamide)s in aqueous solution: a sharp phase transition at the lower critical solution temperature. J Am Chem Soc 128:10030–10031

25. Heugen U, Schwaab G, Bründermann E et al (2006) Solute-induced retardation of water dynamics probed directly by terahertz spectroscopy. Proc Natl Acad Sci 103:12301–12306
26. Hishida M, Tanaka K, Yamamura Y, Saito K (2014) Cooperativity between water and lipids in lamellar to inverted-hexagonal phase transition. J Phys Soc Jpn 83:044801
27. Hishida M, Anjum R, Anada T et al (2022) Effect of osmolytes on water mobility correlates with their stabilizing effect on proteins. J Phys Chem B 126:2466–2475
28. Kabir SR, Yokoyama K, Mihashi K et al (2003) Hyper-Mobile water is induced around actin filaments. Biophys J 85:3154–3161
29. Shiraga K, Ogawa Y, Tanaka K et al (2018) Coexistence of kosmotropic and chaotropic impacts of urea on water as revealed by terahertz spectroscopy. J Phys Chem B 122:1268–1277
30. Yada H, Nagai M, Tanaka K (2008) Origin of the fast relaxation component of water and heavy water revealed by terahertz time-domain attenuated total reflection spectroscopy. Chem Phys Lett 464:166–170
31. Tristram-Nagle S, Nagle JF (2004) Lipid bilayers: thermodynamics, structure, fluctuations, and interactions. Chem Phys Lipid 127:3–14
32. Petrache HI, Gouliaev N, Tristram-Nagle S et al (1998) Interbilayer interactions from high-resolution x-ray scattering. Phys Rev E 57:7014–7024
33. Hishida M, Tanaka K (2011) Long-range hydration effect of lipid membrane studied by terahertz time-domain spectroscopy. Phys Rev Lett 106:158102
34. Yamada T, Seto H (2020) Quasi-elastic neutron scattering studies on hydration water in phospholipid membranes. Front Chem 8:8
35. Yamada T, Takahashi N, Tominaga T et al (2017) Dynamical behavior of hydration water molecules between phospholipid membranes. J Phys Chem B 121:8322–8329
36. Higuchi Y, Asano Y, Kuwahara T, Hishida M (2021) Rotational dynamics of water at the phospholipid bilayer depending on the head groups studied by molecular dynamics simulations. Langmuir 37:5329–5338
37. Grabowska J, Kuffel A, Zielkiewicz J (2021) Revealing the Frank–Evans "iceberg" structures within the solvation layer around hydrophobic solutes. J Phys Chem B 125:1611–1617
38. Laage D, Stirnemann G, Hynes JT (2009) Why water reorientation slows without iceberg formation around hydrophobic solutes. J Phys Chem B 113:2428–2435
39. Yang L, Huang HW (2002) Observation of a membrane fusion intermediate structure. Science 297:1877–1879. https://doi.org/10.1126/science.1074354
40. Emoto K, Umeda M (2000) An essential role for a membrane lipid in cytokinesis. J Cell Biol 149:1215–1224
41. Hishida M, Tanaka K (2012) Transition of the hydration state of a surfactant accompanying structural transitions of self-assembled aggregates. J Phys Condense Matter 24:284113
42. Privalov PL (1997) Thermodynamics of protein folding. J Chem Thermodyn 29:447–474
43. Heinz LP, Grubmüller H (2021) Spatially resolved free-energy contributions of native fold and molten-globule-like Crambin. Biophys J 120:3470–3482
44. Hishida M, Kaneko A, Yamamura Y, Saito K (2023) Contrasting changes in strongly and weakly bound hydration water of a protein upon denaturation. J Phys Chem B 127:6296–6305
45. Singh Negi K, Das N, Khan T, Sen P (2023) Osmolyte induced protein stabilization: modulation of associated water dynamics might be a key factor. Phys Chem Chem Phys
46. Yancey PH, Clark ME, Hand SC et al (1982) Living with water stress: evolution of osmolyte systems. Science 217:1214–1222
47. Auton M, Bolen DW (2005) Predicting the energetics of osmolyte-induced protein folding/unfolding. Proc Natl Acad Sci 102:15065–15068
48. Mukherjee M, Mondal J (2020) Unifying the contrasting mechanisms of protein-stabilizing osmolytes. J Phys Chem B 124:6565–6574
49. Panuszko A, Bruździak P, Kaczkowska E, Stangret J (2016) General mechanism of osmolytes' influence on protein stability irrespective of the type of osmolyte cosolvent. J Phys Chem B 120:11159–11169

50. Higuchi Y, Saleh MA, Anada T et al (2024) Rotational dynamics of water near osmolytes by molecular dynamics simulations. J Phys Chem B 128:5008–5017
51. Tominaga T, Hishida M, Murakami D et al (2022) Experimental evidence of slow mode water in the vicinity of poly(ethylene oxide) at physiological temperature. J Phys Chem B 126:1758–1767
52. Hishida M, Kanno R, Terashima T (2023) Hydration state on poly(ethylene glycol)-bearing homopolymers and random copolymer micelles: in relation to the thermoresponsive property and micellar structure. Macromolecules 56:7587–7596

# Chapter 6
# Dynamic Self-Assembly of Nanospheres Toward Functional Surfaces

**Kai Liu and Junhui He**

**Abstract** In recent decades, nanospheres have demonstrated unique self-assembly behaviors due to their spherical shape, small size, high specific surface area, and tunable surface properties. These characteristics enable the creation of hierarchical or periodic functional surfaces, which play a critical and distinct role in applications such as optics, wettability regulation, catalysis, and environmental sensing. This chapter provides an overview of the principles, methods, and applications of nanospheres dynamic self-assembly in the development of advanced functional surfaces. First, the interactions in the dynamic self-assembly of nanospheres are discussed, including van der Waals forces, electrostatic interaction, hydrogen bonding, coordination, etc. Next, we introduce the methods for inducing the dynamic self-assembly of nanospheres through solvent evaporation, temperature induction, and external force fields and discuss their advantages and limitations. Then, we enumerate the applications of dynamic self-assembly of nanospheres in developing functional surfaces, including photonic crystals, antireflective coatings, wettability control, environmental monitoring, and so on. Finally, the challenges and future development trends in advancing this technology in the field of nanoscience and technology are discussed, aiming to inspire researchers to continue exploring more exciting possibilities in this field.

**Keywords** Nanospheres · Self-assembly · Interactions · Functional surfaces

K. Liu
Functional Nanomaterials Laboratory, Center for Micro/Nanomaterials and Technology, and Key Laboratory of Photochemical Conversion and Optoelectronic Materials, Technical Institute of Physics and Chemistry, Chinese Academy of Sciences, Beijing, China

University of Chinese Academy of Sciences, Beijing, China

J. He (✉)
Functional Nanomaterials Laboratory, Center for Micro/Nanomaterials and Technology, and Key Laboratory of Photochemical Conversion and Optoelectronic Materials, Technical Institute of Physics and Chemistry, Chinese Academy of Sciences, Beijing, China
e-mail: jhhe@mail.ipc.ac.cn


M. Hashizume, Y. Imura (eds.), *Colloids and Interfacial Dynamics*, Physical Chemistry in Action, https://doi.org/10.1007/978-981-96-6728-4_6

## 6.1 Introduction

Nanospheres have attracted widespread attention in the field of nanotechnology due to their unique physical and chemical properties (optical, electronic, magnetic properties, etc. [1–7]), as well as their broad range of applications in scientific research. These particles typically exhibit a spherical shape and range in size from several nm to several hundred nm. The small size, high specific surface area, and modifiable surface characteristics of nanospheres give them a wide range of applications, including drug delivery [8, 9], catalysis [10, 11], energy utilization [12, 13], environmental sensing [14, 15], and other fields [16–19]. Nanospheres can be fabricated from a variety of materials, including polymers, ceramics, metals, and metal oxides. They are often stabilized by surfactants or surface functional groups to prevent aggregation or degradation. In recent years, research on nanospheres has focused on utilizing their characteristics to design advanced functional materials. Notably, the unique surface physicochemical properties of nanospheres enable them not only to function as monodisperse entities but also to act as building units for the assembly of two-dimensional or three-dimensional materials by adjusting the interactions between surface ligands. In this process, nanospheres are affected by forces of different scales within the system, such as attraction, repulsion, or external forces that drive their movement [20, 21] . When influenced by these external forces, nanospheres can be assembled into new intelligent materials with novel functions.

Dynamic self-assembly refers to the process by which nanoparticles spontaneously aggregate and form ordered structures through interactions such as van der Waals forces, electrostatic forces, and hydrogen bonds under specific external conditions [20, 22–24]. This phenomenon is governed by the principles of thermodynamics and kinetics, involving a complex dynamic equilibrium of particle motion, arrangement, and interaction, which ultimately leads to the spontaneous organization of individual particles into ordered structures. These intermolecular interactions form the foundation of the self-assembly process of nanoparticles. Van der Waals forces promote attractive interactions between nanoparticles induced by dipoles, while hydrogen bonds facilitate directional interactions that stabilize the structure. Additionally, electrostatic interactions between charged particles or molecules introduce long-range forces that influence the overall structure [25, 26]. Reversibility is a key feature of dynamic self-assembly. As external conditions change, the already formed ordered structures can dissociate or rearrange to form new aggregates. This reversible self-assembly process provides great flexibility for the application of nanomaterials in biomedical and material sciences [27, 28].

During the dynamic self-assembly process, nanoparticles initially exist in a disordered state and then spontaneously aggregate and assemble under the influence of temperature, pH value, ion strength, or other external factors. From a thermodynamic perspective, the self-assembly process is driven by the tendency of the system to minimize free energy, which is caused by the balance between entropy and enthalpy: The aggregation of nanospheres leads to a decrease in entropy by increasing order, while the mutual attraction between particles results in a decrease in enthalpy, thereby promoting self-assembly [29, 30]. Kinetic factors also play a crucial role, especially in dynamic self-assembly, where the rate of self-assembly is influenced by the diffusion rate of nanoparticles. Rising temperatures typically increase the kinetic energy of particles, thereby increasing their probability of collision and accelerating the self-assembly process. The choice of solvent can also alter the local environment of nanoparticles and change their diffusion kinetics. In addition, the stability of the intermediate state and the size of the energy barrier will significantly affect the path followed by the self-assembly process.

As a key building unit in dynamic self-assembly, nanospheres offer favorable conditions for self-assembly due to their large specific surface area and abundant surface-active sites, enabling the creation of highly ordered and functional surfaces with tunable properties. The individual characteristics of nanospheres, including their size, shape, and surface chemical properties, significantly affect their self-assembly behavior. For example, smaller nanospheres, due to stronger van der Waals interactions, typically form dense, ordered structures, while larger nanospheres exhibit more complex assembly behaviors. Similarly, the shape of nanoparticles influences the type of structure formed—spherical particles tend to promote isotropic stacking. Additionally, the surface chemical properties of nanospheres play a crucial role in controlling their interactions during self-assembly. Surface modification with functional groups or polymers can introduce steric hindrance or promote specific binding interactions, thereby guiding the assembly process [31]. Such modifications can alter assembly dynamics and enhance the formation of advanced properties, such as biocompatibility, targeting ability, or optical properties. Moreover， the surface characteristics of these dynamically assembled structures are highly responsive to external environmental stimuli, enabling them to exhibit excellent adaptive responses while maintaining structural stability and supporting multiple functions. The motivation for studying the dynamic self-assembly of nanospheres arises from the growing demand for innovative functional surfaces in scientific development. For example, in optical applications, dynamically self-assembled nanospheres can modulate light propagation by precisely controlling their periodic structure [32]. In catalysis, dynamic self-assembly can optimize the reaction environments by continuously reorganizing catalytic nanoparticles, thereby improving efficiency and selectivity [33, 34]. In biosensors, dynamically self-assembled and functionalized nanospheres provide a platform for molecular recognition, significantly enhancing sensitivity and specificity [35]. These applications highlight the vast potential of nanosphere dynamic self-assembly technology.

This review aims to systematically summarize the research status of dynamic self-assembly of nanospheres, with a particular focus on their role in the development of functional surfaces. The structure of this review is as follows: The next section will explore the theoretical foundations of dynamic self-assembly, including the driving factors and interactions involved in the process. Following that, we will examine the various methods used to drive dynamic self-assembly of nanospheres, such as solvent evaporation, temperature changes, electric/magnetic fields, and external forces. Subsequently, we will focus on the significance of dynamic self-assembly in functional surface design and emphasize its applications in fields such as photonic crystals, antireflection, wettability regulation, and environmental monitoring. Finally, the article will address current challenges and future outlooks in this technology. Through these discussions, this review seeks to provide researchers with a deeper understanding of the dynamic self-assembly technology of nanospheres, thereby facilitating its further advancement in scientific research.

## 6.2 Interaction During Self-Assembly Process of Nanospheres

The dynamic self-assembly process of nanospheres involves their spontaneous aggregation in response to external environmental influences. When the steady-state conditions of the environment surrounding the nanospheres are altered by external stimuli, driving the system into a nonequilibrium state, the nanospheres will gradually return to thermodynamic equilibrium through energy dissipation, forming a transient, stable self-assembly state [36, 37]. This process ultimately results in the formation of two-dimensional or even three-dimensional ordered structures. The driving factors that influence the self-assembly of nanoparticles represent a key area of research. In this chapter, we introduce the driving forces and typical external triggers of nanoparticle self-assembly, along with an exploration of their modes of action.

The dynamic self-assembly of nanospheres is a complex process in which nanoparticles spontaneously aggregate into ordered structures through various interaction forces under specific stimulus conditions. Typically, nanospheres require energy from the environment to overcome activation barriers, ultimately achieving a delicate balance between attractive and repulsive forces, and this balance enables the particles to reach thermodynamic equilibrium and form ordered structures. The primary forces involved include electrostatic interactions, van der Waals forces, hydrogen bonds, and coordination interactions, etc.

### *6.2.1 van der Waals Forces*

van der Waals interaction is an important short-range force in the self-assembly process of nanospheres, arising from the gravitational forces between instantaneous dipoles and induced dipoles (Fig. 6.1) [38]. Due to their relatively weak nature, van der Waals forces are more effective in smaller nanoparticles. These interactions play a key role in promoting aggregation during the early stages of self-assembly, particularly at high particle concentrations, where they drive particles to form stable aggregates under the influence of thermal motion. However, at low concentrations, the strength of van der Waals interaction rapidly decreases as the distance between the nanospheres increases. The strength of these forces is closely linked to the

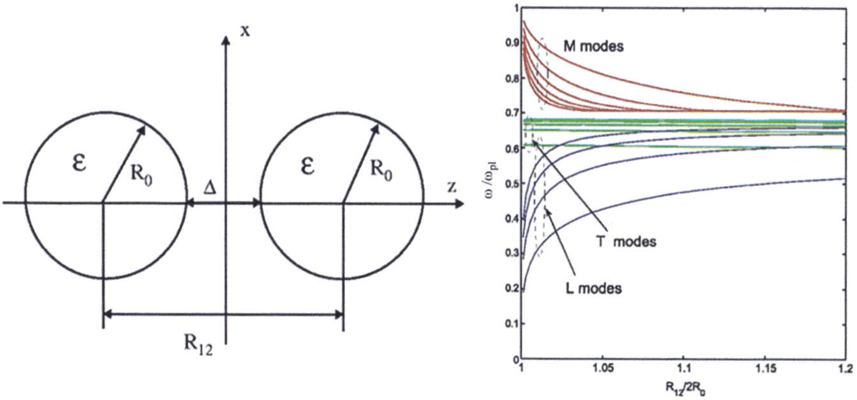


**Fig. 6.1** The function curve of van der Waals force on the normalized distance between identical spheres [38]. Copyright 2008, Springer Nature

surface characteristics of the particles and environmental conditions [39, 40]. The larger the specific surface area of the particles, the stronger the van der Waals forces, as each nanosphere has more neighboring units to interact with. Additionally, the properties and temperature of the solvent can affect the effectiveness of van der Waals forces. In polar solvents, these forces may weaken, while in nonpolar environments, they are significantly enhanced, promoting the aggregation of nanospheres.

## *6.2.2 Electrostatic Interaction*

Electrostatic forces play a crucial role in the self-assembly of nanoparticles, primarily through the interactions between charged particles, including repulsion and attraction. According to Coulomb's law, the strength of the electrostatic force is proportional to the magnitude of the charges and inversely proportional to the distance between the particles [41, 42]. Particles with the same charge will repel each other, while particles with opposite charges will attract. The charge on the particles can be introduced through surface modification with charged groups or by adjusting the pH and electrolyte concentration of the solution. During the initial stage of self-assembly, electrostatic attraction promotes the aggregation of nanospheres. As the structure forms, electrostatic interactions influence the final arrangement and stability of the ordered structure. Proper particle concentration and solution conditions can enhance the efficiency of dynamic self-assembly. Electrostatic repulsion can also help stabilize nanoparticles, as explained by the derjaguin-landau-verwey-overbeek (DLVO) theory [43]. Changes in the dynamic environment, such as pH and particle concentration, can adjust the magnitude of electrostatic forces, thereby controlling the self-assembly process.

### 6.2.3 Hydrogen Bonding

Hydrogen bonding is primarily driven by the attraction between hydrogen atoms and highly electronegative atoms such as O or N, as shown in Fig. 6.2 [44]. The formation of hydrogen bonds significantly enhances the interactions between functionalized nanoparticles, particularly when hydrogen bond donors or acceptors are present on the particle surface, promoting the ordered aggregation of particles. Due to the directivity of hydrogen bonds and their relatively strong interactions, nanoparticles can form stable aggregates, leading to the formation of complex supramolecular structures [45, 46]. Hydrogen bonds are highly sensitive to environmental conditions, such as temperature and pH. As these conditions change, the stability and number of hydrogen bonds may vary, which affects the efficiency of self-assembly and the resulting surface properties. Additionally, hydrogen bonding plays a critical role in the interaction between nanoparticles and biomolecules in biological applications. For example, in drug delivery systems, hydrogen bonding strengthens the binding between drug carriers and target cells, thereby improving targeting and delivery efficiency [47].

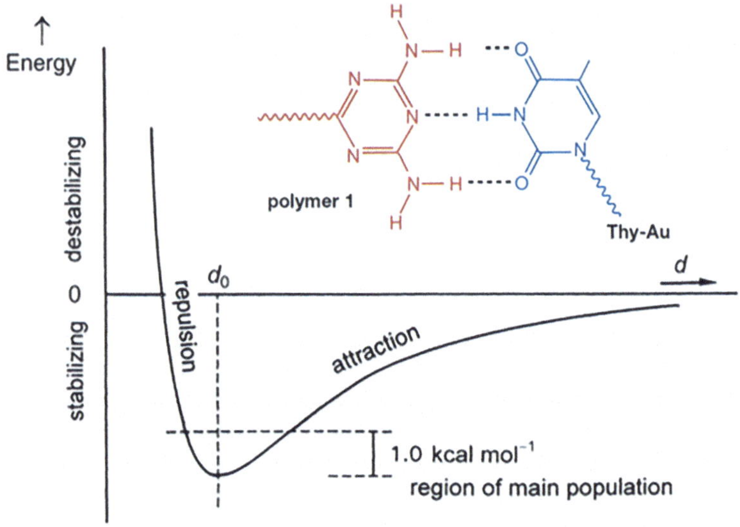


**Fig. 6.2** Schematic representation of a typical hydrogen bond potential [44]. Copyright 2002, John Wiley and Sons

### 6.2.4 Coordination

The coordination effect mainly acts on the self-assembly process of metal-based nanoparticles, where the coordination bonds between metal ions and ligands enhance particle interactions [48, 49]. This effect is typically achieved by introducing coordination-capable molecules, such as amines, carboxylic acids, or phosphates, onto the nanoparticle surface [50, 51]. These ligands form strong coordination complexes with metal ions, increasing the attraction between particles and promoting self-assembly.

Coordination interactions are directional and selective, which allows them to guide nanoparticles in forming highly controllable structures during self-assembly. For example, Wang et al. were the first team to demonstrate the use of metal coordination to guide the self-assembly of patchy nanoparticles into larger chains, as shown in Fig. 6.3a [52]. By functionalizing colloids with metal-coordinated recognition units, the assembly and disassembly of anisotropic particles can be controlled. This method significantly advances the research on the assembly of colloidal particles.

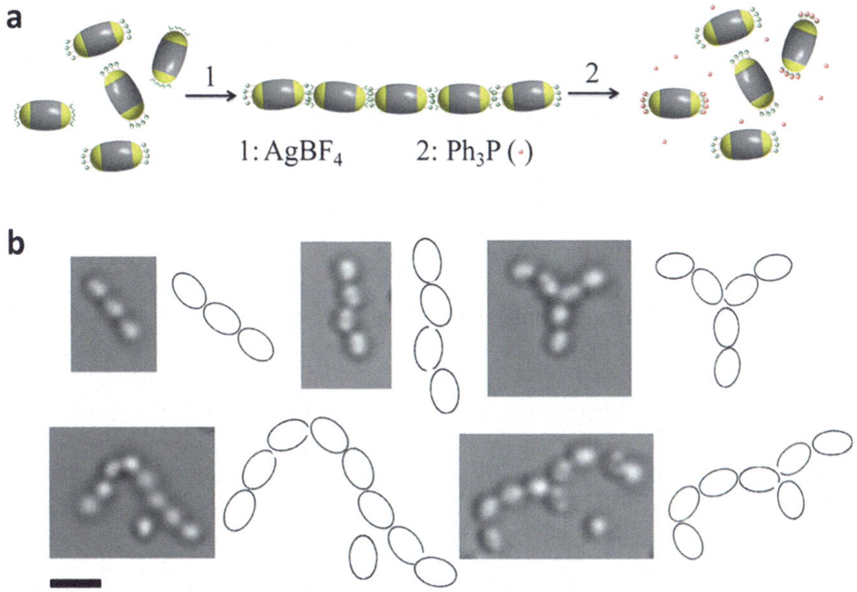


**Fig. 6.3** Self-assembly of two-patch particles via metal coordination. (**a**) Schematic representative showing the triggered assembling and disassembling process of the two-patch particles. (**b**) Scanning electron microscopy (SEM) of chain structures formed by the self-assembly of two-patch particles. Scale bar = 3 μm [52]. Copyright 2013, American Chemical Society

The strength and stability of coordination interactions depend on factors such as the size, structure, recognition units of the nanoparticles and the solvent environment. Under different environmental conditions, the formation and cleavage of coordination bonds can alter the kinetics of self-assembly, thus affecting the final structure (Fig. 6.3b). For example, alkaline medium may cause protonation or deprotonation of ligands, reducing their coordination ability, while increased ionic strength can lead to competition or shielding effects in coordination reactions, which in turn affect self-assembly efficiency [53, 54].

In a typical dynamic self-assembly process, the system environment is subjected to external stimuli, causing the nanospheres to be pushed by one or more of the aforementioned forces, thereby promoting their arrangement into an ordered structure. In the following sections, we will introduce several prominent techniques for nanosphere self-assembly, including wet chemistry methods, template-assisted methods, temperature-induced self-assembly, and electric/magnetic field-induced self-assembly. Each method offers a unique approach to achieving controllable, large-scale, and organized structure of nanospheres on functional surfaces. By gaining a deeper understanding of these dynamic self-assembly processes, researchers can exert precise control over the interactions between nanosphere particles, allowing for better design and selection of the appropriate triggers for dynamic self-assembly on functional interfaces.

## 6.3 Dynamic Self-Assembly Method

### *6.3.1 Wet Chemical Methods*

#### 6.3.1.1 Evaporation-Induced Self-Assembly

Evaporation-induced self-assembly (EISA) is a process that leverages the evaporation of a solvent to drive the self-organization of nanoparticles (such as nanospheres, and nanowires) into ordered structures, facilitating the construction of two-dimensional or three-dimensional nanostructures [55–57]. Before evaporation-induced self-assembly begins, nanoparticles are dispersed in a solvent at a specific concentration to form a stable dispersion. The solvent is typically a volatile liquid, such as water, alcohols, or organic solvents. The evaporation process is crucial in the self-assembly of nanospheres. As the solvent evaporates, the concentration of nanoparticles in the solution increases, leading to their aggregation through interactions such as capillary force, van der Waals forces, and electrostatic forces. The capillary force generated by the liquid's surface tension plays a critical role in the aggregation and self-assembly of nanoparticles [58]. This force causes the particles to migrate toward the evaporation edge and the surface of the solution, where they form arrays or network structures through interactions between the particles (Fig. 6.4a). As evaporation continues, the nanospheres gradually organize into well-defined two-dimensional or three-dimensional structures on the substrate surface

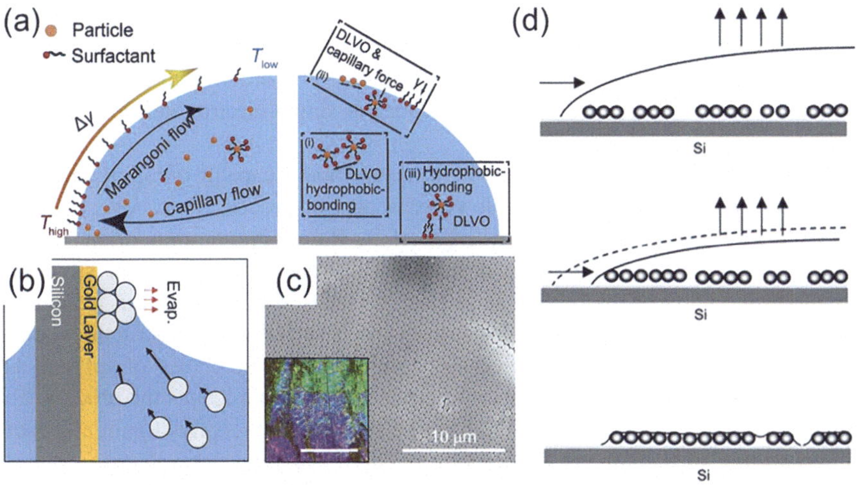


**Fig. 6.4** (**a**) Schematic of the surfactant-dependent flow paradigm (left side) and the particle–particle and particle–interface interactions (right side) in an evaporating surfactant solution droplet containing insoluble particles [58]. Copyright 2020, Elsevier. (**b**) The schematic illustrates the general process of the vertical deposition method. (**c**) Scanning electron microscopy (SEM) and optical diffraction images (inset) of self-assembled nanospheres [59]. Copyright 2019, John Wiley and Sons. (**d**) Schematic illustration of the solvent drying on the silicon surface, inducing the formation of a close-packed array of Au nanoparticles [60]. Copyright 2007, Royal Society of Chemistry

(Fig. 6.4b, c). The surface properties of the substrate significantly influence the self-assembled structure. Hydrophilic substrates tend to promote the uniform assembly of nanospheres, while hydrophobic substrates may cause particle aggregation, as the water film undergoes dehydration before the height of the water layer becomes small enough to generate capillary forces [59]. Additionally, a rapid evaporation rate can result in incomplete particle arrangement, leading to disordered structures. In contrast, a slower evaporation rate provides more time for the particles to self-assemble, resulting in a more ordered arrangement.

Evaporation-induced self-assembly offers a relatively simple experimental procedure. Liu et al. deposited an Au colloidal suspension onto a silicon substrate modified with a monolayer of aminopropyltriethoxysilane (APTES), the Au nanoparticles were then treated with alkylthiol molecules to replace the negatively charged citrate adsorbate on their surface [60]. This modification effectively mitigates the electrostatic repulsion between Au nanoparticles and reduces the electrostatic adsorption between Au nanoparticles and the silicon substrate, thereby facilitating the movement of Au nanoparticles on the surface. As a result, the Au nanoparticles, which were initially randomly dispersed, self-organized into small two-dimensional aggregates on the silicon substrate, forming a submonolayer. Finally, as the evaporation of the alcohol solvent drives the movement of the Au nanoparticle aggregates, the contact line of the solvent is fixed due to the particle–substrate interaction.

Subsequently, the transverse capillary forces, arising from the partial immersion of the nanoparticles, caused further aggregation, leading to the formation of a continuous, ordered array on the silicon surface (Fig. 6.4d). This ordered functional surface opens up new possibilities for applications such as single-electron devices and nonlinear optical response systems.

Evaporation-induced self-assembly (EISA) method is capable of rapidly producing highly ordered nanostructures. However, achieving precise control over the process remains a significant challenge [61, 62]. Rapid evaporation often causes nanospheres to accumulate at the droplet's edge, leading to a circular structure arrangement known as the "coffee ring effect" [63, 64]. Furthermore, this technique faces challenges in achieving consistently ordered self-assembled structures, particularly for complex three-dimensional configurations. The choice of solvent, as well as the dependence on substrate properties, further complicates the process. To enhance the reproducibility and stability of EISA, researchers must gain a deeper understanding of the solvent evaporation dynamics and develop effective control strategies.

#### 6.3.1.2 Langmuir–Blodgett (LB) Method

The Langmuir–Blodgett (LB) method is a well-established self-assembly technique developed by American scientist Irving Langmuir and his student Katherine Blodgett [65]. As shown in Fig. 6.5, this technique utilizes a specialized instrument, the Langmuir–Blodgett trough, to fabricate high-quality single- or multilayer nanomaterial films on solid substrates [66, 67].

In the LB process, the dynamic self-assembly of nanospheres typically occurs at the air–liquid interface. The process generally involves the following steps [68]:

1: Insoluble nanospheres are first dispersed in a solution and then gradually adsorbed onto the gas–liquid interface. Once at the interface, the nanospheres diffuse to minimize surface energy, ultimately forming a stable arrangement.

2: External mechanical devices, such as partition barriers, are employed to compress the nanospheres at the interface, limiting their free movement. This compression increases the density of the nanospheres per unit area, enhancing their orderliness and regular layer formation. Through this process, the nanospheres are tightly arranged, forming regular single- or multilayer structures.

3: Once the desired arrangement of the nanosphere monolayer or multilayer film is achieved, the film is transferred to a solid substrate using a vertical transfer method. During this transfer, the nanospheres remain on the substrate's surface, forming a uniform nanostructured thin film. Depending on the number of transfer cycles, the film can be either a single- or multilayer film.

The second and third steps of the LB method primarily rely on external pressure to tightly arrange nanoparticles. However, during the deposition stage, if the pulling speed is too fast, excessive shear forces may disrupt the ordered monolayer membrane, potentially causing defects. Meanwhile, the compatibility between the

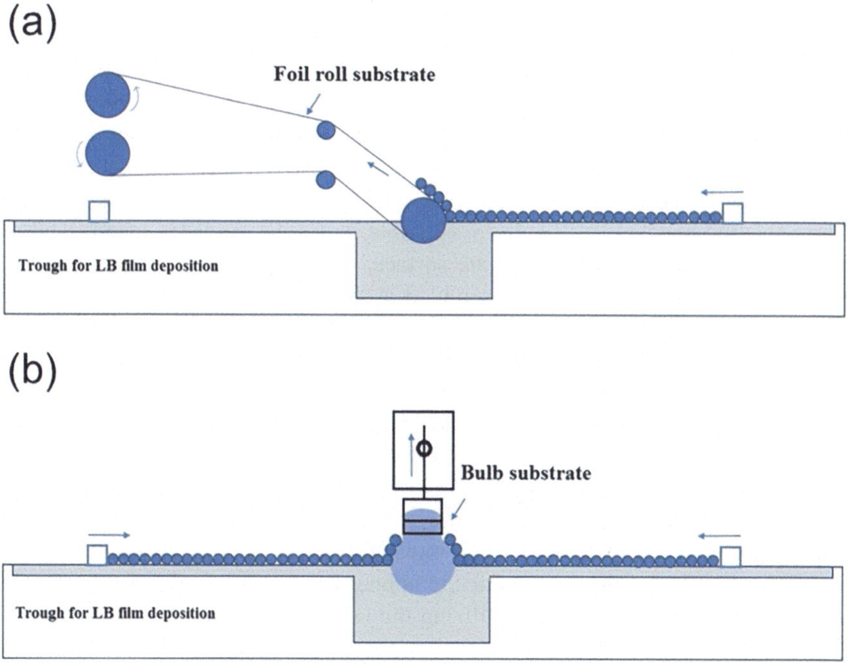


**Fig. 6.5** Schematic diagram of dipping mechanisms on flexible substrate (**a**) and curved object (**b**) in LB trough [68]. Copyright 2020, Elsevier

substrate and the nanospheres plays a crucial role in the self-assembly process, which directly affects the uniformity and stability of the resulting film.

The LB method has the potential to prepare large-area single-layer films due to its simple operation, low cost, and strong controllability. It has found widespread applications in fields such as nanophotonics, sensors, and drug delivery. Although there are still some challenges in terms of stability, with further technological development, LB method is expected to play an increasingly important role in the field of nanotechnology.

### 6.3.2 Template-Assisted Methods

Template-assisted methods assemble nanospheres into desired patterns by using prestructured templates or substrates. The shape, surface properties, and chemical bonds of templates are utilized to control the organization and arrangement of nanospheres across various length scales [69, 70]. Templates are typically classified into hard templates and flexible templates. Special template structures, such as predesigned nanopores and chambers, confine nano-components within a uniform and regular structure. Nanospheres are organized within these predefined templates to form highly ordered structures that replicate the template pattern. This approach allows for the creation of nanosphere arrays with precise geometric shapes that are difficult to achieve using other methods. The template-assisted method first requires designing and manufacturing templates made of metals, silicon, polymers, or other materials suitable for micro- and nanoscales [71]. Surface chemical modifications are then applied to prepare the templates. Subsequently, a solvent containing dispersed nanospheres is transferred onto the surface of the template. As the solution evaporates, the nanospheres begin to assemble under the guidance of physical or chemical interactions on the template surface, forming an ordered arrangement. Once the nanospheres have self-assembled as intended, the template is removed through chemical, physical, or thermal treatment, leaving behind a self-assembled nanosphere array or film (Fig. 6.6a).

Feyter et al. used polystyrene microspheres as photolithography masks and electrochemically covalently deposited aromatic diazonium molecules on graphite substrates to generate regularly separated nanofences with a diameter of 290 nm, which were used as templates for the study of restricted self-assembly of isophthalic acid molecules [72]. After removing the polystyrene bead template through solvent washing, the expected pattern of small molecule assembly structure was obtained (Fig. 6.6b). Similarly, surface-modified templates can also be used to form specific structures with nanoparticles. By utilizing the large surface area of functionalized

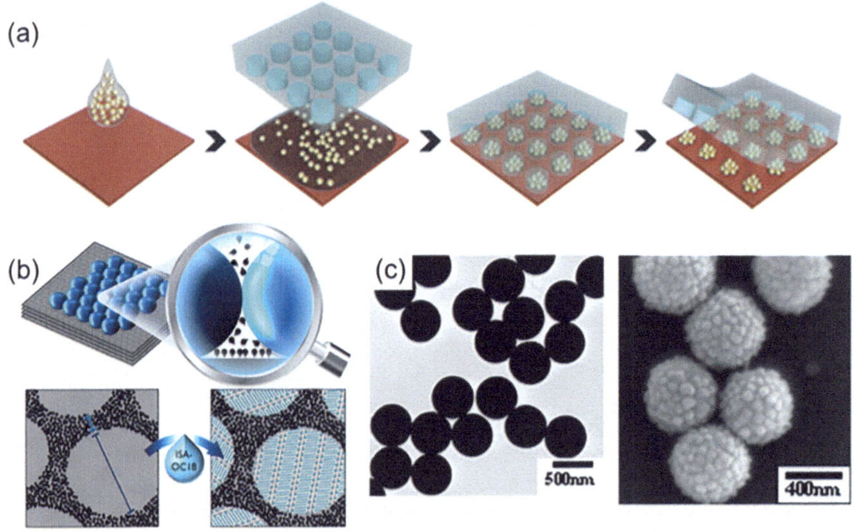


**Fig. 6.6** (**a**) Scheme of the steps in a typical template-assisted self-assembly process [71]. Copyright 2021, John Wiley and Sons. (**b**) Schematic diagram of restricted self-assembly of polystyrene template-assisted 5-octadecoxyisophthalic acid (ISA-OC18), demonstrating the effect of the template on long-range molecular assembly [72]. Copyright 2018, American Chemical Society. (**c**) Small $SiO_2$ is adsorbed onto the surface of larger $SiO_2$ templates for self-assembly to form raspberry-shaped $SiO_2$ nanospheres [73]. Copyright 2007, Elsevier

templates as adsorption sites, nanoparticles can form hierarchical structures. He et al. used larger silica spheres as templates to adsorb small silica nanoparticles, forming nanosphere particles with raspberry-specific structure, as shown in Fig. 6.6c [73]. This silica surface has exciting superhydrophilic and antifog properties.

Although template-assisted self-assembly offers advantages of flexible and generic template design as well as precise control characteristics, it also presents several challenges. One significant issue is the potential incompatibility between the template and the surface properties of the materials, which can lead to uneven stacking or poor alignment of the nanostructures. Furthermore, the production of templates is often costly, and the process of template removal after assembly introduces additional complexity. These factors must be carefully considered when utilizing template-assisted methods for nanofabrication.

### 6.3.3 Thermal-Induced Self-Assembly

Thermally induced self-assembly is a technique that facilitates the spontaneous formation of an ordered nanosphere structure by regulating temperature. This method leverages the impact of temperature variations on both the nanospheres and their surrounding environment. The thermal energy provided by temperature changes can induce chemical or physical transformations in the structural units, triggering movement, chemical interactions, and variations in the solvent evaporation rate. These factors collectively promote the self-assembly of nanospheres into specific, ordered structures [74, 75]. At the level of an individual nanosphere, the molecular collisions in the high-temperature part of the solution environment are more intense, resulting in directional forces caused by local temperature gradient changes, which drive the movement of the nanospheres. On a broader scale, temperature changes typically alter the solvent evaporation rate, facilitating the ordered arrangement of nanospheres on the substrate surface or within the solution. Temperature variations can also change the surface energy of nanospheres, regulate their dispersion state in solution or their interaction with substrates, and induce the spontaneous aggregation of nanospheres. Lin utilizes a spherical mesoporous structure to asymmetrically cover AuNS onto capsules or nanoparticle motors, which drives the movement of the nanospheres [76]. Janus structure is formed by using sputtering coating under vacuum. Monodisperse spherical particles form a single layer on a solid substrate, while the sputtering process only accumulates metal atoms in the upper half of the nanoparticles. Due to the strong surface plasmon resonance (SPR) absorption of AuNS in the near-infrared region, a net thermophoretic force is generated along the asymmetric structure, overcoming Brownian motion and inducing rapid motion of the nanomotor.

Compared with other methods, thermally induced self-assembly can be driven by external forces during the self-assembly process. Notably, the emerging photothermal-driven self-assembly method offers advantages over traditional thermal driving, enabling more precise energy control [77]. However, this method also faces certain challenges. Especially in terms of temperature control accuracy, temperature fluctuations can lead to the formation of irregular structures. Additionally, temperature changes will affect the interaction between the solvent and the nanospheres as well as the evaporation rate of the solvent, which will also impact the final assembly effect. Therefore, more precise experimental control is required to achieve optimal results.

### 6.3.4 Electric/Magnetic-Induced Self-Assembly

The control of nanosphere arrangement and self-assembly through electrical or magnetic interactions is an emerging and promising area of nanotechnology. Electric fields can apply charges or generate force fields to nanoparticles, causing them to align in specific directions and ultimately form ordered or shaped nanostructures [78–80]. This process involves various physical mechanisms, including electric field effects and dielectric polarization. Specifically, when an external electric field is applied to nanoparticles, charged particles experience directional alignment due to the force exerted by the electric field. The resulting charge distribution between the particles induces polarization effects, shifting the surface charge and generating a polarization force.

Yan et al. proposed a universal strategy to reconfigure nanoparticles into various assembly states by introducing imbalanced interactions [81]. The two-dimensional (2D) system of metal dielectric Janus colloids under the influence of direct and alternating electric fields was experimentally validated through computer simulation. Nanoparticles are composed of nano-$SiO_2$ partially covered with a metal layer, and particle motion and colloidal interactions can be controlled in the frequency-dependent dielectric response of deionized water systems. By changing the frequency of the electric field, self-assembly structures of arrays, chains, and clusters were achieved from the same precursor particle (Fig. 6.7a).

Figure 6.7b illustrates the electric field-driven self-assembly device constructed by the Bartlett' team, which integrates an electric field within the rotating frame of a commercial spin-coating machine [82] The electric field applied during the colloidal spin-coating process breaks the axial symmetry and forces the resulting nanospheres to align preferentially at millimeter-level distances. The mechanism behind this alignment involves the application of an alternating electric field during spin coating, which affects fluid flow by exploiting the dielectric contrast between the suspension and the surrounding air, while restricting the suspension through dielectrophoresis. This phenomenon enables the patterning of nanosphere particles (Fig. 6.7c).

The principle of dynamic self-assembly of magnetically driven nanoparticles is similar to that of electrically driven nanoparticles. In this case, an external magnetic field is used to control the arrangement and self-assembly of magnetic nanoparticles. The core of magnetic-driven self-assembly lies in the magnetic properties of

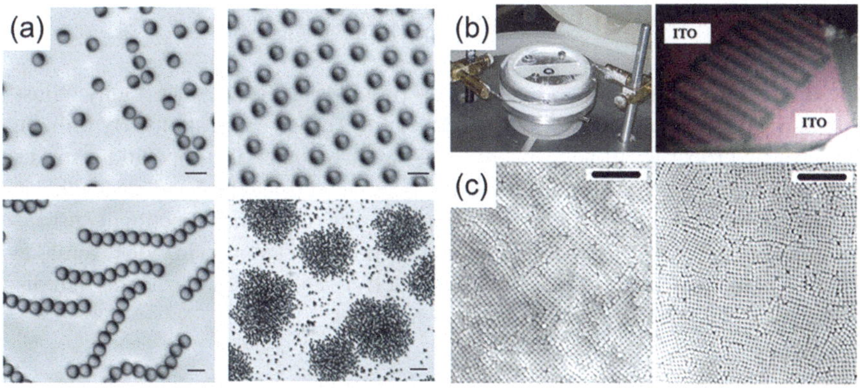


**Fig. 6.7** (**a**) Electrostatic asymmetric nanoparticles programmed with dual charges from the center of a sphere to the opposite hemisphere, and their self-assembly in an electric field to form arrays, chains, and clusters. Scale bar = 5 μm [81]. Copyright 2016, Springer Nature. (**b**) Improved spin-coating chuck with electric field control and patterned indium tin oxide (ITO) substrate (size: 20 × 20 $mm^2$). (**c**) Self-assembly structure of $SiO_2$ nanospheres driven by electric field [82]. Copyright 2012, American Chemical Society

nanospheres. These particles are typically made of ferromagnetic or paramagnetic materials such as ferrite, cobalt, or nickel, or additional magnetic modifications are applied to nonmagnetic particles. Under the influence of an external magnetic field, these particles experience attraction or alignment, with the magnetic dipole forces generated by their magnetism causing them to interact and aggregate in specific configurations [83, 84].

A key advantage of dynamic self-assembly driven by electrical or magnetic fields is its reversibility. By adjusting the strength or direction of the electric or magnetic field, the arrangement of nanoparticles can be dynamically altered, enabling precise, reversible control over their organization and assembly [85]. Additionally, using electric or magnetic effects as triggers offers several benefits, such as instantaneous trigger assembly, repeatability, and the construction of specific patterns. These characteristic makes such methods particularly promising for applications requiring tunable and adaptable nanostructures.

The self-assembly of nanospheres is a multifunctional and powerful technology for fabricating functional surfaces with precise and tunable properties. Each method discussed—wet chemical methods, template-assisted methods, thermal-induced self-assembly, and electric/magnetic-induced self-assembly, provides unique advantages based on application scenarios and material requirements. By strategically selecting the most suitable self-assembly technique, researchers can create self-assembled surfaces with various functions, and these nanostructures can be widely applied in fields including optics, electronics, and biomedicine, showcasing the significant potential of nanosphere self-assembly for advanced technological innovations.

## 6.4 Self-Assembled Functional Surfaces

The dynamic self-assembly of nanospheres provides a versatile and highly adjustable approach to creating functional surfaces for various applications. By utilizing the intrinsic properties of nanospheres, such as size, material composition, surface chemistry, and external stimuli, to drive dynamic self-assembly of nanospheres, researchers have designed optical, wetting, sensing, and other functional surfaces with ideal structures [22, 86–88]. This section provides an overview of several key types of functional surfaces created through self-assembly of nanospheres and discusses their practical applications in different fields.

### *6.4.1 Photonic Crystals*

Photonic crystals are defined as optical materials with periodic structures, which were independently put forward by Yablonovitch and John in 1987 [89, 90]. The periodic arrangement structure of photonic crystals imparts a unique property—the ability to generate a photonic bandgap (PBG) [91, 92]. The PBG structure can effectively block certain frequencies of light waves, preventing photons from propagating through the photonic crystal and producing structural colors. Leveraging this characteristic, the position of the bandgap can be adjusted through designing the crystal structure. For example, modifications to the unit cell size, geometric shape, or refractive index of the photonic crystal can influence the light propagation characteristics and the range of the PBG.

In nature, there are numerous typical structural colors generated by photonic crystals, which are commonly observed in the vibrant and enduring hues of beetle shells, butterfly wings, and certain bird feathers [93]. These organisms utilize structural coloration to convey information or camouflage themselves by adjusting body tissues to modify color (Fig. 6.8a). Inspired by this natural phenomenon, scientists have developed a variety of synthetic photonic crystal structures for applications in fields such as display technology, sensing, and anti-counterfeiting inks. Berger et al. employed the self-assembly properties of short sequences of peptide nucleic acids (PNA) containing guanine to prepare photonic crystal structures with unique optical properties [94]. Specific PNA monomers self-assemble in water to form highly uniform spherical aggregates with an average diameter of 1.7 μm. When a large quantity of these spheres assembles, a colored layer forms at the air–solution interface, presenting a densely packed single-layer structure with hexagonal symmetry under microscopic observation, as shown in Fig. 6.8b. Further research revealed that adding high concentrations of NaCl to the PNA photonic crystal solution significantly alters its color, shifting it from yellow to purple. This color change occurs due to NaCl shielding the electrostatic potential around the nanospheres, leading to their aggregation, the change in interlayer spacing of spheres causes a change in reflection wavelength, and its color-changing principle is similar to the color-changing

**Fig. 6.8** (**a**) Partial photonic crystal structures in nature and their SEM micrographs [93]. Copyright 2007, Springer Nature. (**b**) Optical image of deposited PNA photonic crystal (left) and SEM image of PNA sphere (right) [94]. Copyright 2016, John Wiley and Sons. (**c**) The color of the printed colloidal pattern changes with the ethanol vapor, scale: 0.5 cm [95]. Copyright 2014, American Chemical Society. (**d**) Photos of photonic crystal independent flakes at pH values of 5.0 to 8.0. Scale bar = 3 mm [96]. Copyright 2016, American Chemical Society

mechanism of chameleons, both of which are caused by changes in optical properties due to structural changes. This new type of bioinspired photonic crystal has shown great potential for applications in various fields such as surface-enhanced Raman scattering, diffraction gratings, sensors, and optical coatings. Bai et al. drew inspiration from the color-changing mechanism of long-horned beetles to develop responsive colloidal photonic crystal (CPCs) patterns on both rigid and flexible substrates via inkjet printing of mesoporous colloidal nanoparticle inks [95]. The dual-color responsive CPC patterns (with mesoporous CPCs on leaves and solid $SiO_2$ CPCs on trunks) exhibit a uniform color in $N_2$. However, when exposed to ethanol vapor, the porous nature of mesoporous CPCs causes a pronounced color change and reveals a hidden image (Fig. 6.8c). The mesoporous CPCs demonstrate a rapid response and recovery speed to vapor, as well as strong multi-cycle responsiveness, making them highly suitable for anti-counterfeiting certification applications.

Based on the response mechanism of photonic crystals to various factors (pH, humidity, temperature, etc.), a wide range of functional self-assembly interfaces have been developed [96]. For instance, the change in structural color of photonic crystals in response to varying pH levels can be leveraged to detect the acidity or alkalinity of a system (Fig. 6.8d). These research results provide a theoretical basis and technical support for the further development and application of photonic crystals.

### 6.4.2 Antireflective Surfaces

Reducing the reflection on material surfaces has long been a central focus in the field of modern optics and optoelectronic technology. The dynamic self-assembly technology of nanospheres provides an innovative approach to achieving efficient antireflection [97–99]. Specifically, nanospheres are assembled into an antireflective coating that generates destructive interference by optimizing the refractive index, thereby minimizing light reflection. When the refractive index $n_1$ of the coating is matched with the geometric mean of the refractive indices of air ($n_0$) and substrate ($n_2$), the light passing through this gradient refractive index structure undergoes maximum destructive interference between the reflected light at the coating surface and the coating/substrate interface, thereby reducing reflection and enhancing light transmission. Therefore, selecting appropriate nanosphere materials to achieve optimal refractive index matching is crucial for maximizing the antireflection effect.

The protruding structure on the wings of cicadas creates a refractive index gradient that reduces light reflection, enabling natural camouflage [102]. Inspired by the structure of cicada wings, Li et al. developed a novel antireflective coating by integrating self-assembly methods with shape-memory polymers, as illustrated in Fig. 6.9a [100]. Firstly, the improved Langmuir–Blodgett technique was utilized to self-assemble tightly packed nanosphere arrays, which were then etched to form nanopore arrays. Subsequently, smaller nanosphere particles were induced to self-assemble within these pore arrays, creating a surface structure that generates a refractive index gradient (Fig. 6.9b). This structure demonstrates omnidirectional antireflection properties across the broadband visible light spectrum, achieving an average reflectivity as low as 6% and an average transmittance of 92%. Notably, the coating maintains its antireflection effects at various incident angles (Fig. 6.9c). He's group reported a method for preparing a robust moth-eye-shaped nanostructure antireflective coating on a glass substrate (Fig. 6.9d) [101]. Firstly, silica nanoparticles were self-assembled on the surface of a glass substrate to form a single-layer structure, on the surface of which was then grown a mesoporous silica layer, resulting in a top layer of tightly packed nanodomes, approximately 260 nm in width and 90 nm in height. The lower layer consists of the lower half of the nanosilica and the mesoporous layer. As shown in Fig. 6.9e, using the effective medium theory to calculate the refractive index, the study found that the coating exhibits a quasi-stepwise refractive index change from 0 to 90 nm and from 90 to 180 nm. This design enables the coating to achieve broadband antireflection across the wavelength range of 400–1200 nm, reducing the average reflectivity to 1.88% while maintaining high transmittance at incident angles of up to ±40°. Such structures with excellent antireflection and mechanical properties have enormous potential for application in fields such as solar energy conversion and optical devices.

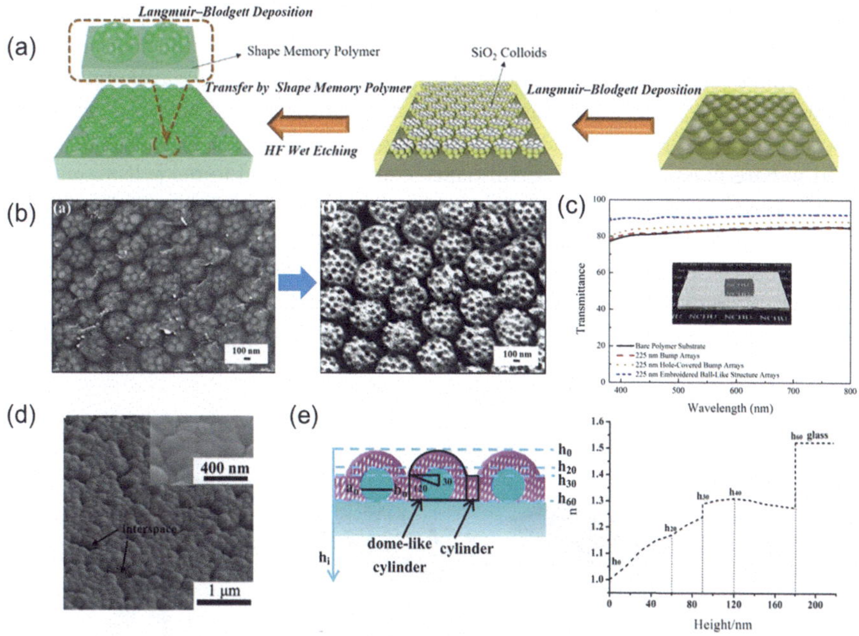


**Fig. 6.9** (**a**) Schematic diagram of the preparation of a self-assembled structure of nanospheres in a nanopore array. (**b**) SEM images of self-assembled surface structures before and after etching. (**c**) Optical properties of embroidered ball-like antireflection structure [100]. Copyright 2021, Elsevier. (**d**) SEM image of the moth-eye-shaped antireflective structure. (**e**) Theoretical diagram and equivalent refractive index curve of moth-eye-shaped antireflection structure [101]. Copyright 2017, American Chemical Society

## 6.4.3 *Superhydrophobic/Superhydrophilic Surfaces*

Surface wettability refers to the ability of solid surface to be wetted by liquid. This property is fundamentally influenced by the surface chemical composition and microstructure. From a chemical perspective, polar groups on the surface often endow it with hydrophilicity, promoting the diffusion of liquids on the surface, while nonpolar groups tend to induce hydrophobicity, making it difficult for liquids to wet the surfaces. The dynamic self-assembly of nanospheres plays a unique and critical role in wetting performance. The chemical groups present on the nanosphere surface determine the surface chemical energy, while the self-assembled structure of the nanospheres can significantly alter the microtopography of the solid surface. According to the Wenzel and Cassie–Baxter models, the roughness of a solid surface can enhance the hydrophilicity of hydrophilic surfaces and increase the hydrophobicity of hydrophobic surfaces.

Ren et al. utilized nanoparticle self-assembly technology to prepare coatings on substrate surfaces with high transparency, superhydrophobicity, low bacterial adhesion, and antibacterial properties [103]. The preparation process is shown in Fig. 6.10a. By mixing hydrophobic $SiO_2$ with antibacterial CuO nanoparticles, self-assembly occurs on the surface of silica sol. The resulting coating exhibited a water contact angle (WCA) as high as 160°, demonstrating superhydrophobic properties. Moreover, the coating maintained high transparency in the wavelength range of 300–2500 nm, with an average transmittance of 93.0%. In bacterial adhesion experiments, its adhesion to *Escherichia coli* was reduced by 3.2 log cells/cm$^2$ compared to bare glass. This high performance can be attributed to the self-assembled superhydrophobic surface, which minimizes the contact area between the surface and bacteria. At the same time, the self-assembled surface releases $Cu^{2+}$ through CuO nanoparticles, generating excessive reactive oxygen species that damage bacterial cells. Functionalized nanosphere units can still exhibit their specific functionality after assembly into coatings. Hydrophobic surfaces can also be obtained by

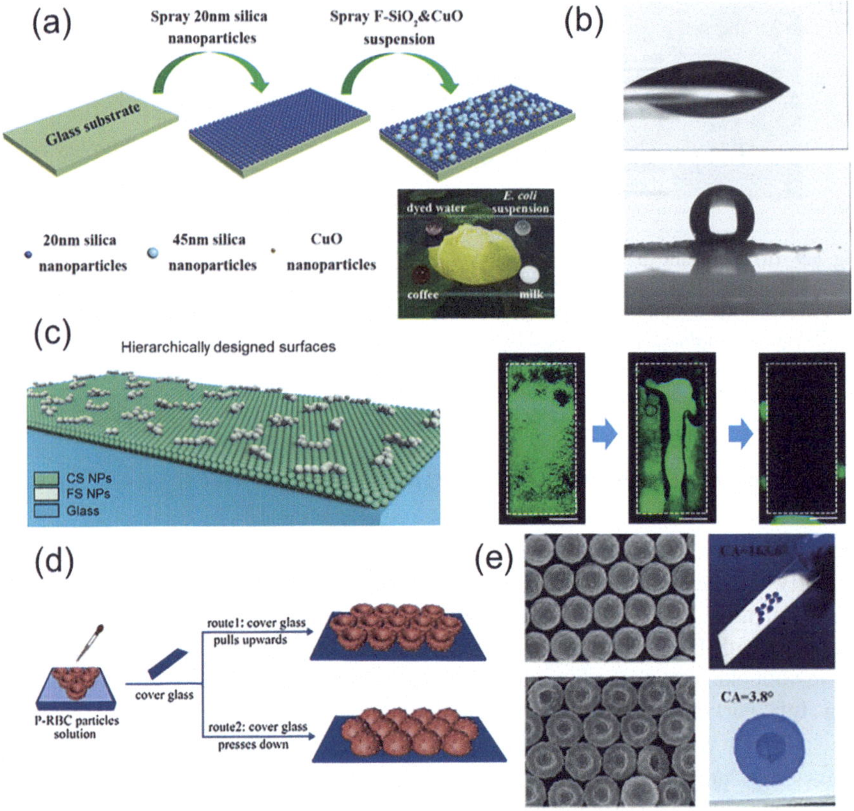


**Fig. 6.10** (**a**) Schematic diagram of self-assembly of F-SiO$_2$ and CuO nanoparticles on substrate surface [103]. Copyright 2018, American Chemical Society. (**b**) Wettability changes of self-assembled surfaces under fluorinated aniline gas phase conditions [104]. Copyright 2018, Elsevier. (**c**) Self-cleaning effect of superhydrophobic surfaces [105]. Copyright 2022, John Wiley and Sons. (**d**) Schematic diagram of self-assembly of amphiphilic/hydrophobic nanoparticles on substrate surface (**e**) SEM image of self-assembled interface (left) and image of hydrophilic/hydrophobic conversion characteristics (right) [106]. Copyright 2019, Elsevier

hydrophobic treatment of hydrophilic self-assembled surfaces. Li's team treated the surface of supramolecular self-assembled microspheres of formylazobenzene and dipeptides in fluorinated aniline vapor, transforming the originally hydrophilic surface into a hydrophobic interface, as shown in Fig. 6.10b, increasing the WCA from 23° to 130° [104].

Superhydrophilic and superhydrophobic functional surfaces have significant implications for a wide range of practical applications. Superhydrophobic surfaces can prevent water droplets from infiltrating functional surfaces and achieve self-cleaning. Even if the surface is covered with dust, water droplets can roll away surface pollutants, greatly reducing maintenance costs for coatings (Fig. 6.10c) [105]. Conversely, the superhydrophilic surface helps to form a continuous water film on its surface, thereby isolating and removing surface pollutants, endowing self-cleaning ability. At the same time, the formation of the water film significantly reduces the visible light transmission loss in transparent materials, making it also highly promising for antifog applications.

In addition, recent studies have shown that nanostructures formed through dynamic self-assembly can respond to external stimuli, enabling reversible switching between hydrophilic and hydrophobic states (Fig. 6.10d, e) [106]. This adaptability opens up new possibilities for intelligent surfaces that respond to environmental conditions.

### *6.4.4 Sensors*

Surface-enhanced Raman spectroscopy (SERS) is a powerful analytical method in biochemical analysis, particularly for ultrasensitive detection. SERS-based analysis can be achieved using self-assembled structures and solid substrates, with SERS performance largely depending on the form of the substrate used [107, 108]. The plasmon coupling strength can be adjusted by controlling the size, shape, and gaps between metal nanoparticles and nanostructures. Research has shown that the incident electromagnetic field is concentrated in a small space; thus, when the gap between adjacent nanoparticles is on the nanoscale, strong field effects can be generated, leading to the amplification of Raman signals near metal nanoparticles (Fig. 6.11a) [109]. Yang et al. utilized an anisotropic layer-by-layer (LBL) self-assembly technique to create a monolayer film of Au@Ag nanorods (NRs) on a nanosphere substrate, forming a broadband-sensitive SERS platform, as shown in Fig. 6.11b [110]. By customizing the sizes of the Au@Ag NRs and polystyrene nanospheres (PS NSs), as well as adjusting the number of layers of the Au@Ag NRs film, they optimized the SERS sensitivity. Among them, the stratified Au@Ag NR-1 MFON-500 substrate exhibits the best SERS performance and achieves a wide quantitative detection range of 4MBA from $1 \times 10^{-8}$ to $1 \times 10^{-6}$ M (Fig. 6.11c–e).

In addition, the ordered interface formed by the dynamic self-assembly process of nanospheres endows them with abundant surface-active sites and enhanced

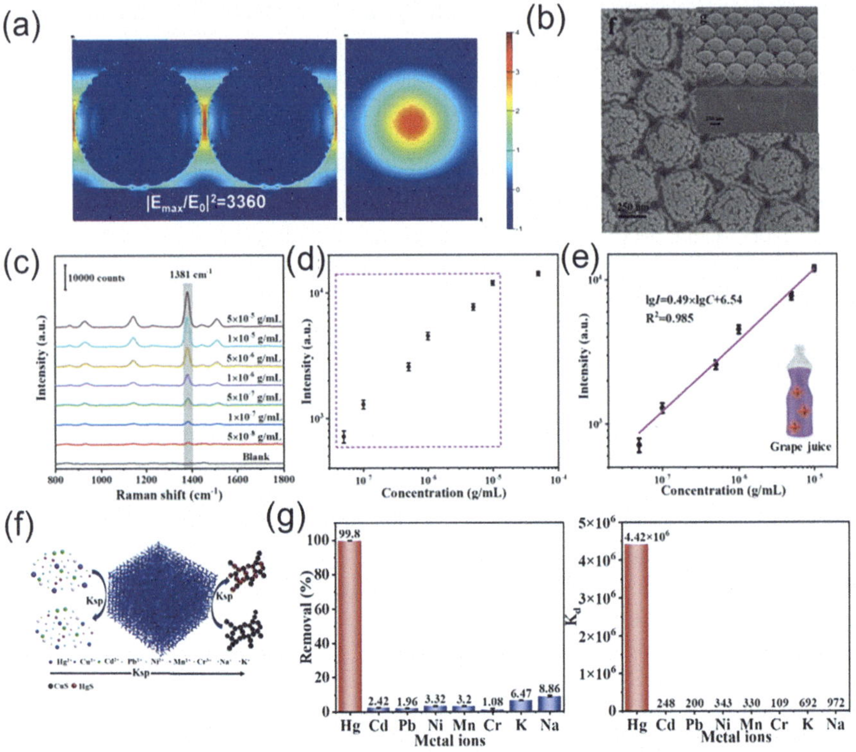


**Fig. 6.11** (**a**) Finite difference time domain (FDTD) simulation images of the contour distribution of the electric field intensity (E) around the nanosphere array in the XZ and YZ planes of the cross-section [109]. Copyright 2021, Elsevier. (**b**) Au@Ag SEM image of NR-1 MFON-500 substrate. The inset is a cross-sectional image. (**c**) Au@Ag SERS spectra of different concentrations of thiamine in grape juice on NR-1 MFON-500 substrate. (**d**) Scatter plot of peak intensity at 1381 $cm^{-1}$ against the concentration of thiram. (**e**) Linear fitting curve of peak intensity at 1381 $cm^{-1}$ for thiram concentrations ranging from $5 \times 10^{-8}$ to $1 \times 10^{-5}$ g/mL [110]. Copyright 2022, Elsevier. (**f**) Uptake mechanism of $Hg^{2+}$ on CuS AG. (**g**) The selectivity of CuS AG for $Hg^{2+}$ in eight metal ions (each concentration of 10 mg•$L^{-1}$)(left) and the $K_d$ values of CuS AG for the eight metal ions (right) [111]. Copyright 2021, Elsevier

adsorption capacity. These surface-active sites can interact with environmental pollutants through physical or chemical adsorption, thereby improving pollutant removal efficiency. Guo et al. developed a novel 3D sponge-like CuS aerogel (CuS AG) by self-assembly of CuS nanocrystals at room temperature and investigated its ability to recognize, capture, and enrich mercury ions [111]. The schematic diagram of its structure is shown in Fig. 6.11f. The highly porous Cu–S network structure has a specific surface area of up to 159 $m^2g^{-1}$ with abundant surface Cu–S sites and ion channels. CuS AG not only exhibits excellent selectivity toward $Hg^{2+}$ but also demonstrates ultrafast capture kinetics (removing 99% within 30 s) (Fig. 6.11g). This self-assembly structure opens a new avenue for the design and fabrication of high-performance materials for metal ion recognition, adsorption, and enrichment.

In addition to the several application fields detailed above, the dynamic self-assembly surface of nanospheres has shown broad application prospects in biomedical, energy catalysis, and other fields. Nanosphere units can achieve functional customization of self-assembled interfaces through precise control (structure, size, properties, etc.) to suit diverse application requirements. Therefore, the dynamic self-assembly surface of nanospheres provides new opportunities and challenges for the development of various fields and is worthy of further research and exploration.

## 6.5 Conclusion and Outlook

The dynamic self-assembly of nanospheres is a technique that spontaneously forms functional surfaces through spherical nanoparticles, which has the characteristics of high efficiency, low cost, and controllability. By leveraging intermolecular interactions and environmental regulation, the structure of the assembled nanospheres can be precisely controlled, resulting in nanostructures with highly ordered surfaces, adjustable porosity, and tailored specific surface areas. These properties make dynamic self-assembly ideal for applications in functional interfaces, optoelectronic devices, detection, and other fields. Furthermore, the dynamic nature of the assembly process allows materials to respond to changes in environmental conditions, providing flexibility in the design of functional materials.

In this review, we systematically list the principles, methods, and construction of functional surfaces for the dynamic self-assembly of nanospheres. The ability to manipulate nanostructures through self-assembly provides enormous opportunities for designing and manufacturing surfaces with specific functions.

Despite exciting achievements in the field of nano self-assembly technology, it is still in the early stages of providing true convenience in practical applications. There are still many unknown factors waiting for researchers to explore in order to truly utilize this technology to promote the progress and development of human society [112]. During the assembly process, the forces between the solution and particles are complex and multifaceted. For more accurate control of the dynamic self-assembly of nanospheres, deeper research is needed into the interactions and assembly dynamics between nanoparticles, particularly in complex, multilevel assembly processes. Additionally, there are still some difficulties and shortcomings in dynamic self-assembly technology, such as the nanostructure formed by self-assembly being easily affected by environmental factors (such as temperature, humidity, and solvents), which may compromise their structural stability. In large-scale production, achieving consistent self-assembled structures and ensuring repeatability and controllability remain significant obstacles. Other issues include visual defects and performance limitations due to poor interface compatibility, as well as challenges related to process costs during large-scale production.

Interdisciplinary collaboration is poised to drive significant breakthroughs in the dynamic self-assembly technology of nanosphere interfaces. The deep integration of multiple disciplines such as biology, physics, and chemistry can provide new insights and solutions for the self-assembly of nanospheres. For example, by combining the self-assembly principles in biology, biocompatible and biologically active nanomaterials can be developed for applications such as drug delivery, tissue engineering, and more. Additionally, the self-assembled nanospheres can be applied in quantum information processing by utilizing quantum effects in physics. Furthermore, chemical synthesis and surface modification techniques enable precise control over nanoparticle interactions and functionality, facilitating the construction of advanced catalytic or sensing platforms. Such interdisciplinary collaboration not only fosters rapid technological development and innovation but also accelerates the transition of dynamic self-assembly technology from the laboratory to practical applications.

The improvement of performance is an inevitable requirement for the development of dynamic self-assembly technology at the interface of nanospheres. Improving the stability, durability, and other performance indicators of self-assembled structures is crucial for practical applications. Achieving this requires a holistic approach, considering factors such as material selection, assembly processes, and structural design, for example, selecting highly stable materials, optimizing self-assembly process parameters, and designing more robust and durable structures. Concurrently, challenges such as nanosphere aggregation and structural inhomogeneity must be addressed. The development of new technologies and methods to overcome these challenges, through ongoing research and innovation, will propel the continued progress of dynamic self-assembly technology.

In addition, sustainability plays a crucial role in the future development of dynamic self-assembly technology for nanosphere interfaces. Developing environmentally friendly nanospheres self-assembly materials will help reduce the impact on the environment and align with the requirements of sustainable development.

**Acknowledgments** This work was supported by the National Natural Science Foundation of China (91963104), the National Key Research and Development Program of China (2019QY(Y)0503, 2017YFA0207102), and the Natural Science Foundation of Beijing (L233012).

## References

1. Yi C, Yang Y, Liu B, He J, Nie Z (2020) Polymer-guided assembly of inorganic nanoparticles. Chem Soc Rev 49:465–508
2. Kim M, Kim J, VanderLaan D, Kubelick KP, Jhunjhunwala A et al (2023) Tunable interparticle connectivity in gold nanosphere assemblies for efficient photoacoustic conversion. Adv Funct Mater 33:2305202
3. Han S, Zal T, Sokolov KV (2021) Fate of antibody-targeted ultrasmall gold nanoparticles in cancer cells after receptor-mediated uptake. ACS Nano 15:9495–9508
4. Niu D, Liu Z, Li Y, Luo X, Zhang J et al (2014) Silica nanospheres: monodispersed and ordered large-pore mesoporous silica nanospheres with tunable pore structure for magnetic functionalization and gene delivery. Adv Mater 26:4910

5. Wu C, Wang J, Zhang X, Kang L, Cao X et al (2022) Hollow gradient-structured iron-anchored carbon nanospheres for enhanced electromagnetic wave absorption. Nanomicro Lett 15:7
6. Adams S, Zhang JZ (2016) Unique optical properties and applications of hollow gold nanospheres (HGNs). Coord Chem Rev 320-321:18–37
7. Li H, Chen L, Li X, Sun D, Zhang H (2022) Recent progress on asymmetric carbon- and silica-based nanomaterials: from synthetic strategies to their applications. Nanomicro Lett 14:45
8. Zhang H, Peng L, Qin W, Jia D, Zhao R et al (2024) Enhanced chemotherapy and anti-metastasis with in situ nanosphere-to-nanofiber transition. Nano Today 54:102122
9. Chen Y, Yang C, Dai Q, Tan J, Dou C, Luo F (2023) Gold-nanosphere mitigates osteoporosis through regulating TMAO metabolism in a gut microbiota-dependent manner. J Nanobiotechnol 21:125
10. Du X, He J (2012) Carrier effect in the synthesis of rattle-type Au@hollow silica nanospheres by impregnation and thermal decomposition method. Microporous Mesoporous Mater 163:201–210
11. Wang D-W, Su D (2014) Heterogeneous nanocarbon materials for oxygen reduction reaction. Energy Environ Sci 7:576–591
12. Yang Y, Zhang W, Xiao Y, Shi Z, Cao X et al (2019) CoNiSe2 heteronanorods decorated with layered-double-hydroxides for efficient hydrogen evolution. Appl Catal B Environ 242:132–139
13. Cepero-Rodríguez E, Sousa-Castillo A, Besteiro LV, Puértolas B, Vázquez-González M, Correa-Duarte MA (2024) Bifunctional Au@UiO-67-bpy-Cu plasmonic nanostructures for the solar-driven CO2 reduction to methanol. Adv Energy Mater n/a:2401887
14. Tian H, Chen J, He J, Liu F (2015) Pd-loaded magnetic mesoporous nanocomposites: a magnetically recoverable catalyst with effective enrichment and high activity for DDT and DDE removal under mild conditions. J Colloid Interface Sci 457:195–202
15. Zhuang W-C, Zhang H-Y, Chen Z-Y, Cheng W-J, Na W-D et al (2024) Visual test paper based on Au/δ-MnO2 hollow nanosphere oxidase-like activity regulation using hexavalent chromium as a smart switch. Rare Metals 43:711–722
16. Fang J, Wang M, Yao K, Zhang T, Krasnok A et al (2021) Dielectric nanospheres: directional modulation of exciton emission using single dielectric nanospheres. Adv Mater 33:2170153
17. Chang L, Liu X, Luo J, Lee C-Y, Zhang J et al (2024) Physiochemical coupled dynamic nanosphere lithography enabling multiple metastructures from single mask. Adv Mater 36:2310469
18. Li J, Hu Y, Yu L, Li L, Ji D et al (2021) Recent advances of nanospheres lithography in organic electronics. Small 17:2100724
19. Zheng K, Sui B, Ilyas K, Boccaccini AR (2021) Porous bioactive glass micro- and nanospheres with controlled morphology: developments, properties and emerging biomedical applications. Mater Horiz 8:300–335
20. Vogel N, Retsch M, Fustin C-A, del Campo A, Jonas U (2015) Advances in colloidal assembly: the design of structure and hierarchy in two and three dimensions. Chem Rev 115:6265–6311
21. Lin Z, Gao C, Chen M, Lin X, He Q (2018) Collective motion and dynamic self-assembly of colloid motors. Curr Opin Colloid Interface Sci 35:51–58
22. van Dommelen R, Fanzio P, Sasso L (2018) Surface self-assembly of colloidal crystals for micro- and nano-patterning. Adv Colloid Interf Sci 251:97–114
23. Ren H, Xiong Z, Wang E, Yuan Z, Sun Y et al (2019) Watching dynamic self-assembly of web buckles in strained MoS2 thin films. ACS Nano 13:3106–3116
24. Giri A, Shreeraj G, Dutta TK, Patra A (2023) Transformation of an imine cage to a covalent organic framework film at the liquid–liquid Interface. Angew Chem Int Ed Engl 62:e202219083

25. Wu L, Wang X, Wang G, Chen G (2018) In situ X-ray scattering observation of two-dimensional interfacial colloidal crystallization. Nat Commun 9:1335
26. Mihut AM, Stenqvist B, Lund M, Schurtenberger P, Crassous JJ (2017) Assembling oppositely charged lock and key responsive colloids: a mesoscale analog of adaptive chemistry. Sci Adv 3:e1700321
27. Ren CZ-J, Solís Muñana P, Dupont J, Zhou SS, Chen JL-Y (2019) Reversible formation of a light-responsive catalyst by utilizing intermolecular cooperative effects. Angew Chem Int Ed Engl 58:15254–15258
28. Martínez-Pedrero F, González-Banciella A, Camino A, Mateos-Maroto A, Ortega F et al (2021) Static and dynamic self-assembly of pearl-like-chains of magnetic colloids confined at fluid interfaces. Small 17:2101188
29. Keating CD, Musick MD, Keefe MH Natan MJ (1999) Kinetics and thermodynamics of au colloid monolayer selfassembly: Undergraduate experiments in surface and nanomaterials chemistry. Journal of Chemical Education 76(7):949
30. Sherman ZM, Swan JW (2016) Dynamic, directed self-assembly of nanoparticles via toggled interactions. ACS Nano 10:5260–5271
31. Wang L, Schubert US, Hoeppener S (2021) Surface chemical reactions on self-assembled silane based monolayers. Chem Soc Rev 50:6507–6540
32. Zornberg LZ, Lewis DJ, Mertiri A, Hueckel T, Carter DJD, Macfarlane RJ (2023) Self-assembling systems for optical out-of-plane coupling devices. ACS Nano 17:3394–3400
33. Zhang Q,Wang W-Z, Yu J-J, Qu D-H, Tian H (2017) Dynamic self-assembly encodes a tri-stable au–TiO photocatalyst. Advanced Materials 29(5):1604948
34. Huang T-H, Espino FKC, Tian X-Y, Widakdo J, Austria HFM et al (2024) Piezocatalytic property of PVDF/graphene self-assembling piezoelectric membrane for environmental remediation. Chem Eng J 487:150569
35. Hu H, Yang C, Li M, Shao D, Mao H-Q, Leong KW (2021) Flash technology-based self-assembly in nanoformulation: fabrication to biomedical applications. Mater Today 42:99–116
36. Aranson IS (2013) Collective behavior in out-of-equilibrium colloidal suspensions. Comptes Rendus. Physique 14:518–527
37. Xu D, Shi R, Sun Z-Y, Lu Z-Y (2021) Mechanism of periodic field driven self-assembly process. J Chem Phys 154
38. Klimov VV, Lambrecht A (2009) Van der Waals forces between plasmonic nanoparticles. Plasmonics 4:31–36
39. Venkataram PS, Hermann J, Vongkovit TJ, Tkatchenko A, Rodriguez AW (2019) Impact of nuclear vibrations on van der Waals and Casimir interactions at zero and finite temperature. 5:eaaw0456
40. Valtiner M, Kristiansen K, Greene GW, Israelachvili JN (2011) Effect of surface roughness and electrostatic surface potentials on forces between dissimilar surfaces in aqueous solution. Advanced Materials 23(20): 2294–2299
41. Ding Y, Cai M, Cui Z, Huang L, Wang L, Lu X, Cai Y (2018) Synthesis of low-dimensional polyion complex nanomaterials via polymerization-induced electrostatic self-assembly. Angewandte Chemie International Edition 57(4):1053–1056
42. Chen Q, Bae SC, Granick S (2011) Directed self-assembly of a colloidal kagome lattice. Nature 469:381–384
43. Batista, C. A., Larson, R. G., & Kotov, N. A. (2015). Nonadditivity of nanoparticle interactions. Science 350(6257), 1242477
44. Steiner, T. (2002) The hydrogen bond in the solid state. Angewandte Chemie International Edition 41(1):48–76
45. Nonappa Ikkala O (2018) Hydrogen bonding directed colloidal self-assembly of nanoparticles into 2d crystals, capsids, and supracolloidal assemblies. Advanced Functional Materials 28(27):1704328

46. Qin Y, Song Z, Miao L, Hu C, Chen Y et al (2023) Hydrogen-bond-mediated micelle aggregating self-assembly towards carbon nanofiber networks for high-energy and long-life zinc ion capacitors. Chem Eng J 470:144256
47. Huang X, Wang P, Li T, Tian X, Guo W et al (2020) Self-assemblies based on traditional medicine Berberine and cinnamic acid for adhesion-induced inhibition multidrug-resistant Staphylococcus aureus. ACS Appl Mater Interfaces 12:227–237
48. Wang M, Wang D, Li Z (2016) Self-assembly of CPO-27-Mg/TiO2 nanocomposite with enhanced performance for photocatalytic CO2 reduction. Appl Catal B Environ 183:47–52
49. Carvalho M-A, Dekkiche H, Nagasaki M, Kikkawa Y, Ruppert R (2019) Coordination-driven construction of porphyrin nanoribbons at a highly oriented pyrolytic graphite (HOPG)/liquid Interface. J Am Chem Soc 141:10137–10141
50. Moradi M, Tulli LG, Nowakowski J, Baljozovic M, Jung TA, Shahgaldian P (2017) Two-dimensional calix[4]arene-based metal–organic coordination networks of tunable crystallinity. Angewandte Chemie International Edition 56(46):14395–14399
51. Kögerler P, Cronin L (2005) Polyoxometalate nanostructures, superclusters, and colloids: From functional clusters to chemical aesthetics. Angewandte Chemie International Edition 44(6):844–846
52. Wang Y, Hollingsworth AD, Yang SK, Patel S, Pine DJ, Weck M (2013) Patchy particle self-assembly via metal coordination. J Am Chem Soc 135:14064–14067
53. Zhang Q, Zhou Z, Li C, Wu P, Sun M (2019) pH-switchable coordinative micelles for enhancing cellular transfection of biocompatible polycations. ACS Appl Mater Interfaces 11:20689–20698
54. Ren X, Wang C-C, Li Y, Wang C-Y, Wang P, Gao S (2022) Ag(I) removal and recovery from wastewater adopting NH2-MIL-125 as efficient adsorbent: a 3Rs (reduce, recycle and reuse) approach and practice. Chem Eng J 442:136306
55. Suh Y, Pham Q, Shao B, Won Y (2019) Self-assembly: the control of colloidal grain boundaries through evaporative vertical self-assembly. Small 15:1970062
56. Zhu P, Kong T, Zhou C, Lei L, Wang L (2018) Engineering microstructure with evaporation-induced self-assembly of microdroplets. Small Methods 2(6):1800017
57. von Baeckmann C, Rubio GMDM, Kählig H, Kurzbach D, Reithofer MR, Kleitz F (2021) Evaporation-induced self-assembly of small peptide-conjugated silica nanoparticles. 60:22700–22705
58. Shao X, Duan F, Hou Y, Zhong X (2020) Role of surfactant in controlling the deposition pattern of a particle-laden droplet: fundamentals and strategies. Adv Colloid Interf Sci 275:102049
59. Suh Y, Pham Q, Shao B, Won Y (2019) The control of colloidal grain boundaries through evaporative vertical self-assembly. 15:1804523
60. Liu ST, Zhu T, Hu RS, Liu ZF (2002) Evaporation-induced self-assembly of gold nanoparticles into a highly organized two-dimensional array. Phys Chem Chem Phys 4:6059–6062
61. Xu Y, He Y, Wu F, Zhou X, Liu M (2024) Formation and application of polymer spherulite-like patterns of halloysite nanotubes by evaporation-induced self-assembly. ACS Appl Mater Interfaces 16:15177–15192
62. Kolegov KS, Barash LY (2020) Applying droplets and films in evaporative lithography. Adv Colloid Interf Sci 285:102271
63. Han W, Lin Z (2012) Learning from "coffee rings": Ordered structures enabled by controlled evaporative self-assembly. Angewandte Chemie International Edition 51(7):1534–1546
64. Marica I, Stefan M, Boca S, Falamaş A, Farcău C (2023) A simple approach for coffee-ring suppression yielding homogeneous drying patterns of ZnO and TiO2 nanoparticles. J Colloid Interface Sci 635:117–127
65. McCullough IIIDH, Regen SL (2004) Don't forget Langmuir–Blodgett films. Chem Commun:2787–2791

66. Zhang Y, Zhang A, He H, Fan Y, Li Y et al (2024) Fabrication of an ultra-thin and ordered SPEEK proton exchange membrane by a Langmuir-Blodgett self-assembly process. J Membr Sci 690:122196
67. Li G, Xu X, Zuo YY (2023) Langmuir-Blodgett transfer from the oil-water interface. J Colloid Interface Sci 630:21–27
68. Kohoutek T, Parchine M, Bardosova M, Pemble ME (2020) Controlled self-assembly of Langmuir-Blodgett colloidal crystal films of monodispersed silica particles on non-planar substrates. Colloids Surf A Physicochem Eng Asp 593:124625
69. Guo X, Duan M, Zhang J, Xi B, Li M, Yin R, Zheng X, Liu Y, Cao F, An X, Xiong S (2022) A general selfassembly induced strategy for synthesizing 2d ultrathin cobalt-based compounds toward optimizing hydrogen evolution catalysis. Advanced Functional Materials 32(51):2209397
70. Wang C, Li J, Paineau E, Laachachi A, Colbeau-Justin C et al (2020) A sol–gel biotemplating route with cellulose nanocrystals to design a photocatalyst for improving hydrogen generation. J Mater Chem A 8:10779–10786
71. Molet P, Passarelli N, Pérez LA, Scarabelli L, Mihi A (2021) Engineering plasmonic colloidal meta-molecules for tunable photonic supercrystals. 9:2100761
72. Van Gorp H, Walke P, Bragança AM, Greenwood J, Ivasenko O et al (2018) Self-assembled polystyrene beads for templated covalent functionalization of graphitic substrates using diazonium chemistry. ACS Appl Mater Interfaces 10:12005–12012
73. Liu X, He J (2007) Hierarchically structured superhydrophilic coatings fabricated by self-assembling raspberry-like silica nanospheres. J Colloid Interface Sci 314:341–345
74. Hwang EY, Lee JS, Lim DW (2019) Oppositely charged, stimuli-responsive anisotropic nanoparticles for colloidal self-assembly. Langmuir 35:4589–4602
75. Shinde A, Huang D, Saldivar M, Xu H, Zeng M et al (2019) Growth of colloidal nanoplate liquid crystals using temperature gradients. ACS Nano 13:12461–12469
76. Xuan M, Wu Z, Shao J, Dai L, Si T, He Q (2016) Near infrared light-powered Janus mesoporous silica nanoparticle motors. J Am Chem Soc 138:6492–6497
77. Fava D, Winnik MA, Kumacheva E (2009) Photothermally-triggered self-assembly of gold nanorods. Chem Commun:2571–2573
78. Yuan J, Takae K, Tanaka H (2022) Impact of inverse squeezing flow on the self-assembly of oppositely charged colloidal particles under electric field br. Phys Rev Lett 129
79. Li Z, Wang W, Yin Y (2020) Colloidal assembly and active tuning of coupled plasmonic nanospheres. Trends Chem 2:593–608
80. Leunissen ME, Vutukuri HR, van Blaaderen A (2009) Directing colloidal self-assembly with biaxial electric fields. Advanced Materials 21(30):3116–3120
81. Yan J, Han M, Zhang J, Xu C, Luijten E, Granick S (2016) Reconfiguring active particles by electrostatic imbalance. Nat Mater 15:1095–1099
82. Bartlett AP, Pichumani M, Giuliani M, González-Viñas W, Yethiraj A (2012) Modified spin-coating technique to achieve directional colloidal crystallization. Langmuir 28:3067–3070
83. Kang N, Zhu J, Zhang X, Wang H, Zhang Z (2022) Reconfiguring self-assembly of photoresponsive hybrid colloids. J Am Chem Soc 144:4754–4758
84. Martín-Roca J, Ortega F, Valeriani C, Rubio RG, Martínez-Pedrero F (2023) Magnetic colloidal currents guided on self-assembled colloidal tracks. Advanced Functional Materials 33(52):2306541
85. Morphew D, Chakrabarti D (2015) Hierarchical self-assembly of colloidal magnetic particles into reconfigurable spherical structures. Nanoscale 7:8343–8350
86. Li Z, Fan Q, Yin Y (2022) Colloidal self-assembly approaches to smart nanostructured materials. Chem Rev 122:4976–5067
87. Milcovich G, Lettieri S, Antunes FE, Medronho B, Fonseca AC et al (2017) Recent advances in smart biotechnology: hydrogels and nanocarriers for tailored bioactive molecules depot. Adv Colloid Interf Sci 249:163–180
88. Karg M, König TAF, Retsch M, Stelling C, Reichstein PM et al (2015) Colloidal self-assembly concepts for light management in photovoltaics. Mater Today 18:185–205
89. Yablonovitch E (1987) Inhibited spontaneous emission in solid-state physics and electronics. Phys Rev Lett 58:2059–2062

90. John S (1987) Strong localization of photons in certain disordered dielectric superlattices. Phys Rev Lett 58:2486–2489
91. Wu S, Xia H, Xu J, Sun X, Liu X (2018) Manipulating luminescence of light emitters by photonic crystals. Advanced Materials 30(47):1803362
92. Li M, Lyu Q, Peng B, Chen X, Zhang L, Zhu J (2022) Bioinspired colloidal photonic composites: Fabrications and emerging applications. Advanced Materials 34(52):2110488
93. Welch VL, Vigneron JP (2007) Beyond butterflies—the diversity of biological photonic crystals. Opt Quant Electron 39:295–303
94. Berger O, Yoskovitz E, Adler-Abramovich L, Gazit E (2016) Spectral transition in bio-inspired self-assembled peptide nucleic acid photonic crystals. Advanced Materials 28(11):2195–2200
95. Bai L, Xie Z, Wang W, Yuan C, Zhao Y et al (2014) Bio-inspired vapor-responsive colloidal photonic crystal patterns by inkjet printing. ACS Nano 8:11094–11100
96. Yetisen AK, Butt H, Yun S-H (2016) Photonic crystal flakes. ACS Sens 1:493–497
97. Liu R, Zhang D, Li Z (2021) Femtosecond laser induced simultaneous functional nanomaterial synthesis, in situ deposition and hierarchical LIPSS nanostructuring for tunable antireflectance and iridescence applications. J Mater Sci Technol 89:179–185
98. Liu J, Tang H, Zhang D, Jiao S, Zhou Z et al (2020) Performance evaluation of the hybrid photovoltaic-thermoelectric system with light and heat management. Energy 211:118618
99. Han Z, Jiao Z, Niu S, Ren L (2019) Ascendant bioinspired antireflective materials: opportunities and challenges coexist. Prog Mater Sci 103:1–68
100. Li P-C, Chen H-Y, Chiang K-T, Yang H (2021) Reversible embroidered ball-like antireflective structure arrays inspired by leafhopper wings. J Colloid Interface Sci 599:119–129
101. Jin B, He J (2017) Self-templated fabrication of robust moth-eye-like nanostructures with broadband and quasi-omnidirectional antireflection properties. ACS Photonics 4:188–196
102. Tseng H-Y, Chen Y-H, Chen R-Y, Yang H (2020) Reversibly erasable broadband omnidirectional antireflection coatings inspired by inclined conical structures on blue-tailed Forest hawk dragonfly wings. ACS Appl Mater Interfaces 12:10883–10892
103. Ren T, Yang M, Wang K, Zhang Y, He J (2018) CuO nanoparticles-containing highly transparent and superhydrophobic coatings with extremely low bacterial adhesion and excellent bactericidal property. ACS Appl Mater Interfaces 10:25717–25725
104. Ma H, Li S, Wei Y, Jiang L, Li J (2018) Fabrication of two-dimensional (2D) ordered microsphere aligned by supramolecular self-assembly of formyl-azobenzene and dipeptide. J Colloid Interface Sci 514:491–495
105. Oh S, Cho J-W, Lee J, Han J, Kim S-K, Nam Y (2022) A scalable haze-free antireflective hierarchical surface with self-cleaning capability. Advanced Science 9(27):2202781
106. Chen X, Song L, Jiang X, Zhang X (2019) Bioinspired superhydrophobic–superhydrophilic convertible film based on anisotropic red blood cell-like particles with protuberances. Colloids Surf A Physicochem Eng Asp 579:123674
107. Cutolo MA, Galeotti F, Spaziani S, Quero G, Calcagno V, Micco A, Irace A, Breglio G, Pisco M, Cusano A (2024) Self-assembled hierarchical nanostructures: Toward engineered sers-active platforms. Laser & Photonics Reviews 18(10):2301056
108. Wu Y, Yu Q, Joung Y, Jeon CS, Lee S et al (2023) Highly uniform self-assembly of gold nanoparticles by butanol-induced dehydration and its SERS applications in SARS-CoV-2 detection. Anal Chem 95:12710–12718
109. Wang K, Li J (2021) Reliable SERS detection of pesticides with a large-scale self-assembled Au@4-MBA@Ag nanoparticle array. Spectrochim Acta A Mol Biomol Spectrosc 263:120218
110. Li X, Lin X, Fang G, Dong H, Li J et al (2022) Interfacial layer-by-layer self-assembly of PS nanospheres and Au@Ag nanorods for fabrication of broadband and sensitive SERS substrates. J Colloid Interface Sci 620:388–398
111. Guo J, Tian H, Yang J, He J (2021) Self-assembly of mercury-ion recognizing CuS nanocrystals into 3D sponge-like aerogel towards superior mercury capturer with outstanding selectivity and efficiency. Chem Eng J 426:130868
112. Jiang S, Van Dyk A, Maurice A, Bohling J, Fasano D, Brownell S (2017) Design colloidal particle morphology and self-assembly for coating applications. Chem Soc Rev 46:3792–3807

# Chapter 7
# Formation of Noble Metal Nanowires and Their Catalytic and Optical Properties

**Yoshiro Imura and Clara Imura**

**Abstract** Anisotropic noble metal nanocrystals are promising novel nanomaterials because of their excellent catalytic and optical properties. In particular, the nanowire structure prepared by an aggregation method acts as a nanocatalyst with high catalytic activity owing to the disordered state of metal atoms within the aggregated domains. To prepare these nanowires, it is necessary to adequately understand their formation dynamics and mechanisms. In this chapter, we describe the nanowire formation dynamics and mechanisms involving the aggregation of nanocrystals, selective adsorption of a capping agent, and their catalytic and optical properties. We also discuss our recent research works on Pd nanowires, dendritic Au nanowires, and ultrathin straight Au–Ag nanowires. These results are useful for preparing functional nanowires with high catalytic activity and unique optical properties.

**Keywords** Nanowires · Diffusion-limited aggregation · Nanocrystals · Formation dynamics · Catalysts

Y. Imura (✉)
Department of Industrial Chemistry, Faculty of Engineering, Tokyo University of Science, Tokyo, Japan
e-mail: imura@rs.tus.ac.jp

C. Imura
Department of Chemistry, Faculty of Science, Ochanomizu University, Tokyo, Japan
e-mail: imura.clara@ocha.ac.jp



M. Hashizume, Y. Imura (eds.), *Colloids and Interfacial Dynamics*, Physical Chemistry in Action, https://doi.org/10.1007/978-981-96-6728-4_7

## 7.1 Introduction

The morphological control of noble metal nanocrystals is important for the development of modern materials, because the physical and chemical properties of these crystals can be easily tuned by tailoring their size and shape [1–4]. Multiple research studies on the fabrication of anisotropic nanocrystals, such as rods, wires, flowers, and cubes, have been performed in the past [5–10]. Anisotropic gold, platinum, and palladium nanocrystals with excellent catalytic properties have been used as catalysts and are easily prepared via a colloidal synthesis in solution using a capping agent to inhibit nanocrystal precipitation [11–13]. According to thermodynamic prediction data, noble metal atoms nucleate and grow into spherical decahedral or icosahedral nanoparticles covered with (111) crystal facets, because the interfacial free energy ($\gamma$) increases in the order $\gamma(111) < \gamma(100) < \gamma(110)$, and the surface area per unit volume of spherical nanoparticles is lower than that of anisotropic nanocrystals [1, 14–16]. Moreover, a truncated octahedron enclosed by a mixture of (111) and (100) crystal facets is the most stable structure for large nanoparticles, because it exhibits the lowest surface area per unit volume [1, 14, 15]. Therefore, anisotropic nanocrystals, such as rods and wires, which are covered with numerous (100) and (110) crystal facets, are fabricated using capping agents with selective adsorption properties [1, 13]. Capping agents with amino groups, such as cetyltrimethylammonium bromide, melamine, and long-chain amines (C18AA, Fig. 7.1a), adsorb more weakly on the Au(111) facet than on the Au(100) and (110) facets [13, 17, 18].

Diffusion-limited aggregation, which controls the reduction rate of metal ions and aggregation rate of metal nanoparticles, is effective in the preparation of nanowires [5, 19, 20]. Such nanowires tend to exhibit a higher catalytic activity than the original spherical nanoparticles owing to their domains (grain boundaries), which

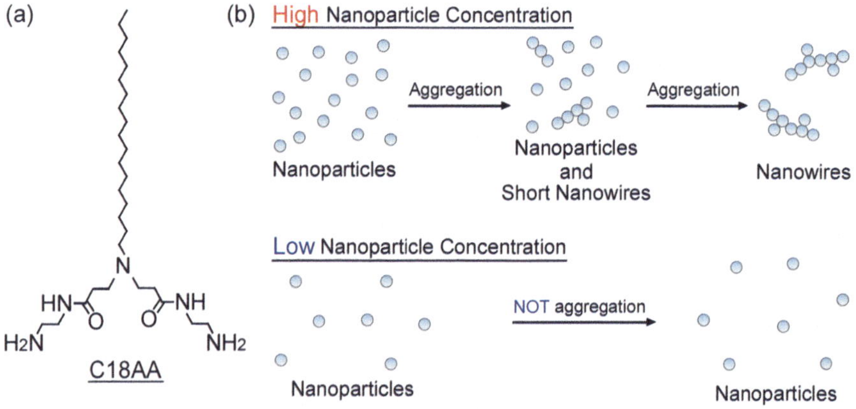


**Fig. 7.1** (**a**) Molecular structure of C18AA. (**b**) Schematic illustration of the nanowire formation process

frequently exhibit high catalytic activity [2, 12, 21]. To prepare nanowires via this method, a high nanoparticle concentration is required because of the difficulty with aggregating nanoparticles at low concentrations (Fig. 7.1b). Hence, the formation dynamics of nanowires must be examined to better understand diffusion-limited aggregation processes.

Ultrathin nanowires with diameters of 2 nm are promising catalysts and optical materials [22–28]. Ultrathin Au nanowires were obtained previously using a soft template and/or oriented attachment method [5, 22–26]. An oriented attachment method involves the aggregation of single-crystalline nanoparticles using capping agents, which are selectively adsorbed on Au(100) and Au(110) facets and weakly adsorbed on Au(111) facets [23]. Compared with two distinct metallic elements, bimetallic nanocrystals represent an emerging class of materials [12, 29, 30]. They include Au–Ag bimetallic nanocrystals, which can be potentially applied as optical materials [31, 32]. The surface plasmon bands of Au–Ag spherical nanoparticles are tunable in the range of 420–520 nm by adjusting the [Au]/[Ag] molar ratio [31, 32]. Therefore, ultrathin Au–Ag bimetallic nanowires are novel optical materials with unique properties, which can be obtained via the oriented attachment method [33]. The formation dynamics of nanowires should be evaluated to elucidate the formation mechanism of ultrathin bimetallic nanowires. In this chapter, we discuss the formation dynamics, catalytic performance, and optical properties of Pd nanowires, dendritic Au nanowires, and ultrathin Au–Ag bimetallic nanowires.

## 7.2 Palladium Nanowires

We previously reported the preparation of Pd nanowires using C18AA as the capping agent and diffusion-limited aggregation [19]. We also found that nanowires exhibited higher catalytic activity during the hydrogenation of *p*-nitrophenol as compared with those of spherical Pd nanoparticles and short Pd nanowires [19]. In this section, we focus on the formation dynamics, mechanism, and catalytic performance of Pd nanowires.

Pd nanowires were prepared by adding $NaBH_4$ (375 μmol) as the reducing agent to an aqueous solution of C18AA and $K_2PdCl_4$ (37.5 μmol). The $[NaBH_4]/[K_2PdCl_4]$ molar ratio was 10. The obtained nanowires had a diameter of ~3.6 nm and a length of a few micrometers (Fig. 7.2e). We examined the nanowire formation dynamics by conducting transmission electron microscopy (TEM) and high-resolution TEM (HR-TEM) observations at various reaction times to elucidate the nanowire formation mechanism (Figs. 7.2 and 7.3). After 1 min of reaction, only spherical nanoparticles were observed (Fig. 7.2a), whereas after 10 min, Pd nanoparticle aggregates and Pd nanowires with an average length of approximately 10 nm were formed (Fig. 7.2b). Upon increasing the reaction time to 3 and 6 h, the length of Pd nanowires increased, although the average diameter remained unchanged (Fig. 7.2c, d). After 24 h, Pd nanowires with diameters of 3.6 nm and lengths of a few micrometers were produced (Fig. 7.2e). The obtained HR-TEM images confirmed that nanowires

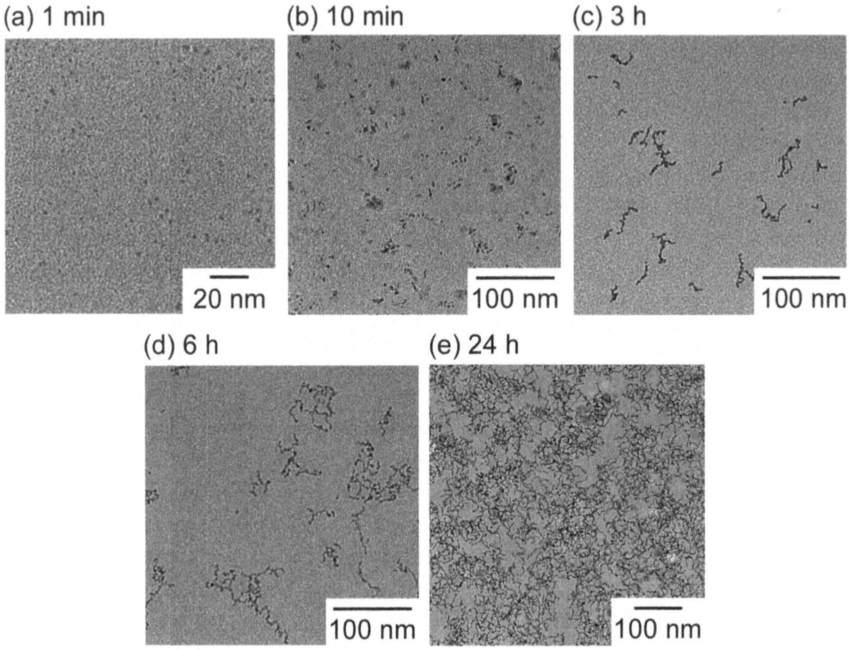


**Fig. 7.2** TEM images of Pd nanocrystals prepared at reaction times of (**a**) 1 min, (**b**) 10 min, (**c**) 3 h, (**d**) 6 h, and (**e**) 24 h after the addition of $NaBH_4$

grew over time (Fig. 7.3). Figure 7.3 also indicates that nanocrystals aggregated to form polycrystalline branched Pd nanowires, because the crystal structure and size of the original nanoparticles were not uniform. Pd nanowires were formed after several hours, and the nanowire aggregation and fusion processes were slow (Figs. 7.2 and 7.3). The concentration of spherical nanoparticles in solution is the main factor affecting nanoparticle aggregation, and the formation of Pd nanowires at low nanoparticle concentrations is suppressed because of the reduced probability of nanoparticle collisions (Fig. 7.1b). To examine this phenomenon in more detail, we prepared smaller Pd nanocrystals by adding 18.75 μmol $K_2PdCl_4$ instead of 37.5 μmol. At $[NaBH_4]/[K_2PdCl_4] = 20$, the TEM images confirm the preparation of short Pd nanowires with diameters of 3.6 nm and lengths of a few hundred nanometers (Fig. 7.4a). Meanwhile, at $[NaBH_4]/[K_2PdCl_4] = 2$, only spherical nanoparticles with an average diameter of 3.4 nm were obtained because of the inhibition of Pd nanocrystal aggregation (Fig. 7.4b). Thus, the formation of nanowires requires a low aggregation rate and the presence of numerous nanocrystals in solution.

Because Pd nanocrystals are effective hydrogenation catalysts, the catalytic performance of long Pd nanowires (Fig. 7.2e), short Pd nanowires (Fig. 7.4a), and spherical Pd nanoparticles (Fig. 7.4b) for the hydrogenation of *p*-nitrophenol to *p*-aminophenol was examined. The reaction rate for the transformation process was determined by monitoring the characteristic ultraviolet–visible (UV–vis) intensities

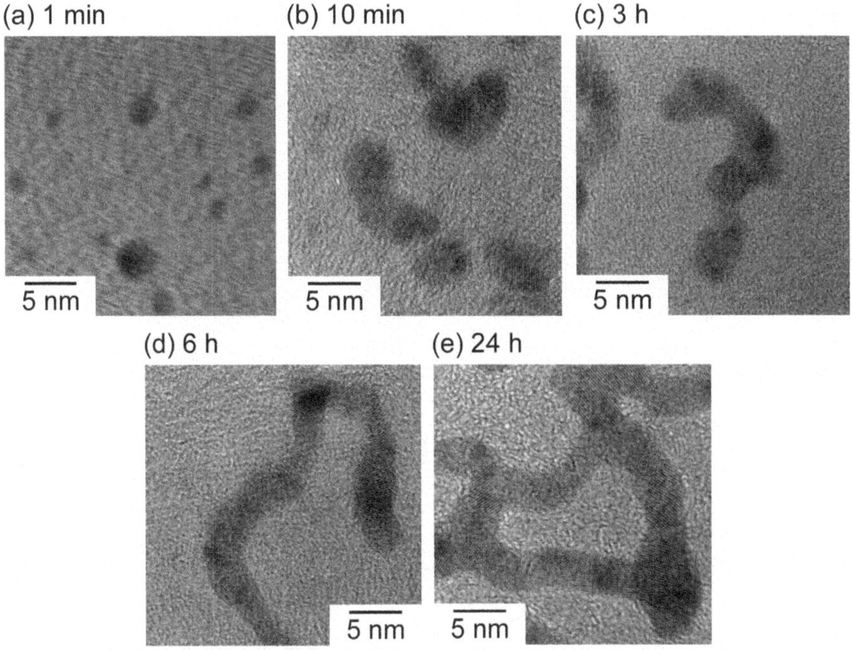


**Fig. 7.3** HR-TEM images of Pd nanocrystals prepared at reaction times of (**a**) 1 min, (**b**) 10 min, (**c**) 3 h, (**d**) 6 h, and (**e**) 24 h after the addition of $NaBH_4$

of *p*-nitrophenol and *p*-aminophenol at 400 and 300 nm, respectively (Fig. 7.4c, d). The absorbance at 400 nm decreased with time in the presence of Pd nanocrystals but remained unchanged in their absence (Fig. 7.4d). In addition, the absorbance decreased in the order of spherical Pd nanoparticles < short Pd nanowires < long Pd nanowires (Fig. 7.4d). The reaction did not proceed in the absence of Pd, and the reactivity order was spherical Pd nanoparticles < short Pd nanowires < long Pd nanowires (Fig. 7.4d). Because the borohydride concentration was initially significantly higher than that of *p*-nitrophenol and the Pd catalyst, pseudo-first-order kinetics were assumed for the reaction. The pseudo-first-order rate constant was evaluated using the expression $-kt = \ln (C_t/C_0)$, where $k$ is the rate constant, and $C_t$ and $C_0$ represent the concentration of *p*-nitrophenol at time $t$ and the initial *p*-nitrophenol concentration at $t = 0$, respectively. All plots of $\ln (C_t/C_0)$ versus time constructed for the three different catalytic species are linear; therefore, $k$-values were calculated from the slopes of the best-fit lines. More specifically, the rate constants of 0.1210, 0.0559, and 0.0175 $s^{-1}$ were calculated for long Pd nanowires, short Pd nanowires, and spherical Pd nanoparticles, respectively (Fig. 7.4e). The surface area per volume of Pd nanowires is smaller than that of spherical Pd nanoparticles because their average diameters are equal to 3.6 and 3.4 nm, respectively. The nanocrystals prepared through the aggregation of spherical nanoparticles exhibited a higher catalytic activity than the original nanoparticles because of the increased

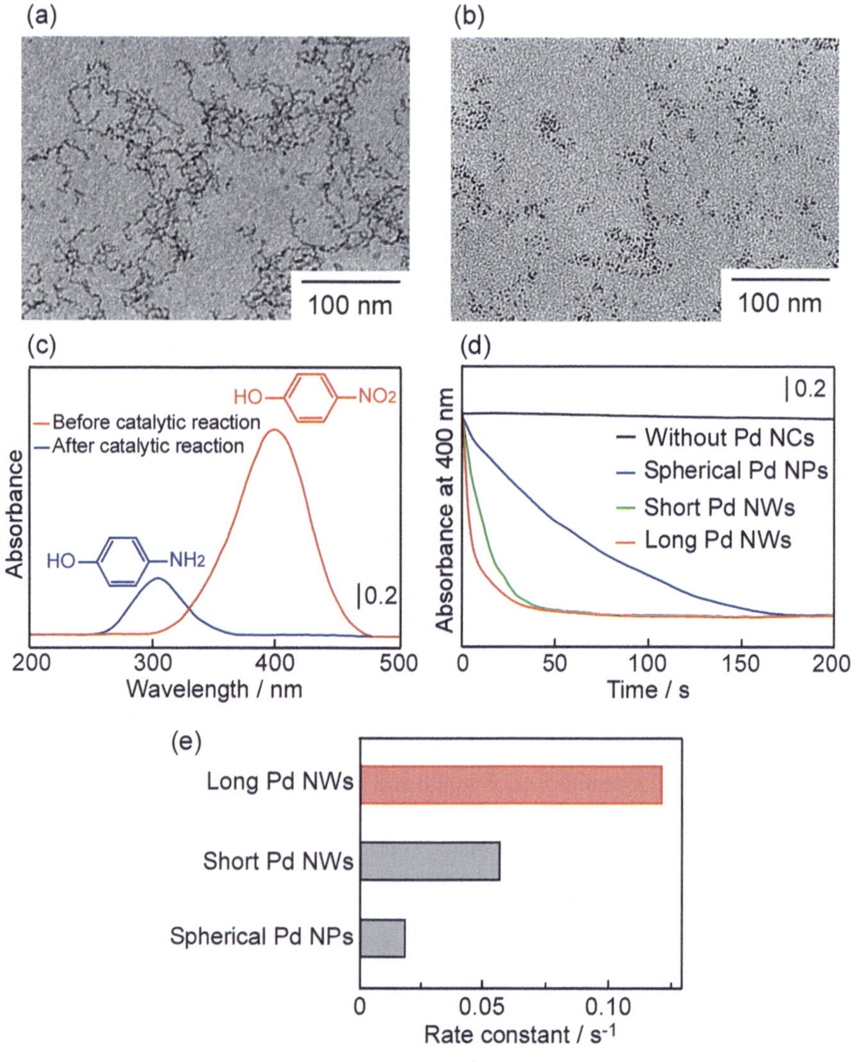


**Fig. 7.4** TEM images of (**a**) short Pd nanowires (NWs) and (**b**) spherical Pd nanoparticles (NPs). (**c**) UV–vis spectra obtained before and after the catalytic reaction in the presence of spherical Pd NPs during the conversion of *p*-nitrophenol to *p*-aminophenol. (**d**) UV–vis spectra recorded at 400 nm for long Pd NWs, short Pd NWs, spherical Pd NPs, and in the absence of Pd nanocrystals (NCs). (**e**) Rate constants *k* were determined for long Pd NWs, short Pd NWs, and spherical Pd NPs

disorder of metal atoms within the aggregated domains. The aggregated domain area of long nanowires is larger than that of short nanowires, which also contributes to their higher activity.

## 7.3 Dendritic Gold Nanowires

Capping agents with amino groups more weakly adsorb on the Au(111) surface than on the Au(100) and (110) surfaces [13, 17, 18]. By using this crystal facet-selective adsorption properties and diffusion-limited aggregation method, the morphology of Au nanocrystals can be effectively controlled [34, 35]. In this section, we describe the formation dynamics, formation mechanism, and catalytic performance of dendritic Au nanowires using C18AA, a capping agent with selective crystal facet adsorption properties, which acts not only as a capping agent but also as a reducing agent [34, 35].

The formation of dendritic Au nanowires upon the addition of an aqueous solution of $HAuCl_4$ (24 mM, 1.0 mL) to an aqueous solution of C18AA (2.0 wt.%, 4.0 mL) at 55 °C is confirmed via TEM. The TEM image shows dendritic Au nanowires comprising branches with sizes of approximately 6.3 nm without spherical Au nanoparticles as byproducts (Fig. 7.5a). The dendritic Au nanowire dispersion exhibits no surface plasmon band at ~530 nm in the UV–vis spectrum because of the absence of spherical Au nanoparticles (Fig. 7.5d). In contrast, only spherical Au nanoparticles with an average diameter of 4.8 nm were prepared by adding an aqueous solution of $HAuCl_4$ (24 mM, 1.0 mL) to an aqueous solution of C18AA

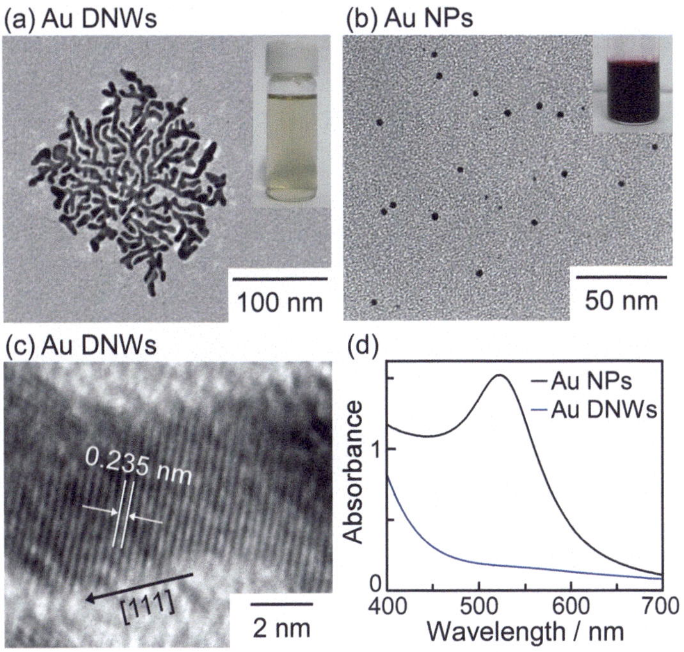


**Fig. 7.5** (**a**, **b**) TEM images, (**c**) HR-TEM image, and (**d**) UV–vis spectra of dendritic Au nanowires (Au DNWs) and spherical Au nanoparticles (Au NPs) obtained after 24 h of the reduction reaction

(0.5 wt.%, 4.0 mL) at 80 °C (Fig. 7.5b). A surface plasmon band appears at 522 nm, indicating the presence of spherical Au nanoparticles (Fig. 7.5d). The HR-TEM image of dendritic Au nanowires depicted in Fig. 7.5c contains fringes with a periodicity of 0.235 nm, which corresponds to the Au(111) lattice spacing along the long axis of the branches, revealing that the growth of dendritic Au nanowires occurs in the Au(111) crystal direction. Because C18AA is preferentially adsorbed on the Au(100) and (110) facets rather than on the Au(111) facet, the Au(100) and (110) crystal facets are covered with high-density layers due to selective adsorption, whereas the Au (111) crystal facet is covered with a low-density layer [18, 22, 34]. Thus, the growth rate on the Au(111) facets is higher than that on the other facets; consequently, the branches of dendritic Au nanowires grow in the [111] direction. The formation of dendritic Au nanowires generally follows the diffusion-limited aggregation mechanism under appropriate reaction conditions. The nanowire formation dynamics were confirmed by conducting TEM observations at various reaction times. The average diameters of dendritic Au nanowires were 30–60 nm after 2 h, 80–160 nm after 6 h, 100–200 nm after 8 h, and 100–200 nm after 14 h of reaction (Fig. 7.6). Similar to Pd nanowires, dendritic Au nanowires were also obtained by slowly aggregating nanocrystals.

The nanowire structure exhibited high catalytic activity for the hydrogenation of *p*-nitrophenol with both Pd and Au. The *k*-values obtained for dendritic Au nanowires and spherical Au nanoparticles were 0.0129 and 0.0064 $s^{-1}$, respectively [35]. The catalytic performance of dendritic Au nanowires is higher than that of spherical Au nanoparticles, although the surface area per volume of dendritic Au nanowires

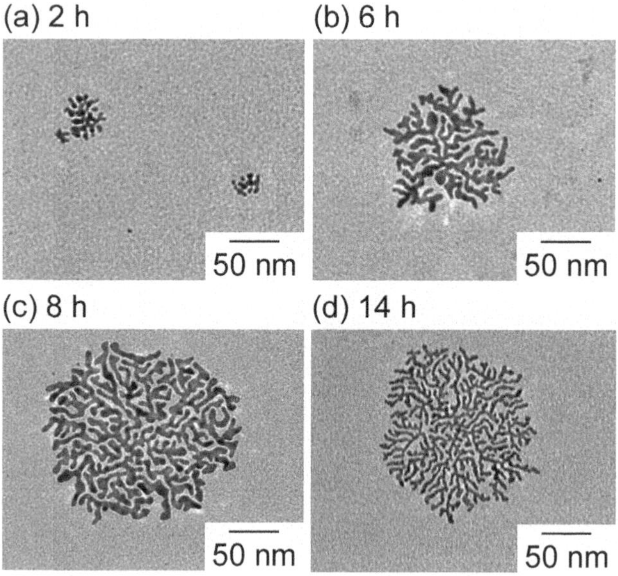


**Fig. 7.6** TEM images of dendritic Au nanowires obtained after (**a**) 2 h, (**b**) 6 h, (**c**) 8 h, and (**d**) 14 h of the reduction reaction

is smaller than that of spherical Au nanoparticles because the average branch size of dendritic Au nanowires and particle size of spherical Au nanoparticles are 6.3 and 4.8 nm, respectively. The higher catalytic activity of dendritic Au nanowires as compared with that of spherical Au nanowires is also due to the larger number of disordered Au atoms within the aggregated domain [35]. These results indicate that the Au nanowire structures prepared through the aggregation of the original Au nanoparticles are novel nanomaterials with high catalytic activity.

## 7.4 Gold and Silver Bimetallic Nanowires

The amino group exhibits selective adsorption properties not only for the Au surface but also for the Ag surface [17]. However, the difference in the adsorption energy between the Ag(111) and (100) crystal facets is smaller than that between the Au(111) and (100) crystal facets [17]. Therefore, anisotropic Ag nanocrystals are more difficult to prepare using a capping agent with an amino group than anisotropic Au nanocrystals. Anisotropic Au–Ag bimetallic nanocrystals are expected to exhibit novel optical and catalytic properties [29]. C18AA can be used to prepare not only Pd nanowires and dendritic Au nanowires but also ultrathin Au–Ag bimetallic nanowires with an average diameter of 2 nm [33]. In this section, we describe the formation dynamics, mechanism, and optical properties of ultrathin Au–Ag bimetallic nanowires.

Ultrathin Au–Ag bimetallic nanowires were prepared by adding ascorbic acid as a reducing agent to a 2 wt.% aqueous C18AA solution containing $HAuCl_4$ and $AgNO_3$ and storing the obtained system for 24 h. The [C18AA]/[Au + Ag] molar ratio was 5.6. The average diameter of ultrathin nanowires was 2.0 nm, and their length was a few micrometers (Fig. 7.7d). TEM–energy-dispersive X-ray spectroscopy experiments revealed that ultrathin nanowires were composed of Au and Ag species. The measured [Au]/[Ag] molar ratio was 1.05, which is in good agreement with the ratio used during preparation. The HR-TEM image of the fabricated nanowires contains fringes with a periodicity of 0.237 nm, which corresponds to the (111) lattice spacing (0.2369 nm) of the alloyed nanocrystals ([Au]/[Ag] = 1). These results indicate that ultrathin Au–Ag bimetallic nanowires grown in the [111] direction were obtained using C18AA as the capping agent. Unlike the use of only Au in the previous section, when Au and Ag were combined, nanowires grew along straight lines. Their formation mechanism was elucidated by evaluating the nanowire formation dynamics via TEM and HR-TEM at various reaction times. The TEM image captured after 1 min reveals the formation of only spherical nanoparticles with a diameter of 2 nm (Fig. 7.7a). The HR-TEM image depicted in Fig. 7.8a indicates that these spherical nanoparticles are single-crystalline nanocrystals. After 2 h of reaction, spherical nanoparticles and short nanowires are observed (Fig. 7.7b). The length and yield of these nanowires increase with increasing reaction time (Fig. 7.7c, d). These results reveal that ultrathin nanowires are formed through the aggregation of single-crystalline nanoparticles with selective growth in the [111]

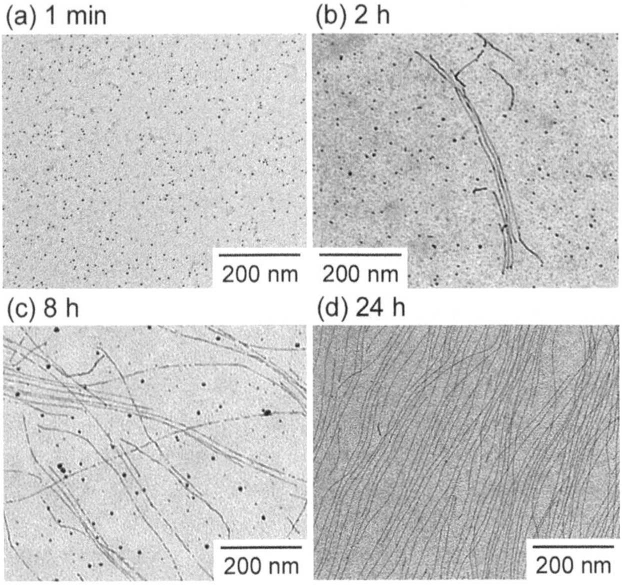


**Fig. 7.7** TEM images of Au–Ag bimetallic nanocrystals obtained after (**a**) 1 min, (**b**) 2 h, (**c**) 8 h, and (**d**) 24 h. The [C18AA]/[Au + Ag] molar ratio is 5.6

direction (Fig. 7.8c) due to the amine groups of C18AA, which are hardly adsorbed on the Au and Ag (111) surfaces as compared with the (100) and (110) surfaces, respectively. In addition, Fig. 7.7 shows that the preparation of ultrathin Au–Ag nanowires requires a low aggregation rate, which is similar to those of Pd and dendritic Au nanowires.

The morphology of nanowires was affected by the molar ratio ([C18AA]/[Au + Ag]). The C18AA concentration was 2 wt.%, and the [Au + Ag] concentration was lowered from 24 to 8 mM. When the molar ratio was 5.6, ultrathin Au–Ag bimetallic nanowires with lengths of a few micrometers were obtained (Fig. 7.9a). The lengths and yields of ultrathin nanowires decreased with increasing the molar ratio. Only spherical Au–Ag bimetallic nanoparticles were obtained at a molar ratio of 16.8, and no ultrathin Au–Ag bimetallic nanowires were observed (Fig. 7.9b). Meanwhile, ultrathin straight nanowires were prepared as the [Au + Ag] concentration decreased from 24 to 12 and 8 mM at a molar ratio of 5.6. The lengths of Au–Ag bimetallic nanowires were controlled by adjusting the molar ratio (Fig. 7.9c). The amount of C18AA on the metal surface increased with increasing molar ratio, and the produced nanocrystals exhibited improved dispersion stability. The formation of these ultrathin nanowires derived through nanocrystal aggregation and fusion was hardly observed because of the improved dispersion stability; consequently, the

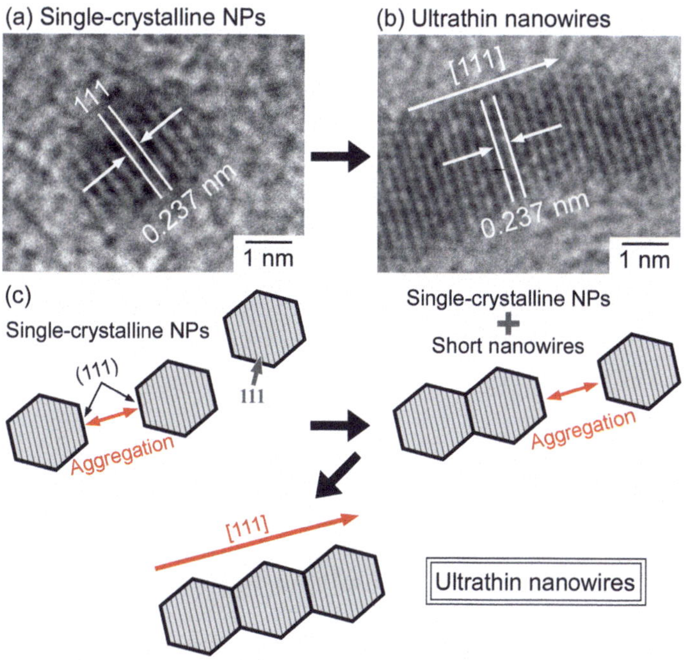


**Fig. 7.8** HR-TEM images of (**a**) single-crystalline Au–Ag bimetallic nanoparticles (NPs) obtained after 1 min and (**b**) ultrathin Au–Ag bimetallic nanowires obtained after 24 h. (**c**) Schematic illustration of the formation mechanism of ultrathin Au–Ag bimetallic nanowires

nanowire length could be controlled by changing the [C18AA]/[Au + Ag] molar ratio.

The optical properties of the prepared nanowires are shown in Fig. 7.10. The Au–Ag bimetallic nanoparticles exhibit the optical properties of both Au nanoparticles and Ag nanoparticles. Alloyed spherical Au–Ag nanoparticles produce plasmon bands between 520 nm (Au nanoparticles) and 420 nm (Ag nanoparticles). In addition, the plasmon bands of the nanowire structures are observed at higher wavelengths than those of spherical nanoparticles. Spherical nanoparticles obtained at a [C18AA]/[Au + Ag] molar ratio of 16.8 exhibit a surface plasmon band at 480 nm (Fig. 7.10). Furthermore, ultrathin Au–Ag bimetallic nanowires generate a surface plasmon band at a higher wavelength than spherical nanoparticles (Fig. 7.10). These results confirm that Au–Ag bimetallic nanowires exhibiting a surface plasmon band in the infrared region were obtained using C18AA as the capping agent. Nanowires with novel properties were formed by controlling the aggregation rate and growth direction of nanocrystals. To prepare nanowires, it is important to understand their formation dynamics and mechanisms.

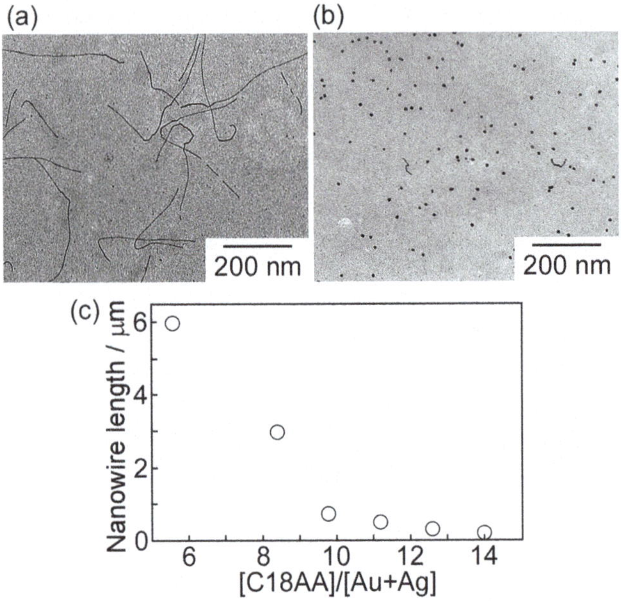


**Fig. 7.9** TEM images of Au–Ag bimetallic nanocrystals obtained at [C18AA]/[Au + Ag] molar ratios of (**a**) 9.8 and (**b**) 16.8. (**c**) Average nanowire length as a function of the [C18AA]/[Au + Ag] molar ratio

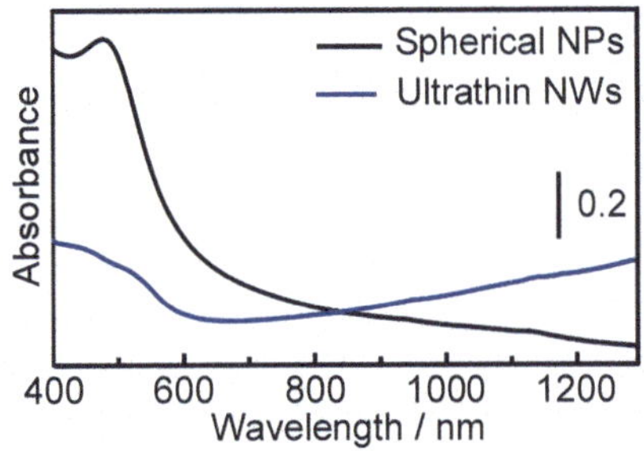


**Fig. 7.10** UV–vis spectra of spherical Au–Ag bimetallic nanoparticles (NPs) and ultrathin Au–Ag bimetallic nanowires (NWs) obtained at [C18AA]/[Au + Ag] molar ratios of 16.8 and 5.6, respectively

## 7.5 Conclusion

In this study, we described the formation dynamics, catalytic performance, and optical properties of Pd nanowires, dendritic Au nanowires, and ultrathin straight Au–Ag nanowires. The preparation of these nanowires required a low aggregation rate and high nanocrystal concentration. The aggregation rate and formation mechanism of the nanowires were determined via TEM and HR-TEM. Furthermore, Pd nanowires and dendritic Au nanowires demonstrated higher catalytic activities for the hydrogenation of *p*-nitrophenol than that of spherical nanoparticles. Ultrathin Au–Ag nanowires, which produce surface plasmon bands in the infrared region, were obtained through the aggregation of single-crystalline nanoparticles with selective growth in the [111] direction. To prepare novel functional anisotropic nanocrystals, it was necessary to evaluate their formation dynamics and elucidate the most feasible formation mechanism. We believe that these results will be useful for preparing anisotropic nanocrystals with superior catalytic and optical properties.

**Acknowledgments** This work was partially supported by JSPS KAKENHI (grant number JP16K17490).

## References


1. Xia Y, Xiong Y, Lim B, Skrabalak SE (2009) Shape-controlled synthesis of metal nanocrystals: simple chemistry meets complex physics? Angew Chem Int Ed 48:60–103
2. Lim B, Xia Y (2011) Metal nanocrystals with highly branched morphologies. Angew Chem Int Ed 50:76–85
3. Shi Y, Lyu Z, Zhao M, Chen R, Nguyen QN, Xia Y (2021) Noble-metal nanocrystals with controlled shapes for catalytic and electrocatalytic applications. Chem Rev 121:649–735
4. Imura Y, Furukawa S, Ozawa K, Morita-Imura C, Kawai T, Komatsu T (2016) Surface clean gold nanoflower obtained by complete removal of capping agents: an active catalyst for alcohol oxidation. RSC Adv 6:17222–17227
5. Imura Y, Maniwa M, Iida K, Saito H, Morita-Imura C, Kawai T (2021) Preparing alumina-supported gold nanowires for alcohol oxidation. ACS Omega 6:16043–16048
6. Morita C, Tanuma H, Kawai C, Ito Y, Imura Y, Kawai T (2013) Room-temperature synthesis of two-dimensional ultrathin gold nanowire parallel array with tunable spacing. Langmuir 29:1669–1675
7. Nguyen QN, Wang C, Shang Y, Janssen A, Xia Y (2023) Colloidal synthesis of metal nanocrystals: from asymmetrical growth to symmetry breaking. Chem Rev 123:3693–3760
8. Huang H, Ruditskiy A, Choi SI, Zhang L, Liu J, Ye Z, Xia Y (2017) One-pot synthesis of penta-twinned palladium nanowires and their enhanced electrocatalytic properties. ACS Appl Mater Interfaces 9:31203–31212
9. Imura Y, Fukuda K, Morita-Imura C, Kawai T (2016) Water and organic solvent dispersible gold nanorods that are pH responsive. ChemistrySelect 1:5404–5408
10. Zhang H, Jin M, Xiong Y, Lim B, Xia Y (2012) Shape-controlled synthesis of Pd nanocrystals and their catalytic applications. Acc Chem Res 46:1783–1794
11. Imura Y, Kan R, Akiyama R, Saito H, Morita-Imura C, Kawai T (2020) Magnetic $Fe_3O_4$-supported gold nanoflowers with lattice-selected surfaces: preparation and catalytic performance. ACS Omega 5:15755–15760
12. Imura Y, Tsujimoto K, Morita C, Kawai T (2014) Preparation and catalytic activity of Pd and bimetallic Pd–Ni nanowires. Langmuir 30:5026–5030
13. Yang TH, Shi Y, Janssen A, Xia Y (2020) Surface capping agents and their roles in shape-controlled synthesis of colloidal metal nanocrystals. Angew Chem Int Ed 59:15378–15401
14. Barnard AS, Young NP, Kirkland AI, van Huis MA, Xu H (2009) Nanogold: a quantitative phase map. ACS Nano 3:1431–1436
15. Imura Y, Tanaka M, Kasuga A, Akiyama R, Ogawa D, Sugimori H, Morita-Imura C, Kawai T (2023) Nanoarchitectonics and catalytic performance of au-Pd nanoflowers supported on $Fe_2O_3$. J Oleo Sci 72:1055–1061
16. Imura Y, Saito H, Jino Y, Akiyama R, Morita-Imura C, Kawai T (2024) Nanoarchitectonics for improving catalytic performance of α-alumina-supported gold nanoflower by water extraction and ultraviolet–ozone treatment. J Oleo Sci 73:1241–1247

17. You H, Liu X, Liu H, Fang J (2016) Theoretical description of the role of amine surfactant on the anisotropic growth of gold nanocrystals. CrystEngComm 18:3934–3941
18. Imura Y, Morita C, Endo H, Kondo T, Kawai T (2010) Reversible phase transfer and fractionation of au nanoparticles by pH change. Chem Commun 46:9206–9208
19. Imura Y, Kataoka H, Akiyama R, Morita-Imura C, Kawai T (2018) Preparation and reconstruction of long branched palladium nanowires exhibiting high catalytic activities. ChemistrySelect 3:13387–13390
20. Witten TA, Sander LM (1981) Diffusion-limited aggregation, a kinetic critical phenomenon. Phys Rev Lett 47:1400–1403
21. Ruan L, Zhu E, Chen Y, Lin Z, Huang X, Duan X, Huang Y (2013) Biomimetic synthesis of an ultrathin platinum nanowire network with a high twin density for enhanced electrocatalytic activity and durability. Angew Chem Int Ed 52:12577–12581
22. Imura Y, Tanuma H, Sugimoto H, Ito R, Hojo S, Endo H, Morita C, Kawai T (2011) Water-dispersible ultrathin Au nanowires prepared using a lamellar template of a long-chain amidoamine derivative. Chem Commun 47:6380–6382
23. Halder A, Ravishankar N (2007) Ultrafine single-crystalline gold nanowire arrays by oriented attachment. Adv Mater 19:1854–1858
24. Lu X, Yavuz MS, Tuan HY, Korgel BA, Xia Y (2008) Ultrathin gold nanowires can be obtained by reducing polymeric strands of Oleylamine–AuCl complexes formed via aurophilic interaction. J Am Chem Soc 130:8900–8901
25. Huo Z, Ck T, Huang W, Zhang X, Yang P (2008) Sub-two nanometer single crystal au nanowires. Nano Lett 8:2041–2044
26. Feng H, Yang Y, You Y, Li G, Guo J, Yu T, Shen Z, Wu T, Xing B (2009) Simple and rapid synthesis of ultrathin gold nanowires, their self-assembly and application in surface-enhanced Raman scattering. Chem Commun:1984–1986
27. Wang Y, Choi SI, Zhao X, Xie S, Peng HC, Chi M, Huang CZ, Xia Y (2014) Polyol synthesis of ultrathin Pd nanowires via attachment-based growth and their enhanced activity towards formic acid oxidation. Adv Funct Mater 24:131–139
28. Zhu W, Zhang YJ, Zhang H, Lv H, Li Q, Michalsky R, Peterson AA, Sun S (2014) Active and selective conversion of $CO_2$ to CO on ultrathin au nanowires. J Am Chem Soc 136:16132–16135
29. Imura Y, Akiyama R, Furukawa S, Kan R, Morita-Imura C, Komatsu T, Kawai T (2019) Au–ag nanoflower catalysts with clean surfaces for alcohol oxidation. Chem Asian J 14:547–552
30. Wang D, Li Y (2011) Bimetallic nanocrystals: liquid-phase synthesis and catalytic applications. Adv Mater 23:1044–1060
31. Ferrando R, Jellinek J, Johnston RL (2008) Nanoalloys: from theory to applications of alloy clusters and nanoparticles. Chem Rev 108:845–910
32. Mallin MP, Murphy CJ (2002) Solution-phase synthesis of Sub-10 nm au–ag alloy nanoparticles. Nano Lett 2(11):1235–1237
33. Imura Y, Mori T, Morita-Imura C, Kataoka H, Akiyama R, Kurata H, Kawai T (2018) Preparation and length control of water-dispersible ultrathin gold and silver bimetallic nanowires. Colloids Surf A Physicochem Eng Asp 543:9–14
34. Imura Y, Maezawa A, Morita C, Kawai T (2012) Neuron-shaped gold nanocrystals and two-dimensional dendritic gold nanowires fabricated by use of a long-chain amidoamine derivative. Langmuir 28:14998–15004
35. Imura Y, Kataoka H, Saito H, Minakawa M, Morita-Imura C, Kawai T (2019) Morphological stability and catalytic performance of supported and unsupported dendritic gold nanowire catalysts. ChemistrySelect 4:9908–9914

# Chapter 8
# Photoexcited Carrier Dynamics and Photocatalytic Activity on Titanium Dioxide Surfaces

**Kenichi Ozawa, Ryu Yukawa, Keita Hiromori, and Nobuo Nakajima**

**Abstract** The mechanism of superb photocatalytic activity of $TiO_2$ has been studied by synchrotron radiation photoelectron spectroscopy. Photocatalytic activity was evaluated by X-ray photoelectron spectroscopy (XPS), and photogenerated carrier lifetimes were measured by pump–probe time-resolved XPS (TR-XPS). Single-crystal rutile and anatase $TiO_2$ with well-defined surfaces were investigated, and the correlation between photocatalytic activity and the carrier lifetime has been discussed based on the experimental results. The carrier lifetime was determined by measuring the spontaneous change of the electron binding energies, which corresponded to generation and quenching of the surface photovoltage. Photocatalytic activity was evaluated by the rate of decrease in the C 1s XPS peak intensity due to photocatalytic decomposition and desorption of adsorbed acetic acid molecules. There is a linear and positive correlation between the carrier lifetime and photocatalytic activity; the longer the carrier lifetime, the higher the photocatalytic activity. Since the carrier lifetime is sensitive to the surface potential, photocatalytic activity can be tuned by changing the surface potential and controlling the carrier behavior. This was demonstrated by introducing a rutile/anatase boundary on the $TiO_2$ surface.

K. Ozawa (✉)
Institute of Materials Structure Science, High Energy Accelerator Research Organization, Tsukuba, Ibaraki, Japan

SOKENDAI (The Graduate University for Advanced Science), Hayama, Japan
e-mail: kenichi.ozawa@kek.jp

R. Yukawa
International Center for Synchrotron Radiation Innovation Smart (SRIS), Tohoku University, Sendai, Japan

K. Hiromori · N. Nakajima
Graduate School of Advanced Science and Engineering, Hiroshima University, Higashihiroshima, Japan



M. Hashizume, Y. Imura (eds.), *Colloids and Interfacial Dynamics*, Physical Chemistry in Action, https://doi.org/10.1007/978-981-96-6728-4_8

**Keywords** Time-resolved X-ray photoelectron spectroscopy · Photocatalysis · Carrier lifetime · Band bending · Titanium dioxide

## 8.1 Introduction

Photocatalysis is a phenomenon in which chemical reactions are accelerated by certain substances under irradiation with light. Such substances are called photocatalysts and are mainly metal oxides. Among them, titanium dioxide ($TiO_2$) has been studied for a long time as a substance that exhibits a very strong photocatalytic phenomenon. One of the pioneering works was already done in the 1940s when Jacobsen discussed the mechanism of chalking behavior of $TiO_2$-containing pigments [1]. Weyl and Förland are also pioneers and have discussed photochemistry of rutile $TiO_2$ in conjunction with the role of point defects and trace doping atoms [2]. A turning point in photocatalyst research was the discovery of Honda–Fujishima effect, in which electrochemical photolysis of water proceeds on the $TiO_2$ electrode [3, 4]. Due to the background of the oil crisis in the 1970s, research into photocatalysts for producing hydrogen fuel expanded explosively.

The photocatalytic activity exhibited by $TiO_2$ is not limited to photolysis of water, but also includes decomposition and removal of harmful substances [5] and sterilization of bacteria and viruses [6]. Although photoinduced superhydrophilicity is not exactly "photocatalysis," it is one of the important phenomena of $TiO_2$ [7]. Taking advantage of these photoinduced properties, $TiO_2$ has been widely used in various products [8] and is the most commercially successful photocatalyst so far.

What both catalysis and photocatalysis have in common is that a series of chemical reactions proceed spontaneously, and the substance that catalyzes the reaction itself does not change before or after the reaction. On the other hand, the difference between the two is whether carriers, i.e., electrons ($e^-$) and holes ($h^+$), participate in the reaction or not. In photocatalysis, carriers generated upon photon absorption migrate in the photocatalyst and are transported to the surface where they interact with adsorbed species to initiate chemical reactions (Fig. 8.1). In ambient conditions, where $O_2$ and $H_2O$ are available, electrons and holes interact with these species first to form so-called reactive oxygen species, like $O_2^{\bullet -}$, $HO_2^{\infty}$, $\bullet OH$, and $O^{\bullet}$, by the following reactions [9, 10]:

$$e^- + O_2 \rightarrow O_2^{\bullet -},$$
$$O_2^{\bullet -} + H^+ \rightarrow HO_2^{\bullet},$$
$$h^+ + H_2O \rightarrow \bullet OH + H^+,$$
$$h^+ + O_2^{\bullet -} \rightarrow 2O^{\bullet}.$$

These radicals interact with organic compounds and initiate decomposition reactions that produce $CO_2$ and $H_2O$ as final products. Even in the absence of $O_2$ or $H_2O$, chemical reactions proceed as electrons and holes interact directly with organic compounds [11, 12].

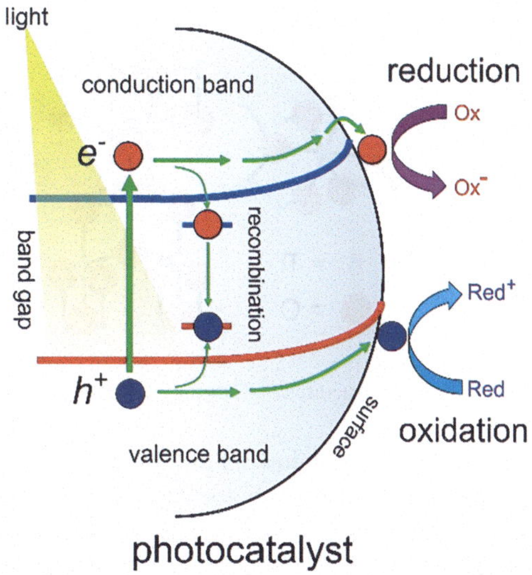


**Fig. 8.1** Photocatalytic processes from photoabsorption to redox reactions

Because of the important role of the carriers in photocatalysis, much effort has been devoted to clarifying the carrier dynamics from generation to interaction with chemical compounds. The time scale of the carrier dynamics ranges from femtoseconds to microseconds [10, 13, 14]. As carriers migrate to the surface, they are often trapped by traps such as point defects and/or grain boundaries, and only the surviving carriers have the opportunity to reach the surface and come into contact with chemicals. The main driving force of the carrier migration is drift along the potential gradient of the internal electric field, i.e., band bending. Therefore, it is indispensable to correlate the carrier dynamics with band bending of photocatalysts [14, 15] for a complete understanding of photocatalysis.

This chapter shows how photogenerated carriers behave in rutile and anatase $TiO_2$ crystals investigated by time-resolved X-ray photoelectron spectroscopy (TR-XPS) utilizing a laser-pump and synchrotron-radiation-probe technique. Additionally, we show how photocatalytic activity depends on the crystalline phase and surface orientation of $TiO_2$. Photocatalytic activity was evaluated by X-ray photoelectron spectroscopy (XPS) under an ultrahigh vacuum condition.

## 8.2 Photocatalytic Activity of $TiO_2$

$TiO_2$ is the most extensively studied photocatalyst, with many books and review articles available. Naturally producing $TiO_2$ are rutile ($P4_2/mnm$), anatase ($I4_1/amd$), and brookite (*Pbca*) polymorphs (Fig. 8.2). The rutile polymorph is thermodynamically the most stable phase, followed by brookite and anatase [16]. However, this order changes when the particle size is reduced; anatase becomes more stable than

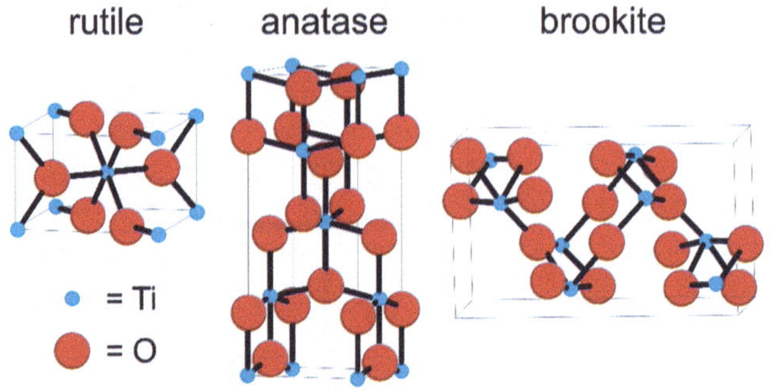


**Fig. 8.2** Crystal structures of $TiO_2$ polymorphs: rutile, anatase, and brookite

rutile as the size is below 50 or 14 nm [16]. Anatase and rutile crystal polymorphs are commonly used as photocatalysts in the form of nanometer-sized powders to increase surface area. Thus, the $TiO_2$ photocatalysts in practical use are the anatase polymorph that is advantageous for photocatalytic activity, because anatase $TiO_2$ is known to have higher photocatalytic activity than rutile. One of the earliest studies reporting this phenomenon was conducted by Tanaka et al. in 1991 [17], who showed that the higher the anatase content of titanium dioxide, the higher the rate of photocatalytic degradation of some organic compounds in aqueous solution. Xu et al. compared the reaction rate of photocatalytic CO oxidation on single-crystalline anatase and rutile surfaces and found that anatase has ten times higher activity than rutile [18]. Luttrell et al. reported that the decomposition rate of methyl orange was two times higher for the anatase surface than the rutile surface [19]. Anatase also shows higher activity than rutile in the production of formaldehyde through photocatalytic methanol oxidation [20].

Many hypotheses have been proposed to explain this structural dependence: a difference of the band-gap energy (3.1–3.2 eV for anatase and 2.9–3.0 eV for rutile [21]), a difference in the properties of polarons (electron–photon coupling) [22–24], a difference of electron and hole effective masses [25, 26], a difference in the lifetime of photogenerated carriers [18, 26–28], etc. A "synergistic effect" will provide a hint for identifying the factors essential for the photocatalytic activity of $TiO_2$. The synergistic effect of $TiO_2$ is the effect that a mixture (not a physical mixture) of rutile and anatase particles exhibits a higher activity than the pure phases [29–32].

A rutile/anatase junction is considered to have the effect of suppressing fast electron–hole recombination and promoting carrier transport [33, 34], resulting in a prolonged carrier lifetime and an increased chance to reach the photocatalyst surface. Therefore, to enhance the photocatalytic activity of $TiO_2$, extending the lifetime of photogenerated electrons and holes seems to be a good strategy.

## 8.3 Time-Resolved X-Ray Photoelectron Spectroscopy

The carrier dynamics is a very fast phenomenon from femtoseconds to microseconds [10, 13, 14]. Various techniques have been employed to evaluate carrier lifetime: time-resolved transient absorption spectroscopy, time-resolved photoluminescence, time-resolved photoconductivity, time-resolved photoelectron spectroscopy, etc. The carrier lifetimes of $TiO_2$ photocatalysts determined by these time-resolved techniques are summarized in Ref. 14. The table summarizes only some of the values reported in the literature, but the range of values is wide, from 1 ps to 33 μs. The situation was rather complicated, because the reported values were obtained from different forms of $TiO_2$ of different sizes under different environments. Clearly, determining carrier lifetime alone is not sufficient to understand carrier dynamics and relate it to photocatalytic activity.

Time-resolved photoelectron spectroscopy is a powerful tool for directly studying temporal changes in the electronic structure of materials. This technique can be broadly divided into two types: time-resolved *two-photon* photoelectron spectroscopy (TR-2PPE) and time-resolved *X-ray* photoelectron spectroscopy (TR-XPS). Both measurements employ a pump–probe technique with a pump light for the excitation of a material to be measured and a probe light for the determination of a decay process to the ground state. Ultrashort pulses are required for the pump and probe lights, whose pulse widths determine the time resolution. For TR-2PPE, femtosecond laser pulses are commonly used for both pump and probe lights, allowing femtosecond dynamics to be investigated by direct detection of photoexcited electrons [35, 36]. For TR-XPS, while the femtosecond laser is also used for the pump light to make an excited state, a pulse light of an X-ray synchrotron radiation, an X-ray free electron laser, or a high harmonic generation light is used as the probe light [37]. Here, the core electrons of constituent elements are examined instead of excited electrons. In other words, TR-XPS indirectly investigates the excited states of the system. Unlike TR-2PPE, TR-XPS captures a localized view of the excited system, because the core electrons are more localized within the atom than the excited electrons in the conduction band. This is a great advantage when one wants to track electron transport from one location to another. Another advantage is that not only electron but also hole transport can be detected, because the binding energy of the core electrons is equally sensitive to the electric field due to excited electrons and holes.

One of the examples of TR-XPS experiments is shown in Fig. 8.3, in which the time evolution of the excited states of copper phthalocyanine (CuPc) and $C_{60}$ in CuPc/$C_{60}$/$TiO_2$(110) and $C_{60}$/CuPc/$TiO_2$(110) is examined by measuring C 1s spectra [38]. The C 1s spectra of these systems are composed of $C_{60}$ and CuPc components. Both components move toward higher binding energies upon irradiation of pump laser pulse (the photon energy of 3.11 eV). The magnitude of the shift is between 10 and 20 meV at the maximum. The peak shifts are interpreted by charge transfer of the photoexcited electrons from CuPc or $C_{60}$ to the $TiO_2$ substrate, leaving the holes in CuPc or $C_{60}$. Although the shift is very small, different curves of

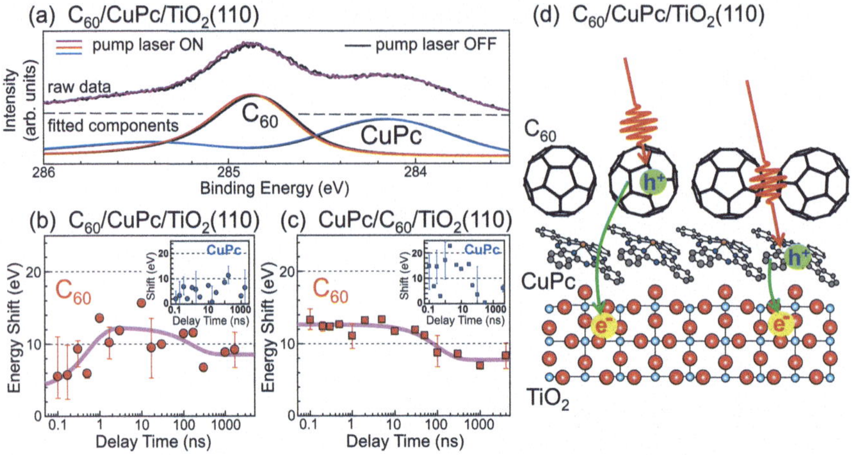


**Fig. 8.3** TR-XPS research on model organic solar cell systems: $C_{60}$/CuPc/$TiO_2$(110) and CuPc/$C_{60}$/$TiO_2$(110) [38]. (**a**) C 1s spectra for $C_{60}$ (0.5 nm)/CuPc (1.0 nm)/rutile $TiO_2$(110) acquired with and without irradiation of a 3.11-eV pump laser. The spectrum with the laser was measured at a delay time of 10 ns using a 385-eV photon energy. (**b**, **c**) Laser-induced shifts of the C 1s components of $C_{60}$ and CuPc in two different stacked systems. (**d**) Schematic drawing of charge transfer from organic layers to the $TiO_2$ substrate

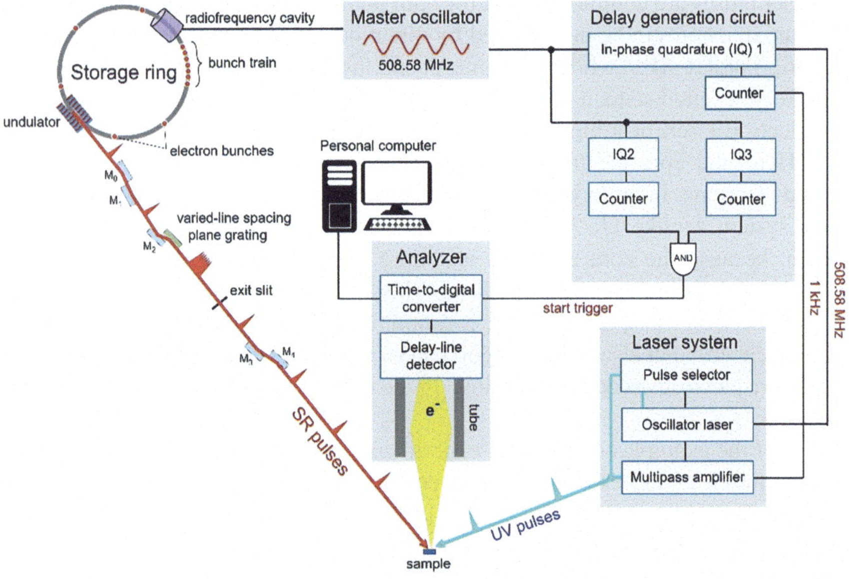


**Fig. 8.4** Block diagram of the timing control for the TR-XPS measurement system at SPring-8 BL07LSU

energy shift as a function of time after excitation are observed between $C_{60}$ and CuPc for two systems with different stacking orders. This result shows that it is in principle possible to identify where the excited carriers are at each time, demonstrating the advantageous feature of TR-XPS.

Figure 8.4 shows a schematic illustration of a TR-XPS system, which was installed and operated on beamline BL07LSU [39] of SPring-8, a synchrotron radiation facility in Japan. The system consisted of a femtosecond laser system for pump pulses, a synchrotron radiation beamline to provide probe pulses, a time-of-flight type spectrometer for TR-XPS measurements, and a timing control unit. The most important part of the system is the timing control unit that provides the laser and synchrotron radiation pulses in a synchronous manner, and the time-resolved measurements were performed by controlling the time difference (delay time) between the laser and synchrotron radiation pulses reaching the sample surface. At SPring-8, a sine wave of 508.58 MHz from a master oscillator was used to regulate a radio frequency cavity of the storage ring, and the same sine wave was used as a reference signal and sent to the laser and the analyzer systems at BL07LSU. The time resolution is determined by the wider pulse width of the pump or probe light. For the TR-XPS system at BL07LSU, the width of the synchrotron radiation pulse was about 50 ps, which was larger than that of the femtosecond laser pulse (~35 fs). Thus, the time resolution was about 50 ps. More details about this measurement system are found elsewhere [40, 41].

## 8.4 Band Bending and Surface Photovoltage

### *8.4.1 Surface Photovoltage*

The light-induced temporal shift of the core levels observed in TR-XPS, as shown in Fig. 8.3, is caused by the change in the valence state of an element from neutral to partially charged states.TR-XPS can also be used to track photovoltage changes

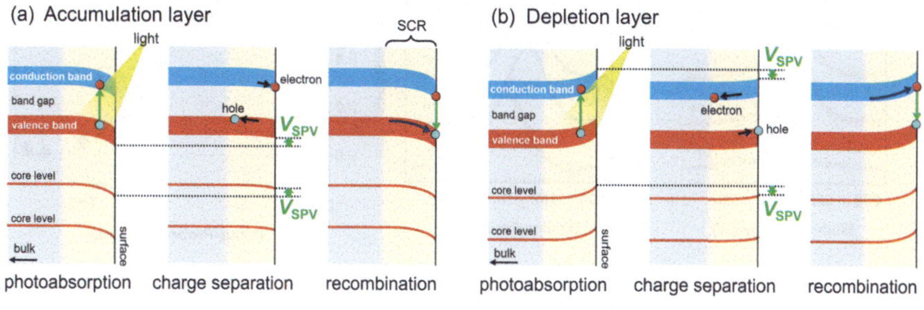


**Fig. 8.5** Surface photovoltage process for accumulation (**a**) and depletion (**b**) layers. The energy positions of the conduction and valence bands, as well as all core levels, change at the surface upon charge separation

at semiconductor surfaces or semiconductor heterojunctions. Surface photovoltage (SPV) occurs when light with an energy higher than the band-gap energy is irradiated onto semiconductor surfaces. A necessary condition for the SPV generation is that an electron accumulation layer or a depletion layer is formed on the semiconductor surface, namely, that downward or upward band bending is formed on the surface. Figure 8.5 shows how the SPV is generated and quenched on the two types of surfaces.

The region where band bending is formed is called a space-charge region (SCR). As photoabsorption occurs and weakly bound electron–hole pairs, i.e., excitons, are formed within the SCR, electrons and holes sense an electric field by the built-in potential of the SCR and experience forces in opposite directions along the potential gradient. This facilitates dissociation of excitons into electrons and holes. The electrons and holes are drifted toward the surface and the bottom of the SCR in the accumulation layer and vice versa in the depletion layer. Such charge separation always generates a counter-electric field that reduces the intrinsic built-in potential, thereby reducing the magnitude of initial band bending. The SPV corresponds to the magnitude of band bending that decreases with light irradiation $V_{SPV}$ (Fig. 8.5). Maximum $V_{SPV}$ is achieved when a flat band is formed and, thus, equals the magnitude of initial band bending.

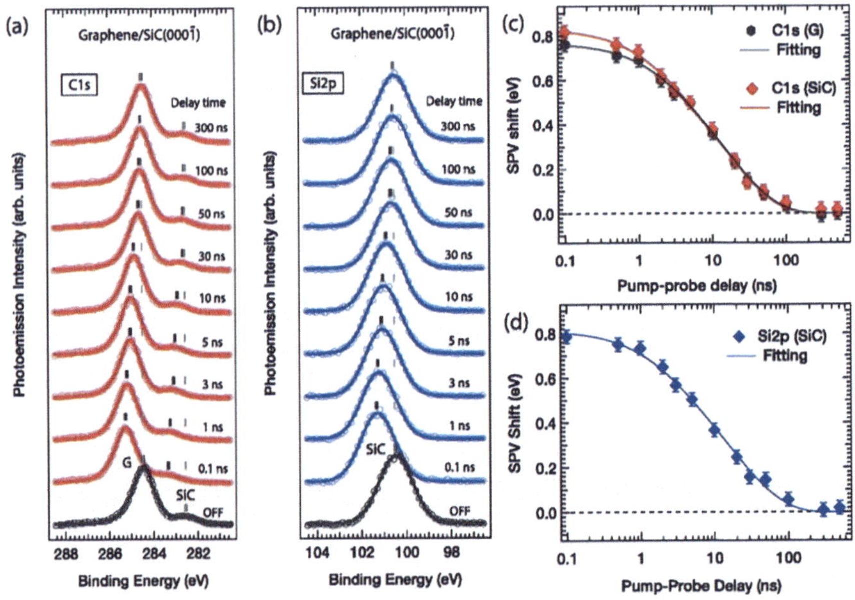


**Fig. 8.6** TR-XPS research on graphene/SiC$(000\bar{1})$ [42]. Time-resolved photoemission spectra for (**a**) C 1s and (**b**) Si 2p core levels. The SPV shift measured from the peak position of the dark condition was extracted and shown as a function of the delay time for (**c**) C 1s and (d) Si 2p core levels. (Reprinted from Someya et al. [42]. Copyright 2018 AIP Publishing)

When excitons are formed in a flat band region (e.g., deep in the bulk), fast electron–hole recombination should proceed since two charged particles are not significantly separated spatially. Even though excitons are dissociated, electrons and holes move randomly toward either the surface or bulk sides by thermal diffusion. Therefore, an additional electric field is not formed upon light illumination, and the flat-band structure is preserved.

TR-XPS allows tracking temporal changes of the SPV on semiconductor surfaces, because the valence states and the core levels of the semiconductor surface undergo an energy shift due to SPV. An example of the TR-XPS measurement is shown in Fig. 8.6. C 1s and Si 2p core levels of the graphene/SiC(000$\overline{1}$) surface shifted to the higher binding energies by ~0.8 eV, as the surface was illuminated by a 3.1-eV laser, and the magnitude of the shift was diminished with a time constant of 40–50 ns [42]. The core-level shift is attributed to the change of band bending due to the SPV, because the two curves of the C 1s and Si 2p energy shifts are identical. The fact that the SPV-induced shift is on the high binding energy side indicates that a depletion layer is formed on the graphene/SiC(000$\overline{1}$) surface. As in this example, even if we do not know the nature of the SCR, SPV measurements can reveal which type of band bending (upward or downward) is formed.

Kronik and Shapira have published a comprehensive review on the SPV [43]. The details of the theoretical and experimental aspects of the SPV are provided in their review.

### 8.4.2 *Band Bending Profile*

A band bending profile of the SCR determines a dissociation rate of photoexcited excitons and the behavior of electrons and holes within the SCR. Here, a calculation method for evaluating upward and downward band bending profiles is described. At a semiconductor/semiconductor heterojunction, the material is discontinuously connected to the other substance. A semiconductor surface is a type of an interface between the vacuum and the semiconductor. The potential structure inside the semiconductor is bent to some extent in a direction perpendicular to the interface under the influence of the electric field applied from the interface. If the crystal structure in the bulk is uniform, the band bending structure largely depends on the electric field from the interface, the density of carriers that screen the electric field, and the dielectric constant of the crystal.

Here, we present the simplest model to calculate the band bending structure. The depth $z$ variation of the potential energy $V(z)$ can be obtained by solving the following Poisson equation [44]:

$$\frac{\mathrm{d}^2V(z)}{\mathrm{d}z^2} = -\frac{e^2}{\varepsilon\varepsilon_0}\left[N_\mathrm{D} - N_\mathrm{A} - n(z) + p(z)\right], \tag{8.1}$$

where $e$ is the elementary charge, $\varepsilon$ and $\varepsilon_0$ are the relative and vacuum dielectric constants, respectively, $N_D$ is the bulk donor density, $N_A$ is the bulk acceptor density, $n(z)$ is the electron density in the conduction band, and $p(z)$ is the hole density in the valence band. The position $z = 0$ is set at the surface or the interface of the crystal. In the case of an n-type semiconductor with a sufficiently large band gap, the value $p(z) - N_A$ can be neglected. Therefore, Eq. (8.1) is simplified to

$$\frac{d^2V(z)}{dz^2} = -\frac{e^2}{\varepsilon\varepsilon_0}\left[N_D - n(z)\right]. \tag{8.2}$$

From here, simplified calculation methods are presented for each case where the bands bend upwardly (Fig. 8.7a) and downwardly (Fig. 8.7b). When the band bends upwardly to energies higher enough than the thermal excitation energy of electrons ($k_BT$; $k_B$ is Boltzmann constant and $T$ is the temperature of the material), the number

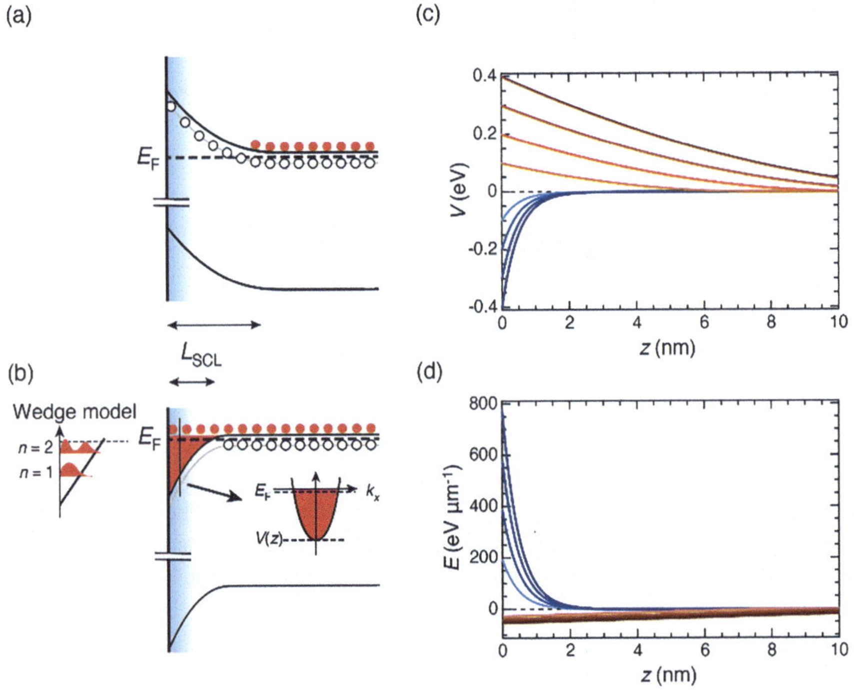


**Fig. 8.7** Band bending models for (**a**) upward bending and (**b**) downward bending. Red circles indicate the thermally excited electrons in the conduction band, and the red-colored area in panel (**b**) represents the energy region where the conduction band is filled with carrier electrons. (**c**) Band bending structures for the anatase $TiO_2$ surface. The magnitudes of band bending $V_s$ are 0.1, 0.2, 0.3, and 0.4 eV at the surface ($z = 0$) for both upward (blue) and downward (brown) band bending curves. $V = 0$ eV is the potential energy at the deep inside the bulk crystal. (**d**) Electronic fields, or potential gradients of band bending, are calculated from the curves in panel (**c**)

of carrier electrons in the conduction band near the surface becomes negligibly small, and the positively charged ionized donors primarily act to screen the electric field (Fig. 8.7a). In this case, Eq. (8.2) becomes

$$\frac{\mathrm{d}^2 V(z)}{\mathrm{d}z^2} = -\frac{e^2 N_\mathrm{D}}{\varepsilon\varepsilon_0}. \tag{8.3}$$

Since the right-hand side is constant, this differential equation is easily solved. Integrating twice and applying appropriate boundary conditions, one obtains a quadratic function against $z$:

$$V(z) = -\frac{e^2 N_\mathrm{D}}{2\varepsilon\varepsilon_0}\left(z - W_\mathrm{SCR}\right)^2, \tag{8.4}$$

where $W_\mathrm{SCR} = \left[2\varepsilon\varepsilon_0 V_\mathrm{s} / (e^2 N_\mathrm{D})\right]^{\frac{1}{2}}$ is the width of the SCR, and $V_\mathrm{s} = V(0)$ is the energy at the surface/interface. $V(z)$ is set to zero at $z = W_\mathrm{SCR}$.

When the band bends downward and a part of the conduction band goes below the Fermi level, carrier electrons accumulate in the conduction band. Since the density of accumulated electrons, $n(z)$, is typically much larger than $N_\mathrm{D}$, accumulated electrons are the prime charges that screen the electric field from the surface/interface. Thus, the contribution of $N_\mathrm{D}$ can be neglected, and Eq. (8.2) is simplified to

$$\frac{\mathrm{d}^2 V(z)}{\mathrm{d}z^2} = -\frac{e^2 n(z)}{\varepsilon\varepsilon_0}. \tag{8.5}$$

$n(z)$ is obtained by integrating the density of states of the conduction electrons in the energy range of conduction band minimum (CBM) to Fermi level $E_\mathrm{F}$:

$$n(z) = \frac{1}{3\pi^2}\left(\frac{\sqrt{2m^*}}{\hbar}\right)^3 \left|V(z)\right|^{\frac{3}{2}}. \tag{8.6}$$

Here, $\hbar$ is the reduced Planck constant, $m^*$ is the isotropic effective mass of conduction electrons, and $E_\mathrm{F}$ is set to zero. Moreover, a single parabolic band is assumed to consist of the conduction band bottom. Since Eq. (8.5) includes $V(z)$ on the right-hand side, the differential equation cannot be solved analytically. Generally, the potential curve $V(z)$ is obtained by solving the equation self-consistently with the boundary condition $V(0) = V_\mathrm{s}$.

In an actual material, due to the quantum confinement effect between the surface and the potential $V(z)$, electrons in the accumulation layer are quantized along the depth direction. To quantitatively evaluate this quantization, the wedge model, i.e., approximating the potential shape with a wedge, is used as the simplest and analytically solvable potential structure (inset of Fig. 8.7b) [45, 46]. Using the wedge model, $W_{SCR}$ can be simply expressed as [47].

$$W_{SCR} = \left( \frac{15\pi^2 \hbar^3 \varepsilon\varepsilon_0}{4e^2 \sqrt{m^{*3}} |V_s|} \right)^{\frac{1}{2}}. \tag{8.7}$$

The $W_{SCR}$ values of both depletion and accumulation layers increase proportionally with $\sqrt{\varepsilon}$ . Interestingly, although $W_{SCR}$ is proportional to $\sqrt{V_s}$ in the depletion layer, it becomes small as the function of $V_s^{-1/4}$ in the accumulation layer. The latter relation comes from the fact that larger $V_s$ leads to a steeper potential gradient.

Figures 8.7c, d show, respectively, calculated $V(z)$, i.e., the band bending profile, and the electric field $E(z) = dV(z)/dz$, i.e., the potential gradient, for the surface of anatase $TiO_2$. Here, the following parameters were used: $\varepsilon = 31$, $N_D = 0.6 \times 10^{19}$ $cm^{-3}$ (used to calculate the depletion-layer profile) [48], $m^* = 0.5m_e$ ($m_e$ is the mass of a free electron) [49]. Both $V(z)$ in the deep bulk side and $E_F$ were set to zero. The width of the SCR, $W_{SCR}$, is much shorter for the accumulation layer than the depletion layer. For example, $W_{SCR}$ are 15 and 2 nm for the accumulation and depletion layers, respectively, at $|V_s| = 0.4$ eV. This is because the charge density that screens the electric field is fixed at $N_D$ in the depletion layer, whereas the charge density becomes significantly higher near the surface region ($z < 1$ nm), reaching a value around $10^{21}$ $cm^{-3}$, in the accumulation layer. Consequently, the electric field, i.e., the potential gradient, moderately relaxes in the depletion layer but sharply in the accumulation layer (Fig. 8.7b).

## 8.5 Determination of Carrier Lifetime from SPV Decay Curves

When a semiconductor surface develops an SCR with a magnitude of band bending $V_s$ and is irradiated by light, the SPV $V_{SPV}$ is generated. $V_{SPV}$ is determined by initial band bending $V_s$ and a photogenerated carrier density $n_{ph}$ as [50].

$$\frac{V_{SPV}}{k_B T} \exp\left( \frac{V_{SPV}}{k_B T} \right) = \frac{n_{ph}}{n_0} \exp\left( \frac{V_s}{k_B T} \right), \tag{8.8}$$

where $n_0$ is the density of intrinsic carriers. $n_0$ equals a bulk donor density $N_D$ when the conduction band is not occupied. On the other hand, if the CBM crosses $E_F$ and the conduction band is partially occupied, $n_0$ is approximated by $n(z)$ given in Eq. (8.6). Under the condition that both $V_s$ and $V_{SPV}$ are sufficiently larger than $k_B T$, Eq. (8.8) is solved to give the following equation:

$$V_{\mathrm{SPV}} = \eta k_{\mathrm{B}} T ln\left(1+\gamma \frac{n_{\mathrm{ph}}}{n_0}\right), \tag{8.9}$$

where $\eta$ and $\gamma$ are constants.

The value of $V_{\mathrm{SPV}}$ is reduced with time because of decreasing $n_{\mathrm{ph}}$ due to electron–hole recombination. Therefore, these values are functions of time $t$, i.e., $V_{\mathrm{SPV}}(t)$ and $n_{\mathrm{ph}}(t)$. $V_{\mathrm{SPV}}(t)$ is maximized when the electrons and holes are separated to the both sides of the SCR (the surface and the bottom of the SCR). Then, electron–hole recombination proceeds via a thermionic emission process [51], which is characterized by the carriers at the bottom of the SCR overpassing the surface potential barrier (Fig. 8.5). A differential equation of $V_{\mathrm{SPV}}(t)$ is derived from Eq. (8.9):

$$\frac{\mathrm{d}V_{\mathrm{SPV}}(t)}{\mathrm{d}t} = \eta k_{\mathrm{B}} T \frac{\gamma \frac{\mathrm{d}n_{\mathrm{ph}}(t)}{\mathrm{d}t}}{n_0 + \gamma n_{\mathrm{ph}}(t)}.$$

$n_{\mathrm{ph}}(t)$ is given using the carrier lifetime $\tau$ as

$$\frac{\mathrm{d}n_{\mathrm{ph}}(t)}{\mathrm{d}t} = -\frac{n_{\mathrm{ph}}(t)}{\tau}, \tag{8.10}$$

and $\tau$ is given as follows [52]:

$$\tau = \tau_{\infty} \exp\left[-\frac{V_{\mathrm{SPV}}(t)}{\eta k_{\mathrm{B}} T}\right], \tag{8.11}$$

where $\tau_{\infty}$ is the carrier lifetime at $V_{\mathrm{SPV}}(t \to \infty) = 0$. Using Eqs. (8.9), (8.10), and (8.11), Eq. (8.8) can be rewritten as

$$\frac{\mathrm{d}V_{\mathrm{SPV}}(t)}{\mathrm{d}t} = -\frac{\eta k_{\mathrm{B}} T}{\tau_{\infty}}\left[\exp\left(\frac{V_{\mathrm{SPV}}}{\eta k_{\mathrm{B}} T}\right) - 1\right].$$

This differential equation can be solved analytically, and the expression of $V_{\mathrm{SPV}}(t)$ is obtained:

$$V_{\mathrm{SPV}}(t) = -\eta k_{\mathrm{B}} T \ln\left\{1 - \left[1 - \exp\left(-\frac{V_0}{\eta k_{\mathrm{B}} T}\right)\right]\exp\left(-\frac{t}{\tau_{\infty}}\right)\right\}. \tag{8.12}$$

Here, $V_0$ is a SPV at $t = 0$ or the maximum SPV.

Figure 8.8 shows $V_{\mathrm{SPV}}$ curves as a function of $t$. $\tau_{\infty}$ is the parameter that determines how quickly or slowly the SPV decreases and is an experimentally determinable carrier lifetime. The constant $\eta$ is called an ideality factor [51], which is an

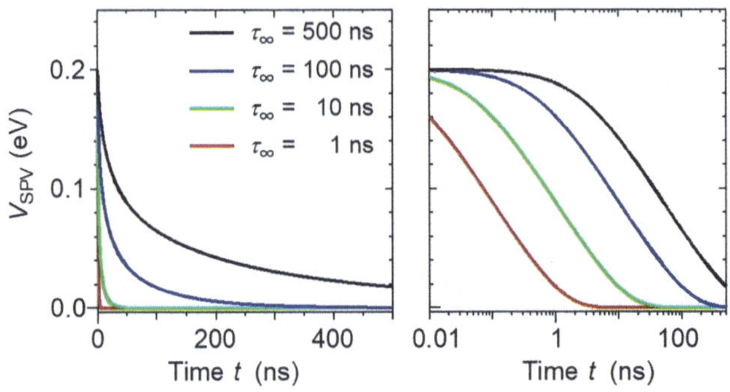


**Fig. 8.8** Change of $V_{SPV}$ as a function of time $t$ in linear and log scales, calculated by Eq. (8.12). Parameters are $T = 300$ K, $V_0 = 0.2$ eV, $\eta = 1.5$, and $\tau_\infty = 1$, 10, 100, and 500 ns

indicator of the rate of electron–hole recombination before the carriers overpass the surface potential barrier and typically has a value between 1 and 2.

## 8.6 A Case Study of $TiO_2$ Photocatalyst

### *8.6.1 Carrier Lifetime*

TR-XPS measurements of $TiO_2$ photocatalysts were carried out at BL07LSU of SPring-8 [39]. The $TiO_2$ photocatalysts were anatase and rutile $TiO_2$ single crystals with well-defined surface orientations, i.e., anatase $TiO_2$(101), anatase $TiO_2$(001), rutile $TiO_2$(001), rutile $TiO_2$(100), rutile $TiO_2$(110), and rutile $TiO_2$(011). The surfaces were cleaned by cycles of $Ar^+$ sputtering (1–2 kV, $1 \times 10^{-4}$ Pa Ar, 10 min) and annealing in an $O_2$ atmosphere [900 K, (1–2.5) $\times 10^{-4}$ Pa, 10 min] until no carbon contaminants were observed by XPS and sharp low-energy electron diffraction (LEED) spots were observed (Fig. 8.9) [14].

For the pump–probe TR-XPS measurements, a second harmonic of an amplifier Ti:sapphire pulsed laser was used for the pump light. The photon energy was 3.06 eV (405 nm), and the pulse duration was 35 fs with a repetition rate of 1 kHz. For the probe light, 600-eV synchrotron radiation was used to measure the Ti 2p core-level peak. The pulse duration was about 50 ps, and this determined the time resolution of the TR-XPS measurements. Photoelectron energies were analyzed by a time-of-flight electron spectrometer (ARTOF 10 k, VG Scienta) [40, 41].

Figure 8.10a shows a series of Ti $2p_{3/2}$ peaks of anatase $TiO_2$(001) and rutile $TiO_2$(110) as a function of the time difference between the pump and probe light, i.e., a delay time. The peak moves to the lower binding energy side at the delay time $t = 0.2$ ns by 0.22 eV on anatase $TiO_2$(001) and 0.13 eV on rutile $TiO_2$(110). The shift is caused by the SPV. The magnitude of the shift, which corresponds to $V_{SPV}$, is diminished with $t$ and goes back to the initial position at $t = 100$ ns. The direction of

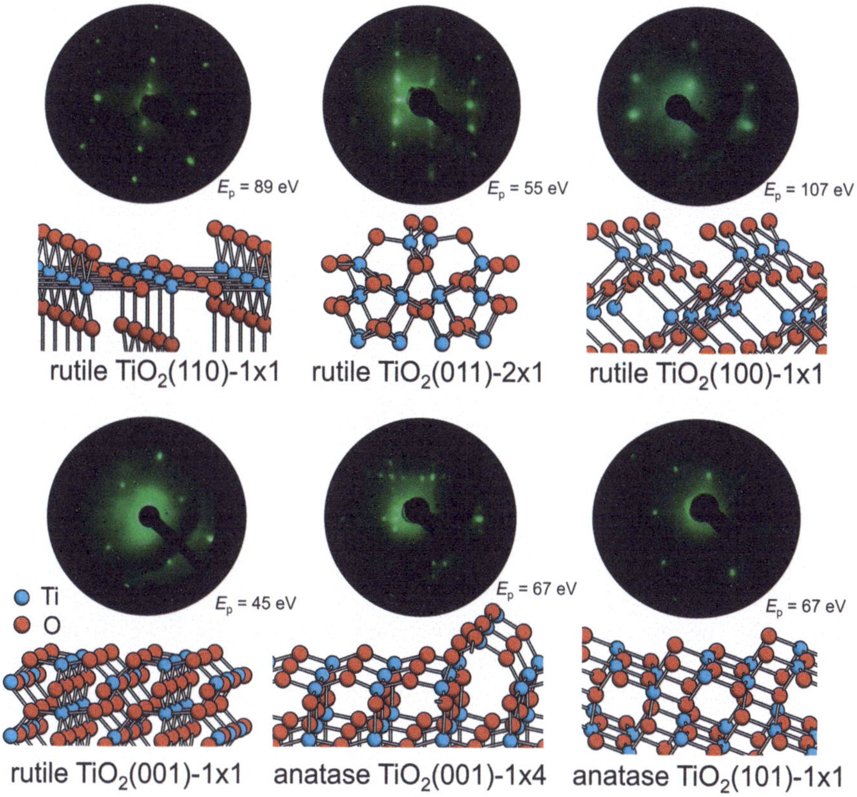


**Fig. 8.9** LEED patterns and structural models of $TiO_2$ surfaces. $E_p$ is the primary energy of the electron beam

the peak shift, i.e., the lower binding energy side, suggests that accumulation layers were formed on both surfaces (Fig. 8.5). This is a reasonable result, because the $TiO_2$ surfaces cleaned in the ultrahigh vacuum usually develop an accumulation layer caused by unavoidable surface oxygen vacancies.

Figure 8.10b shows a delay-time-dependent change of $V_{SPV}$ obtained from anatase $TiO_2$(001), rutile $TiO_2$(110), rutile $TiO_2$(011), and sputtered rutile $TiO_2$(110). Although there are only four data points, the SPV on rutile $TiO_2$(011) is diminished quickly within 1 ns after the photoirradiation. On the other hand, the SPV survives for longer times on the pristine and sputtered rutile $TiO_2$(110) surfaces. The SPV on anatase $TiO_2$(001) is somewhere in between, although the largest initial $V_{SPV}$ of 0.22 eV is achieved. Solid lines are results of least-squares fitting of the plotted data with Eq. (8.12). In the fitting procedure, $\tau_\infty$ and $V_0$ were treated as variables, while $\eta$ and $T$ were fixed parameters ($\eta = 1.4$ and $T = 300$ K). Please see Ref. 53 for the choice of the $\eta$ value. The obtained $\tau_\infty$ values are 90, 150, 1.3, and 230 ns for anatase $TiO_2$(001), rutile $TiO_2$(110), rutile $TiO_2$(011), and sputtered rutile $TiO_2$(110),

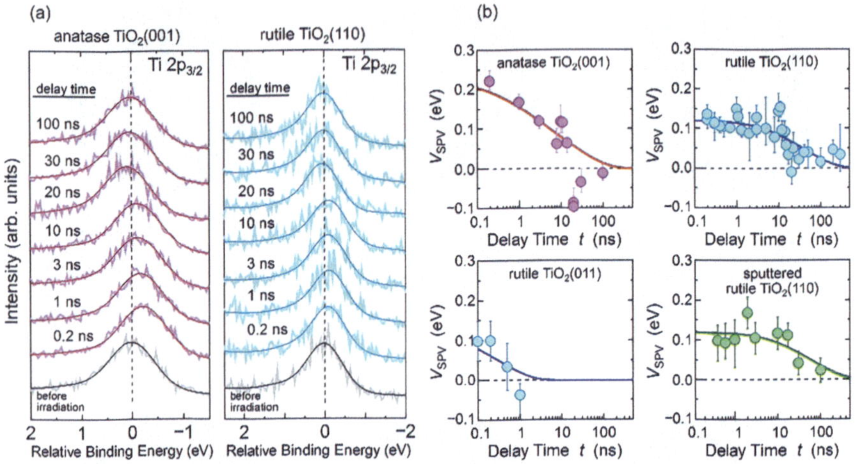


**Fig. 8.10** TR-XPS research on $TiO_2$ surfaces. (**a**) Ti $2p_{3/2}$ spectra of anatase $TiO_2(001)$ and rutile $TiO_2(110)$ at delay times between 0.2 and 100 ns. The pump and probe light energies were 3.06 and 600 eV, respectively. (**b**) Change of $V_{SPV}$ of four $TiO_2$ surfaces as a function of the delay time

respectively. The carrier lifetime is the longest on sputtered rutile $TiO_2(110)$ and the shortest on rutile $TiO_2(011)$. This result gives an important message: the lifetime of photogenerated carriers depends on surface orientation [(110) versus (011)] and the surface condition (pristine versus sputtered) rather than the crystal structure (anatase versus rutile).

Many factors affect $\tau_\infty$ [53]: initial band bending $V_s$, a capture cross section of the carriers, a thermal velocity of the carriers, a density of carrier traps, etc. Among them, $V_s$ is the most influential factor to determine $\tau_\infty$ through the following equation [53]:

$$\tau_\infty = \frac{W_{SCR}}{S_p} \exp\left(\frac{V_s}{\eta k_B T}\right). \qquad (8.13)$$

Here, $S_p$ is the surface recombination velocity. When the surface condition is changed and $V_s$ is changed, the value of $\tau_\infty$ is altered significantly. Changes of $W_{SCR}$ and $S_p$ have less effect on $\tau_\infty$ compared to $V_s$. Therefore, the error will not be very large even if we only focus on the change in $V_s$ and ignore other parameters to discuss the change in $\tau_\infty$.

Figure 8.11 shows valence-band XPS spectra of four $TiO_2$ surfaces. A comparison is made between the clean surface and acetic-acid-saturated surfaces. The valence band maximum (VBM) position moves toward the higher binding energy side by acetic-acid adsorption on the $TiO_2$ surfaces, except for sputtered rutile $TiO_2(110)$. Considering that the accumulation layers are formed on the clean $TiO_2$ surfaces, $V_s$ is enlarged on three surfaces, whereas it is slightly diminished on sputtered rutile $TiO_2(110)$. The changes in $V_s$ are 0.05, 0.04, 0.06, and −0.05 eV for

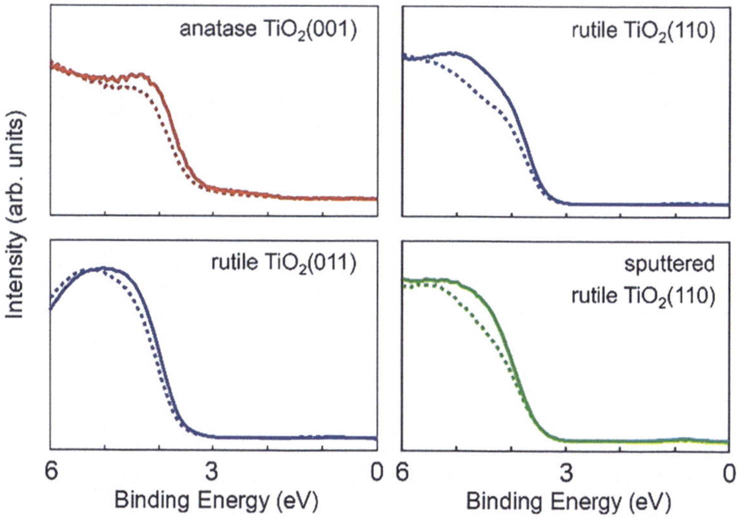


**Fig. 8.11** Valence band spectra of the clean (solid lines) and acetic-acid-covered (dashed lines) $TiO_2$ surfaces. The photon energy was 100 eV

anatase $TiO_2(001)$, rutile $TiO_2(110)$, rutile $TiO_2(011)$, and sputtered rutile $TiO_2(110)$, respectively. Although we did not measure the carrier lifetime on acetic-acid-saturated surfaces ($\tau_{\infty,Ac}$), the values are estimated by Eq. (8.13). Assuming that $W_{SCR}$ and $S_p$ are the same before and after adsorption, $\tau_{\infty,Ac}$ on anatase $TiO_2(001)$, rutile $TiO_2(110)$, rutile $TiO_2(011)$, and sputtered rutile $TiO_2(110)$ are 360, 450, 7, and 60 ns, respectively. As this example shows, the carrier lifetime is sensitive to $V_s$. Therefore, it is necessary to compare the carrier lifetimes of different surfaces with the same $V_s$ value. One of the conclusions drawn from the TR-XPS study is that the carrier lifetime is longer for anatase $TiO_2(001)$ than for rutile $TiO_2(110)$ at the same $V_s$ value [54].

### 8.6.2 Correlation Between Photocatalytic Activity and Carrier Lifetime

The excellent photocatalytic activity of $TiO_2$ has given rise to many hypotheses to explain the phenomenon. The most likely and experimentally supported hypothesis is the length of the carrier lifetime [18, 26–28]; the longer the carrier lifetime, the higher the photocatalytic activity. This hypothesis has also been proved directly by a combination of TR-XPS and XPS measurements [55].

Figure 8.12a shows C 1s XPS spectra of acetic-acid-saturated $TiO_2$ surfaces. The characteristic two-peak structure is attributed to methyl carbon (the low-binding-energy component) and carboxylate carbon (the high-binding-energy component).

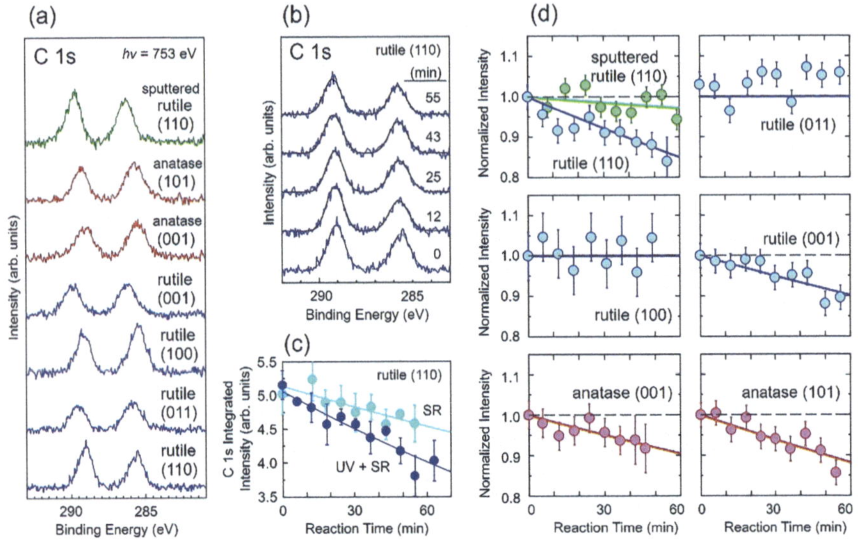


**Fig. 8.12** (**a**) C 1s spectra of acetic-acid-covered $TiO_2$ surfaces, acquired with a 753-eV photon energy. (**b**) Change of the C 1s spectrum of acetic-acid-covered rutile $TiO_2$(110) as a function of the irradiation time of the UV laser (3.31 eV). (**c**) Plots of the C 1s intensities against the reaction time (the UV irradiation time). The intensity decreases even without UV irradiation (the plots indicated by SR). The difference in the intensity decrease is the effect of UV. (**d**) C 1s intensity decreases due to the UV irradiation effect. Intensities are normalized to 1.0 before UV irradiation, and only the UV-induced decrease is taken into account

A slight difference in the peak binding energies depending on the surface reflects the magnitude of band bending. Acetic acid adsorbs on the $TiO_2$ surface either molecularly or dissociatively, but it was difficult to distinguish between these two forms from our XPS data.

To evaluate the photocatalytic activity of the $TiO_2$ surfaces, we monitored the decrease in the C 1s peak intensity under irradiation of an ultraviolet (UV) laser. The measurements were carried out at BL-13B of The Photon Factory, a synchrotron radiation facility [56], using a concentric hemispherical electron energy analyzer (SES200, Gamma Data/Scienta). A continuous-wave UV laser with a wavelength of 375 nm (3.31 eV) was irradiated on the acetic-acid-saturated $TiO_2$ surfaces while measuring the C 1s XPS spectra. The C 1s peak intensity decreased slightly with time as shown in Fig. 8.12b, which shows an example of a series of C 1s spectra as a function of the reaction time. The peak intensity decrease was observed, even when the UV laser was not irradiated, meaning that the synchrotron radiation induced decomposition and desorption of adsorbed acetic acid. However, UV laser irradiation enhanced the decreasing rate (Fig. 8.12c). Thus, the effect of the photocatalytic reaction corresponds to the difference in the C 1s intensity decrease with and without UV laser irradiation. A summary of the UV-induced decreases in the C 1s peak intensity on seven $TiO_2$ surfaces is shown in Fig. 8.12d. A large decrease was observed on the rutile $TiO_2$(011) and (001) surfaces and the anatase $TiO_2$(101) and (001) surfaces, whereas the rutile $TiO_2$(100) and (011) surfaces are found to be insensitive to the UV laser. Although photocatalytic reactions were not examined in our study, the following reaction scheme is assumed [11]:

$CH_3COOH \rightarrow CH_3COO^-(a) + H^+(a)$ (or $CH_3COOH(a)$).
$TiO_2 + h\nu \rightarrow h^+ + e^-$,
$CH_3COO^-(a) + h^+ \rightarrow CH_3(a) + CO_2\uparrow$.
(or $CH_3COOH(a) + h^+ \rightarrow CH_3(a) + H^+(a) + CO_2\uparrow$),
$CH_3(a) + H^+(a) + e^- \rightarrow CH_4\uparrow$.

Here, (a) means adsorbed species and $h\nu$ stands for a photon. This reaction scheme is what would occur in the absence of $O_2$ or $H_2O$.

Solid lines in Fig. 8.12d are results of least-squares fit of the experimental data with a first-order reaction equation: $I = \exp(-k_{ph}t)$, where $I$ and $k_{ph}$ are the normalized C 1s peak intensity and the reaction rate constant, respectively. A faster decrease in the C 1s intensity gives a larger $k_{ph}$ value and means higher photocatalytic activity. The evaluated $k_{ph}$ values on seven surfaces are summarized in Table 8.1 together with coverages of acetic acid and the calculated carrier lifetimes on the acetic-acid-saturated surfaces. There is no clear correlation between $k_{ph}$ and the coverage (1 monolayer (ML) corresponds to $2.6 \times 10^{14}$ cm$^{-2}$ [55]), indicating that the number of adsorbed species does not contribute to the activity. On the other hand, a linear and positive correlation is easily recognized between $k_{ph}$ and $\tau_{\infty,Ac}$ (Fig. 8.13). This is strong evidence that the carrier lifetime is a prime factor in determining photocatalytic activity. As the free carriers live longer after their generation upon photoadsorption, they have more chance to encounter adsorbed species on the surface. Therefore, the carrier lifetime is a good indicator to describe the surface dependence

**Table 8.1** Rate constant $k_{ph}$, acetic-acid coverage, and carrier lifetime on acetic-acid-covered surfaces $\tau_{\infty,Ac}$

| Surface | $k_{ph}$ (min$^{-1}$) | Coverage (ML) | $\tau_{\infty,Ac}$ (ns) |
|---|---|---|---|
| Rutile (110) | $2.7 \times 10^{-3}$ | 0.50 | 450 |
| Anatase (101) | $2.1 \times 10^{-3}$ | 0.56 | – |
| Rutile (001) | $1.7 \times 10^{-3}$ | 0.45 | – |
| Anatase (001) | $1.7 \times 10^{-3}$ | 0.54 | 360 |
| Sputtered rutile (110) | $0.5 \times 10^{-3}$ | 0.66 | 60 |
| Rutile (011) | 0 | 0.41 | 7 |
| Rutile (100) | 0 | 0.63 | – |

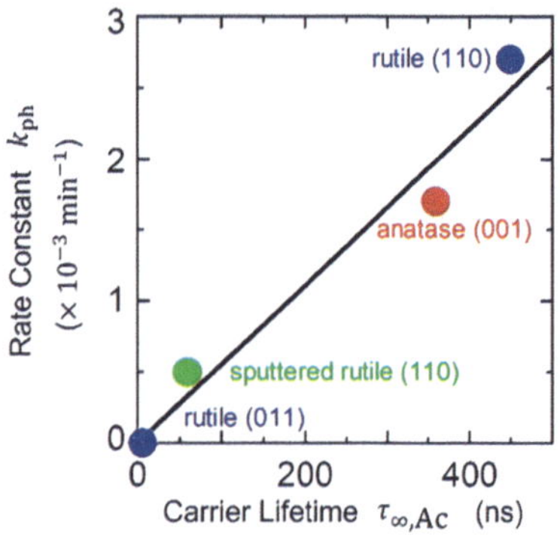


**Fig. 8.13** Relation between the rate constant of photocatalytic degradation and desorption of acetic acid and the lifetime of photoexcited carriers on four $TiO_2$ surfaces

of photocatalytic activity. The result also suggests that the activity can be controlled by changing the surface potential to extend the carrier lifetime.

## 8.7 Photocatalytic Activity at Rutile/Anatase Boundary on $TiO_2$ Surfaces

### *8.7.1 Band Alignment at the Boundary*

The synergistic effect of the $TiO_2$ photocatalysts [29–32] is considered to result from the boundary between the rutile and anatase phases [33, 34], as described in Sect. 8.3. It is proposed that a discontinuous energy band alignment is formed at the boundary, reflecting the difference in the binding energy of electronic levels between these two phases. From the XPS measurements of $TiO_2$ nanoparticles with rutile/anatase heterojunctions [33], the energy levels of the rutile phase have been determined to be lower (in binding energy) than those of the anatase phase by 0.39 eV at the VBM. The energy difference between the two phases has also been reproduced by theoretical calculations [33, 34]. Thus, a potential gradient should exist at the rutile/anatase boundary, which promotes exciton dissociation and carrier transport, eventually leading to enhanced photocatalytic activity. However, it has not been directly confirmed experimentally how steep the potential gradient is at the boundary. Furthermore, photocatalytic activity of the rutile/anatase boundary alone has not been selectively determined experimentally. XPS using microfocused beams allows selective measurements to determine the energy levels and the photocatalytic activity at the boundary [57].

Figure 8.14a shows a two-dimensional (2D) distribution of the anatase and rutile phases on the $TiO_2$ surface. The rutile/anatase boundary was prepared by heating a part of an anatase $TiO_2(001)$ surface using a high-power green laser (527 nm, 10 W) and by inducing an anatase-to-rutile phase transition [57]. The 2D mapping was made by utilizing a characteristic shape of the $e_g$ peak in the Ti $L_3$ edge X-ray absorption spectroscopy (XAS) spectra (Fig. 8.14b). The ratio of the absorption

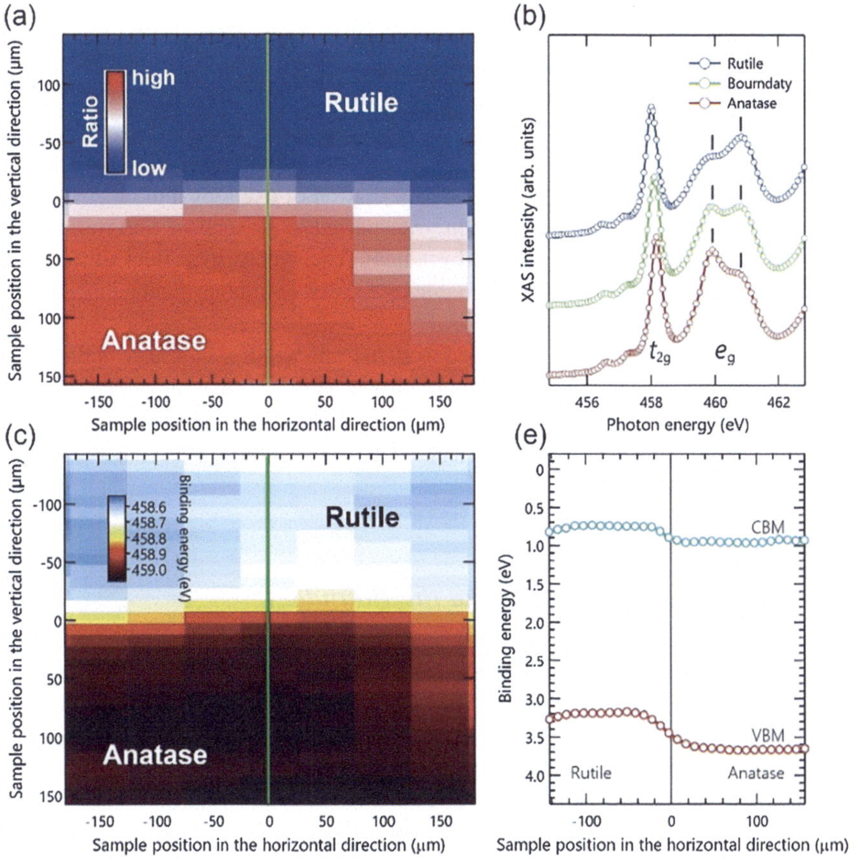


**Fig. 8.14** (**a**) 2D distribution of the rutile and anatase phases. The ratio of the 460- and 461-eV intensities of the Ti $L_3$ $e_g$ XAS peak is used to distinguish between the different phases. (**b**) Ti $L_3$ $e_g$ XAS spectra of rutile, anatase, and rutile/anatase boundary regions. (**c**) 2D distribution of the rutile and anatase phases mapped by the Ti $2p_{3/2}$ XPS peak energy. (**d**) CBM and VBM profiles along the rutile/anatase boundary (green lines in panels (**a**) and (**c**)), deduced from the XPS and XAS measurements. The energy levels are discontinuously connected macroscopically but are continuous at the micrometer order

intensity (height) at 460 and 461 eV, i.e., $R_{eg} = I_{460\ eV}/I_{461\ eV}$, depends on the crystal structure, with $R_{eg} > 1$ for the anatase phase and $R_{eg} < 1$ for the rutile phase [58]. The same phase distribution image was obtained (Fig. 8.14c) when the Ti $2p_{3/2}$ core-level binding energy was used; the Ti $2p_{3/2}$ XPS peak is located at 458.8 eV on the rutile surface, and it shifts to 459.1 eV on the anatase phase. A similar binding energy shift is also observed in the VBM, where the VBM binding energies are 3.2 and 3.65 eV on the rutile and anatase surfaces, respectively [57].

The XAS and XPS results indicate that the binding energy shift is caused by band bending and that the crystalline phase and the electron binding energy are correlated with each other. Therefore, a potential gradient should be formed at the rutile/anatase boundary. Figure 8.14d shows VBM and CBM profiles across the boundary, derived from the VBM position of the valence band XPS spectra (for the VBM) and the Ti $2p_{3/2}$ XPS peak position and the edge position of the Ti $L_3$ XAS spectra (for the CBM). The energy levels of the rutile and anatase phases are smoothly connected at the boundary, but the band alignment is macroscopically discontinuous, and this should facilitate dissociation of excitons into electrons and holes. The electrons and holes are, then, transported to the anatase and rutile sides from the boundary along the potential gradient. The synergistic effect of the $TiO_2$ photocatalysts [29–32] is, thus, explainable by the discontinuous band connection at the rutile/anatase boundary.

The width of the boundary, estimated from the full-width at half maximum of the first derivatives of the band alignment curves (Fig. 8.14d), is about 30 μm. Considering the spatial resolution of 11 μm, the footprint size in the vertical direction, the boundary width is estimated to be 28 μm. The value is, however, larger than the actual width, as discussed below.

## 8.7.2 *Photocatalytic Activity at the Boundary*

The $TiO_2$ sample with the rutile/anatase boundary on the surface is an excellent platform to verify the synergistic effect. That is, one can examine whether the boundary has higher photocatalytic activity than the anatase and rutile surfaces. By employing the same method explained in Sect. 8.6.2, i.e., measuring the C 1s XPS intensity decrease caused by UV-induced decomposition and desorption of adsorbed acetic acid, photocatalytic activity has been compared among three regions. The microfocused beam with a beam footprint of 11 μm (vertical) × 100 μm (horizontal) allowed to assess the photocatalytic activity at the rutile/anatase boundary

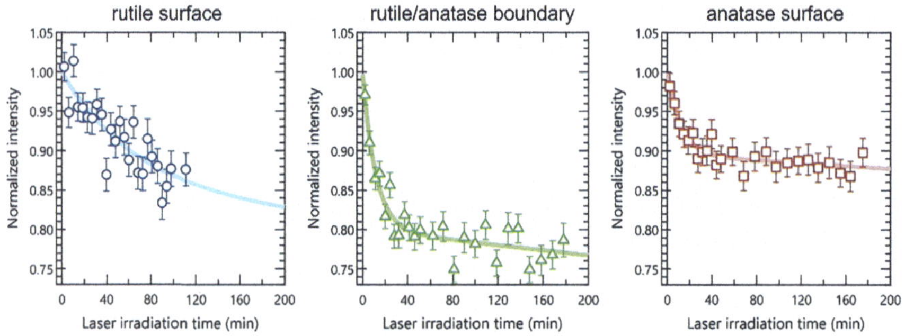


**Fig. 8.15** C 1s XPS intensity decreases due to the UV irradiation effect. An acetic-acid-covered $TiO_2$ surface having a rutile/anatase boundary was used to evaluate position-dependent photocatalytic activity. The C 1s intensities are normalized to 1.0 before UV irradiation, and only the UV-induced decrease is taken into account

exclusively. Figure 8.15 shows that the C 1s peak intensity decreases largely during the first 30 min on the anatase surface and at the rutile/anatase boundary, followed by a slower decrease at longer times. Such a two-step decrease is not obvious, but it is also present on the rutile surface. The origin of the two-step decrease is not understood at present, but two scenarios are envisaged; one is that there are two reactions to produce $CH_4$ and $CO_2$ [11] and $C_2H_6$ [12, 59, 60] as reaction products, and the other is that monodentate and bidentate forms of acetic acid (molecularly adsorbed species and acetate, respectively) involve the reaction. Further studies are necessary to understand this interesting behavior. The absence of the two-step decrease in Fig. 8.12d is due to the small number of measurement points and because only the first 60 min were examined.

The C 1s intensity decreases by ~20% during the first 1 h at the rutile/anatase boundary, but the decreasing amounts are only ~10% on both rutile and anatase surfaces. In this period, the fast reaction mainly contributes to the removal of the adsorbates from the surface. Under the assumption that decomposition and desorption of adsorbed species proceed in a first-order reaction, the two-step decrease is reproduced by a two-component exponential equation; $I = m \exp(-k_1 t) + (1 - m) \exp(-k_2 t)$, where $I$, $m$, $t$, and $k$ are the C 1s intensity, the ratio of the exponential functions, the reaction time (the UV irradiation time), and the rate constants of the fast ($k_1$) and slow ($k_2$) reactions, respectively. Solid lines in Fig. 8.15 are the results of least-squares fit of the measured data using the two-component exponential equation. The values of $k_1$ are $8.6 \times 10^{-2}$ min$^{-1}$ at the rutile/anatase boundary and $8.3 \times 10^{-2}$ and $1.6 \times 10^{-2}$ min$^{-1}$ on the anatase and rutile surfaces, respectively [58]. Higher photocatalytic activity at the boundary compared to the single-phase surfaces is directly evidenced by this analysis. Moreover, the $k_1$ values of the single-phase surfaces are consistent with a generally accepted view that anatase $TiO_2$ is more active than rutile $TiO_2$ [17–20].

### 8.7.3 Carrier Behavior at the Boundary

The origin of the high photocatalytic activity at the rutile/anatase boundary is highly attributed to the potential gradient at the boundary, which should promote dissociation of excitons and transportation of electrons and holes away from the boundary. From the width of the boundary (28 μm) and the energy differences between the rutile and anatase phases (0.45 and 0.2 eV for the VBM and CBM, respectively), the average potential gradients at the boundary are 0.016 and 0.007 eV μm$^{-1}$ in the valence and conduction bands, respectively. However, these values should be underestimated. This is because the rutile/anatase boundary was not a straight line, but curved and winding, as confirmed by Raman spectroscopy measurements [58] and X-ray photoemission electron microscopy (X-PEEM) measurements. Figure 8.16a shows a 2D distribution of the rutile and anatase phases deduced from the peak position of the Ti $L_3$ $t_{2g}$ XAS edge. Similar to the Ti $2p_{3/2}$ XPS peak (Fig. 8.14c), the $t_{2g}$ peak position depends on the crystalline structure with lower and higher photon

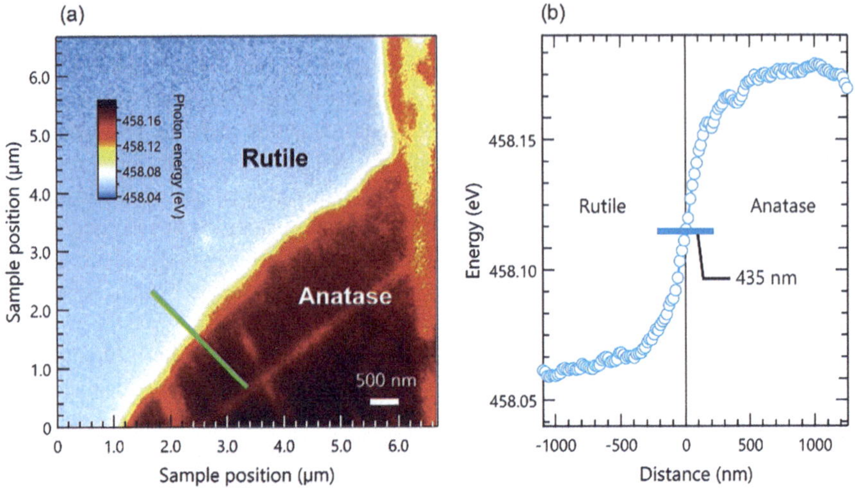


**Fig. 8.16** Results of the X-PEEM measurements. (**a**) 2D distribution of the rutile and anatase phases mapped by the Ti $L_3$ $t_{2g}$ XAS peak energy. (**b**) Line profile of the $t_{2g}$ peak energy along the rutile/anatase boundary indicated by a green line in panel (**a**)

energies for the rutile and anatase phases, respectively. It is obvious that the rutile/anatase boundary is curved and winding. Therefore, the microfocused beam used for the XPS measurements was too large to accurately evaluate the boundary width. X-PEEM with a high spatial resolution is a more suitable measurement method.

Figure 8.16b shows a $t_{2g}$ peak position as a function of the distance from the boundary. A step-like structure is observed, and a full-width at half maximum of the first derivative of the step-like curve is 0.44 μm. It is unfortunate that the spatial resolution of X-PEEM under the measurement condition employed was not evaluated. Thus, 0.44 μm is the maximum possible boundary width, and the actual width should be much narrower.

Taking 0.44 μm as a boundary width, the potential gradients are calculated to be 1.0 and 0.45 eV $\mu m^{-1}$ in the valence and conduction bands, respectively. These values are, however, smaller than the expected potential gradient in the accumulation layer (Fig. 8.7d); the potential gradient is 389 eV $\mu m^{-1}$ at the surface when 0.2-eV downward bending is formed. Therefore, the electrons and holes experience a strong electric field in the direction perpendicular to the surface and a weak electric field in the direction parallel to the surface. Photogenerated electrons and holes move in this complex electric field as follows; the electrons and holes are transported to the surface and to the bulk sides, respectively, along the downward band bending, and these carriers are further transported to the anatase and rutile phases, respectively, along the potential gradient at the rutile/anatase boundary. Figure 8.17 shows a schematic diagram of carrier behavior. Our result suggests that even weak electric fields below 1 eV $\mu m^{-1}$ are effective in charge separation and carrier lifetime elongation. This is an important message that photocatalytic activity can be enhanced by

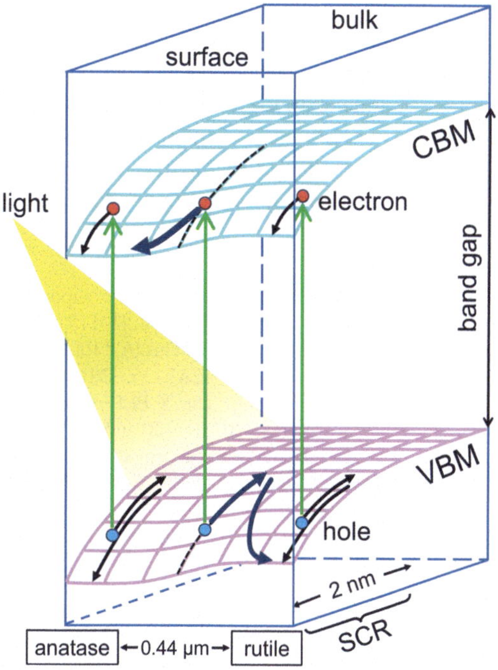


**Fig. 8.17** Schematic band structure of the $TiO_2$ surface near the rutile/anatase boundary and behavior of the photoexcited electrons and holes

introducing a weak potential gradient onto the photocatalyst surface, such as by forming heterojunctions or adsorbing charged particles.

## 8.8 Conclusion

Photocatalytic activity was evaluated by XPS and photogenerated carrier lifetimes were measured by pump–probe TR-XPS. Single-crystal rutile and anatase $TiO_2$ with well-defined surfaces were investigated. The carrier lifetime was determined by measuring the spontaneous change of the electron binding energies, which corresponded to generation and quenching of the surface photovoltage. The 3.06-eV pulsed laser was used as a pump light and synchrotron radiation was used as a probe light. Determined lifetimes were $\tau_\infty = 90$, 150, 1.3, and 230 ns on anatase $TiO_2(001)$, rutile $TiO_2(110)$, rutile $TiO_2(011)$, and sputtered rutile $TiO_2(110)$, respectively. Photocatalytic activity was evaluated by the rate of decrease in the C 1s XPS peak intensity due to photocatalytic decomposition and desorption of adsorbed acetic acid molecules. The C 1s spectra were measured using synchrotron radiation, while a continuous UV laser with the energy of 3.31 eV was irradiated to induce photocatalytic reaction. The reaction rate constant $k_{ph}$ varied between 0 and $2.7 \times 10^{-3}$ min$^{-1}$ depending on the surfaces. There is a linear and positive correlation between the carrier lifetime and the photocatalytic activity; the longer the carrier lifetime, the higher the photocatalytic activity. Since the carrier lifetime is sensitive to the surface potential, photocatalytic activity can be tuned by changing the surface potential and controlling the carrier behavior. This was demonstrated by introducing a rutile/anatase boundary on the $TiO_2$ surface, where the band discontinuity is introduced. Microfocused XPS measurements revealed that the photocatalytic activity is higher at the boundary than in the anatase and rutile regions, implying that the potential gradient at the boundary should promote exciton dissociation and extend the carrier lifetime.

**Acknowledgments** This work was partly supported by JSPS KAKENHI Grant Numbers 16H03867, 16H06027, 21H01805, and 22H01157. The TR-XPS measurements at SPring-8 were performed using facilities of the Synchrotron Radiation Research Organization, The University of Tokyo (Proposal Numbers 2012A7426, 2012B7433, 2013A744, 2014A7463, 2015A7487, 2016A7503, and 2017A7533). The X-PEEM experiment was performed with the approval of SPring-8 (Prolosal Numbers 2023B1720 and 2019A1092). The XPS measurements at the Photon Factory were performed under the approval of the Photon Factory Advisory Committee (Proposal Numbers 2012S2-006, 2015S2-008, 2016G529, 2017G525, and 2021S2-003).

## References

1. Jacobsen AE (1949) Titanium dioxide pigments: correlation between photochemical reactivity and chalking. Ind Eng Chem 41(3):523–526. https://doi.org/10.1021/ie50471a018
2. Weyl WA, Förland T (1950) Photochemistry of rutile Ind Eng Chem 42(2):257–263. https://doi.org/10.1021/ie50482a019
3. Fujishima A, Honda K (1971) Electrochemical evidence for the mechanism of the primary stage of photosynthesis. Bull Chem Soc Jpn 44(4):1148–1150. https://doi.org/10.1246/bcsj.44.1148
4. Fujishima A, Honda K (1972) Electrochemical photolysis of water at the semiconductor electrode. Nature 238:37–38. https://doi.org/10.1038/238037a0
5. Li R, Li T, Zhou Q (2020) Impact of titanium dioxide ($TiO_2$) modification on its application to pollution treatment–a review. Catalysts 10(7):804-1-32. https://doi.org/10.3390/catal10070804
6. Bono N, Ponti F, Punta G, Candiani G (2021) Effect of UV irradiation and $TiO_2$-photocatalysis on airborne bacteria and viruses: an overview. Materials 14(5):1075-1-19. https://doi.org/10.3390/ma14051075
7. Banerjee S, Dionysiou DD, Pillai SC (2015) Self-cleaning applications of $TiO_2$ by photo-induced hydrophilicity and photocatalysis. Appl Catal B Environ 176-177:396–428. https://doi.org/10.1016/j.apcatb.2015.03.058
8. Parrino F, Palmisano L (eds) (2020) Titanium dioxide ($TiO_2$) and its applications, 1st edn. Elsevier, pp 311–479. https://doi.org/10.1016/C2019-0-01050-3
9. Hashimoto K, Irie H, Fujishima A (2005) $TiO_2$ photocatalysis: a historical overview and future prospects. Jpn J Appl Phys 44:8269–8285. https://doi.org/10.1143/JJAP.44.8269
10. Schneider J, Matsuoka M, Takeuchi M, Zhang J, Horiuchi Y, Anpo M, Bahnemann DW (2014) Understanding $TiO_2$ photocatalysis: mechanisms and materials. Chem Rev 114(19):9919–9986. https://doi.org/10.1021/cr5001892
11. Quah EL, Wilson JN, Idriss H (2010) Photoreaction of the rutile $TiO_2$(011) single-crystal surface: reaction with acetic acid. Langmuir 26(9):6411–6417. https://doi.org/10.1021/la9040985

12. Kraeutler B, Bard AJ (1977) Photoelectrosynthesis of ethane from acetate ion at an n-type titanium dioxide electrode. The photo-Kolbe reaction. J Am Chem Soc 99(23):7729–7731. https://doi.org/10.1021/ja00465a065
13. Hoffmann MR, Martin ST, Choi W, Bahnemann DW (1995) Environmental applications of semiconductor photocatalysis. Chem Rev 95(1):69–96. https://doi.org/10.1021/cr00033a004
14. Ozawa K, Yamamoto S, Mase K, Matsuda I (2019) A surface science approach to unveiling the $TiO_2$ photocatalytic mechanism: correlation between photocatalytic activity and carrier lifetime. e-J Surf Sci Nanotechnol 17:130–147. https://doi.org/10.1380/ejssnt.2019.130
15. Zhang Z, Yates JT Jr (2012) Band bending in semiconductors: chemical and physical consequences at surfaces and interfaces. Chem Rev 112(10):5520–5551. https://doi.org/10.1021/cr3000626
16. Ranade MR, Navrotsky A, Zhang HZ, Banfield JF, Elder SH, Zaban A, Borse PH, Kulkarni SK, Doran GS, Whitfield HJ (2002) Energetics of nanocrystalline $TiO_2$. Proc Natl Acad Sci USA 99(suppl. 2):6476–6481. https://doi.org/10.1073/pnas.251534898
17. Tanaka K, Capule MFV, Hisanaga T (1991) Effect of crystallinity of $TiO_2$ on its photocatalytic action. Chem Phys Lett 187(1–2):73–76. https://doi.org/10.1016/0009-2614(91)90486-S
18. Xu M, Gao Y, Moreno EM, Kunst M, Muhler M, Wang Y, Idriss H, Wöll C (2011) Photocatalytic activity of bulk $TiO_2$ anatase and rutile single crystals using infrared absorption spectroscopy. Phys Rev Lett 106:138302-1–138302-4. https://doi.org/10.1103/PhysRevLett.106.138302
19. Luttrell T, Halpegamage S, Tao J, Kramer A, Sutter E, Batzill M (2014) Whyis anatase a better photocatalyst than rutile?–model studies on epitaxial $TiO_2$ film. Sci Rep 4:4043-1-8. https://doi.org/10.1038/srep04043
20. Ahmed AY, Kandiel TA, Oekermann T, Bahnemann D (2011) Photocatalytic activities of different well-defined single crystal $TiO_2$ surfaces: anatase versus rutile. J Phys Chem Lett 2(19):2461–2465. https://doi.org/10.1021/jz201156b
21. Tang H, Lévy F, Berger H, Schmid PE (1995) Urbach tail of anatase $TiO_2$. Phys Rev B 52:7771–7774. https://doi.org/10.1103/PhysRevB.52.7771
22. Deskins AN, Dupuis M (2007) Electron transport via polaron hopping in bulk $TiO_2$: a density functional theory characterization. Phys Rev B 75:195212-1-10. https://doi.org/10.1103/PhysRevB.75.195212
23. Setvin M, Franchini C, Hao X, Schmid M, Janotti A, Kaltak M, Van de Walle CG, Kresse G, Diebold U (2014) Direct view at excess electrons in $TiO_2$ rutile and anatase. Phys Rev Lett 113:086402-1–086402-5. https://doi.org/10.1103/PhysRevLett.113.086402
24. Zhukov VP, Chulkov EV (2014) Ab initio calculations of the electron-phonon interaction and characteristics of large polarons in rutile and anatase. Phys Solid State 56:1302–1309. https://doi.org/10.1134/S1063783414070361
25. Lee H-S, Woo C-S, Youn B-K, Kim S-Y, Oh S-T, Sung Y-E, Lee H-I (2005) Bandgap modulation of $TiO_2$ and its effect on the activity in photocatalytic oxidation of 2-isopropyl-6-methyl-4-pyrimidinol. Top Catal 35:255–260. https://doi.org/10.1007/s11244-005-3832-2
26. Zhang J, Zhou P, Liu J, Yu J (2014) New understanding of the difference of photocatalytic activity among anatase, rutile and brookite $TiO_2$. Phys Chem Chem Phys 16:20382–20386. https://doi.org/10.1039/c4cp02201g
27. Yamada Y, Kanemitsu Y (2012) Determination of electron and hole lifetimes of rutile and anatase $TiO_2$ single crystals. Appl Phys Lett 101(13):133907-1-4. https://doi.org/10.1063/1.4754831
28. Yamakata A, Vequizo JJM, Matsunaga H (2015) Distinctive behavior of photogenerated electrons and holes in anatase and rutile $TiO_2$ powders. J Phys Chem C 119(43):24538–24545. https://doi.org/10.1021/acs.jpcc.5b09236
29. Sclafani A, Herrmann JM (1996) Comparison of the photoelectronic and photocatalytic activities of various anatase and rutile forms of titania in pure liquid organic phases and in aqueous solutions. J Phys Chem 100(32):13655–13661. https://doi.org/10.1021/jp9533584
30. Ohno T, Tokieda K, Higashida S, Matsumura M (2003) Synergism between rutile and anatase $TiO_2$ particles in photocatalytic oxidation of naphthalene. Appl Catal A General 244(2):383–391. https://doi.org/10.1016/S0926-860X(02)00610-5

31. Kho YK, Iwase A, Teoh WY, Mädler L, Kudo A, Amal R (2010) Photocatalytic $H_2$ evolution over $TiO_2$ nanoparticles. The synergistic effect of anatase and rutile. J Phys Chem C 114(6):2821–2829. https://doi.org/10.1021/jp910810r
32. Su R, Bechstein R, Sø L, Vang RT, Sillassen M, Esbjörnsson B, Palmqvist A, Besenbacher F (2011) How the anatase-to-rutile ratio influences the photoreactivity of $TiO_2$. J Phys Chem C 115(49):24287–24292. https://doi.org/10.1021/jp2086768
33. Deák P, Aradi B, Frauenheim T (2011) Band lineup and charge carrier separation in mixed rutile-anatase systems. J Phys Chem C 115(8):3443–3446. https://doi.org/10.1021/jp1115492
34. Scanlon DO, Dunnill CW, Buckeridge J, Shevlin SA, Logsdail AJ, Woodley SM, Catlow CRA, Powell MJ, Palgrave RG, Parkin IP, Watson GW, Keal TW, Sherwood P, Walsh A, Sokol AA (2013) Band alignment of rutile and anatase $TiO_2$. Nat Mater 12:798–801. https://doi.org/10.1038/nmat3697
35. Bauer M, Marienfeld A, Aeschlimann M (2015) Hot electron lifetimes in metals probed by time-resolved two-photon photoemission. Prog Surf Sci 90(3):319–376. https://doi.org/10.1016/j.progsurf.2015.05.001
36. Onda K, Li B, Petek H (2004) Two-photon photoemission spectroscopy of $TiO_2$(110) surfaces modified by defects and $O_2$ or $H_2O$ adsorbates. Phys Rev B 70:045415-1-11. https://doi.org/10.1103/PhysRevB.70.045415
37. Milne CJ, Penfold TJ, Chergui M (2014) Recent experimental and theoretical developments in time-resolved X-ray spectroscopies. Coordination Chem Rev 277-278:44–68. https://doi.org/10.1016/j.ccr.2014.02.013
38. Ozawa K, Yamamoto S, Miyazawa T, Yano K, Okudaira K, Mase K, Matsuda I (2021) Influence of stacking order of phthalocyanine and fullerene layers on the photoexcited carrier dynamics in model organic solar cell. J Phys Chem C 125(25):13963–13970. https://doi.org/10.1021/acs.jpcc.1c03584
39. Yamamoto S, Senba Y, Tanaka T, Ohashi H, Hirono T, Kimura H, Fujisawa M, Miyawaki J, Harasawa A, Seike T, Takahashi S, Nariyama N, Matsushita T, Takeuchi M, Ohata T, Furukawa Y, Takeshita K, Goto S, Harada Y, Shin S, Kitamura H, Kakizaki A, Oshima M, Matsuda I (2014) New soft X-ray beamline BL07LSU at SPring-8. J Synchrotron Radiat 21(2):352–365. https://doi.org/10.1107/S1600577513034796
40. Ogawa M, Yamamoto S, Kousa Y, Nakamura F, Yukawa R, Fukushima A, Harasawa A, Kondoh H, Tanaka Y, Kakizaki A, Matsuda I (2012) Development of soft X-ray time-resolved photoemission spectroscopy system with a two-dimensional angle-resolved time-of-flight analyzer at SPring-8 BL07LSU. Rev Sci Instrum 83:023109-1–023109-7. https://doi.org/10.1063/1.3687428
41. Yamamoto S, Matsuda I (2013) Time-resolved photoelectron spectroscopies using synchrotron radiation: past, present, and future. J Phys Soc Jpn 82(2):021003-1-18. https://doi.org/10.7566/JPSJ.82.021003
42. Someya T, Fukidome H, Endo N, Takahashi K, Yamamoto S, Matsuda I (2018) Interfacial carrier dynamics of graphene on SiC, traced by the full-range time-resolved core-level photoemission spectroscopy. Appl Phys Lett 113(5):051601-1-4. https://doi.org/10.1063/1.5043223
43. Kronik L, Shapira Y (1999) Surface photovoltage phenomena: theory, experiment, and applications. Surf Sci Rep 37(1–5):1–206. https://doi.org/10.1016/S0167-5729(99)00002-3
44. Mönch W (2001) Semiconductor surfaces and interfaces, 3rd edn. Springer, Berlin/Heidelberg, pp 21–31. https://doi.org/10.1007/978-3-662-04459-9
45. Lüth H (2015) Solid surfaces, interfaces and thin films, 6th edn. Springer, Cham, pp 337–391. https://doi.org/10.1007/978-3-319-10756-1
46. Yukawa R, Ozawa K, Yamamoto S, Liu R-Y, Matsuda I (2015) Anisotropic effective mass approximation model to calculate multiple subband structures at wide-gap semiconductor surfaces: application to accumulation layers of $SrTiO_3$ and ZnO. Surf Sci 641:224–230. https://doi.org/10.1016/j.susc.2015.07.007

47. Yukawa R, Yamamoto S, Akikubo K, Takeuchi K, Ozawa K, Kumigashira H, Matsuda I (2016) Tailoring photovoltage response at $SrRuO_3/SrTiO_3$ heterostructures. Adv Mater Interfaces 3(22):1600527-1-5. https://doi.org/10.1002/admi.201600527
48. Kavan L, Grätzel M, Gilbert SE, Klemenz C, Scheel HJ (1996) Electrochemical and photoelectrochemical investigation of single-crystal anatase. J Am Chem Soc 118(28):6716–6723. https://doi.org/10.1021/ja954172l
49. Rödel TC, Fortuna F, Bertran F, Gabay M, Rozenberg MJ, Santander-Syro AF, Le Fèvre P (2015) Engineering two-dimensional electron gases at the (001) and (101) surfaces of $TiO_2$ anatase using light. Phys Rev B 92:041106(R)-1–041106(R)-5. https://doi.org/10.1103/PhysRevB.92.041106
50. Bröcker D, Gießel T, Widdra W (2004) Charge carrier dynamics at the $SiO_2$/Si(100) surface: a time-resolved photoemission study with combined laser and synchrotron radiation. Chem Phys 299(2–3):247–251. https://doi.org/10.1016/j.chemphys.2003.11.028
51. Sah C-T, Noyce RN, Shockley W (1957) Carrier generation and recombination in p-n junctions and p-n junction characteristics. Proc IRE 45(9):1228–1243. https://doi.org/10.1109/JRPROC.1957.278528
52. Hamers RJ, Cahill DG (1991) Ultrafast time resolution in scanned probe microscopies: surface photovoltage on Si(111)-(7×7). J Vac Sci Technol B 9(2):514–518. https://doi.org/10.1116/1.585559
53. Ozawa K, Yamamoto S, Yukawa R, Liu R, Emori M, Inoue K, Higuchi T, Sakama H, Mase K, Matsuda I (2016) What determines the lifetime of photoexcited carriers on $TiO_2$ surfaces? J Phys Chem C 120(51):29283–29289. https://doi.org/10.1021/acs.jpcc.6b10136
54. Ozawa K, Emori M, Yamamoto S, Yukawa R, Yamamoto S, Hobara R, Fujikawa K, Sakama H, Matsuda I (2014) Electron-hole recombination time at $TiO_2$ single-crystal surfaces: influence of surface band bending. J Phys Chem Lett 5(11):1953–1957. https://doi.org/10.1021/jz500770c
55. Ozawa K, Yamamoto S, Yukawa R, Liu R-Y, Terashima N, Natsui Y, Kato H, Mase K, Matsuda I (2018) Correlation between photocatalytic activity and carrier lifetime: acetic acid on single-crystal surfaces of anatase and rutile $TiO_2$. J Phys Chem C 122(17):9562–9569. https://doi.org/10.1021/acs.jpcc.8b02259
56. Ozawa K, Aiura Y, Wakabayashi D, Tanaka H, Kikuchi T, Toyoshima A, Mase K (2022) Beamline commissioning for microscopic measurements with ultraviolet and soft X-ray beam at the upgraded beamline BL-13B of the photon factory. J Synchrotron Radiat 29(2):400–408. https://doi.org/10.1107/S160057752200090X
57. Hiromori K, Nakajima N, Hasegawa T, Wada S-i, Takahashi O, Ohkochi T, Mase K, Ozawa K (2024) Electronic origin of enhanced photocatalytic activity at the anatase/rutile boundary: a case of acetic acid on $TiO_2$ surface. J Phys Chem C 128(51):21767–21775. https://doi.org/10.1021/acs.jpcc.4c05630
58. Ruus R, Kikas A, Saar A, Ausmees A, Nõmmiste E, Aarik J, Aidla A, Uustare T, Martinson I (1997) Ti 2p and O 1s X-ray absorption of $TiO_2$ polymorphs. Solid State Commun 104(4):199–203. https://doi.org/10.1016/S0038-1098(97)00300-1
59. Henderson MA (2011) A surface science perspective on $TiO_2$ photocatalysis. Surf Sci Rep 66(6–7):185–297. https://doi.org/10.1016/j.surfrep.2011.01.001
60. Wilson JN, Idriss H (2003) Effect of surface reconstruction of $TiO_2$(001) single crystal on the photoreaction of acetic acid. J Catal 214(1):46–52. https://doi.org/10.1016/S0021-9517(02)00172-0

# Chapter 9
# Carrier Transfer from Photocatalyst to Cocatalyst

**Ke-Hsuan Wang, Masaaki Yoshida, and Chechia Hu**

**Abstract** Photocatalytic water splitting, the process of using light energy to split water molecules into hydrogen and oxygen, has garnered significant attention for its potential in clean energy production. This process offers a promising solution to the global energy crisis by utilizing solar energy to produce hydrogen fuel, a clean alternative to fossil fuels. However, the efficiency of photocatalytic water splitting heavily depends on the use of suitable photocatalysts, which are materials that absorb light and facilitate the necessary reactions. Cocatalysts play a crucial role in enhancing the performance of photocatalysts, by accelerating reaction kinetics, improving selectivity, and stabilizing the overall process. This chapter provides a comprehensive overview of the role of cocatalysts in photocatalytic water splitting. We discuss the mechanisms through which cocatalysts function, their effects on reaction rates, and their contribution to optimizing the efficiency of photocatalytic systems. Furthermore, we examine the key materials used as cocatalysts, their interactions with photocatalysts, and recent advancements in this rapidly evolving field.

**Keywords** Photocatalyst · Cocatalyst · Carrier transfer · Hydrogen production · In situ analysis · Infrared absorption · X-ray absorption

K.-H. Wang (✉)
Department of Applied Chemistry, Sanyo-Onoda City University, Sanyo-Onoda, Japan
e-mail: wang@rs.socu.ac.jp

M. Yoshida
Department of Applied Chemistry, Yamaguchi University, Ube, Japan

C. Hu
Department of Chemical Engineering, National Taiwan University of Science and Technology, Taipei, Taiwan



M. Hashizume, Y. Imura (eds.), *Colloids and Interfacial Dynamics*, Physical Chemistry in Action, https://doi.org/10.1007/978-981-96-6728-4_9

## 9.1 Significance of Water Splitting

Water splitting plays a crucial role in achieving clean hydrogen production, serving as a key technology in building a sustainable energy society. Water is abundantly available on Earth and can be readily sourced from oceans, rivers, and lakes, enabling a sustainable energy cycle while minimizing environmental impacts. Furthermore, hydrogen produced from water splitting returns to water after combustion, creating a closed energy cycle without introducing toxic by-products. These unique characteristics highlight hydrogen as a clean energy carrier capable of replacing fossil fuels. However, the reaction that splits water into hydrogen and oxygen involves a significant positive Gibbs free energy change:

$$2H_2O \rightarrow 2H_2 + O_2 \quad \Delta G^0_{298} = 237\,\mathrm{kJ/mol} \tag{9.1}$$

To drive this reaction, an external energy supply is required. Achieving sustainability will rely heavily on the adoption of renewable energy sources. While various renewable energy options are available, many are inherently intermittent due to their reliance on weather conditions and time of day. This intermittency can be challenging for civilian applications and disrupt energy grid stability. To address this, excess electricity generated during periods of high production and low demand can be redirected to produce hydrogen—offering an alternative or complement to traditional storage methods like pumped hydroelectric storage or flywheels. The hydrogen can then be efficiently stored and utilized when needed, enhancing energy supply stability and maximizing the effective use of renewable resources.

Currently, the primary hydrogen production processes utilizing renewable energy include:

1. Electrolysis: Water is electrolyzed using electricity generated from wind, hydro, or solar power to produce hydrogen.
2. Biomass Reforming: Hydrogen is produced via steam reforming of biomass.
3. Photocatalytic Water Splitting: Water is directly split into hydrogen and oxygen using light energy.

Each of these technologies has its own advantages and drawbacks. Among them, photocatalytic water splitting is particularly noteworthy for its relatively straightforward system design. For example, it does not require specialized reactors like electrolysis cells and can be readily scaled to large reaction areas. However, challenges of photocatalytic water splitting remain, including the separation of hydrogen and oxygen gas and its limited efficiency. Overcoming these challenges will unlock the potential of photocatalytic water splitting technology, positioning it as a cornerstone for achieving a sustainable energy society via the efficient and clean production of hydrogen gas.

## 9.2 Photocatalytic Water Splitting

A photocatalyst is a material that absorbs light energy and utilizes it to drive chemical reactions. Photocatalysts have been extensively investigated for their applications in environmental purification and energy conversion processes, including water splitting and pollutant decomposition. Materials that function as photocatalysts include semiconductors, complexes, and dyes. Among them, only semiconductor photocatalysts are currently known to facilitate the complete water splitting reaction (the reaction that generates hydrogen and oxygen from water in a 2:1 ratio) [1–4].

The research on semiconductor photocatalysts dates back to 1972 when Akira Fujishima and Kenichi Honda discovered the Honda–Fujishima effect [5]. They conducted experiments by immersing $TiO_2$ electrodes and Pt electrodes in an electrolyte and connecting them to an external power supply while being irradiated with light containing ultraviolet rays. In this study, it was observed that oxygen gas was generated from the $TiO_2$ electrode, and hydrogen gas was generated from the Pt electrode. Notably, this reaction occurs at a potential lower than 1.23 V vs. RHE (the theoretical potential required for complete water splitting), highlighting an important achievement that demonstrated the feasibility of water splitting using light energy.

Semiconductors have an energy gap, called the bandgap, between the valence band and the conduction band. The relationship between this bandgap and light energy is expressed by the following equation:

$$\text{Bandgap}(\text{eV}) = 1240 / \text{wavelength}(\text{nm}) \tag{9.2}$$

When light with energy greater than the bandgap is irradiated onto a semiconductor photocatalyst, electrons in the valence band absorb the light energy and are excited to the conduction band, as illustrated in Fig. 9.1. Meanwhile, holes are created in the valence band as electron vacancies. These excited electrons and the

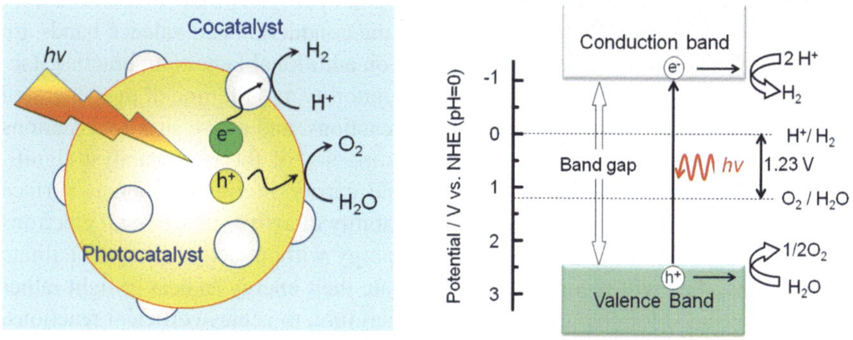


**Fig. 9.1** Schematics showing the reaction of water splitting using photocatalysts

generated holes migrate through the photocatalyst and reach the surface, where they trigger the following chemical reactions:

**Oxidation Reaction (Oxygen Evolution Reaction/OER)**

The holes ($h^+$) remaining in the valence band of the photocatalyst oxidize water molecules on the photocatalyst surface, generating oxygen gas and protons.

$$2H_2O + 4h^+ \rightarrow O_2 + 4H^+ \quad (9.3)$$

**Reduction Reaction (Hydrogen Evolution Reaction/HER)**

The excited electrons ($e^-$) that move to the conduction band react with protons to generate hydrogen gas. The generated hydrogen can be utilized as an energy carrier.

$$4H^+ + 4e^- \rightarrow 2H_2 \quad (9.4)$$

For these oxidation and reduction reactions to occur, the band structure of the semiconductor photocatalyst must meet specific conditions. First, the bottom of the conduction band must be at an energy level lower than the redox potential of $H^+/H_2$ (0 V vs. NHE), ensuring that the excited electrons have enough energy to reduce protons and generate hydrogen gas. Conversely, the top of the valence band must be at an energy level higher than the redox potential of $O_2/H_2O$ (1.23 V vs. NHE), allowing holes to oxidize water molecules and generate oxygen gas. Thus, the energy levels of the conduction band and the valence band must appropriately span these redox potentials, fulfilling the thermodynamic prerequisites for water splitting.

$ZnO_2$ photocatalysts meet the aforementioned energy requirements and are capable of facilitating complete water splitting without the aid of cocatalysts, as their excited electrons and holes possess sufficiently high energy levels. For other common photocatalysts such as $TiO_2$ and $SrTiO_3$, their energy level of the conduction band is located just above the reduction potential of water. This results in a limited driving force for electron-mediated water reduction, thereby impeding hydrogen production. To overcome this limitation, cocatalysts such as Pt, NiO, and $RuO_2$ can be employed to enhance the catalytic activity and hydrogen production efficiency.

However, even when the energy levels of the conduction and valence bands are appropriate, efficient water splitting depends on additional factors. In practice, factors such as charge separation properties, the mobility and lifetime of the generated electrons and holes, overpotentials in redox reactions, and active sites for reactions all influence the reaction efficiency. Bulk properties of the photocatalyst significantly affect charge separation, mobility, and carrier lifetimes, whereas surface properties dictate overpotentials and the availability of active sites. Even if electrons and holes reach the surface with sufficient energy without active sites to facilitate the reactions, they will recombine and dissipate their energy as heat or light rather than being utilized for chemical reactions. Therefore, to achieve efficient reactions, it is crucial to incorporate cocatalysts on the surface. Cocatalysts can provide the necessary active sites, help prevent charge recombination, and optimize the overall efficiency of the water splitting process. Their presence is indispensable for achieving high-performance photocatalytic water splitting.

## 9.3 Role of Cocatalysts in Enhancing Photocatalytic Water Splitting

Cocatalysts are critical components in improving the efficiency of photocatalytic water splitting [6]. While they are not required for the fundamental process, they significantly enhance the reaction rates by addressing the inefficiencies inherent in photocatalytic systems.

### *9.3.1 Enhancing Charge Separation and Suppressing Charge Recombination*

Photocatalytic water splitting begins with the absorption of photons possessing energy greater than the bandgap of the photocatalyst material. This absorption generates photoexcited electron–hole pairs within the bulk of the material. For these charge carriers to participate in redox reactions, they must migrate from the bulk to the surface of the photocatalyst. However, this migration is often hindered by imperfections in the material. The crystallinity of the photocatalyst and the presence of defects significantly impact this process. Defects, in particular, serve as recombination centers, allowing electrons and holes to recombine with each other before reaching the active sites, leading to decreased photocatalytic efficiency.

To address these inherent limitations, cocatalysts are employed to promote effective charge separation and suppress recombination. Cocatalysts act as electron sinks or mediators, ensuring that photogenerated electrons are efficiently transferred to the desired reaction sites. For example, Pt, a widely used cocatalyst, effectively captures electrons from the conduction band of photocatalysts. Pt thus prevents their recombination with holes and thus enhances the availability of electrons for the HER. This mechanism significantly accelerates hydrogen production by maximizing the utilization of photogenerated charge carriers.

In addition to capturing electrons, cocatalysts play a critical role in directing charge carriers to appropriate reaction sites on the photocatalyst surface. This targeted flow reduces internal resistance and optimizes the kinetics of charge transfer. By suppressing recombination and facilitating charge separation, cocatalysts enhance the efficiencies of both the HER and the OER, the two essential half-reactions in water splitting.

Recent advancements have also highlighted the importance of material engineering in improving charge separation. Optimizing the crystallinity of photocatalysts and minimizing defects through synthesis techniques or postsynthetic treatments can substantially reduce recombination rates. Furthermore, integrating nanostructured cocatalysts enhances surface interactions, providing additional pathways for efficient charge carrier migration.

### 9.3.2 Providing Active Sites and Enhancing Catalytic Activity

Cocatalysts also enhance the catalytic activity of photocatalytic systems by providing active sites for redox reactions. Transition metal oxides, such as NiO and $RuO_2$, are commonly employed for the OER due to their ability to stabilize oxygen intermediates. For example, NiO facilitates the oxidation of water molecules by stabilizing oxygen intermediates on its surface, thereby promoting oxygen generation. Similarly, noble metals like Pt and $RuO_2$ provide active sites for proton reduction, significantly improving hydrogen production rates.

By lowering the overpotentials required for the OER and HER, cocatalysts minimize energy losses associated with these processes. This dual functionality of stabilizing intermediates and lowering energy barriers makes cocatalysts indispensable for efficient photocatalytic water splitting.

### 9.3.3 Synergistic Effects of Cocatalysts

The integration of multiple cocatalysts can produce synergistic effects that significantly enhance photocatalytic performance. For example, combining noble metals like Pt with transition metal oxides such as NiO or $RuO_2$ allows the system to leverage the superior electron transfer capabilities of noble metals and the stability and catalytic activity of metal oxides. This approach not only enhances reaction kinetics but also reduces the reliance on expensive noble metals, thereby improving the economic viability of the process.

Composite cocatalysts, such as Pt–Ni or Pt–Co alloys, have demonstrated remarkable improvements in photocatalytic efficiency. These materials combine the unique properties of their constituents to enhance charge transfer and catalytic activity, leading to higher hydrogen production rates and enhanced overall water splitting efficiency.

### 9.3.4 Role and Function of Noble-Metal/Cr-Layer Core/Shell Structure Cocatalysts

While synergistic combinations of cocatalysts demonstrate the potential to optimize performance by leveraging the complementary properties of distinct materials, further refinement is needed to address critical inefficiencies such as back-reactions and long-term stability. The development of core/shell structures, such as noble-metal/Cr-layer cocatalysts, represents a significant advancement in overcoming these limitations [7, 8]. These systems build upon the synergistic principles by introducing an additional layer of functionality—a protective shell that enhances selectivity and durability.

Core/shell structures offer a unique mechanism for addressing specific challenges inherent in photocatalytic water splitting. The integration of a noble metal core with a $Cr_2O_3$ shell not only facilitates efficient charge transfer and reaction site availability but also suppresses undesired back-reactions, such as the recombination of hydrogen and oxygen. These innovative structures exemplify the potential of engineered cocatalysts to overcome the limitations of traditional designs, thereby paving the way for more efficient and robust photocatalytic systems.

Noble-metal/Cr-layer core/shell cocatalysts have emerged as a highly effective candidate [9]. Figure 9.2 illustrates the significant enhancement in photocatalytic activity achieved by $Rh/Cr_2O_3$ and $Pt/Cr_2O_3$ cocatalysts compared to bare Rh or Pt on the $(Ga_{1-x}Zn_x)(N_{1-x}O_x)$ system under visible-light irradiation. The $Cr_2O_3$ shell serves multiple functions, including the suppression of undesired back-reactions and the enhancement of hydrogen evolution efficiency. In particular, systems

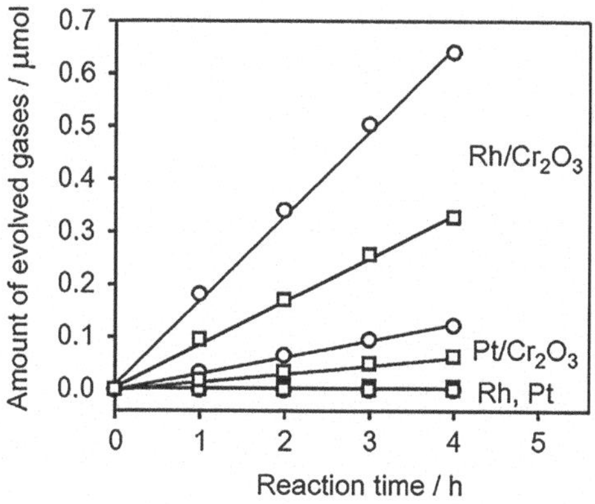


**Fig. 9.2** Photocatalytic activity ($H_2$, circles; $O_2$, squares) of $(Ga_{1-x}Zn_x)(N_{1-x}O_x)$ modified with Rh, Pt, $Pt/Cr_2O_3$, or $Rh/Cr_2O_3$ cocatalysts for overall water splitting under visible irradiation (>400 nm). Photocatalysts were prepared according to the method described in previous reports. Reactions were performed using 0.15 g of sample dispersed in pure water (370 mL) under irradiation from a high-pressure mercury lamp (450 W) via a UV filter (aqueous 2 M $NaNO_2$ solution). (After Yoshida et al. [9]. Reproduced with permission from the American Chemical Society)

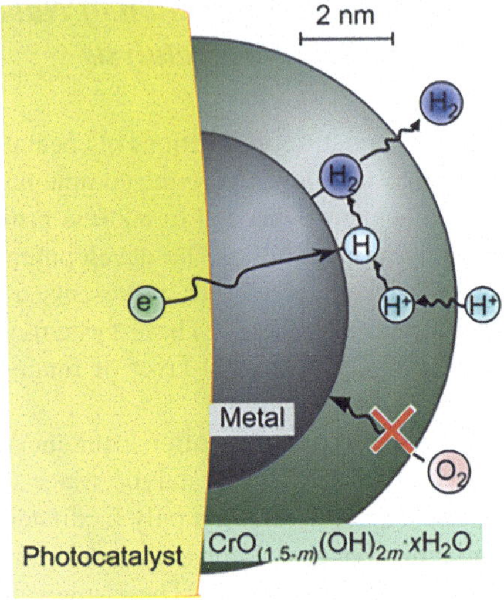


**Fig. 9.3** Schematic model of $H_2$ evolution reaction on core/shell noble-metal/$Cr_2O_3$ particulate system as a cocatalyst for photocatalytic overall water splitting. (After Yoshida et al. [9]. Reproduced with permission from the American Chemical Society)

utilizing $(Ga_{1-x}Zn_x)(N_{1-x}O_x)$ with cocatalysts like Rh/$Cr_2O_3$ and Pt/$Cr_2O_3$ have been shown to outperform their bare noble metal counterpart under visible-light irradiation.

The $Cr_2O_3$ shell plays a critical role as a selective barrier, allowing the passage of protons and hydrogen while blocking oxygen permeation. This selective gas permeability prevents the recombination of evolved hydrogen and oxygen, a prevalent issue on bare noble metals. Electrochemical analyses reveal that the $Cr_2O_3$ layer forms a passivated structure, probably $CrO_{(1.5-m)}(OH)_{2m}{\cdot}xH_2O$, at the interface. This structure retains high catalytic activity while suppressing back-reactions, ensuring efficient overall water splitting. Figure 9.3 in the referenced text schematically illustrates this mechanism, showing how protons penetrate the $Cr_2O_3$ shell to the noble metal core, where they are reduced to hydrogen molecules.

In addition to suppressing undesired reactions, the noble-metal/Cr-layer structure enhances the stability and longevity of cocatalysts. The $Cr_2O_3$ shell minimizes interaction between the noble metal and oxygen in the reaction medium, thereby extending the cocatalyst's operational lifetime.

### 9.3.5 *Role and Function of Lanthanoid-Oxide/Noble-Metal Cocatalysts Electroless-Deposited on GaN:ZnO*

Building on the advancements offered by noble-metal/Cr-layer core/shell structures, the incorporation of lanthanoid oxides as surface modifiers introduces a new dimension to cocatalyst design [10]. While $Cr_2O_3$ layers have demonstrated effective suppression of undesired back-reactions, lanthanoid oxides bring additional functionalities that align with the goals of enhanced stability and selectivity in photocatalytic systems. This section explores the role of lanthanoid oxides, electroless-deposited on noble-metal cocatalysts, in enhancing the performance of $(Ga_{1-x}Zn_x)(N_{1-x}O_x)$-based photocatalysts. By expanding on the concepts discussed in the previous section, this study examined how lanthanoid oxides contribute to suppressing back-reactions and improving the efficiency of hydrogen and oxygen evolution under visible-light irradiation.

Lanthanoid-oxide/noble-metal cocatalysts deposited on $(Ga_{1-x}Zn_x)(N_{1-x}O_x)$ are investigated for photocatalytic overall water splitting under visible light. As shown in Fig. 9.4, lanthanoid oxides, including La, Pr, Sm, Gd, and Dy, significantly enhance hydrogen and oxygen evolution compared to unmodified Rh-metal/$(Ga_{1-x}Zn_x)(N_{1-x}O_x)$, which is prone to severe back-reactions. The deposition of these oxides suppresses the recombination of evolved hydrogen and oxygen on noble metals, improving overall efficiency.

The proposed mechanism, illustrated in Fig. 9.5, reveals that lanthanoid oxide layers selectively permit hydrogen diffusion while blocking oxygen, a property attributed to the structural characteristics of the oxide layers. Among the various oxides studied, Pr oxide demonstrates remarkable performance, particularly at a deposition of 13 wt%. Although the back-reactions are not completely eliminated, the presence of lanthanoid oxides significantly enhances both stability and efficiency of the photocatalyst.

This study positions lanthanoid oxides as viable alternatives to $Cr_2O_3$ for mitigating back-reactions in photocatalytic systems. These findings underline the critical role of lanthanoid oxides in advancing cocatalyst functionality, paving the way for innovative approaches to sustainable hydrogen production. By integrating

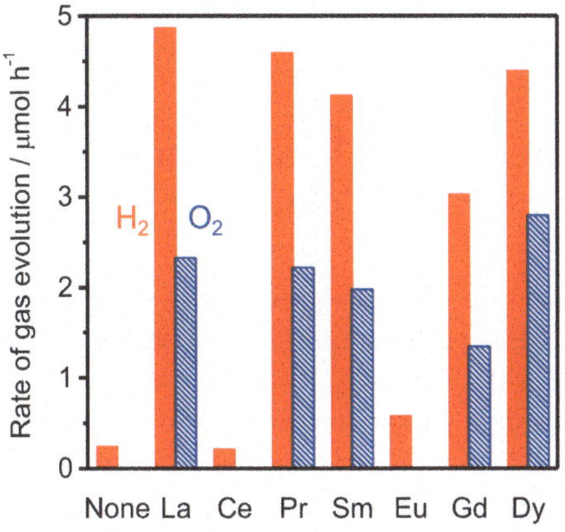


**Fig. 9.4** Photocatalytic activities of Rh-metal/$(Ga_{1-x}Zn_x)(N_{1-x}O_x)$ after $NaBH_4$ treatment with various lanthanoid nitrates for overall water splitting under visible light irradiation. The reaction was performed using 0.10 g of sample in an aqueous solution (150 ml) with a 300 W Xe lamp with a cutoff filter (>400 nm). The lanthanoid nitrates as precursors were used at 13 wt% amount as a metal base. The activities were estimated at an average of 3 h (Dy: average of 2 h). (After Yoshida et al. [13]. Reproduced with permission from the American Chemical Society)

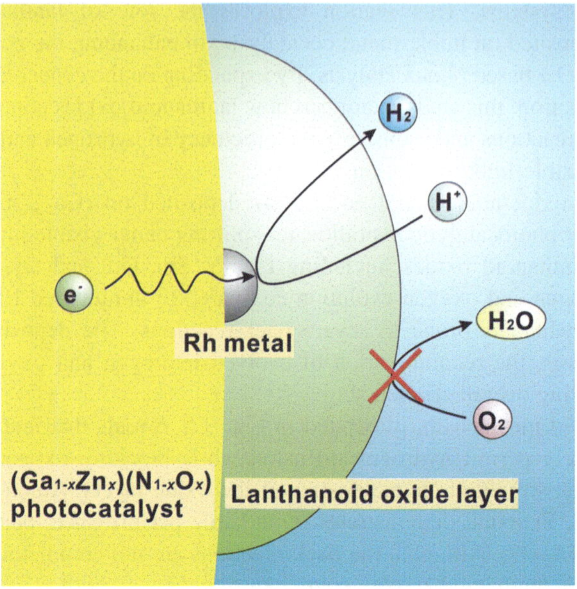


**Fig. 9.5** Schematic model of $H_2$ evolution reaction on $(Ga_{1-x}Zn_x)(N_{1-x}O_x)$ photocatalyst electroless-deposited with lanthanoid-oxide/noble-metal cocatalysts for photocatalytic overall water splitting. (After Yoshida et al. [13]. Reproduced with permission from the American Chemical Society)

lanthanoid oxides, photocatalysts gain enhanced selectivity and operational stability, marking a significant step forward in the design of efficient and robust systems for visible-light-driven water splitting.

## 9.4 In Situ Monitoring of Charge Transfer from Photocatalyst to Cocatalyst

Recent advancements in photocatalytic water splitting have emphasized the importance of understanding charge transfer from photocatalyst to cocatalyst. These advancements highlight the necessity for direct monitoring of charge transfer to refine photocatalyst design. In situ monitoring methodologies have significantly deepened our understanding of charge transfer processes, providing critical insights into the migration pathways and reaction mechanisms of charge carriers [11]. These techniques reveal key factors influencing photocatalytic performance under operational conditions. For instance, Asakura et al. employed femtosecond time-resolved X-ray absorption fine structure (XAFS) spectroscopy with an X-ray free-electron laser to capture the localization of excited electrons in $WO_3$ photocatalysts, revealing structural and electronic changes [12]. Similarly, Kamat et al. documented charge migration in $TiO_2$ photocatalysts using UV–vis spectrophotometry, showing the impact on Au nanoparticles within thiol-crosslinked systems [13]. Durrant and collaborators utilized transient absorption spectroscopy to demonstrate how Rh reduction in $SrTiO_3$ extends electron lifetimes [14]. Yamakata's team applied time-resolved visible-to-mid-infrared spectroscopy to explore the role of $CoO_x$ cocatalysts on $LaTiO_2N$ in capturing holes and extending electron lifetimes [15]. Additionally, Ozawa et al. showed using time-resolved photoelectron spectroscopy that potential barriers on $TiO_2$ surfaces influence carrier dynamics [16].

Research on photocatalytic systems tends to focus more on the generation of electron–hole pairs after photoexcitation and the migration of carriers to the catalyst surface, while relatively few studies have investigated the detailed redox reactions at the photocatalyst–cocatalyst interface. This is mainly because the redox reactions at this interface involve intricate electrochemical dynamics that are challenging to study in situ. Although methods such as Attenuated Total Reflection Surface-Enhanced Infrared Absorption Spectroscopy (ATR-SEIRAS) [17] and XAFS provide valuable insights into interfacial reactions, their use is not as widespread as techniques for studying carrier generation or migration. Therefore, this chapter mainly introduces ATR-SEIRAS and XAFS as key methods for exploring the detailed redox reactions at the photocatalyst–cocatalyst interface.

### 9.4.1 Carrier Transfer to Hydrogen Evolution Cocatalysts

ATR-SEIRAS has emerged as an indispensable tool for investigating charge transfer processes at the molecular level, particularly within the domain of material science and catalyst research. By offering in situ, real-time insights into electronic and structural dynamics, ATR-SEIRAS has significantly advanced our understanding of HER and related phenomena.

In photocatalytic systems, cocatalysts deposited on photocatalysts enhance hydrogen generation by optimizing the catalytic efficiency at the cocatalyst/water interface. This process is heavily influenced by the electrochemical potential at the cocatalyst, which must be sufficiently negative compared to the $H^+/H_2$ redox potential under irradiation. Monitoring such potential changes is critical to optimizing catalytic performance. The vibrational frequency of adsorbed CO molecules on metal surfaces serves as a sensitive probe of electronic density and local potential at adsorption sites [18]. These frequencies shift based on interactions with the supporting materials and surrounding environments, providing valuable information about charge transfer processes critical to hydrogen evolution cocatalysts.

In situ ATR-SEIRAS studies, such as those conducted on Pt cocatalysts supported on GaN photocatalysts, have revealed shifts in the Fermi level during hydrogen evolution under UV irradiation [19]. As shown in Fig. 9.6, the experimental setup involves a spectroelectrochemical cell with Pt nanoparticles (200–300 nm, Fig. 9.6) deposited via electroless and photodeposition methods. These studies confirm efficient hydrogen evolution in methanol solutions, underscoring the system's validity for studying charge transfer mechanisms. ATR-SEIRAS measurements

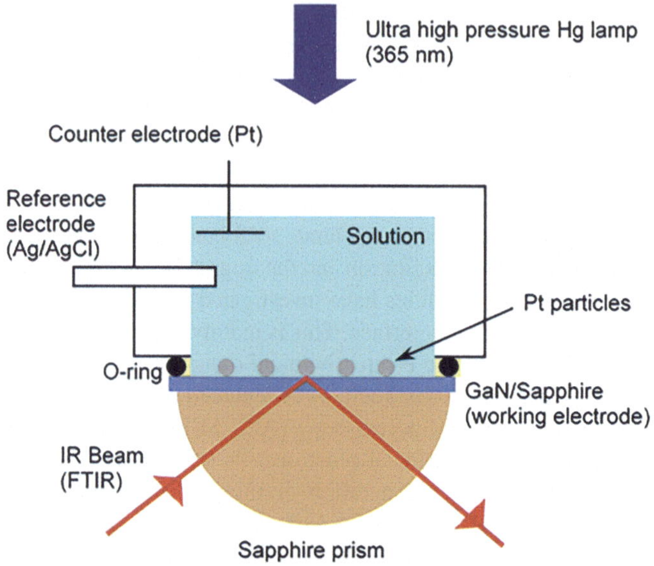


**Fig. 9.6** Schematic illustration of optical geometry and spectroelectrochemical cell for in situ ATR-SEIRAS measurements. A glass electrochemical cell was used with a Pt mesh counter electrode, an Ag/AgCl (saturated KCl) reference electrode, and a prepared sample, which was connected to a Cu wire with In paste, as a working electrode in an 0.1 M $Na_2SO_4$ aqueous solution with bubbling Ar or CO. A high-pressure mercury lamp was used for UV irradiation (365 nm). (Figure provided by Yoshida M., Department of Applied Chemistry, Yamaguchi, Japan)

demonstrate dynamic changes in CO vibrational frequencies during UV irradiation, reflecting charge separation at the GaN–Pt interface. Positive frequency shifts correspond to hole migration, while subsequent negative shifts indicate electron transfer from GaN to Pt. These results highlight the influence of band bending at the interface, as indicated by a flat-band potential of approximately −0.2 V. Time-resolved SEIRA spectra (Fig. 9.7) reveal a Fermi level shift to −0.35 V during irradiation, which is consistent with hydrogen evolution potentials. However, slower hole migration creates localized imbalances at some Pt sites, acting as recombination centers. This reversible charge transfer process underscores the need for optimizing electron transfer rates at GaN – Pt interfaces.

Further studies utilizing visible-light-responsive oxynitride photocatalysts, such as GaN:ZnO, have expanded the understanding of carrier dynamics [20]. Infrared spectra of CO adsorbed on Pt in Pt/GaN:ZnO photocatalysts show frequency shifts under irradiation, indicating the migration of photoexcited electrons from GaN:ZnO to Pt. Conversely, Pt/$LaTiO_2N$ systems exhibit minimal frequency shifts, highlighting differences in interfacial carrier transfer and hydrogen evolution potential. These findings, supported by band bending mechanisms, emphasize that carrier separation efficiency varies across photocatalyst materials, with GaN:ZnO outperforming TaON and $LaTiO_2N$ due to reduced recombination rates.

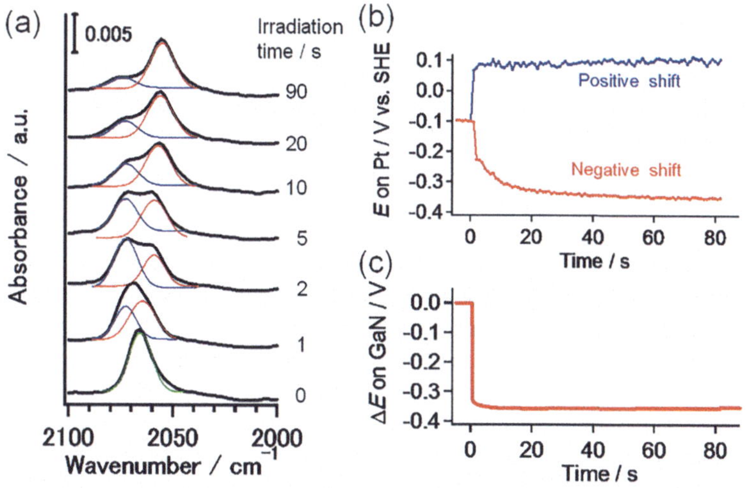


**Fig. 9.7** (**a**) Irradiation-time-dependent SEIRA spectra of adsorbed CO on Pt cocatalyst at the rest potential of the GaN photocatalyst electrode in CO-saturated 0.1 M $Na_2SO_4$ aqueous solution at light intensities of $8.6 \times 10^{18}$ photons $cm^{-2}$ $h^{-1}$. The curve-fitting results were shown in each spectrum using Gaussian functions. Reference spectrum was measured under bubbling Ar prior to CO bubbling. Time dependence of the estimated potential on the Pt cocatalyst with deconvolution processing (**b**) and the rest potential of the GaN photocatalyst electrode (**c**) during in situ ATR-SEIRAS measurement under irradiation. (After Yoshida et al. [9]. Reproduced with permission from the American Chemical Society)

### *9.4.2 Carrier Transfer to Oxygen Evolution Cocatalysts*

XAFS is a powerful analytical technique used to study the electronic properties and local structure of materials by examining how X-rays are absorbed by atoms [21]. It focuses specifically on the absorption behavior at energies near and above the core-level binding energy of an atom. Due to this energy dependence, XAFS spectrum is typically divided into two regions: X-ray absorption near-edge structure (XANES) and extended X-ray absorption fine structure (EXAFS). XANES is highly sensitive to the formal oxidation state of the absorbing atom, enabling the estimation of valence states through the half-edge method, which correlates the energy at half the normalized absorption with oxidation states [22–24]. In contrast, EXAFS provides information about the distances, coordination numbers, and species of the neighboring atoms around the absorbing atom. In photocatalytic systems, excited carriers transferring from the photocatalyst to the cocatalyst can alter the valence state of the cocatalyst. This change is observable as a shift in the absorption edge energy of the metal elements in the cocatalyst, as detected by XAFS. Such shifts offer valuable insights into whether excited carriers have successfully migrated to the cocatalyst, thereby enhancing our understanding of charge transfer mechanisms in photocatalytic systems.

For instance, photoexcited carrier transfer from $SrTiO_3$ to MnOx during oxygen evolution has been investigated using in situ XAFS [25]. As illustrated in Fig. 9.8, the experimental setup involves irradiating a MnOx/$SrTiO_3$ sample in a 0.04 M $AgNO_3$ solution, enabling the observation of carrier dynamics. The Mn K-edge XAFS spectra (Fig. 9.9) reveal a shift in the Mn absorption edge from 6557 to 6560 eV during irradiation, signifying the oxidation of $Mn^{3+}$ to $Mn^{4+}$. This transition is facilitated by hole transfer from $SrTiO_3$ to $MnO_x$, which enhances oxygen evolution by increasing the local electrode potential of the cocatalyst. In the absence of $AgNO_3$, recombination reactions dominate, leaving the Mn oxidation state unchanged. Figure 9.10 schematically shows how $Ag^+$ reduction prevents recombination, enabling efficient hole transfer to $MnO_x$.

Another study employs conversion-electron-yield soft X-ray absorption spectroscopy (CEY-SXAS) to explore carrier dynamics in Al-doped $SrTiO_3$ (STO:Al) loaded with Rh/$Cr_2O_3$ and CoOOH cocatalysts under UV–visible irradiation [26]. As depicted in Fig. 9.11, this setup enables real-time monitoring of surface reactions under ambient pressure, with water vapor playing a critical role in stabilizing photoinduced holes at the CoOOH OER site. The mechanistic schematic (Fig. 9.12) shows how Rh/$Cr_2O_3$ facilitates electron capture while CoOOH oxidizes water with photoinduced holes. Without water vapor or Rh/$Cr_2O_3$, holes cannot efficiently reach CoOOH, severely limiting oxygen evolution.

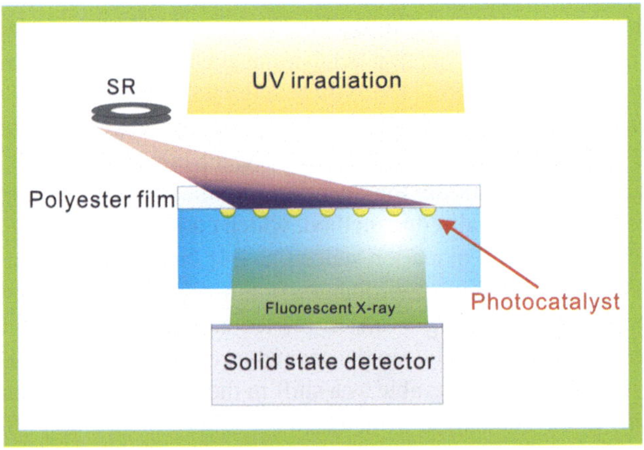


**Fig. 9.8** Schematic illustration of in situ XAFS measurements during photocatalytic reactions. The sample is fixed on a polyester film over a Teflon cell filled with distilled water or a 0.04 M aqueous $AgNO_3$ solution under UV irradiation from a Xe lamp. (After Yoshida et al. [25]. Reproduced with permission from the Chemical Society of Japan)

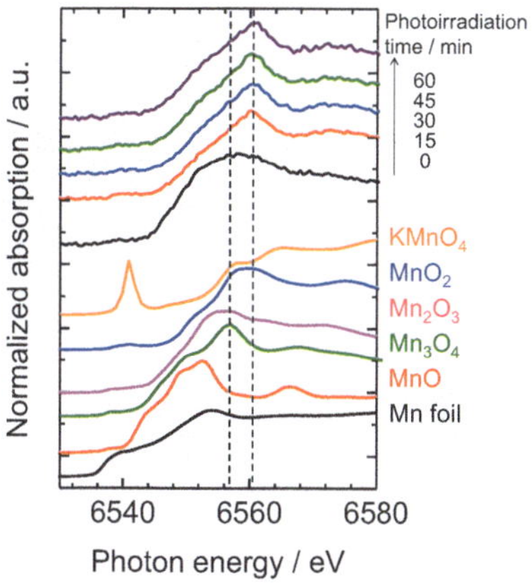


**Fig. 9.9** In situ Mn K-edge XAFS spectra of a $MnO_x/SrTiO_3$ sample in 0.04 M aqueous $AgNO_3$ solution under UV irradiation from a Xe lamp equipped with an IR cutoff filter and neutral density filters. XAFS spectra of Mn foil, MnO, $Mn_3O_4$, $Mn_2O_3$, $MnO_2$, and $KMnO_4$ are shown for reference. (After Yoshida et al. [25]. Reproduced with permission from the Chemical Society of Japan)

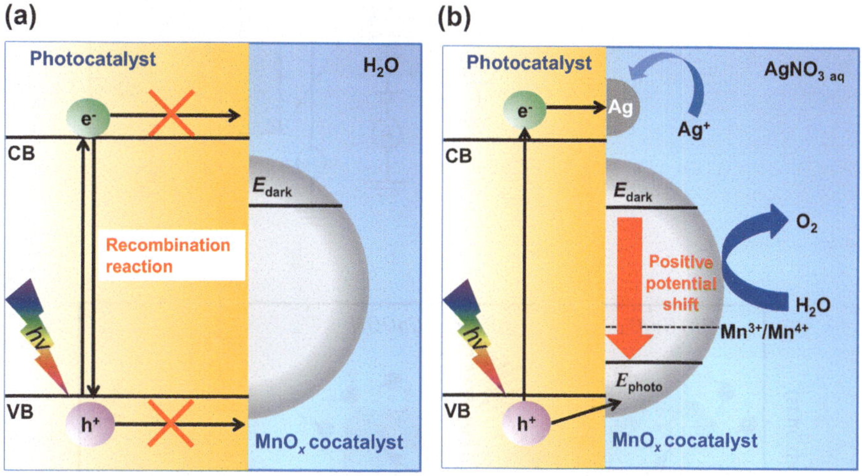


**Fig. 9.10** Schematic models of photoexcited carrier transfer in the $MnO_x/SrTiO_3$ photocatalyst in (**a**) distilled water and (**b**) aqueous $AgNO_3$ solution under UV irradiation. Edark and Ephoto denote the local electrode potentials before and during UV irradiation, respectively. (After Yoshida et al. [14]. Reproduced with permission from the Chemical Society of Japan)

## 9.5 Conclusion

Cocatalysts are critical for enhancing photocatalytic water splitting by improving charge separation, suppressing recombination, and providing active redox sites for HER and OER. Advances in cocatalyst design, such as composite cocatalysts, which integrate complementary materials like noble metals and transition metal

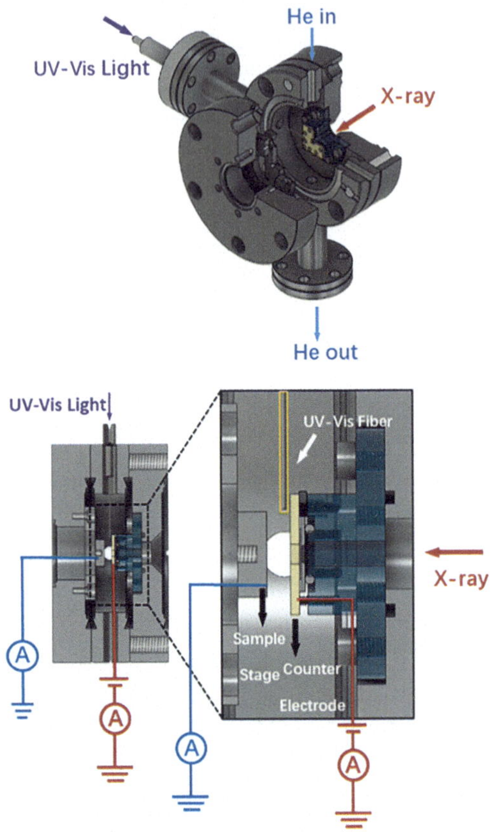


**Fig. 9.11** A CEY-SXAS cell modified for introduction of UV–vis lights. The UV–vis lights are introduced via a fiber whose exit is opened to the sample stage. (After Kondoh et al. [14]. Reproduced with permission from the Chemical Society of Japan)

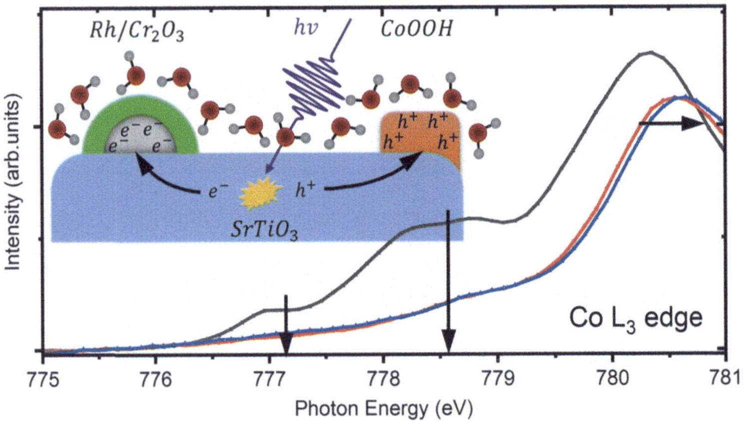


**Fig. 9.12** Schematics showing the $Rh/Cr_2O_3$ HER site facilitating electron capture and the CoOOH OER site oxidizing water with photoinduced holes. (After Kondoh et al. [14]. Reproduced with permission from the Chemical Society of Japan)

oxides, contribute synergistically to suppressing back-reactions and improving catalytic activity. These combinations leverage the superior charge transfer properties of noble metals and the catalytic robustness of oxides, further enhancing the efficiency of hydrogen and oxygen evolution. Additionally, the integration of lanthanoid oxides as surface modifiers introduces new functionalities, including enhanced selectivity and durability, paving the way for innovative cocatalyst designs.

Simultaneously, in situ monitoring techniques, such as ATR-SEIRAS and XAFS, have revolutionized our understanding of charge transfer at photocatalyst–cocatalyst interfaces. These methods provide detailed insights into band bending, carrier separation efficiency, and oxidation state transitions critical for HER and OER. For instance, ATR-SEIRAS highlights electronic density changes in Pt systems, while XAFS captures oxidation dynamics in systems like $SrTiO_3$–$MnO_x$. These findings emphasize the importance of optimizing interfacial dynamics for efficient and sustainable water splitting.

Despite these advances, challenges remain in fully understanding complex interfacial redox reactions and their sensitivity to operational conditions. Current in situ methodologies often fall short of capturing the full complexity of real-world systems, requiring further refinement and integration of complementary techniques.

## References


1. Hisatomi T, Kubota J, Domen K (2014) Recent advances in semiconductors for photocatalytic and photoelectrochemical water splitting. Chem Soc Rev 43:7520–7535
2. Maeda K (2011) Photocatalytic water splitting using semiconductor particles: history and recent developments. J Photochem Photobiol C: Photochem Rev 12:237–268
3. Takata T, Domen K (2019) Particulate photocatalysts for water splitting: recent advances and future prospects. ACS Energy Lett. 4:542–549
4. Kudo A, Miseki Y (2009) Heterogeneous photocatalyst materials for water splitting. Chem Soc Rev 38:253–278
5. Fujishima A, Honda K (1972) Electrochemical photolysis of water at a semiconductor electrode. Nature 238:37
6. Yang J, Wang D, Han H, Li C (2013) Roles of cocatalysts in photocatalysis and photoelectrocatalysis. Acc Chem Res 46:1900–1909
7. Maeda K, Teramura K, Lu D, Saito N, Inoue Y, Domen K (2007) Roles of Rh/Cr O (core/shell) nanoparticles photodeposited on visible-light-responsive $(Ga_{1-x}Zn_x)(N_{1-x}O_x)$ solid solutions in photocatalytic overall water splitting. J Phys Chem C 111:7554–7560
8. Maeda K, Teramura K, Lu D, Saito N, Inoue Y, Domen K (2006) Noble-metal/Cr O core/shell nanoparticles as a cocatalyst for photocatalytic overall water splitting. Angew Chem Int Ed Engl 45:7806–7809
9. Yoshida M, Takanabe K, Maeda K, Ishikawa A, Kubota J, Sakata Y, Ikezawa Y, Domen K (2009) Role and function of noble-metal/Cr-layer core/shell structure cocatalysts for photocatalytic overall water splitting studied by model electrodes. J Phys Chem C 113:10151–10157
10. Yoshida M, Maeda K, Lu D, Kubota J, Domen K (2013) Lanthanoid oxide layers on rhodium-loaded $(Ga_{1-x}Zn_x)(N_{1-x}O_x)$ photocatalyst as a modifier for overall water splitting under visible-light irradiation. J Phys Chem C 117:14000–14006
11. Cowan AJ, Durrant JR (2013) Long-lived charge separated states in nanostructured semiconductor photoelectrodes for the production of solar fuels. Chem Soc Rev 42:2281–2293

12. Uemura Y, Kido D, Wakisaka Y, Uehara H, Ohba T, Niwa Y, Nozawa S, Sato T, Ichiyanagi K, Fukaya R, Adachi S, Katayama T, Togashi T, Owada S, Ogawa K, Yabashi M, Hatada K, Takakusagi S, Yokoyama T, Ohtani B, Asakura K (2016) Dynamics of photoelectrons and structural changes of tungsten trioxide observed by femtosecond transient XAFS. Angew Chem Int Ed Engl 55:1364–1367
13. Subramanian V, Wolf EE, Kamat PV (2004) Catalysis with TiO /gold nanocomposites. Effect of metal particle size on the Fermi level equilibration. J Am Chem Soc 126:4943–4950
14. Moss B, Wang Q, Butler KT, Grau-Crespo R, Selim S, Regoutz A, Hisatomi T, Godin R, Payne DJ, Kafizas A, Domen K, Steier L, Durrant JR (2021) Linking in situ charge accumulation to electronic structure in doped SrTiO reveals design principles for hydrogen-evolving photocatalysts. Nat Mater 20:511–517
15. Yamakata A, Kawaguchi M, Nishimura N, Minegishi T, Kubota J, Domen K (2014) Behavior and energy states of photogenerated charge carriers on Pt- or CoO -loaded LaTiO N photocatalysts: time-resolved visible to mid-infrared absorption study. J Phys Chem C 118:23897–23906
16. Ozawa K, Emori M, Yamamoto S, Yukawa R, Yamamoto S, Hobara R, Fujikawa K, Sakama H, Matsuda I (2014) Electron-hole recombination time at TiO single-crystal surfaces: influence of surface band bending. J Phys Chem Lett 5:1953–1957
17. Osawa M (1997) Dynamic processes in electrochemical reactions studied by surface-enhanced infrared absorption spectroscopy (SEIRAS). Bull Chem Soc Jpn 70:2861–2880
18. Samjeské G, Komatsu K, Osawa M (2009) Dynamics of CO oxidation on a polycrystalline platinum electrode: a time-resolved infrared study. J Phys Chem C 113:10222–10228
19. Yoshida M, Yamakata A, Takanabe K, Kubota J, Osawa M, Domen K (2009) ATR-SEIRAS investigation of the Fermi level of Pt cocatalyst on a GaN photocatalyst for hydrogen evolution under irradiation. J Am Chem Soc 131:13218–13219
20. Lu X, Bandara A, Katayama M, Yamakata A, Kubota J, Domen K (2011) Infrared spectroscopic study of the potential change at cocatalyst particles on oxynitride photocatalysts for water splitting by visible light irradiation. J Phys Chem C 115:23902–23907
21. Crozier ED (1997) A review of the current status of XAFS spectroscopy. Nucl Instr Meth Phys Res B 133:134–144
22. Araki Y, Tsunekawa S, Sakai A, Harada K, Nagatsuka R, Suzuki–Sakamaki M, Amemiya K, Wang KH, Kawai T, Yoshida M (2022) Development of a hemispherical cavity cobalt electrocatalyst for water oxidation based on a polystyrene colloidal template electrodeposition method. ChemistrySelect 7:e202200600
23. Miura T, Tsunekawa S, Onishi S, Ina T, Wang KH, Watanabe G, Hu C, Kondoh H, Kawai T, Yoshida M (2021) Assessing nickel oxide electrocatalysts incorporating diamines and having improved oxygen evolution activity using operando UV/visible and X-ray absorption spectroscopy. Phys Chem Chem Phys 23:23280–23287
24. Tsunekawa S, Yamamoto F, Wang KH, Nagasaka M, Yuzawa H, Takakusagi S, Kondoh H, Asakura K, Kawai T, Yoshida M (2020) Operando observations of a manganese oxide electrocatalyst for water oxidation using hard/tender/soft X-ray absorption spectroscopy. J Phys Chem C 124:23611–23618
25. Yoshida M, Gon N, Maeda S, Mineo T, Nitta K, Kato K, Nitani H, Abe H, Uruga T, Kondoh H (2014) In situ XAFS study of the photoinduced potential shift of a $MnO_X$ cocatalyst on a $SrTiO_3$ photocatalyst. Chem Lett 43:1725–1727
26. Wang Z, Toyoshima R, Yoshida M, Mase K, Kondoh H (2024) In situ observation of charge transfer in the cocatalyst-loaded $SrTiO_3$ : Al photocatalyst under UV irradiation by conversion-electron-yield soft–X–ray absorption spectroscopy. J Phys Chem C 128:9193–9201

# Chapter 10
# Dynamics of Polymer Chains Via Interfacial Viscoelasticity Analysis

**Yoshihisa Fujii, Kenji Yamaoka, Shintaro Yasuda, and Akira Uchiyama**

**Abstract** This chapter depicts two research findings about nondestructive mechanical stimulation for analyzing interfacial molecular motion by a quartz crystal microbalance (QCM), a micro-mass analysis technique, highlighting polymer dynamics observed at the solid interface. The gelation behavior at the interface and the gelation temperature of methylcellulose (MC) aqueous solution were elucidated as a function of distance from the electrode surface. MC chains were restricted its mobility and aggregation during the sol–gel transition due to adsorption on the electrode surface. Secondly, the glass transition temperature, specifically on buried interfaces, was employed to elucidate the thermal molecular motion of polystyrene. These studies strengthened the understanding of interfacial dynamics and enabled the nondestructive and precise evaluation of molecular motion characteristics at interfaces using a cost-effective, versatile technology known as QCM, consequently allowing interface research for various researchers.

**Keywords** Polymer · Viscoelasticity · Glass transition · Interfacial gelation · Adsorption · Polymer thin film · Quartz crystal microbalance · Neutron reflectivity

## 10.1 Introduction

Since the 1940s, polymeric materials have been extensively utilized in several commercial products, and they are now essential for a variety of uses in everyday life [1]. In recent years, polymeric materials have been progressively used in

Y. Fujii (✉) · S. Yasuda · A. Uchiyama
Department of Chemistry for Materials, Graduate School of Engineering, Mie University, Tsu, Japan
e-mail: fujii@chem.mie-u.ac.jp

K. Yamaoka
Department of Macromolecular Science, Graduate School of Science, The University of Osaka, Suita, Japan



M. Hashizume, Y. Imura (eds.), *Colloids and Interfacial Dynamics*, Physical Chemistry in Action, https://doi.org/10.1007/978-981-96-6728-4_10

smaller-scale microelectronic circuits and other applications. On the other hand, as the scale decreases, the surface area-to-volume ratio markedly increases, making it necessary to take into account the effects of the enlarged interfacial region on material properties [2]. Moreover, understanding the influence of surfaces and interfaces on functional properties may result in exceptional behaviors.

The chemical and physical properties of polymers at surfaces and interfaces are closely related to the expression of their functionality. The surfaces and interfaces of materials are crucial to many scientific processes due to their interaction with the external environment. Due to the absence of neighboring molecules on the surface, a decrease in molecular mass density and changes in dynamics are observed. Conversely, at the material interface, the binding of molecular chains due to contact with the surface can lead to physical properties that differ from those in the bulk. Consequently, regulating the surface and interface properties is expected to enhance the performance and functionality of materials.

On the other hand, evaluating local material properties at surfaces and interfaces is challenging. Near the surface, in reality, due to the progress of modern nanotechnology, it is becoming technically less difficult to accurately evaluate the structure and properties of polymer film surfaces. Despite having such a situation, near the interface with the substrate, accurate measurements are difficult due to the influence of the substrate material. Moreover, applying direct force near the solid interface presents challenges, which is why extensive study continues in this area. Many studies on polymer surfaces and interfaces have been conducted using thin films. The nanoscale treated by thin films effectively reflects the macromolecular properties of each region, as the surface-to-volume ratio is significantly larger.

The glass transition temperature ($T_g$) is often discussed in relation to the kinetic properties of polymer interfaces. In 1991, McKenna et al. first reported a decrease in the $T_g$ of organic liquids confined within nanopores [3]. This finding prompted further studies reporting both increases and decreases in $T_g$ under various experimental conditions. On the other hand, thin films may exhibit different physical properties compared to bulk materials, with the large contribution of surface and interface effects playing a significant role. In 1993, Reiter demonstrated the unusual dynamics of polystyrene (PS) ultrathin films through the kinetics of dewetting [4]. The films exhibited dewetting behavior, collapsing at temperatures below the $T_g$ of the bulk material. Furthermore, the growth of nuclei during dewetting of ultrathin films was found to occur much faster than in thicker films, suggesting that the $T_g$ of ultrathin films is lower than that of thicker films. The decrease in $T_g$ is often explained by free volume theory and is attributed to changes in the motion of polymer chain segments near $T_g$. Furthermore, $T_g$ substantially affects the viscoelasticity at the interface. In particular, at the interface of polymeric materials, changes in viscoelasticity affect the mobility and mechanical properties of the polymer segments, which in turn influence the overall material properties. Therefore, understanding the relationship between $T_g$ and viscoelasticity at the interface is crucial in material design.

In 1994, Keddie et al. employed spectroscopic ellipsometry to study the thickness dependence of $T_g$ in polymethylmethacrylate (PMMA) thin films and reported that hydrogen bonding at the substrate interface restricts segment mobility, leading to an increase in $T_g$ compared to the bulk [5, 6]. Prucker et al. utilized optical waveguide spectroscopy to observe a reduction in $T_g$ as the thickness of PMMA thin films supported on hydrophobic substrates decreased [7]. In 2001, Wu et al. discovered that based on thermal expansion measurements using an X-ray reflectivity measurement, the $T_g$ of thin films increases relative to the bulk as the thickness decreases [8]. Using the layer model, numerous studies have measured local regions at surfaces and interfaces. The first study of $T_g$ gradients in thin films was conducted by the fluorescence measurement by Torkelson et al. [9–11]. They placed dye-labeled molecules at known positions in the thin films and measured the fluorescence intensity as a function of film density. This method successfully collected information on $T_g$ at various depths within the film. Kajiyama et al. also showed, from temperature-dependent X-ray photoelectron spectroscopy (TDXPS) and angle-dependent XPS (ADXPS) measurements, that the $T_g$ of the polymer becomes smaller than the bulk value as it moves toward the photosensitive surface [12]. Xie et al. measured PS thin films using positron annihilation lifetime spectroscopy (PALS), with the near-surface region as a probe, and reported that the surface $T_g$ is comparable to the bulk $T_g$ [13]. On the other hand, Jean et al. also measured the near-surface $T_g$ of PS thin films using PALS and reported a decrease from the bulk value when the probe depth exceeded 40 nm [14]. Thus, there are examples where different results have been reported even for similar measurement methods, highlighting the need for more precise analysis to measure local areas at surfaces and interfaces. The effect of thin film surfaces on $T_g$ is often attributed to free surface effects [9–11]. Recently, surface $T_g$ has been measured using near-edge X-ray absorption fine structure (NEXAFS) spectroscopy and scanning probe microscopy (SPM). Tanaka et al. used lateral force microscopy (LFM) and scanning viscoelasticity microscopy (SVM) to study the surface molecular motion of PS thin films and found that the films were in a glass-rubber transition state even at 293 K [15–17]. They concluded that this was due to the surface segregation of the chain end groups.

On the other hand, measuring $T_g$ at the interface with the support is a limited approach. Roth et al. measured the $T_g$ of individual layers supported by different polymers by selectively placing fluorescent dyes on the polymer bilayer [18]. PS thin films supported on bulk PS showed a decrease in $T_g$, while PS thin films supported on bulk PMMA did not, suggesting that the interface between the two polymers affects the dynamics of the nanoscale layers. On the basis of inelastic neutron scattering measurements, Kanaya, Inoue, and co-workers claimed the existence of a dead layer with a thickness of ca. 5 nm [19]. So far, several groups further looked into the substrate effect and illuminated the role of a physically adsorbed (i.e., via physisorption) polymer layer of several nanometers thickness ("adsorbed nanolayer") at the polymer–substrate interface in the deviation of $T_g$ in polymer thin films from that in the bulk. Tsui et al. clarified the existence of a polymer residual layer [20] (also called dead layer [21]) is a very thin interfacial polymer layer at the substrate, and in most cases, this thin layer has no thermal expansion coefficient or molecular mobility [22].

Tanaka et al. investigated the segment mobility at the interface between PS and inorganic solids using fluorescence lifetime measurements with evanescent wave excitation and reported that the $T_g$ at the interface is higher than in the bulk [23, 24]. Thus, many spectroscopic approaches have been studied to investigate $T_g$ variations at interfaces. However, direct characterization of the structure and properties of buried interfaces continues to pose significant obstacles.

Consequently, we proposed a methodology for assessing the dynamics of polymer chains near solid interfaces by mechanical measurements utilizing a quartz crystal microbalance (QCM) and examined the impact of viscoelastic alterations on material properties. We believe that this technique will strengthen the understanding of the dynamical properties of buried interfaces.

## 10.2 Viscoelastic Measurements for Soft Matter by Using Quartz Crystal Microbalance

The piezoelectric oscillation of a quartz resonator has been used as an ultra-sensitive mass sensor, utilizing the Sauerbrey relationship between the resonance frequency and the mass per unit area deposited on the crystal. This relationship has enabled the quartz crystal microbalance (QCM) to be a mainstay of vacuum science by applying thickness measurements of deposition films and a gas sensor. The AT-cut quartz crystal resonator is a typical quartz crystal resonator comprising an extremely thin quartz crystal cut along the AT plane with thin metal film electrodes attached to both sides. The AT-cut quartz resonator vibrates in the thickness-shear type at a certain frequency, which is called a resonance frequency, by applying an alternating-current (AC) voltage to electrodes. The resonance frequency of the quartz resonator depends on the thickness of the crystal and is typically in the order of $10^6$ Hz. If a thin film deposited on one surface of the resonator has the exact same acoustic properties as the resonator, the thickness of the crystal increases by the thickness of the deposited thin film, the wavelength of the vibration increases, and the resonance frequency decreases accordingly. Therefore, the change in mass on the surface can be calculated from the resonance frequency change. It measures an mass density in the units of mass per unit area with high sensitivity in the range of ng/cm$^2$. The relationship between the mass per unit area and the resonance frequency changes is reported as the following by Sauerbrey [25]:

$$\Delta f = -\frac{2nf_0^2}{Z_q}\Delta m \tag{10.1}$$

where $\Delta f$ and $\Delta m$ represent the resonance frequency change and the change in mass per unit on the quartz resonator, respectively. $n$ is an overtone number of the quartz crystal, which resonates at odd harmonics. $f_0$ is the fundamental frequency of the quartz crystal depending on the thickness of the crystal. $Z_q$ is a constant called the acoustic impedance of AT-cut quartz crystal ($8.8 \times 10^6$ kg•m$^{-2}$•s$^{-1}$). Eq. (10.1) means that the mass sensitivity of QCM increases with increasing $n$ and $f_0$.

Although QCM has been used primarily in air or vacuum, the application of QCM has expanded in the electrochemical and biological field, such as an electric double layer, adsorption of proteins, polymers, and surfactants in solution, antigen–antibody reactions, and molecular recognition because it is possible to oscillate a quartz resonator in liquids. Kanazawa and Gordon demonstrated that $\Delta f$ depends not only on the mass change of the deposition but also on the viscosity in the liquid by evaluating the viscosity of sucrose and glucose aqueous solution with each concentration by using QCM. The viscosity of the liquid in which the vibration of the resonator propagates acts as a load on the resonator and reports the relationship between $\Delta f$ and the viscosity of Newtonian liquid and the penetration as follows [26, 27]:

$$|\Delta f| = |\Delta\Gamma| = \sqrt{\frac{nf_0^3}{\pi Z_q^2}}\sqrt{\eta\rho} \tag{10.2}$$

$$\delta = \sqrt{\frac{\eta}{\pi f_0 \rho}} \tag{10.3}$$

where $\Delta\Gamma$ is the dissipation defined by the half-bandwidth at half-height of a resonance peak and means the imaginary part of a resonance frequency. The absolute values of $\Delta f$ and $\Delta\Gamma$ are equal, moreover, $\Delta f$ and $\Delta\Gamma$ are reciprocals of each other. $\eta$ and $\rho$ represent the viscosity and density of the Newtonian liquid. The viscosity of Newtonian liquids is independent of a shear rate, so it is the same at the higher frequency measured by QCM. $\delta$ is the penetration depth of the vibration of the resonator in Newtonian liquid. The vibration amplitude of the quartz resonator attenuates exponentially from the interface. It is defined as the distance at which the amplitude is $1/e$ and means the analytical depth of the resonator. When the quartz resonator has $f_0$ of 9 MHz, the $\delta$ in water is approximately 190 nm. The equation is extended to dynamic modulus to evaluate viscoelastic materials as follows:

$$G' = \frac{\pi^2 Z_q^2}{\rho f_0^2}\left(\Delta\Gamma^2 - \Delta f^2\right) \tag{10.4}$$

$$G'' = -\frac{2\pi^2 Z_q^2}{\rho f_0^2}\left(\Delta f \circ \Delta\Gamma\right) \tag{10.5}$$

where $G'$ and $G''$ are storage and loss moduli, respectively. Those equations are held according to the simple model, which assumes that the quartz resonator is in contact with a homogeneous viscoelastic material over an area sufficiently larger than the propagation constant of the resonator (Fig. 10.1a). In viscoelastic materials, the penetration depth is expressed as follows [28]:

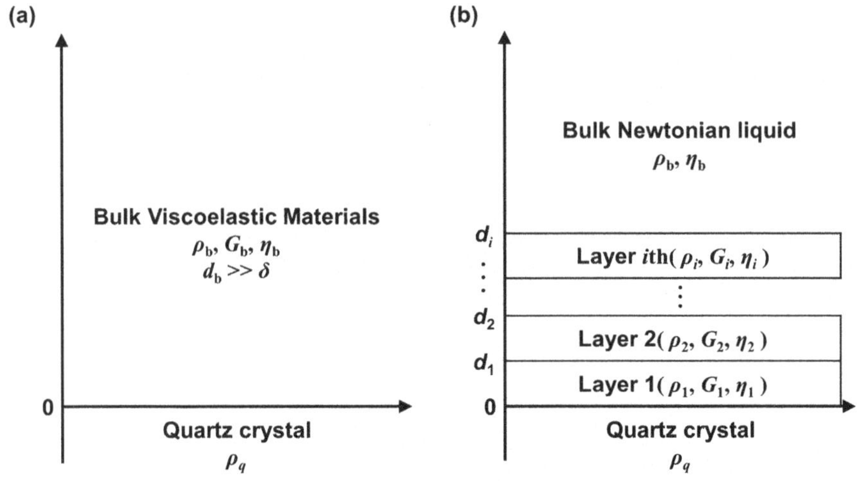


**Fig. 10.1** Geometry of the quartz sensor (**a**) in contact with a homogeneous viscoelastic material and (**b**) covered by the multilayer viscoelastic film in a bulk Newtonian liquid

$$\delta = -\frac{1}{2\pi n f_0 \sqrt{\rho}} \frac{|G^*|}{\sqrt{G''}} \tag{10.6}$$

where $G^*$ is a complex modulus ($G^* = G' + iG''$). The shear modulus as determined with the eqs. (10.4) and (10.5) is the shear modulus at the respective frequency. It must not be confused with the low-frequency shear modulus. The interfacial structure is generally heterogeneous due to adsorption, orientation, and segregation to the vertical direction of an interface. Assuming the sample is homogeneous in the resonator plane, it can be modeled as stratified layer systems vertically from the interface with the quartz resonator. Figure 10.1b shows the geometry of the quartz sensor covered by the $i$th multilayer viscoelastic film with the thickness of $d_i$. The covered quartz sensor oscillates in a bulk Newtonian liquid. In case the thickness of multilayer viscoelastic film is thin, which is smaller than δ, $\Delta f$, and $\Delta\Gamma$ are indicated as follows [29]:

$$\Delta f \approx -\frac{f_0}{2n\pi\rho_q}\left\{\frac{\eta_b}{\delta_b} + \sum_i \left[2n\pi f_0 d_i \rho_i - 2d_i \left(\frac{\eta_b}{\delta_b}\right)^2 \frac{G_i''}{|G_i^*|^2}\right]\right\} \tag{10.7}$$

$$\Delta\Gamma \approx \frac{1}{2n^2\pi\rho_q}\left\{\frac{\eta_b}{\delta_b} + \sum_i \left[2d_i \left(\frac{\eta_b}{\delta_b}\right)^2 \frac{G_i'}{|G_i^*|^2}\right]\right\} \tag{10.8}$$

where $\rho_q$ is the density of the quartz. $\eta_b$ is the viscosity of the bulk Newtonian liquid, and $\delta_b$ is the penetration depth in the bulk. $\rho_i$ $d_i$, $G'_i$, $G''_i$, and $G^*_i$ are density, thickness, storage modulus, loss modulus, and complex modulus, respectively. From eqs. (10.7) and (10.8), it follows that the contribution of the ultrathin films ($d_i \sim 0$) is small in comparison with the bulk liquid acoustic response as shown in term 1. However, a thin layer with a finite thickness will demonstrate a different acoustic response depending on the viscoelasticity. Considering the slip in the boundary makes the analysis more confusing. The QCM excels at measuring viscoelasticity by applying mechanical vibration to a sample on an electrode substrate, which is difficult to measure using other viscoelasticity measurement methods. In addition, it has excellent spatial resolution in the vertical direction of the interface on the order of nanometers.

## 10.3 Viscoelastic Measurement of Adsorbed and Grafted Polymer Monolayer

Adsorbed and grafted polymers onto a solid substrate functionalized to the solid surfaces and applied to the materials in the biomedical and industrial fields, such as inhibitory adsorbing proteins and thrombocytes, and dispersibility of solid particles. Control of the surface functionalities requires an understanding of the structure and dynamics of the adsorbed and grafted polymers, which differ from the bulk properties because a part of the chains is bound on the solid substrate. The structure and dynamics of the bound chains have been evaluated by scanning probe microscopy, which applies force from the surface, and spectroscopic techniques using an evanescent field and a neutron beam. This chapter presents studies that evaluated the viscoelastic properties of adsorbed and grafted polymer monolayers by using QCM.

Polymer adsorption at the solid/liquid interface is influenced not only by the chemical composition of polymers but also by the arrangement of monomers within the polymer chain. Designing polymers with precise sequences has been challenging, restricting investigations of sequence effects on adsorption behavior to molecular simulations. Composto et al. utilize polypeptide synthesis, which allows for controlled sequence design, to examine how small changes in sequence affect adsorption at the solid/liquid interface [30]. They study the adsorption behavior of three polypeptides, consisting of 90% lysine and 10% cysteine, onto a gold surface by using a quartz crystal microbalance with dissipation monitoring (QCM-D) and total internal reflection ellipsometry. The viscoelastic properties of poly peptides with different adsorption conformations were evaluated in the relationship between the resonance frequency shift ($\Delta f$) and the dissipation shift ($\Delta D$), which effect relates the ratio of energy loss to the total energy stored in the system. The frequency shifts indicate changes in mass per unit area accompanying the adsorption. The dissipation factor provides insight into the conformation of the adsorbed polymer layer. A large shift in the dissipation indicates a loosely bound polymer layer, while a tightly bound layer results in less energy loss and thus a smaller shift in the dissipation (Fig. 10.2a). The polypeptides containing evenly spaced cysteines ($(K_5CK_9CK_4)_2$) showed a smaller shift in $\Delta D$, despite having the largest $\Delta f$ (the red arrow in Fig. 10.2b), suggesting that the adsorbed polypeptides layer adopted a more tightly bound train conformation, in which the segments are laid on the gold surface.

The adsorption isotherms and kinetics of polymers in solution were frequently measured by QCM. However, current methods for analysis of QCM data often assume that the adsorbed layer is a homogeneous medium, which does not accurately describe solution-adsorption physics. The liquid solvent including the adsorbed layer is included in the mass, and the mass per unit area of adsorptions is overestimated. Furthermore, they do not consider friction between the adsorbed polymers and the free solvent flowing in the adsorbed layer. Radke et al. developed a two-phase (2P) continuum model by considering the contributions of adsorbate and liquid solvent based on self-consistent, which includes friction between the adsorbed polymers and free solvent, and the 2P model was compared with Sauerbrey for a rigid adsorbed layer and the multilayer model proposed by Voinova et al. The 2P model agrees with the multilayer model within experimental error at the finite thickness, but they diverge at large adsorbed thickness by increasing the contribution of friction between the adsorbed polymers and free solvent. The 2P model can be applied to various adsorption scenarios, which are not limited to polymers [31].

Gillissen et al. simultaneously determine the viscoelasticity of irreversible adsorbed polymers, reversible adsorbed polymers on the substrates, and bulk polymers in solutions by using QCM-D and successfully decouple these viscoelastic contributions to quantify the kinetic and conformation of adsorbed polymers [32].

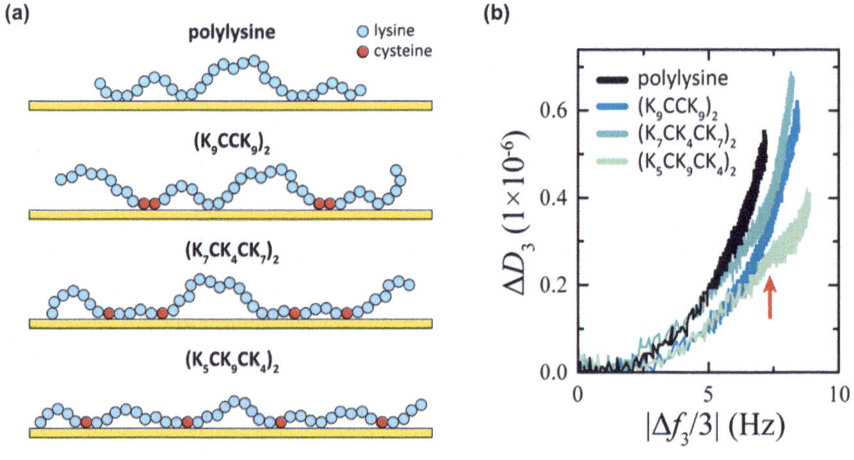


**Fig. 10.2** (**a**) Schematic images of the adsorbed polypeptide chains of the different polymer sequences. (**b**) Dissipation ($\Delta D$) vs resonance frequency shift ($\Delta f$) plots show changes in the viscoelastic behavior of the adsorbed polypeptide layer with increasing mass. (Reprinted with permission from *Langmuir* **2023**, *39*, 1740–1749. Copyright 2023 American Chemical Society)

The irreversible adsorbed polymer has significant changes in the conformation with many adhesion sites per polymer at sufficient adhesion energy. The reversible adsorbed polymer is in dynamic equilibrium with bulk polymers and has fewer adhesion sites per polymer and less significant conformational changes. The viscoelasticity of bulk polymers ($\nu_{bulk}$) was evaluated from the resonance frequency shift during polyethylene glycol (PEG) solution loading on the coated sensor to avoid the adsorption and agree reasonably well with the Zimm model, which predicts $\nu_{bulk} \sim \omega^{-1/3}$ at high ω when the molecular weight of PEG is 0.2 kg/mol or greater. The viscoelasticity of the adsorbed polymers suggested that the reversible adsorbed polymers have a similar viscoelasticity as the bulk polymers because reversible adsorption does not change the polymer conformation and is similar conformation as in the bulk. On the other hand, the irreversible adsorbed polymers are less flexible than the bulk polymer due to the small loops of the adsorbed chains compared to the coils in the bulk. They demonstrated that QCM can evaluate the dynamics of polymers with different conformations adsorbed on the solid surface.

Modification of the quartz sensor surface enables QCM to measure the surface adsorption of active species onto different substrates. Hodges et al. evaluated the adsorption activity of sodium polystyrene sulfonate (NaPSS) on different substrates [33]. Figure 10.3 shows that differences in both $\Delta f$ and $\Delta D$ are measured with low NaPSS concentrations when the sensor surface is changed from silica (no NaPSS adsorption) to alumina (NaPSS adsorption). $\Delta f$ below 1 wt% for alumina rising to a plateau frequency equivalent to a 0.9-nm thick Sauerbrey film. $\Delta f$ changes represent the adsorption isotherm of NaPSS at 20 °C. $\Delta D$ changes fit well the Kanazawa–Gordon model adjusted by considering the finite viscoelastic layer on the sensor. The adsorption isotherm of NaPSS on alumina was demonstrated to fit with a Langmuir isotherm by changing the quartz sensor surface.

Polymer brushes, which consist of polymer chains grafted onto a solid substrate, are widely used as a tool for surface modification and provide characteristic

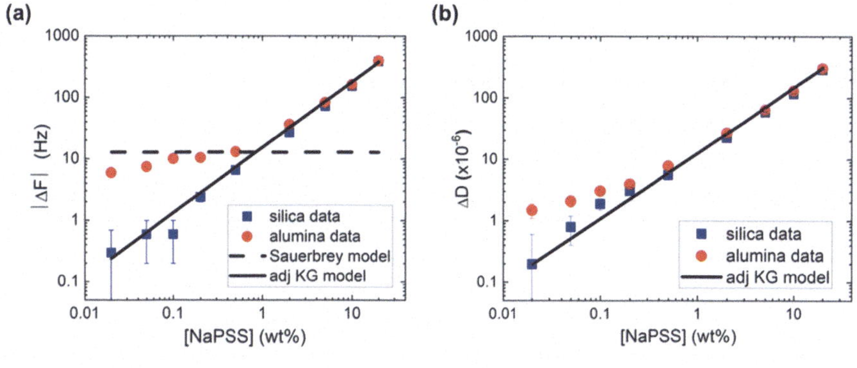


**Fig. 10.3** NaPSS concentration dependence of (**a**) the resonance frequency shift ($\Delta f$) and (**b**) dissipation shift($\Delta D$) on the silica (no NaPSS adsorption) and alumina (NaPSS adsorption) surface. (Reprinted with permission from *ACS Omega* **2020**, *5*, 22,081–22,090. Copyright 2020 American Chemical Society)

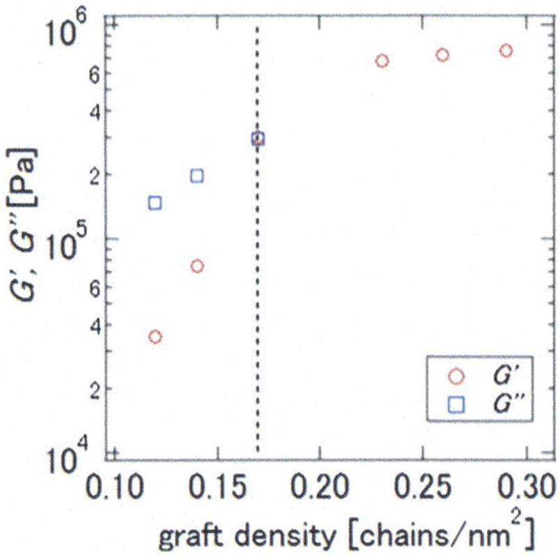


**Fig. 10.4** Storage ($G'$, red circles) and loss ($G''$, blue square) moduli of polymer brush depending on the graft density. Black dashed vertical line denotes the graft density at which $\Delta\Gamma$ reached its peak. (Reprinted with permission from *Langmuir* **2017**, *33*, 5166–5172. Copyright 2017 American Chemical Society)

properties such as preventing adhesion and lubrication. The graft density is an important factor in determining the properties of polymer brushes. Yokoyama et al. evaluated the formation kinetics of polymer brushes prepared by the grafting-to method using QCM and neutron reflectivity (NR) [34]. $G'$ and $G''$ of polymer brushes were calculated based on the three-layer model (first; quartz, second; visco-elastic film, third; Newtonian liquid) from $\Delta f$ and $\Delta D$ in the process of polymer brush formation. Once a certain brush density was reached, the crossover between $G'$ and $G''$ was observed (Fig. 10.4), and the brush layer changed from a viscous layer to an elastic layer like hydrogels. The brush layer behaving as an elastic layer is expected to work as a layer to prevent the adhesion of foreign bodies with its elastic repulsion. Therefore, the characteristic brush density determined by QCM may be defined as an antifouling threshold brush density. Effects of graft density on the characteristic properties of polymer brushes, such as the coil-globule transition of the thermoresponsive polymer [35–37] and the hydration state [38, 39], evaluated by QCM.

## 10.4 Surface and Interfacial Gelation of Thermoresponsive Polymer

Thermoresponsive polymers are applied as surface modifications for the temperature switching of hydrophilic and hydrophobic properties through adsorption and grafting on solid substrates. The polymer chains adsorbed on a solid interface take various adsorption conformations (e.g., loop, train, and tail) unlike in the bulk. The mobility of the polymer chains is greatly reduced by bounding to the interface and influence the transition behavior. In addition, the interfacial gelation phenomenon of thermo responsive polymers has long been known, and the thermoresponsive sol–gel transition behavior at solid interfaces is complex. So, the current understanding of the influence of polymer chains bound to the solid surface on the transition behavior of thermoresponsive polymers is rather restricted. We evaluated the thermoresponsive sol–gel transition of aqueous methylcellulose (MC) solutions to clarify the effects of substrates [40, 41], segregation [42] of MC, and adsorbed chains [43] by using QCM and NR.

MC is a cellulose derivative with a part of hydroxy groups of cellulose modified to hydrophobically methoxy groups. When the average degree of substitution (DS) of a methoxy group is between 1.7 and 2.0 per anhydroglucose repeat unit, it becomes soluble in water at low temperatures because the substituted methoxy groups break the hydrogen bonds between the hydroxy groups of cellulose. The aqueous MC solution transforms into a turbid hydrogel upon heating because dehydrated hydrophobic groups aggregate and form physical crosslinks, and it returns to the sol state when cooled. It shows the thermoreversible sol–gel transition behavior.

Shear modulus changes of aqueous MC solutions with the sol–gel transition were determined by the frequency and dissipation changes of quartz resonators. The QCM measurement was conducted with the homemade QCM measurement system using a vector network analyzer (DG8SAQ VNWA, SDR-Kits) as shown in Fig. 10.5. $\Delta f$ and $\Delta\Gamma$ were monitored with the temperature changes in the aqueous MC solutions, and temperature dependences of $\Delta f$ and $\Delta\Gamma$ of the aqueous MC solution with 10 times the concentration of an overlap concentration of MC ($c^* = 0.58$ wt%) are shown in Fig. 10.6. The gradual increase in $\Delta f$ (decrease in $\Delta\Gamma$) between 10 and 40 °C was caused by a decrease in the solution viscosity associated with increasing temperature. $\Delta f$ rapidly decreased (increase in $\Delta\Gamma$) at temperatures above 45 °C because the gelation of the aqueous rapidly increased the solution viscosity and elasticity. Subsequently, the gelation was complete at 60 °C and higher; therefore, $\Delta f$ and $\Delta\Gamma$ displayed constant values. When the MC gel was cooled from 70 to 10 °C, the $\Delta f$ and $\Delta\Gamma$ values did not change until near 40 °C, displaying

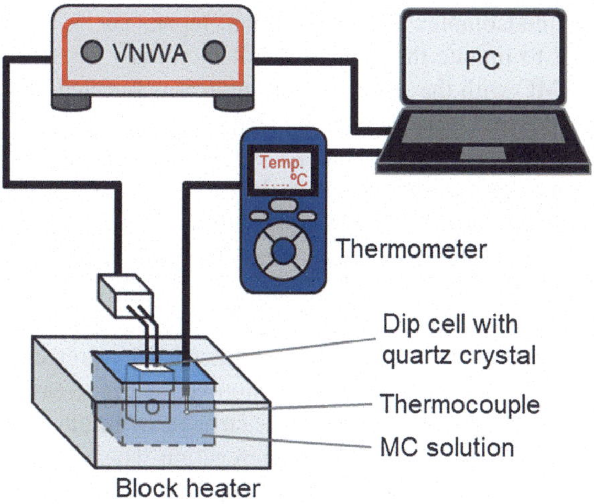


**Fig. 10.5** Schematic image of the homemade QCM equipment

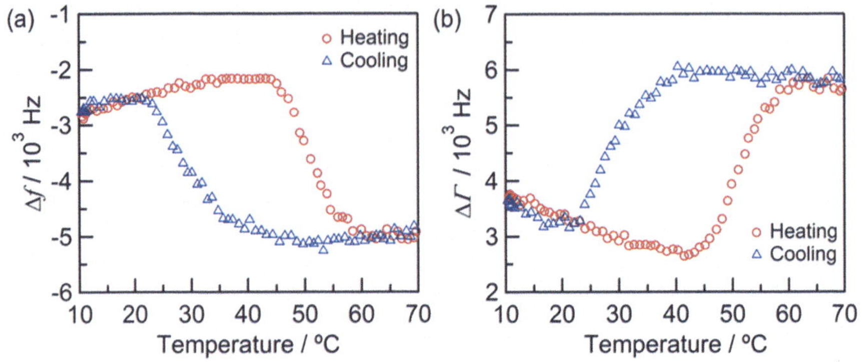


**Fig. 10.6** Temperature dependence of (**a**) $\Delta f$ and (**b**) $\Delta\Gamma$ of the aqueous MC solution with the concentration of 10$c$*

constant. From approximately 35 °C, $\Delta f$ rapidly increased ($\Delta\Gamma$ decreased) to a value similar to the preheating value at 20 °C and below. Hysteresis and thermal reversibility of the aqueous MC solution were observed as changes in $\Delta f$ and $\Delta\Gamma$ during the QCM measurements, empirically demonstrating that QCM can be used to evaluate the sol–gel transition. In addition, the gelation of the aqueous MC solution was measured by changing the substrate of quartz sensors to gold (Au), silicone (Si), and silicone oxide ($SiO_2$). The hydrophobic-treated Si substrates lowered the gelation temperature because MC adsorbed on the hydrophobic surface and increased the concentration of MC at the interface between the MC solution and solid substrate of the quartz sensor [40, 41].

The interfacial structure between the aqueous MC solution and substrate of the quartz sensor was observed by neutron reflectometry with the Soft Interface Analyzer (SOFIA) at BL16, Materials and Life Science Facility, Japan Proton Accelerator Research Complex (J-PARC), Tokai, Japan. The Si substrate sputtered with Au was used to imitate the Au surface of the quartz sensor. The deuterium oxide solution of MC with the 10$c$* concentration was injected into the dedicated module and attached to the Au substrate to make the solid/liquid interface. The neutron reflectivity curves were measured at 25–70 °C. The compositional distribution of MC perpendicular to the interface with the Au substrate was evaluated by the fitting analysis based on the model neutron scattering density ($b/V$) profile to discuss the adsorption and interfacial segregation. The volume fraction of MC ($\Phi_{MC}$) depending on the distance from the Au substrate ($z$) is shown in Fig. 10.7. The $\Phi_{MC}$ significantly increased from around 10 nm at 25 °C. It is suggested that the MC chains are adsorbed onto the Au interface. The interfacial segregation with a gradient in the depth direction was observed and improved with increasing the temperature. The hydrophobic interaction of MC was stronger with the dehydration at the high temperature and caused the interfacial segregation. The gradient spread greater than the adsorption thickness, suggesting that the interfacial segregation occurred

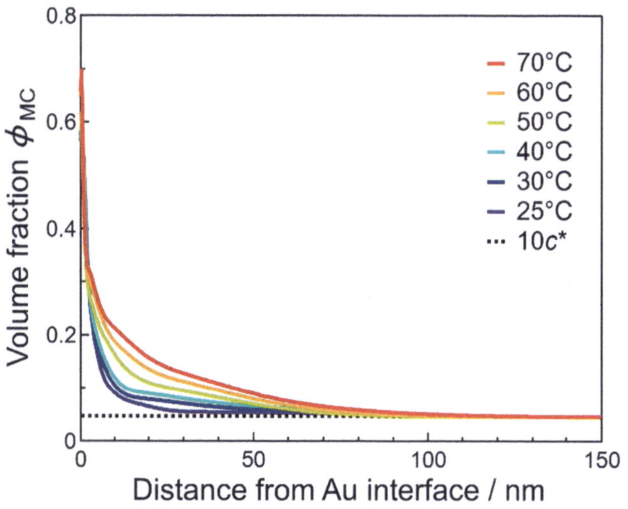


**Fig. 10.7** Volume fraction of MC at the interface between the Au substrate and the deuterium oxide solution of MC with the 10$c$* concentration

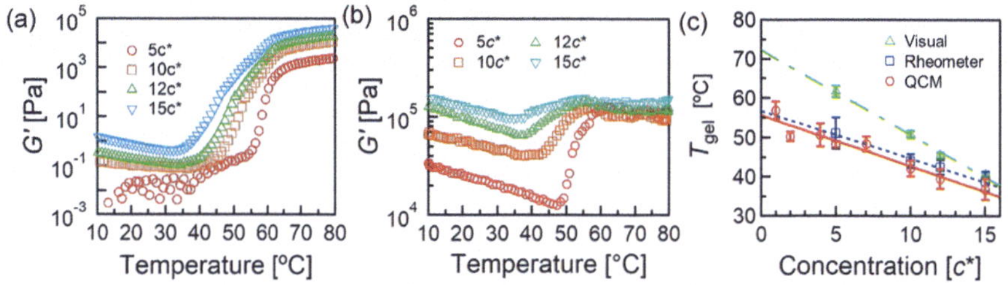


**Fig. 10.8** MC concentration dependence of the storage modulus ($G'$) accompanying the gelation during the heating process (**a**) evaluated by rheology measurements and (**b**) estimated via the QCM. (**c**) MC concentration dependence of the gelation temperature decided from visual observations ( $T_{\text{gel}}^{\text{vis}}$ ) in the bulk ( $T_{gel}^{b}$ ) evaluated by rheology measurements, and at the interface of the gold substrate ( $T_{\text{gel}}^{\text{Q}}$ ) evaluated by the QCM. Each color line shows the linear fitting

not only because of the hydrophobic interaction but also because of entanglements of MC and long-length van der Waals interaction between MC and Au [42].

The shear modulus of the aqueous MC solution was calculated from $\Delta f$ and $\Delta\Gamma$ based on the Kanazawa–Gordon (KG) model (Eqs. 10.4 and 10.5). Although the MC concentration gradient at the Au interface was confirmed from the interfacial structure analysis via NR, the analysis here assumed that the MC solution layer is homogeneous to simplify the calculation. Temperature dependence of a storage modulus ($G'$) of aqueous MC solutions evaluated by a rheometer and QCM is shown in Fig. 10.8a, b, respectively. Both figures showed a rapid increase in the $G'$ caused by the gelation. The onset temperature of the increase in $G'$ was defined as the gelation temperature ($T_{\text{gel}}$), and it decreased with increasing the MC concentration. The

$G'$ value evaluated by QCM was larger than by the rheometer measured at 1 Hz. The shear modulus, evaluated by QCM, reflected the local motion, such as a segmental motion of a polymer in the glass-to-rubber transition region because QCM measurements are conducted at high frequency (9 MHz). The MC concentration dependence of $T_{\mathrm{gel}}$ evaluated by each technique is shown in Fig. 10.8c. $T_{\mathrm{gel}}$ in the bulk was defined as the onset temperature of the $G'$ change evaluated by rheology measurements ($T_{gel}^{b}$). Conversely, $T_{\mathrm{gel}}$ at the interface with the Au substrate was defined as that estimated via QCM ($T_{\mathrm{gel}}^{\mathrm{Q}}$). $T_{\mathrm{gel}}^{\mathrm{vis}}$ was determined by the inclination experiment based on a visual observation. All $T_{\mathrm{gel}}$ decreased monotonically with increasing concentration because the liquid–liquid phase separation, which induces the gelation, tends to occur with the concentration increase. The slope of $T_{\mathrm{gel}}^{\mathrm{vis}}$ was the largest of the others. It was considered that the MC chain network is rearranged during inclination observation because the characteristic time is sufficiently longer than the relaxation time of the network rearrangement. Hence, $T_{\mathrm{gel}}^{\mathrm{vis}}$ was especially higher than other $T_{\mathrm{gel}}$ at the low concentration. Compared with $T_{gel}^{b}$ and $T_{\mathrm{gel}}^{\mathrm{Q}}$, the slope to the MC concentration was almost equivalent. Nevertheless, $T_{\mathrm{gel}}^{\mathrm{Q}}$ was smaller than $T_{gel}^{b}$ by approximately 2–4 °C within the applied concentrations in this work. Here, we are not denying the possibility that gelation at the QCM time scale occurs at a much lower temperature. We considered that the origins of the gelation evaluated by the rheometer and by the QCM might be different due to the determination of $T_{\mathrm{gel}}$. Further experiments are needed to clarify the origin of fast dynamics evaluated by the QCM. However, assuming that $T_{\mathrm{gel}}$ reflects an increase of the $G'$ due to the gelation, the $T_{\mathrm{gel}}$ could decrease owing to the aggregation near the interface.

We performed interfacial selective viscoelastic analysis by combining a simple KG model and controlling the analytical depth ($\delta$) of a QCM sensor to elucidate the effect of the bound polymer chains at the interface on the thermoresponsive sol–gel transition behavior of the aqueous MC solution [43]. The $\delta$ of QCM was changed by the overtone number ($n$) based on eq. 10.6. $T_{\mathrm{gel}}^{\mathrm{Q}}$ of aqueous MC solutions was measured by changing the $\delta$ and plotted by $\delta$ for each MC concentration as shown in Fig. 10.9. For the larger $\delta$, $T_{\mathrm{gel}}^{\mathrm{Q}}$ was independent of $\delta$ and constant for each

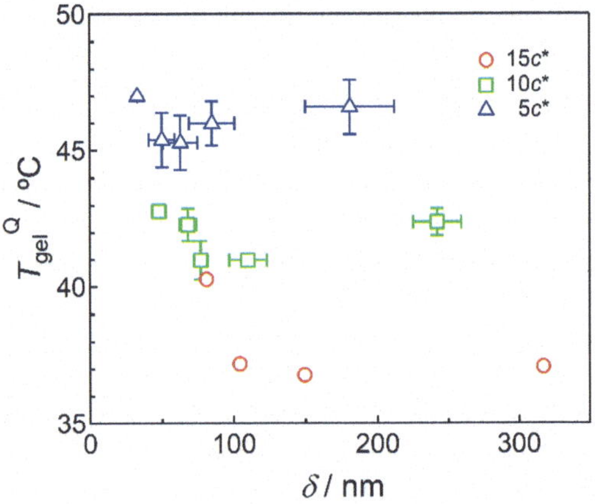


**Fig. 10.9** Analytical depth ($\delta$) dependence of the gelation temperature ($T_{\mathrm{gel}}^{\mathrm{Q}}$) for the aqueous MC solutions at each concentration

concentration. $T_{gel}^{Q}$ at a larger δ decreased with an increase in the MC concentration. The tendency of decreasing $T_{gel}^{Q}$ at the largest δ ($n = 1$) was consistent with the concentration dependence of the bulk. By contrast, $T_{gel}^{Q}$ increased near the solid interface when δ became shorter. The adsorption of the polymer at the surface slowed the chain mobility. As a consequence, the reduction of the MC chain mobility at the interface prevented the chain aggregation accompanying the sol–gel transition, causing the $T_{gel}^{Q}$ to increase. The increasing trend of $T_{gel}^{Q}$ near the interface changed with the MC concentration. This trend was observed below 50, 75, and 100 nm at $5c^*$, $10c^*$, and $15c^*$, respectively. The adsorbed thickness in the concentrated solution is not expected to change. The distance at which $T_{gel}^{Q}$ began to increase is larger than the extension of MC chains ($2R_g \simeq 30$ nm) and suggested nonmonolayer adsorption. The entanglement of MC chains increased with the concentration and propagated the interfacial interaction between the substrate and polymers to be longer.

## 10.5 Glass Transition Temperature

The previous section established using QCM that the sol–gel transition behavior of thermoresponsive polymers at solid interfaces exhibits interfacial properties that differ from those of the bulk. Furthermore, it was revealed that this behavior originates from the structure of polymer chains in contact with the interface and their interactions with the solid. This finding indicates that evaluating interfacial viscoelasticity by QCM could work as an effective approach for elucidating the viscoelastic properties of polymers at buried interfaces.

The introduction highlights extensive discussion on the effects of air surface and substrate interface on the $T_g$ of polymer thin films. At the air interface, $T_g$ decreases due to the activation of molecular chain thermal mobility prompted by free volume. At the interface, $T_g$ increases due to the restriction of the thermal molecular motion of polymer chains by the confinement effect. Many studies of $T_g$ in these polymer thin films have considered spectroscopic measurement approaches. However, differing from the bulk and surface, there are very few studies in which $T_g$ near the substrate interface has been directly evaluated by a mechanical approach. While there are limited studies reporting $T_g$ measurements of polymer thin films in air using QCM [44–47], these investigations have predominantly focused on microscale thick films, with no consideration of the "interfacial effect" in thin films. In this section, we investigated the $T_g$ of polymers in contact with heterogeneous solid interfaces using QCM, focusing on nanoscale polymer thin films. Notably, we successfully measured the $T_g$ of PS thin films near the Au substrate interface, and this achievement is highlighted.

PS thin films with thicknesses ranging from 30 to 400 nm were prepared by spin-coating on plasma-etched Au electrodes using toluene as the solvent. After spin-coating, thermal annealing in vacuo was applied at 150 °C for 12 h to remove the residual solvent. The stiffness (shear modulus) and density of quartz crystals are known to decrease with increasing temperature. The relationship between the resonance frequency of AT-cut quartz and temperature can be expressed by the following cubic polynomial equation [48]:

$$\Delta f_{\mathrm{T}} = a_0 + a_1 T + a_2 T^2 + a_3 T^3 \tag{10.9}$$

where $T$ is the temperature of the quartz crystal and $a_0$–$a_3$ are temperature coefficients. These coefficients vary depending on slight differences in the cutting angle of the quartz and the heating rate. In order to precisely assess the frequency variation caused by changes in the physical properties of the thin film, we focused solely on measuring the $\Delta f_{\mathrm{T}}$ of the quartz crystal under the same conditions as the sample measurement. Subsequently, we analyzed the temperature dependence of $\Delta f$ after subtracting it as a reference.

Accurately identifying the temperature of the quartz crystal is crucial when assessing the temperature dependency of $\Delta f$. Previous studies have reported measurements of first-order phase transition phenomena, such as polymer crystallization and metal melting, based on changes in $\Delta f$ [49, 50]. To achieve precise temperature calibration, a standard material with a known melting point was deposited over the quartz crystal oscillator, and its melting point was measured through the observed frequency shift. This method improved the accuracy of temperature calibration, therefore guaranteeing the reliability of $T_{\mathrm{g}}$ evaluation for PS thin films. Figure 10.10 shows the relationship between $\Delta f$ and temperature for indium during the heating process. The melting point ($T_{\mathrm{m}}$) of indium is 156.6 °C. As the temperature increased, $\Delta f$, which was initially kept constant, showed a dramatic decrease at a specific temperature before stabilizing at a low constant value. In the solid state, indium exhibits high rigidity, leading to negligible vibrational energy dissipation at the contact interface. On the other hand, when the temperature exceeds the melting point and the material transitions to a liquid state, the dissipation of vibrational energy increases due to the liquid's viscosity, resulting in an immediate reduction in $\Delta f$. This indicates that the rapid decrease in $\Delta f$ resulting from the temperature increase denotes the melting of indium and that the surface temperature of the QCM

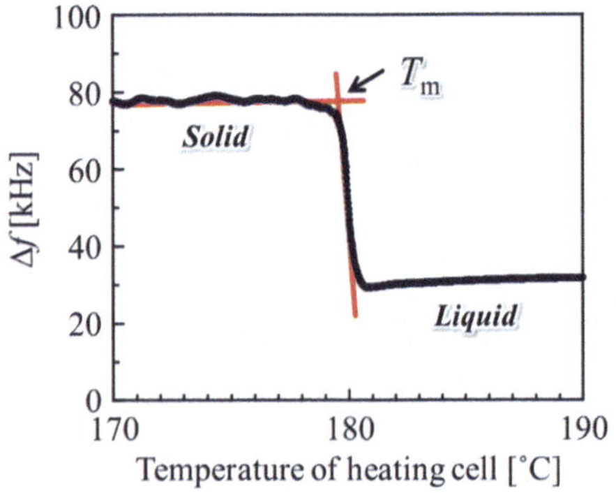


**Fig. 10.10** The temperature dependence of $\Delta f$ of indium

sensor attained 156.6 °C. The evaluation of the melting points of several reference materials demonstrates that QCM can be reliably evaluated across a wide temperature range.

Figure 10.11 shows the temperature dependence of $\Delta f$ for a PS thin film during the heating process, as measured using QCM. The molecular weight of PS is 96.4 kg/mol, and the film thickness is about 200 nm. The dashed line indicates the bulk $T_g$ of PS determined by differential scanning calorimetry (DSC). As the temperature increased, $\Delta f$ decreased, and two distinct changes in slope were observed at certain temperatures. An inflection point was noted at a temperature comparable to $T_g^{DSC}$, while another inflection point was observed at a significantly higher temperature. The response of a quartz crystal oscillator coated with a viscoelastic film can be described using an equivalent circuit model. The relationship between $\Delta f$ and the complex modulus can be expressed by the following equation [51]:

$$\Delta f_n = \frac{f_0}{\pi Z_q} \sin\left(\frac{\phi_n}{2}\right) \sqrt{\rho_f \left|G_n^*\right|} \tag{10.10}$$

where $\Delta f_n$ represents the frequency shift, $n$ is the overtone number, $Z_q$ is the complex impedance of the quartz, $f_0$ is the fundamental resonance frequency of the quartz, $\phi_n$ is the phase angle, $\rho_f$ is the density of the thin film, and $G_n^*$ is the complex modulus. This equation indicates that variations in $\Delta f$ are associated with changes in the density and complex modulus of the film. The decrease in $\Delta f$ observed at temperatures below $T_g^{DSC}$ is attributed to the reduction in storage modulus and the density decrease of the PS thin film due to thermal expansion as the temperature increases. The dramatic change in the slope of $\Delta f$ at a specific temperature indicates the shift of the PS thin film from a glassy to a rubbery state, leading to considerable change in modulus and thermal expansion coefficient. Therefore, the temperatures at which the slope changed were defined as $T_g^{low}$ and $T_g^{high}$, from the lower temperature side, respectively. It is considered that the molecular chains near the solid interface are restricted in their thermal motion due to confinement effects from the substrate, leading to an increase in $T_g$. Consequently, it can be interpreted that $T_g^{low}$ reflects the

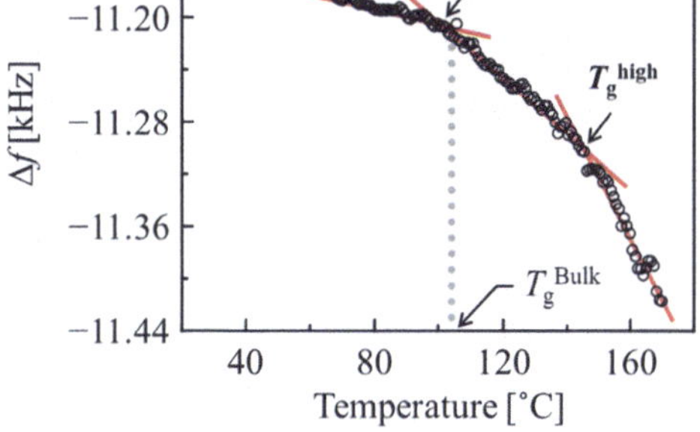


**Fig. 10.11** The temperature dependence of $\Delta f$ of PS thin films. The temperature increase rate is 1 °C/min, and the data are for the second run, which removes the effect of thermal history

$T_g$ in the bulk and $T_g^{high}$ reflects the $T_g$ near the substrate interface. It is assumed that the hierarchical glass transition phenomenon of the PS thin film was observed as the temperature increased.

In the next step, we examined the effect of film thickness on $T_g$. Figure 10.12 shows the thickness dependence of $T_g^{low}$ and $T_g^{high}$ for PS thin films, as measured by QCM. The dotted line in the figure represents the $T_g^{DSC}$ (the bulk $T_g$ measured by DSC).

The shaded gray area indicates the range of $T_g$ values for PS thin films on silicon substrates that have been reported using techniques such as ellipsometry, X-ray reflectivity, and positron annihilation lifetime spectroscopy [52–55]. In general, the thickness dependence of $T_g$ in thin films is known to be described by the following Keddie's empirical equation [5]:

$$T_g(d) = T_g^{\infty}\left[1-\left(\frac{A}{d}\right)^{\delta}\right] \tag{10.11}$$

where $d$ is the film thickness, $T_g^{\circ}$ is the bulk $T_g$, $A$ is a characteristic length, and δ is an exponent. Firstly, $T_g^{low}$ exhibited values comparable to $T_g^{DSC}$ for film thicknesses above 100 nm, with no significant differences observed in the range of 100–400 nm. However, when the film thickness was reduced to below 100 nm, $T_g^{low}$ showed a decreasing trend in relation to the decreasing film thickness. This trend is consistent with the shaded region, suggesting that $T_g^{low}$ corresponds to the bulk $T_g$ of PS thin films. It is particularly noteworthy that when the film thickness decreased, $T_g^{low}$ exhibited a value inferior to that of the shaded gray area. This outcome indicates that, in contrast to measuring techniques such as ellipsometry, QCM can evaluate $T_g$ by eliminating the effects of the substrate interface. On the other hand, $T_g^{high}$ exhibited almost no variation with changes in film thickness and did not show a decreasing trend as the film thickness decreased. This result indicates that $T_g^{high}$

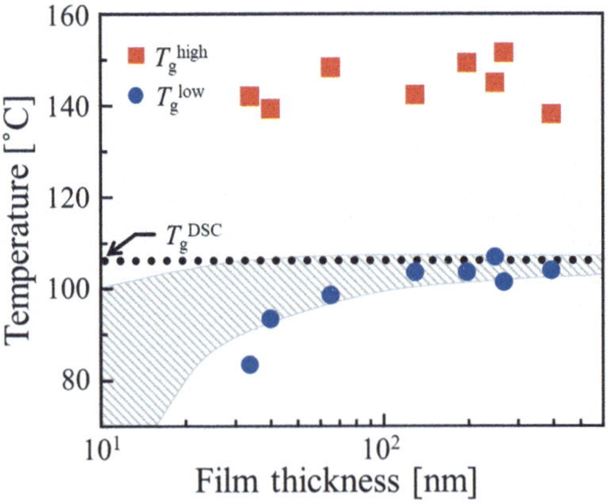


**Fig. 10.12** The thickness dependence of $T_g$ of PS thin film with QCM. The dash line indicates $T_g^{DSC}$, and the shaded area indicates the region where published data from other study [52–55]

mainly reflects the $T_g$ in the region where the surface effect is small, that is, near the substrate interface of the PS thin film.

## 10.6 Conclusion

This chapter presents a method to investigate the molecular motion properties at interfaces by means of QCM, a technique that, while not widely utilized, serves as a micro-mass analysis method. The outcomes obtained recently have been addressed as well. The investigation into the gelation of methylcellulose at the interface revealed that the adsorption of the polymer at the surface slowed chain mobility. Consequently, this reduction in chain mobility at the interface inhibited chain aggregation during the sol–gel transition, resulting in an increase in the gelation temperature. In the evaluation of the thermal mobility of molecular chains near the substrate interface using polystyrene, it was successfully clarified the glass transition behavior at the interface and that of the rest of the thin film separately. In conclusion, various methods for analyzing buried interfaces exist, including fluorescence lifetime measurement and the application of quantum beams like X-rays and neutron beams, which have been mentioned in numerous publications. Even so, it may be a minor interest of our own, but we believe that by sticking to "mechanical stimulation and reaction," even at the interface, as in the traditional and classical dynamic viscoelasticity measurement historically, we could contribute a new perspective to the science of interfacial dynamics, which has been developed by outstanding researchers far beyond our reach.

**Acknowledgments** This research was supported by a Grants-in-Aid for Scientific Research (JSPS KAKENHI) (grant numbers JP16K05926 and JP19H05720), JST SPRING, Grant Number JPMJSP2137, OKASAN-KATO FUNDATION, and Alumni Association research fund of the Faculty of Engineering at Mie University. The neutron scattering experiment was approved by the Neutron Science Proposal Review Committee of J-PARC/MLF (Proposal No. 2018A0271) and supported by the Inter-University Research Program on Neutron Scattering of IMSS, KEK.

## References

1. Karim A, Kumar S (eds) (2000) Polymer surfaces, interfaces and thin films. World Scientific, Singapore/New Jersey/London/Hong Kong
2. Tsui OKC, Russell TP (2008) Polymer thin films. World Scientific, New Jersey/London/Singapore/Beijing/Shanghai/Hong Kong/Taipei/Chennai
3. Jackson CL, McKenna GB (1991) The glass transition of organic liquids confined to small pores. J Non-Cryst Solids 131-133:221–224. https://doi.org/10.1016/0022-3093(91)90305-P
4. Reiter G (1993) Mobility of polymers in films thinner than their unperturbed size. Europhys Lett 23:579–584. https://doi.org/10.1209/0295-5075/23/8/007
5. Keddie JL, Jones RAL, Cory RA (1994) Size-dependent depression of the glass transition temperature in polymer films. Europhys Lett 27:59–64. https://doi.org/10.1209/0295-5075/27/1/011

6. Keddie JL, Jones RAL, Cory RA (1994) Interface and surface effects on the glass-transition temperature in thin polymer films. Faraday Discuss 98:219–230. https://doi.org/10.1039/FD9949800219
7. Prucker O, Christian S, Bock H, Rühe J, Frank CW, Knoll W (1998) On the glass transition in ultrathin polymer films of different molecular architecture. Macromol Chem Phys 199:1435–1444. https://doi.org/10.1002/(SICI)1521-3935(19980701)199:7%3C1435::AID-MACP1435%3E3.0.CO;2-%23
8. Fryer DS, Peters RD, Kim EJ, Tomaszewski JE, White CC, Wu W (2001) Dependence of the glass transition temperature of polymer films on interfacial energy and thickness. Macromolecules 34:5624–5634. https://doi.org/10.1021/ma001932q
9. Ellison CJ, Torkelson JM (2003) The distribution of glass-transition temperatures in nanoscopically confined glass formers. Nat Mater 2:695–700. https://doi.org/10.1038/nmat980
10. Ellison CJ, Torkelson JM (2002) Sensing the glass transition in thin and ultrathin polymer films via fluorescence probes and labels. J Polym Sci B 40:2745–2758. https://doi.org/10.1002/polb.10343
11. Ellison CJ, Kim SD, Hall DB, Torkelson JM (2002) Confinement and processing effects on glass transition temperature and physical aging in ultrathin polymer films: novel fluorescence measurements. Eur Phys J E 8:155–166. https://doi.org/10.1140/epje/i2001-10057-y
12. Kajiyama T, Tanaka K, Takahara A (1995) Depth dependence of the surface glass transition temperature of a poly(styrene-block-methyl methacrylate) diblock copolymer film on the basis of temperature-dependent x-ray photoelectron spectroscopy. Macromolecules 28:3482–3484. https://doi.org/10.1021/ma00113a059
13. Xie L, DeMaggio GB, Frieze WE, DeVries J, Gidley DW (1995) Positronium formation as a probe of polymer surfaces and thin films. Phys Rev Lett 74:4947–4950. https://doi.org/10.1103/PhysRevLett.74.4947
14. Jean YC, Zhang R, Cao H, Yuan JP, Huang CM, Nielsen B, Kumar PA (1997) Glass transition of polystyrene near the surface studied by slow-positron-annihilation spectroscopy. Phys Rev B 56:R8459–R 8462. https://doi.org/10.1103/PhysRevB.56.R8459
15. Kajiyama T, Tanaka K, Takahara A (1997) Surface molecular motion of the monodisperse polystyrene films. Macromolecules 30:280–285. https://doi.org/10.1021/ma960582y
16. Tanaka K, Takahara A, Kajiyama T (2000) Rheological analysis of surface relaxation process of monodisperse polystyrene films. Macromolecules 33:7588–7593. https://doi.org/10.1021/ma000406w
17. Kajiyama T, Tanaka K, Takahara A (1998) Study of the surface glass transition behaviour of amorphous polymer film by scanning-force microscopy and surface spectroscopy. Polymer 39:4665–4673. https://doi.org/10.1016/S0032-3861(98)00049-4
18. Roth CB, McNerny KL, Jager WF, Torkelson JM (2007) Eliminating the enhanced mobility at the free surface of polystyrene: fluorescence studies of the glass transition temperature in thin bilayer films of immiscible polymers. Macromolecules 40:2568–2574. https://doi.org/10.1021/ma062864w
19. Inoue R, Kanaya T, Nishida K, Tsukushi I, Shibata K (2005) Inelastic neutron scattering study of low energy excitations in polymer thin films. Phys Rev Lett 95:056102. https://doi.org/10.1103/PhysRevLett.95.056102
20. Fujii Y, Yang Z, Leach J, Atarashi H, Tanaka K, Tsui OKC (2009) Affinity of polystyrene films to hydrogen-passivated silicon and its relevance to the $T_g$ of the films. Macromolecules 42:7418–7422. https://doi.org/10.1021/ma901851w
21. Koga T, Jiang N, Gin P, Endoh MK, Narayanan S, Lurio LB, Sinha SK (2011) Impact of an irreversibly adsorbed layer on local viscosity of nanoconfined polymer melts. Phys Rev Lett 107:225901. https://doi.org/10.1103/PhysRevLett.107.225901
22. Napolitano S, Pilleri A, Rolla P, Wübbenhorst M (2010) Unusual deviations from bulk behavior in ultrathin films of poly(tert-butylstyrene): can dead layers induce a reduction of $T_g$? ACS Nano 4:841848. https://doi.org/10.1021/nn9014517

23. Tanaka K, Tsuchimura Y, Akabbori K, Ito F, Nagamura T (2006) Time- and space-resolved fluorescence study on interfacial mobility of polymers. Appl Phys Lett 89:061916. https://doi.org/10.1063/1.2335593
24. Tanaka K, Tateishi Y, Okada Y, Nagamura T, Doi M, Morita H (2009) Interfacial mobility of polymers on inorganic solids. J Phys Chem B 113:4571. https://doi.org/10.1021/jp810370f
25. Sauerbrey G (1959) Verwendung von schwingquarzen zur wägung dünner schichten und zur mikrowägung. Zeitschrift fuer Physik A: Hadrons and Nuclei 155:206–222. https://doi.org/10.1007/BF01337937
26. Kanazawa KK, Gordon JG (1985) Frequency of a quartz microbalance in contact with liquid. Anal Chem 57:1770–1771. https://doi.org/10.1021/ac00285a062
27. Johannsmann D (ed) (2015) The quartz crystal microbalance in soft matter research. Fundamentals and modeling. Springer, Cham/Heidelberg/New York/Dordrecht/London
28. Johannsmann D (2021) Studying soft interfaces with shear waves: principles and applications of the quartz crystal microbalance (QCM). Sensors 21:3490. https://doi.org/10.3390/s21103490
29. Voinova MV, Rodahl M, Jonson M, Kasemo B (1999) Viscoelastic acoustic response of layered polymer films at fluid-solid interfaces: continuum mechanics approach. Phys Scr 59:391–396. https://doi.org/10.1238/Physica.Regular.059a00391
30. O'Bryan CS, Murdoch TJ, Strickland DJ, Rose KA, Bendejacq D, Lee D, Composto RJ (2023) Investigating the sequence specific adsorption behavior of polypeptides at the solid/liquid interface. Langmuir 39:1740–1749. https://doi.org/10.1021/acs.langmuir.2c02292
31. Berlinger SA, Chen X, Yutkin M, Radke CJ (2022) A two-phase model for adsorption from solution using quartz crystal microbalance with dissipation. Langmuir 38:10114–10127. https://doi.org/10.1021/acs.langmuir.2c00998
32. Gillissen JJJ, Jackman JA, Sut TN, Cho NJ (2020) Disentangling bulk polymers from adsorbed polymers using the quartz crystal microbalance. Appl Mater Today 18:100460. https://doi.org/10.1016/j.apmt.2019.100460
33. Hodges CS, Harbottle D, Biggs S (2020) Investigating adsorbing viscoelastic fluids using the quartz crystal microbalance. ACS Omega 5:22081–22090. https://doi.org/10.1021/acsomega.0c02100
34. Tanoue H, Yamada NL, Ito K, Yokoyama H (2017) Quantitative analysis of polymer brush formation kinetics using quartz crystal microbalance: viscoelasticity of polymer brush. Langmuir 33:5166–5172. https://doi.org/10.1021/acs.langmuir.7b00961
35. Koenig M, Rodenhausen KB, Rauch S, Bittrich E, Eichhorn KJ, Schubert M, Stamm M, Uhlmann P (2018) Salt sensitivity of the thermoresponsive behavior of PNIPAAm brushes. Langmuir 34:2448–2454. https://doi.org/10.1021/acs.langmuir.7b03919
36. Moghaddam SZ, Thormann E (2017) Hofmeister effect on PNIPAAm in bulk and at an interface: surface partitioning of weakly hydrated anions. Langmuir 33:4806–4815. https://doi.org/10.1021/acs.langmuir.7b00953
37. Schweigerdt A, Heinen S, Stöbener DD, Weinhart M (2021) Grafting density-dependent phase transition mechanism of thermoresponsive poly(glycidyl ether) brushes: a comprehensive QCM-D study. Langmuir 37:7087–7096. https://doi.org/10.1021/acs.langmuir.1c00695
38. Furusawa H, Sekine T, Ozeki T (2016) Hydration and viscoelastic properties of high- and low-density polymer brushes using a quartz-crystal microbalance based on admittance analysis (QCM-A). Macromolecules 49:3463–3470. https://doi.org/10.1021/acs.macromol.6b00035
39. Lin CH, Luo SC (2021) Combination of AFM and electrochemical QCM-D for probing zwitterionic polymer brushes in water: visualization of ionic strength and surface potential effects. Langmuir 37:12476–12486. https://doi.org/10.1021/acs.langmuir.1c02230
40. Yamaoka K, Fujii Y, Torikai N (2021) Evaluation of local gelation behavior of aqueous methylcellulose solution using quartz crystal microbalance. J Japan Inst Met Mater 85:23–29. https://doi.org/10.2320/jinstmet.JD202007

41. Yamaoka K, Fujii Y, Torikai N (2021) Evaluation of local gelation behavior of aqueous methylcellulose solution using quartz crystal microbalance. Mater Trans 62:647–654. https://doi.org/10.2320/matertrans.MT-M2020392
42. Yamaoka K, Yamada NL, Hori K, Fujii Y, Torikai N (2022) Interfacial selective study on the gelation behavior of aqueous methylcellulose solution via a quartz crystal microbalance. Langmuir 38:4494–4502. https://doi.org/10.1021/acs.langmuir.1c02728
43. Yamaoka K, Fujii Y, Torikai N (2024) Sol–gel transition of a thermo-responsive polymer at the closest solid interface. Langmuir 40:17405–17409. https://doi.org/10.1021/acs.langmuir.4c01522
44. Forrest JA, Svanberg C, Révész K, Rodahl M, Torell LM, Kasemo B (1998) Relaxation dynamics in ultrathin polymer films. Phys Rev E 58:R1226-R 1229. https://doi.org/10.1103/PhysRevE.58.R1226
45. Forrest JA, Mattsson J, Börjesson L (2002) Using adhesion to probe viscoelasticity of polymer film surfaces: a quartz crystal microbalance study. Eur Phys J E 8:129–136. https://doi.org/10.1140/epje/i2000-10102-5
46. Du X, Tang X, Fang J, Zhu DM (2017) Quartz crystal resonator study of glass transitions in polyvinylbutyral (PVB) films. J Appl Polym Sci 134:45433. https://doi.org/10.1002/app.45433
47. Shull KR, Taghon M, Wang Q (2020) Investigations of the high-frequency dynamic properties of polymeric systems with quartz crystal resonators. Biointerphases 15:0210012. https://doi.org/10.1116/1.5142762
48. Wang D, Mousavi P, Hauser PJ, Oxenham W, Grant CS (2005) Quartz crystal microbalance in elevated temperature viscous liquids: temperature effect compensation and lubricant degradation monitoring. Colloids surf. A 268:30–39. https://doi.org/10.1016/j.colsurfa.2005.05.075
49. Kim BC, Maeda K, Kim YH (2011) Monitoring nonisothermal crystallization of thermoplastic polymers using a quartz crystal resonator. J Appl Polym Sci 120:3370–3380. https://doi.org/10.1002/app.33552
50. Dawson BD, Lee SM, Krim J (2009) Tribo-induced melting transition at a sliding asperity contact. Phys Rev Lett 103:205502. https://doi.org/10.1103/PhysRevLett.103.205502
51. Denolf GC, Sturdy LF, Shull KR (2014) High-frequency rheological characterization of homogeneous polymer films with the quartz crystal microbalance. Langmuir 30:9731–9740. https://doi.org/10.1021/la502090a
52. Cui Z, Fujii Y, Tanaka K (2012) Effect of long range interactions on the glass transition temperature of thin polystyrene films. ACS Macro Lett 1:1317–1320. https://doi.org/10.1021/mz300391g
53. Fukao K, Miyamoto Y (2000) Glass transitions and dynamics in thin polymer films: dielectric relaxation of thin films of polystyrene. Phys Rev E 61:1743–1754. https://doi.org/10.1103/PhysRevE.61.1743
54. Alcoutlabi M, McKenna GB (2005) Effects of confinement on material behaviour at the nanometre size scale. J Phys: Condens Matter 17:R461–R524. https://doi.org/10.1088/0953-8984/17/15/R01
55. Forrest JA, Dalnoki-Veress K (2001) The glass transition in thin polymer films. Adv Colloid Interf Sci 94:167–195. https://doi.org/10.1016/S0001-8686(01)00060-4

# Chapter 11
# Design and Applications of Stimuli-Responsive Polymer Interfaces

**Takashi Miyata**

**Abstract** Stimuli-responsive polymer interfaces, encompassing polymer particles, self-assemblies, layers, films, and gels, exhibit remarkable adaptability to environmental changes such as pH, temperature, and light. Their polymer interfaces undergo reversible structural and/or property changes in response to external stimuli, enabling a wide range of innovative applications in diverse fields. The ability to engineer such interfaces with molecular precision enables precise control over interfacial phenomena, paving the way for innovative applications in drug delivery, biosensing, cell cultures, chemical reactions, and environmental remediation. This chapter delves into the fundamental principles governing the design of stimuli-responsive polymer interfaces, such as particles, self-assemblies, layers, films, and hydrogels, explores cutting-edge synthetic methodologies including controlled/living polymerizations, supramolecular assembly, and crosslinking strategies, and delineates their physical and chemical behavior in response to external stimuli such as pH, temperature, light, and biomolecular interactions. Furthermore, this chapter highlights their applications in areas such as drug delivery, sensing, cell cultures, chemical reactions, and environmental remediation. The versatility and tunability of these smart polymer interfaces make them highly attractive for developing adaptive systems that can respond dynamically to changing conditions.

**Keywords** Stimuli-responsive polymer · Particle · Self-assembly · Layer · Film · Hydrogel

---

T. Miyata (✉)
Department of Chemistry and Materials Engineering, Faculty of Chemistry, Materials and Bioengineering, Kansai University, Suita, Japan
e-mail: tmiyata@kansai-u.ac.jp



M. Hashizume, Y. Imura (eds.), *Colloids and Interfacial Dynamics*, Physical Chemistry in Action, https://doi.org/10.1007/978-981-96-6728-4_11

## 11.1 Introduction

Stimuli-responsive polymer materials, which undergo reversible changes in their structures and properties in response to specific external stimuli, such as pH, temperature, light, and so on, have opened new frontiers in various fields, including drug delivery, sensors, cell culture, and environmental monitoring [1–5]. Among these materials, stimuli-responsive polymer interfaces, including polymer particles, self-assemblies, films, and gels, have garnered significant attention due to their unique ability to exhibit dynamic behavior under changing environmental conditions [6–10]. The dynamic interfaces designed using various stimuli-responsive polymers are central to the design of smart materials capable of responding to stimuli in a controlled and predictable manner. Stimuli-responsive polymer interfaces are often constructed from polymers that possess functional groups or moieties capable of responding to the environment. This stimulus-responsiveness can lead to changes in the structures and properties, such as swelling, dissolution, or conformational rearrangements. The most commonly studied stimuli-responsive systems are based on polymers containing functional groups that respond to pH changes (e.g., poly(acrylic acid)), temperature (e.g., poly(*N*-isopropylacrylamide) (PNIPAAm)), light (e.g., photo-responsive azobenzene-based polymers), and so on. These polymers can be employed in the fabrication of stimuli-responsive polymer interfaces that exhibit high selectivity, tunability, and responsiveness.

Polymeric particles have emerged as one of the most promising platforms for stimuli-responsive polymer interfaces. The stimuli-responsive polymer particles, typically produced via heterogeneous radical polymerization such as emulsion polymerization or precipitation polymerization, can undergo significant changes in size, shape, and surface properties upon exposure to external stimuli [6, 9, 10]. For instance, temperature-responsive PNIPAAm-based particles that exhibit reversible changes in the size and dispersibility in aqueous media in response to changes in temperature were prepared by the precipitation polymerization [9]. Such temperature-responsive polymer particles have been designed as ideal candidates for drug delivery applications, where the release rate of the drug can be controlled by temperature changes. Additionally, pH-responsive polymer particles made from poly(acrylic acid) or its derivatives have been widely investigated for targeted drug delivery, where the particles can release encapsulated drugs in specific environments, such as the acidic pH of tumor tissues or the stomach [5]. Self-assembly of amphiphilic polymers represents another crucial approach in the development of stimuli-responsive polymer interfaces. Self-assembled structures can be engineered by exploiting the inherent properties of amphiphilic block copolymers, which can form well-defined nanostructures in response to changes in environmental conditions [7, 8]. These structures, such as micelles, vesicles, and capsules, can serve as carriers for drugs, proteins, or nucleic acids, with the self-assembly process providing an elegant and efficient method for their formation. Moreover, the stimuli-responsive behavior of polymer self-assemblies can be fine-tuned by varying the block composition, molecular weight, and type of stimulus. For example, polymeric micelles formed from pH-responsive block copolymers, which change from hydrophobic to hydrophilic under acidic conditions, can be used as drug carriers for controlled drug release in response to the localized pH of a target tissue [5, 8]. Polymer layers and films also play a pivotal role in the development of stimuli-responsive interfaces. Stimuli-responsive polymer layers and films can undergo significant changes in their chemical and physical surface properties when exposed to external stimuli, making them suitable for applications ranging from sensors to cell cultures [1, 7]. For instance, film surfaces coated with PNIPAAm can switch the hydrophilic/hydrophobic properties reversibly in response to temperature, which has been harnessed for applications in cell regulation [11]. Similarly, hydrogels, which are crosslinked networks of hydrophilic polymers, can undergo large changes in volume when exposed to changes in pH, temperature, or ionic strength. These properties make them ideal candidates for use in tissue engineering, drug delivery, and wound healing, where they can respond to the changing environment of the body.

Recent advances in the design and fabrication of stimuli-responsive polymer interfaces have expanded the scope of their applications of metal nanoparticles, such as gold nanoparticles (AuNPs) and silver nanoparticles (AgNPs). The integration of various stimuli-responsive polymers into hybrid systems, such as polymer-metal nanoparticle hybrid materials, has led to the development of hybrid materials with enhanced performance and functionality of metal nanoparticles [12]. For example, stimuli-responsive polymer-AuNPs hybrids can exhibit enhanced sensing capabilities, with the ability to detect specific molecules or environmental changes in real time. Moreover, the development of multi-responsive polymer interfaces that can respond to more than one type of stimulus has further broadened their applicability, enabling the design of advanced systems with complex and tailored responses [13, 14]. The ability to engineer polymer interfaces with precise control over their stimuli-responsive behavior has opened new ways in fields such as controlled drug delivery, tissue engineering, and environmental monitoring. However, significant challenges remain in the design and optimization of polymer interfaces such as particles, self-assemblies, layers, and films. Achieving the desired level of responsiveness while maintaining the stability and biocompatibility of the polymer interfaces is a key consideration, particularly in biomedical applications. This chapter aims to provide a comprehensive overview of the design principles, fabrication techniques, and applications of stimuli-responsive polymer interfaces, including polymer particles, self-assemblies, layers, and films. The chapter also focuses on the synthesis of polymer particles, self-assembled structures, layers, and films, highlighting the various stimuli that can trigger their responses. In addition, the current challenges are explored in the field, as well as the future directions for the development of stimuli-responsive polymer interfaces.

## 11.2 Stimuli-Responsive Polymers

Stimuli-responsive polymers are a class of functional polymers that undergo significant changes in their physical or chemical properties in response to external stimuli, including temperature, pH, light, and so on (Fig. 11.1) [15–20]. These innovative polymers have garnered substantial attention in various fields, particularly in biomedical applications such as drug delivery, diagnostics, and tissue engineering. Their ability to intelligently respond to environmental signals makes them suitable for enhancing therapeutic efficacy and minimizing side effects in medical treatments, positioning them at the forefront of polymer science and engineering. The development of stimuli-responsive polymers began in the late twentieth century, coinciding with the rise of smart materials, and has evolved significantly due to advancements in nanotechnology and polymer synthesis techniques. Research has revealed that the molecular structures of these polymers can be finely tuned to achieve desired responsiveness, leading to multifunctional applications, particularly in theranostics, integrating therapeutic and diagnostic functions in a single platform. The versatility of stimuli-responsive polymers has led to their implementation in innovative drug delivery systems that exploit physiological changes in disease states, enhancing localized treatment and improving patient outcomes. Despite their potential, the integration of stimuli-responsive polymers into clinical practice is not without challenges. Concerns about biocompatibility, stability under physiological conditions, and the potential for adverse immune responses have been identified as critical barriers to their widespread application. Moreover, the molecular and material designs of these stimuli-responsive polymers necessitate ongoing research to improve their feasibility and safety in their practical applications.

The synthesis of stimuli-responsive polymers involves various methodologies that cater to the desired properties and functionalities of the final material. Among these, traditional radical polymerization and controlled radical polymerization techniques are prominently utilized. Controlled radical polymerizations, such as reversible addition-fragmentation chain transfer (RAFT) and atom transfer radical polymerization (ATRP), allow for better control over polymer architecture and molecular weight distribution, which are crucial for achieving specific responsiveness to external stimuli [20]. Stimuli-responsive polymers can be designed to respond to a variety of external stimuli, including chemical, physical, and biological

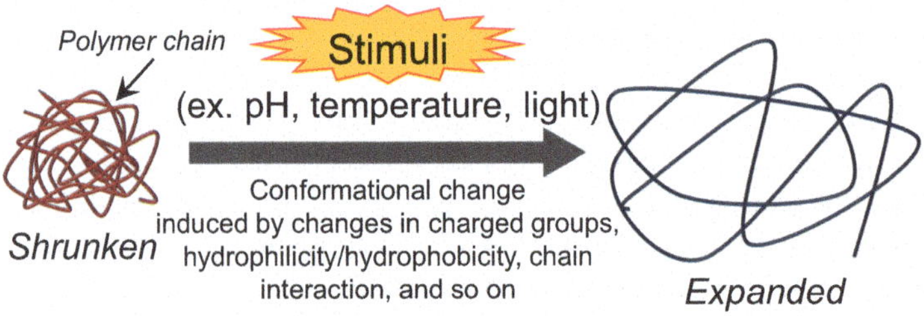


**Fig. 11.1** Conformational changes of stimuli-responsive polymers upon environmental stimuli

triggers. pH-Responsive polymers exhibit changes in their physical or chemical properties in response to variations in pH levels, making them particularly useful in biomedical applications [17]. The human body contains several environments with distinct pH values, such as the highly acidic stomach (pH 1–3) and the more neutral intestinal environment (pH 5–8). These polymers typically contain ionizable groups that can either be weakly acidic or basic, which respond to changes in pH by ionizing and inducing significant structural changes. Upon ionization, the electrostatic repulsions between the charged groups lead to the expansion of the polymer chains, enhancing their solubility and interaction with the surrounding environments. Temperature-responsive polymers are classified into two types, lower critical solution temperature (LCST)-type and upper critical solution temperature (UCST)-type polymers [18]. While LCST-type polymers are soluble in solvents at temperatures below the LCST, they become insoluble above the LCST because of a drastic change in the polymer chain from hydrophilic to hydrophobic. In contrast, UCST-type polymers exhibit opposite temperature-responsive behavior below and above the UCST. Importantly, the LCST-type polymers have both hydrophilic and hydrophobic moieties in a monomer unit or a polymer chain. Representative temperature-responsive polymer is poly(*N*-isopropylacrylamide) (PNIPAAm), which has an LCST at 32 °C. The LCST can be tuned by the balance between the hydrophilic and hydrophobic moieties. For instance, the introduction of hydrophobic moieties to the LCST-type polymers resulted in lowering the LCST. UCST-type polymers have functional groups that exhibit attractive interactions, such as hydrogen bonds and ionic bonds, sensitive to temperature changes. Photo-responsive polymers are engineered to respond to light stimuli, providing high precision and control in their activation [19]. These polymers utilize photo-responsive chromophores, such as azobenzene, spiropyran, or nitrobenzyl groups, which change their structure upon exposure to specific wavelengths of light. The ideal wavelength for activation is often tuned to the near-infrared region, as it penetrates biological tissues more effectively and minimizes absorption by cells, allowing for targeted applications without significant harm to surrounding tissues. Thus, the versatility of stimuli-responsive polymers extends to the development of smart materials and dynamic interfaces, including stimuli-responsive polymer particles, self-assemblies, and layers, films, as described in the following sections.

## 11.3 Responsive Polymer Particles

Stimuli-responsive polymer particles are a class of engineered polymer interfaces capable of undergoing significant changes in size, dispersibility, or other properties in aqueous media in response to various external stimuli, including pH, temperature, and light (Fig. 11.2). These innovative particles have attracted considerable interest in fields such as drug delivery, sensing, tissue engineering, and environmental applications due to their ability to release therapeutic agents or alter their functionality in a controlled manner. Their unique responsiveness allows for targeted

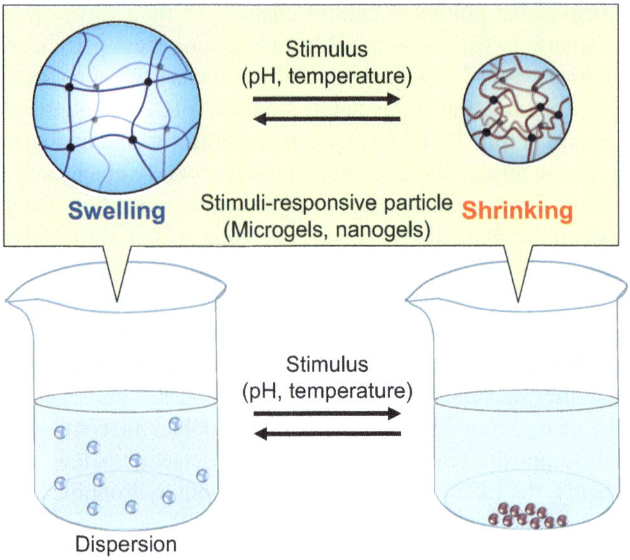


**Fig. 11.2** Stimuli-responsive polymer particles that undergo changes in size in response to environmental stimuli

and controlled drug delivery, enhancing the effectiveness of pharmaceuticals while minimizing side effects, which makes them notable in advancing biomedical technologies [6, 9, 10, 21]. Among the various types of stimuli-responsive polymer particles, pH-responsive polymer particles are designed to respond to changes in acidity, enabling targeted drug delivery in acidic tumor microenvironments. Temperature-responsive polymer particles, like PNIPAAm-based microgels, exhibit phase transitions that regulate drug release in response to temperature changes, particularly useful in localized hyperthermia treatments. Additionally, photo-responsive polymer particles utilize photosensitive components to control their behavior upon exposure to specific wavelengths of light, offering precision in therapeutic applications [22, 23]. Magnetic field-responsive polymer particles further enhance targeted drug delivery by combining magnetic nanoparticles with therapeutic agents, enabling localized heating and drug release at desired sites, such as tumors [23]. Despite their promising applications, the development of stimuli-responsive polymer particles is not without challenges. Future research aims to address these limitations by exploring the use of endogenous stimuli, such as biological signals and conditions, which could lead to more robust and safer intelligent drug delivery systems [4, 24]. Such stimuli-responsive polymer particles hold the potential to revolutionize various fields by enabling precise control over drug release and material properties, thereby paving the way for innovative solutions to complex medical and environmental challenges.

pH-Responsive polymer particles are designed to respond to changes in the acidity or basicity of their environment. The polymer particles typically undergo changes in size and dispersibility in aqueous media when exposed to different pH levels [6, 25, 26]. For instance, cationic polymers, such as poly(2-(dimethylamino)ethyl methacrylate) and poly(4-vinyl pyridine), change from hydrophobic to hydrophilic states as the pH decreases, enhancing their interaction with aqueous environments and enabling targeted drug release [5, 27, 28]. Conversely, anionic polymers like poly(acrylic acid) and poly(aspartic acid) exhibit an opposite behavior, becoming hydrophobic at lower pH levels. These unique properties make pH-responsive polymer particles particularly useful for targeting acidic tumor microenvironments, thereby facilitating localized drug delivery. Temperature-responsive particles undergo changes in size and dispersibility in aqueous media in response to changes in temperature, often characterized by a specific phase transition temperature, known as the lower critical solution temperature (LCST) or upper critical solution temperature (UCST) of their constituent polymers [29–32]. A prominent example is PNIPAAm, which exhibits an LCST around 32 °C, close to human body temperature. While PNIPAAm remains soluble in aqueous media at temperatures below the LCST, it becomes hydrophobic and insoluble at temperatures above the LCST. Using the temperature-dependent solubility of PNIPAAm, the temperature-responsive microgels, which exhibit a volume phase transition in response to temperature changes, have been typically prepared via precipitation polymerization of *N*-isopropylacrylamide (NIPAAm) in aqueous media [30, 31]. Recently, a novel strategy for the fabrication of stimuli-responsive core–shell microgels has been proposed, offering significant potential for applications across a range of fields [33]. On this strategy, core–shell microgels composed of zwitterionic poly(2-methacryloyloxyethyl phosphorylcholine) (PMPC) hydrogel core and temperature-responsive poly[oligo(ethylene glycol) methacrylate-*co*-2-(2′-methoxyethoxy) ethyl methacrylate] (P(OEGMA-*co*-$MEO_2MA$)) shell were synthesized through inverse miniemulsion RAFT polymerization. The synthetic protocol comprises several critical steps: (i) the synthesis of a water-soluble amphiphilic block copolymer emulsifier, P(OEGMA-*co*-$MEO_2MA$)-*b*-PMPC, via RAFT polymerization; (ii) the generation of stable water-in-oil (W/O) emulsions, utilizing the block copolymer as an emulsifier; and (iii) the inverse miniemulsion RAFT copolymerization of the zwitterionic monomer, 2-methacryloyloxyethyl phosphorylcholine (MPC), and the crosslinker, *N,N′*-methylenebisacrylamide, with polymerization initiated from the emulsifier localized at the emulsion interfaces. The resulting core–shell microgels, dispersed in aqueous medium, displayed a pronounced reduction in optical transmittance when heated above 38 °C. This temperature-responsive change in the transmittance was attributed to the lower critical solution temperature (LCST) behavior of the shell composed of P(OEGMA-*co*-$MEO_2MA$), which transitions to a hydrophobic state above 38 °C. This temperature-responsive collapse of the shell significantly alters the microgel's optical properties. This work provides a robust and versatile synthetic framework for the construction of core–shell microgels integrating zwitterionic hydrophilicity with tunable temperature-responsive behavior. These temperature-responsive properties are advantageous in applications requiring drug release in response to localized hyperthermia or changes in body temperature.

Photo-responsive polymer particles utilize light as a trigger for their responsive behavior. Incorporating photo-responsive moieties allows these particles to undergo reversible changes in their size and dispersibility in aqueous media when exposed to specific wavelengths of light. This property can be harnessed for applications requiring precise control over drug release or particle behavior. For example, Photo-responsive particles can be designed to swell or collapse upon light irradiation, which can be exploited in targeted therapy applications where light can be precisely controlled and directed [34–37]. Photo-crosslinkable and photodegradable gels, capable of modulating their physical properties such as stiffness and viscoelasticity upon light exposure, have emerged as photo-responsive scaffolds for the dynamic regulation of cellular behaviors [38–42]. In parallel, photo-responsive micro- and nanogels, whose structures and properties can be precisely controlled by external light stimuli, have attracted attention as fundamental units for constructing cell-adhesive or non-adhesive surfaces. Recently, zwitterionic polymer particles functionalized with photodimerizable groups on their surfaces have been synthesized via surfactant-free emulsion polymerization followed by surface modification, resulting in bioinert colloidal building blocks that exhibit excellent biocompatibility, resistance to nonspecific biofouling, and responsiveness to environmental changes [43] (Fig. 11.3). Upon exposure to ultraviolet light (UV) of suitable wavelength, the photodimerizable groups on the particle surfaces undergo covalent crosslinking, thereby inducing particle–particle assembly in aqueous media. This photo-induced aggregation leads to the formation of zwitterionic particle-based films, whose surfaces effectively prevent protein adsorption and cellular adhesion due to the inherent antifouling properties of the zwitterionic functionalities. Consequently, these zwitterionic polymer particles bearing photodimerizable groups present a promising strategy for fabricating scaffolds with tunable mechanical and surface properties suitable for tissue engineering applications. Moreover, their light-triggered assembly behavior can be harnessed to construct well-organized nanostructures with potential utility not only in cellular regulation but also

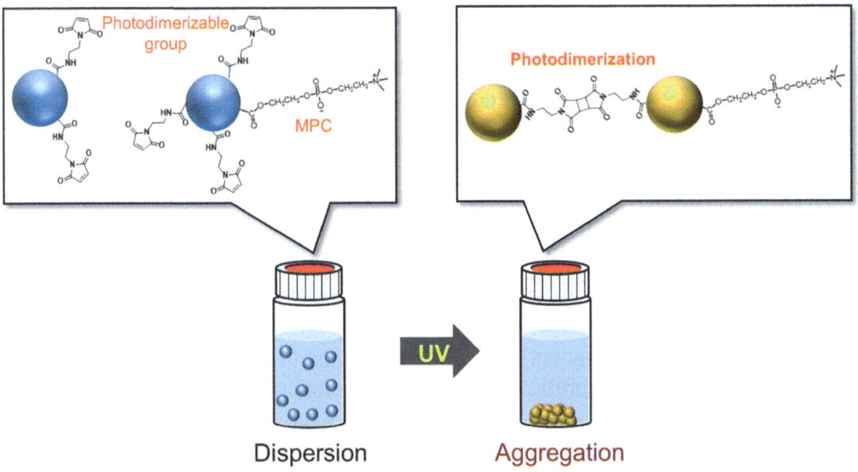


**Fig. 11.3** Photo-assembling of zwitterionic polymer particles with photodimerizable groups under UV exposure

in advanced fields such as photonic materials and nanoelectronics. The integration of zwitterionic antifouling properties with spatiotemporal control via photoactivation thus provides a versatile and robust platform for the development of smart, multifunctional materials.

Biomolecules play vital roles in maintaining homeostasis and mediating various physiological processes, often serving as biomarkers that reflect dynamic changes in biological systems. These molecular signals provide crucial diagnostic indicators and represent promising targets for the development of smart drug delivery systems (DDSs) capable of autonomous, feedback-driven drug release. Therefore, molecularly stimuli-responsive gels, which undergo changes in volume upon recognizing specific biomolecular stimuli, are also emerging as powerful platforms for diagnostics and controlled therapy [44–51]. Unlike the rapid evolution of pH- and temperature-responsive polymer particles, molecularly stimuli-responsive polymer particles remain comparatively underexplored. Molecularly stimuli-responsive micro- and nanogels with selective molecular binding sites offer significant advantages in biomedical applications, such as targeted drug delivery and biosensing [52–56]. Their high sensitivity, modularity, and responsiveness to molecular signals such as glucose and proteins render them attractive for personalized medicine. One prominent application lies in the management of diabetes, where the self-regulation of insulin release based on real-time glucose sensing remains a therapeutic goal. Glucose-responsive gels that swell or shrink in the presence of glucose have been widely studied for constructing closed-loop insulin delivery systems [57, 58]. The glucose-responsive systems fall into three main categories based on the use of glucose oxidase (GOx), phenylboronic acid (PBA), and concanavalin A (Con A). GOx and Con A provide high sensitivity and specificity but suffer from low stability under physiological conditions. In contrast, synthetic PBA-based systems exhibit enhanced robustness but often compromise on selectivity and biodegradability. Hybrid approaches that strategically integrate natural and synthetic components can offset these individual limitations, making glucose-responsive systems the most extensively explored class of molecularly stimuli-responsive materials. For example, glucose-responsive gel particles that undergo glucose-triggered volume changes have also been engineered for responsive insulin release [59, 60]. Traditional designs combine GOx with pH-sensitive polymers to mediate insulin release through gluconic acid generation [61–64]. Volpatti et al. encapsulated insulin, GOx, and catalase within acetalized dextran nanoparticles to fabricate systems exhibiting both fast and sustained insulin release profiles, dictated by the acetal content. These nanoparticles successfully regulated blood glucose levels in diabetic mouse models [65]. In parallel, fully synthetic glucose-responsive gel particles have been developed using temperature-responsive polymers such as PNIPAAm functionalized with PBA groups. The PBA-modified PNIPAAm gel particles selectively bind glucose through reversible covalent interactions, leading to charge inversion or swelling under physiological conditions [66–70]. Hoare and Pelton designed amphoteric PNIPAAm-based gel particles with glucose-responsive behaviors at body temperature and pH [71]. Zhang et al. investigated the swelling kinetics of PBA-introduced PNIPAAm microgels in aqueous glucose solutions [72], while Guo et al. synthesized RAFT-polymerized nanogels containing PBA, 2-(acrylamido)glucopyranose, and a fluorescent boron dipyrromethene moiety, facilitating both glucose detection and insulin delivery [73].

Innovative designs of molecularly stimuli-responsive polymer materials, such as gels and particles, have further enabled direct transduction of molecular recognition into changes in the structures. Since the swelling properties of gels are strongly dependent on crosslinking density, dynamic crosslinks formed via biomolecular complexes, such as carbohydrate–lectin or antigen–antibody complexes, can endow gels with specific, reversible volume changes upon recognition [74–79]. Molecular imprinting has also yielded gels that shrink upon binding target molecules like tumor marker glycoproteins or endocrine disruptors such as bisphenol A (BPA) [80, 81]. Molecular complexes as dynamic crosslinks have also been employed in the synthesis of molecularly stimuli-responsive particles. For example, surfactant-free emulsion polymerization using a complex of polymerizable group-introduced lectin Con A and a methacrylate monomer bearing a glucose moiety (2-glucosyloxyethyl methacrylate: GEMA) produced biohybrid particles with GEMA-ConA complex as dynamic crosslinks [82]. Upon glucose exposure, competitive binding induces dissociation of the GEMA-ConA complex crosslinks, leading to rapid swelling of the biohybrid particles (Fig. 11.4). In contrast, the biohybrid particles with

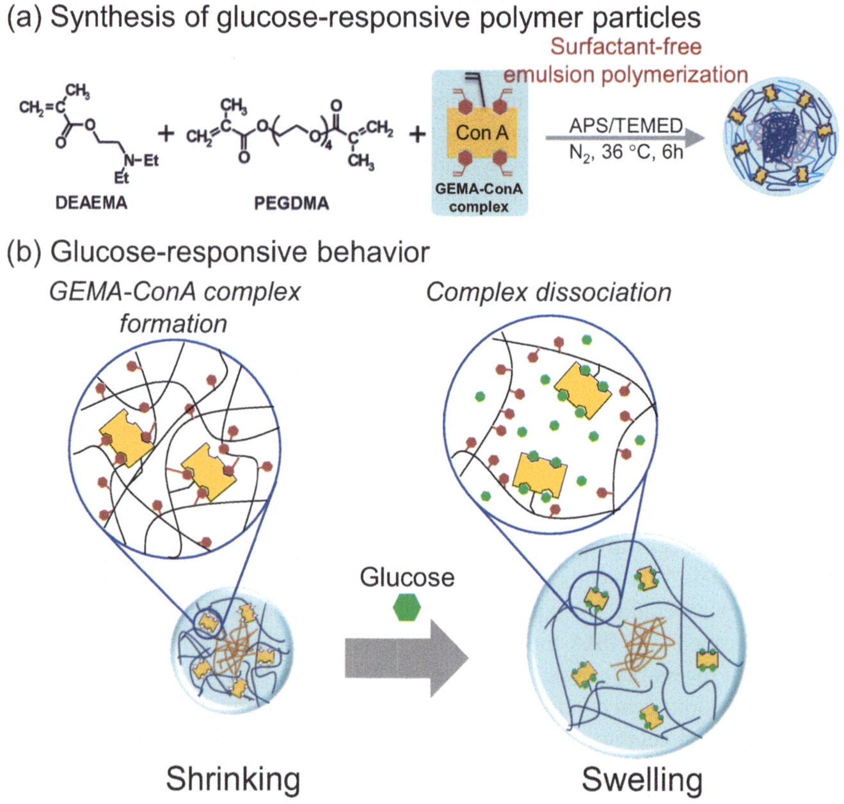


**Fig. 11.4** Glucose-responsive behavior of biohybrid polymer particles with GEMA-ConA complexes as dynamic crosslinks

GEMA-ConA complex crosslinks exhibit no response to galactose, which does not bind to Con A, demonstrating selective recognition. Beyond traditional applications, molecularly stimuli-responsive polymer particles have been used to fabricate bioresponsive microlenses. Lyon et al. assembled PNIPAAm-based microgels bearing biotin groups into arrays on glass substrates, enabling the optical detection of avidin–biotin interactions via changes in lens properties under bright-field microscopy [44, 83, 84]. In another example, hybrid gel particles were synthesized by coating silica ($SiO_2$) particles with cyclodextrin (CD)-containing polyacrylamide gel layers via surface-initiated atom transfer radical polymerization (SI-ATRP) [85]. Upon exposure to bisphenol A (BPA) as a target molecule, the gel layer formed host–guest CD-BPA-CD complexes, increasing crosslinking density and inducing particle shrinkage (Fig. 11.5). These organic–inorganic hybrids exemplify how molecular recognition can be harnessed to modulate material properties for biosensing.

Stimuli-responsive particles incorporating catalytic components represent a unique class of smart polymer interfaces that function as "nano-reactors," offering enhanced and externally controllable catalytic performance [12, 86–89]. These soft, solvent-swollen particles can encapsulate metal nanoparticles (e.g., Au, Ag, Pd, Cu), bimetallic systems (e.g., Cu/Pd) [90], or enzymes and enzyme mimics [91], thereby integrating responsive behavior with catalytic activity. Hybrid systems composed of metal nanoparticles embedded within stimuli-responsive polymer matrices, especially those using gold nanoparticles (AuNPs) or silver nanoparticles (AgNPs), allow precise modulation of particle dispersibility in aqueous media and optical properties. In addition, AuNPs have been extensively employed in catalytic platforms due to their favorable catalytic properties [12, 86, 92–95]. A combination of AuNPs with temperature-responsive polymers, such as PNIPAAm, yields hybrid nanostructures capable of temperature-regulated catalysis. For example, RAFT polymerization has been employed to form well-defined PNIPAAm shells around AuNPs, enabling the fabrication of systems responsive to temperature [96]. Stimuli-responsive hybrid particles not only enhance catalytic efficiency but also facilitate recycling and separation, which are critical for sustainable catalysis. A key advantage of these hybrid particles is their ability to maintain dispersion, prevent nanoparticle aggregation, and remain catalytically active under controlled stimuli. Despite their promise, temperature-responsive polymer–AuNP hybrids face several

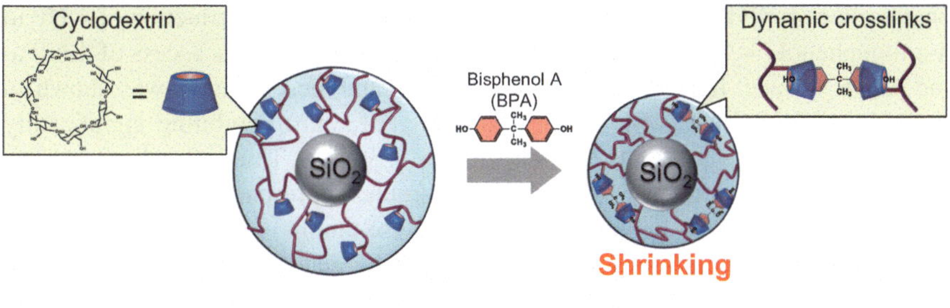


**Fig. 11.5** Target molecule-responsive behavior of organic–inorganic hybrid particles composed of $SiO_2$ core and gel layer with CD ligands

limitations. Hybridization of AuNPs as catalysts with stimuli-responsive polymers often suffers from inefficient anchoring or limited thermal robustness. The presence of a polymeric shell may also prevent substrate diffusion, thereby reducing catalytic activity. AuNP-based hybrids have garnered much attention for use in sensing and catalysis due to their high surface reactivity and stability [12, 92–95]. Their utility spans various organic reactions, including coupling, hydrogenation, oxidation, and reduction processes [97–100]. In addition, to enhance stability in aqueous environments, AuNP surfaces are often modified using thiolate surfactants or thiol-functionalized polymers.

Temperature-responsive PNIPAAm–AuNP hybrids can be fabricated via three main methods: (1) surface grafting of thiol-terminated PNIPAAm chains to AuNP ("grafting-to"), (2) polymerization of NIPAAm directly on AuNPs ("grafting-from"), and (3) in situ AuNP formation within PNIPAAm-based self-assemblies or gels. Methods (1) and (2) produce core–shell structures where the polymer shell undergoes reversible collapse above the LCST, modulating the reactant access to the AuNP core. For example, PNIPAAm–AuNP core–shell hybrid particles were synthesized via "grafting-to" methodology, demonstrating temperature-dependent catalytic reduction of 4-nitrophenol (4-NP) with excellent recyclability [101]. Another approach involves the non-covalent association of temperature-responsive polymers, such as hyperbranched polyethylenimine, with citrate-stabilized AuNPs. Their LCST and catalytic activity can be finely tuned by adjusting polymer molecular weight, isobutyramide content, and pH [102]. Additionally, AuNPs embedded within PNIPAAm-based block copolymer micelles or microgels have demonstrated switchable catalytic behavior, active below LCST and inactive above, due to altered substrate accessibility [103]. This reversible control facilitates efficient catalyst recovery and reuse, making these systems promising candidates for "on-demand" catalysis. However, the catalytic regulation in these hybrid particles, particularly those synthesized via methods (1) and (2), remains partially limited due to incomplete surface coverage and weak polymer–AuNP interactions. Moreover, in method (3), the physical entrapment of AuNPs in polymer matrices can lead to particle leakage over time.

A recent novel method enables the covalent incorporation of AuNPs within PNIPAAm microgels via copolymerization of a polymerizable AuNP monomer with NIPAAm (Fig. 11.6) [104, 105]. This method ensures stable AuNPs immobilization within the gel particles, enabling reversible "on–off" switching of catalytic activity based on temperature. Below the LCST (~34 °C), the hydrated PNIPAAm chains allow 4-NP to access the AuNP surface, facilitating the reduction of 4-NP to 4-aminophenol (4-AP). Above the LCST, chain collapse restricts access of 4-NP to the AuNP surface, inhibiting the reaction. The catalytic efficiency was shown to depend on AuNP size, with smaller particles (~13 nm) exhibiting significantly higher apparent rate constants than larger ones (~36 nm). Thus, stimuli-responsive hybrid particles embedded with metal nanoparticles represent a versatile platform for modulating catalytic behavior. Through precise tuning of polymer composition and environmental conditions, these hybrid particles enable spatiotemporal control over reaction kinetics, recyclability, and substrate selectivity. Advances in covalent

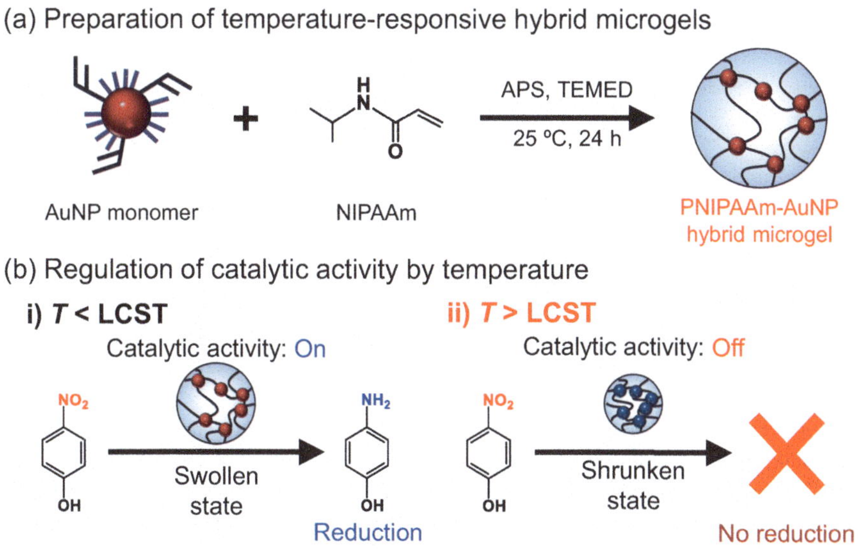


**Fig. 11.6** (a) Preparation of temperature-responsive polymer–gold nanoparticle hybrid microgels using an AuNP monomer and (b) regulation of their catalytic activity in response to temperature changes

integration and stimuli-responsive systems continue to overcome current challenges, expanding their applicability in green chemistry, sensing, and therapeutic technologies.

## 11.4 Responsive Polymer Self-Assemblies

Amphiphilic polymers composed of both hydrophilic and hydrophobic parts form self-assemblies, such as micelles, vesicles, and other nanostructures, in aqueous media because the hydrophobic parts are aggregated by hydrophobic interactions [7, 8, 10, 16, 106–112]. Stimuli-responsive polymer self-assemblies are sophisticated materials that respond dynamically to external environmental changes, such as temperature, pH, and light (Fig. 11.7). These self-assembled structures are notable for their versatile applications in fields such as drug delivery, bioimaging, and therapeutic diagnostics, where their ability to encapsulate and release active substances in response to specific stimuli enhances the efficacy and precision of treatments [113]. Various types of stimuli-responsive polymer self-assemblies, such as micelles, vesicles, and other nanostructures, stand out due to their unique structural properties and functional advantages. Polymeric micelles and vesicles, formed from amphiphilic block copolymers, can encapsulate hydrophobic or hydrophilic compounds, allowing for controlled release in response to changes in pH or temperature. Stimuli-responsive polymer self-assemblies, characterized by their hydrophilic and

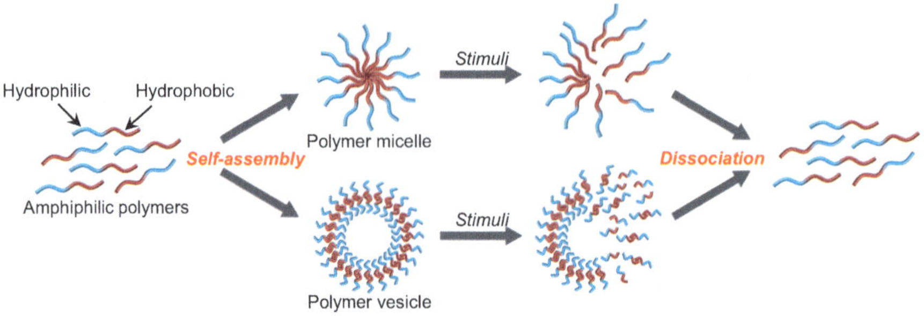


**Fig. 11.7** Formation and dissociation of stimuli-responsive polymer self-assemblies such as micelles and vesicles

hydrophobic blocks, offer enhanced stability in aqueous media and tunable properties, making them suitable for transporting a diverse range of therapeutic agents, including anticancer drugs and imaging agents. Furthermore, dendritic systems and smart nanocarriers have emerged, capable of sophisticated drug delivery mechanisms that utilize stimuli such as enzymatic activity or light to trigger the release of the drugs [4, 114]. Despite the promising applications of stimuli-responsive polymer self-assemblies, challenges such as premature drug release, potential toxicity, and the complexity of synthesizing well-designed polymers with precise responsiveness pose significant hurdles. Moreover, the integration of stimuli-responsive polymer self-assemblies into clinical settings raises questions regarding their long-term stability, biocompatibility, and effectiveness in real-world applications. However, the development of stimuli-responsive polymer self-assemblies represents a cutting-edge frontier in materials science, offering new possibilities for targeted therapeutic delivery and advanced biomedical applications.

In general, polymeric micelles are formed from amphiphilic block copolymers composed of both hydrophilic and hydrophobic blocks [8, 108, 109]. The polymeric micelles can respond to various stimuli, including temperature and pH. For instance, LCST-type temperature-responsive polymeric micelles dissociate with a decrease in temperature, leading to the release of their encapsulated drugs. Their structure allows them to encapsulate hydrophobic drugs within a hydrophobic core, making them versatile carriers for drugs and imaging agents. Polymer vesicles, also known as polymersomes, are another type of self-assembled structure, characterized as spherical shells with an aqueous core surrounded by a bilayer membrane of amphiphilic block copolymers [7, 110–112]. They offer several advantages over traditional liposomes, such as enhanced mechanical stability and tunable membrane properties. Polymersomes can transport a wide range of compounds, including anticancer drugs and imaging agents, and can function as controlled drug release systems that respond to stimuli such as pH or ionic strength.

Biological membranes, constituted by self-assembled phospholipid bilayers, serve not only as compartmentalizing barriers but also as dynamic interfaces for selective molecular recognition and transmembrane transport. Inspired by their structural complexity and responsiveness, researchers have developed biomimetic polymer self-assemblies, such as micelles [115–117], vesicles [118], and liposomes [119–121], for targeted drug delivery and biosensing applications. These self-assembled nanostructures offer spatial and temporal control over therapeutic release profiles. A central strategy in controlled drug release exploits environmentally triggered structural changes in polymeric self-assemblies. However, conventional stimuli-responsive polymer self-assemblies often undergo dissociation upon external stimuli (e.g., pH or temperature changes), resulting in premature burst release. Crosslinking techniques can improve the stability of the assembled structures but frequently compromise release kinetics. Thus, designing robust, reversible self-assemblies that maintain structural integrity under stimuli remains a formidable challenge.

Stimuli-responsive polymers can alter their physical and chemical properties in response to environmental changes. Among them, liquid crystalline polymers (LCPs) are particularly compelling due to their anisotropic yet dynamic molecular ordering, inspired by biological membranes [122, 123]. Thermotropic LCPs exhibit reversible phase transitions between liquid crystalline and isotropic phases, and their architecture, whether main-chain or side-chain, critically influences transition temperatures. Side-chain LCPs incorporating mesogens onto flexible polysiloxane backbones display tunable liquid crystalline–isotropic phase transition temperature ($T_i$) at relatively lower temperatures compared to the liquid crystalline polymers with C–C backbones. Their flexibility and the cooperative behavior of pendant mesogenic units facilitate phase transitions at physiological temperatures [124–126]. For example, membranes prepared using LCPs with $T_i$ can modulate the permeability of solutes through a liquid crystalline–isotropic phase transition near body temperature [127–129]. However, such membranes' macroscopic size limits their application in nanoscale drug delivery systems. Recently, temperature-responsive polymeric micelles were prepared as reversibly responsive drug carriers using amphiphilic side-chain LCPs. The amphiphilic liquid crystalline polymers (LCP-*g*-OEGs) feature hydrophobic mesogenic units and hydrophilic oligo(ethylene glycol) (OEGMA) chains grafted onto a poly(methylsiloxane) (PMS) backbone [130, 131]. The amphiphilic LCP-*g*-OEGs exhibit reversible liquid crystalline–isotropic phase transitions near physiological temperature and form stable micellar nanostructures by their self-assembly in aqueous media. The temperature-responsive behavior characterized by differential scanning calorimetry (DSC), polarized optical microscopy (POM), and X-ray diffraction (XRD) revealed a nematic–isotropic phase transition temperature ($T_{NI}$) at 36.2 °C for LCP-*g*-OEG, matching physiological conditions. The POM images displayed loss of birefringence at $T_{NI}$, confirming the nematic–isotropic phase transition. The amphiphilic architecture of LCP-*g*-OEG promoted spontaneous micelle formation in water, driven by hydrophobic interactions among mesogens (Fig. 11.8a). The hydrophilic oligo(ethylene glycol) (OEG) corona ensured colloidal stability and reduced cytotoxicity [132]. Pyrene fluorescence assays showed a low critical micelle concentration (CMC) of $3.6 \times 10^{-4}$ mg/mL at 25 °C and $1.5 \times 10^{-3}$ mg/mL at 40 °C, indicating high thermodynamic stability even under dilution, a critical property for systemic delivery. The dynamic light scattering (DLS) measurements and transmission electron microscope (TEM)

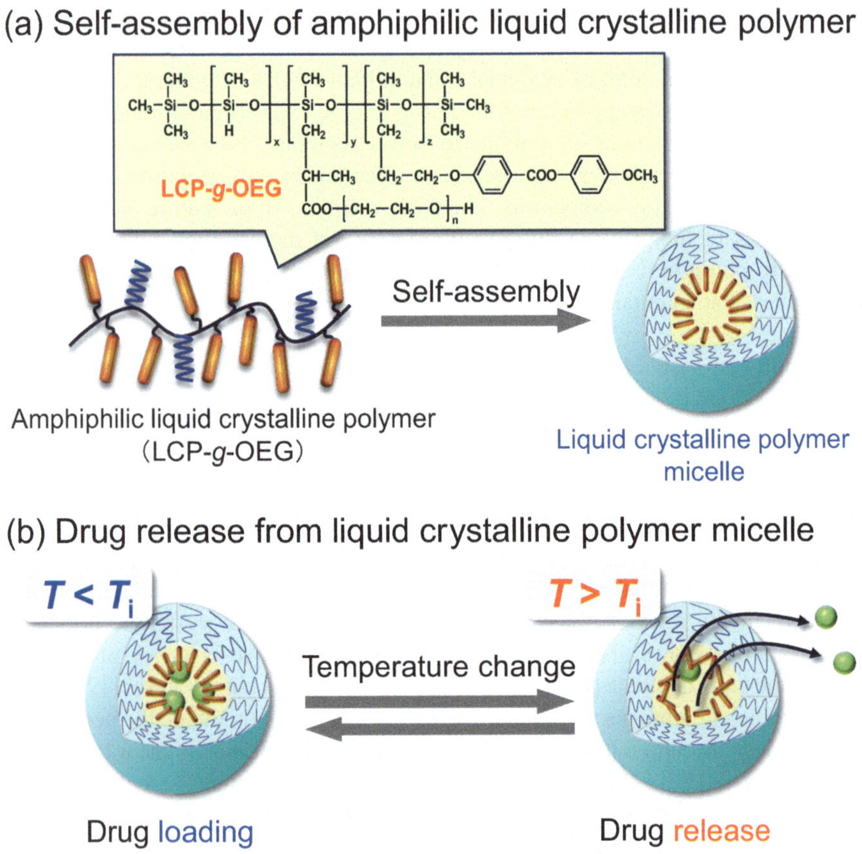


**Fig. 11.8** (a) Self-assembly of amphiphilic liquid crystalline polymers in aqueous media and (b) regulation of drug release from their self-assemblies in response to temperature changes

observations revealed spherical micelles (~130 nm diameter), maintaining size and low polydispersity index over 28 days and across a broad temperature range. In contrast, LCPs lacking OEG chains aggregated due to poor water compatibility, underscoring the importance of hydrophilic OEG chains in dispersion and cytocompatibility. Fluorescein as a model drug remained largely retained within LCP-*g*-OEG micelles at 25 °C, whereas a significant increase in release rate was observed at temperatures above $T_{NI}$ (Fig. 11.8b). Arrhenius plots of release rate constants showed a marked discontinuity at ~36.2 °C, confirming that the liquid crystalline–isotropic phase transition governs drug diffusivity within the micellar core. Reversible ON–OFF regulation of drug release in response to alternating the temperature between 25 and 50 °C repeatedly was achieved using the LCP-g-OEG micelles without micelle dissociation, in contrast to conventional self-assembled micelles that are often dissociated irreversibly under stimuli. The fluorescein-loaded LCP-*g*-OEG micelles also demonstrated selective cellular uptake by tumor cells

(HeLa, HepG2) at 30 °C, with negligible uptake in normal cells, suggesting that they have potential applications as temperature-responsive drug carriers for tumor-specific targeting. Thus, LCP-*g*-OEG micelles represent a new class of stimuli-responsive DDSs that integrate liquid crystalline ordering, reversibly temperature-responsiveness, and biological functionality. Their ability to modulate drug release via reversible liquid crystalline–isotropic phase transitions, without structural dissociation, positions them as highly promising nanocarriers for precision medical applications.

## 11.5 Responsive Polymer Layers

Surface plasmon resonance (SPR) sensors are extensively employed in biochemical and biomedical research to elucidate dynamic biomolecular interactions owing to their high sensitivity and quantitative capabilities [133–140]. SPR operates by detecting variations in the local refractive index near gold sensor chip surfaces, which arise from the binding of analytes to ligands immobilized on the surface, manifesting as shifts in the resonance angle. Conventionally, ligands such as proteins are immobilized directly onto SPR chip surfaces via standard chemical techniques, including amine coupling. However, the two-dimensional confinement of the chip surface imposes a limit on the quantity of immobilized ligands, thereby restricting sensor performance. On the other hand, hydrogels, composed of water-swollen, crosslinked polymer networks, offer a three-dimensional (3D) framework with interconnected pores that enable solute diffusion and cooperative polymer chain dynamics. These features render hydrogels attractive as immobilization matrices for biomolecules, facilitating molecular recognition and responsiveness. For instance, hydrogels incorporating antibody–antigen or lectin–carbohydrate interactions as dynamic crosslinks have exhibited stimuli-responsive volume changes in the presence of target biomolecules [74, 78, 80]. Critically, the 3D architecture based on highly swollen networks allows for greater ligand loading compared to flat surfaces, positioning hydrogel-coated chips as promising platforms for amplifying SPR signals. For example, antibody-immobilized hydrogel layers were fabricated on SPR sensor chips via SI-ATRP. SI-ATRP offers precise control over polymer chain length and composition, allowing tunable layer thickness and ligand content by adjusting reaction parameters such as monomer concentration and polymerization time [141–143]. Hydrogel layers were formed by SI-ATRP of acrylamide (AAm) and an acryloyl-functionalized antibody after the introduction of ATRP initiators onto the SPR chip surfaces [144]. Scanning electron microscopy (SEM) demonstrated that the gel layer thickness increased proportionally with polymerization time, consistent with the time-dependent growth of polymer molecular weight during ATRP. The hydrogel layer formation allows for higher antibody loading than achievable with direct surface immobilization of antibody. The hydrogel layer chips exhibited significantly larger resonance angle shifts upon exposure to rabbit IgG compared to the directly immobilized chips, indicating enhanced signal output. In contrast, exposure to bovine serum albumin (BSA) elicited negligible responses, confirming the specificity of the sensor. Analysis of binding interactions revealed that the association constant ($K_a$) for rabbit IgG binding to the hydrogel layer chip was $1.3 \times 10^5$ $M^{-1}$ while that of the directly antibody-immobilized chip was $7.3 \times 10^4$ $M^{-1}$. The resonance angle shift increased with layer thickness up to ~225 nm, indicating improved signal generation with greater antibody content. Beyond this threshold, however, the signal decreased and approached the level of directly immobilized chips. This phenomenon is attributed to the physical limitation of SPR detection, which is confined to changes in permittivity within approximately 300 nm from the gold surface. Although thicker hydrogel layers contain more antibody and can bind more antigen, the diffusion of analyte molecules into deeper regions of the gel becomes increasingly hindered. Consequently, the contribution of antigen binding far from the sensor surface to the SPR signal diminishes due to limited evanescent field penetration and reduced local permittivity changes.

Molecular imprinting is a robust methodology for fabricating synthetic receptors with high specificity for target molecules [145–148]. In molecular imprinting, after functional monomers interact with the template molecule through non-covalent or covalent interactions and are copolymerized with crosslinkers, the template is removed from the resulting networks to form complementary binding cavities. These molecularly imprinted polymers (MIPs) exhibit selectivity in size, shape, and chemical functionality, rendering them suitable for diverse applications including biosensing, catalysis, and separation technologies. While molecular imprinting of small molecules has been extensively studied, imprinting of large, structurally delicate biomolecules such as proteins remains challenging due to their conformational complexity and sensitivity to processing conditions [149–153]. For example, SPR sensor chips featuring protein-imprinted hydrogel layers with specific affinity toward a target protein, concanavalin A (ConA), were fabricated using molecular imprinting with minimal crosslinker content [154]. Traditional imprinting approaches employ substantial crosslinker quantities to rigidify recognition sites; however, this approach leveraged the flexibility and spacious architecture of hydrogel networks to accommodate the larger dimensions of proteins while preserving conformational features essential for molecular recognition. Hydrogel layers with molecular binding sites were synthesized on SPR sensor chips via SI-ATRP using GEMA as the glucose-containing ligand monomer, acrylamide (AAm), and a small amount of *N,N′*-methylenebisacrylamide (MBAA) as a crosslinker in molecular imprinting using lectin ConA as the template. Following polymerization, template ConA was removed in aqueous glucose solution to yield ConA-imprinted hydrogel layers. Hydrogel layer thickness increased proportionally with polymerization time, consistent with the ATRP growth mechanism in which polymer molecular weight scales with reaction time and monomer availability. The ConA-imprinted hydrogel layer (~289.0 nm) produced a significantly larger resonance angle shift compared to the nonimprinted hydrogel layer (~291.5 nm), indicating enhanced specific binding. Affinity analysis revealed an association constant of $2.02 \times 10^7$ $M^{-1}$ for the imprinted hydrogel layer versus $5.26 \times 10^5$ $M^{-1}$ for the control, validating the successful formation of recognition sites via molecular imprinting with GEMA as the ligand monomer. Increasing layer thickness resulted in a gradual decline in resonance angle shift. This inverse relationship contrasts with previous observations for antibody-functionalized hydrogels, where signal intensity initially increased with thickness up to ~300 nm before declining. Given that the effective penetration depth of SPR is approximately 300 nm, hydrogel layers exceeding this limit likely hinder diffusion of analyte molecules toward the sensor surface, attenuating signal response. In thicker layers, restricted molecular diffusion reduces the interaction of ConA with the gold interface, thereby decreasing the detectable permittivity change. Thus, precise control of hydrogel thickness is critical for optimizing SPR sensor performance. The combination of SI-ATRP and molecular imprinting is useful in designing hydrogel-based SPR sensors with tunable sensitivity and specificity.

Biomolecules such as antibodies or lectins can also be used as ligands in the fabrication of hydrogel layers on SPR chips by molecular imprinting. For example, molecularly imprinted hydrogel layers in the form of brush-type and chemically crosslinked gel-type architectures were fabricated on SPR sensor chips via SI-ATRP, employing a lectin (ConA) and an antibody (anti-IgG) as biomolecular ligands for the template glycoprotein (rabbit immunoglobulin G: IgG), in the absence and presence of a chemical crosslinker, respectively [155]. Although the IgG-imprinted brush-type sensor chip exhibited a binding affinity for IgG comparable to that of the nonimprinted brush-type sensor chip, the IgG-imprinted gel-type layer demonstrated an approximately twofold increase in the affinity constant relative to the nonimprinted gel-type layer. These results underscore the critical role of chemical crosslinking in the formation of well-defined molecular recognition sites through molecular imprinting. Consequently, the combination of biomolecular ligands and crosslinkers facilitates the construction of hydrogel layer sensor interfaces with enhanced specificity and affinity toward target biomolecules. The resulting gel-type molecularly imprinted layers incorporating both lectins and antibodies as ligands represent a promising platform for the development of SPR-based biosensors aimed at the selective detection of proteins.

In addition, bioinert components such as poly(ethylene glycol) (PEG) and poly(2-methacryloyloxyethyl phosphorylcholine) (PMPC) have been widely used for the suppression of nonspecific protein adsorption onto the SPR sensor chip surfaces. For example, molecularly imprinted hydrogel layers possessing specific recognition sites for the lectin ConA were fabricated on SPR sensor chips via SI-ATRP of 2-methacryloyloxyethyl phosphorylcholine (MPC), GEMA, and *N,N′*-methylenebisacrylamide (MBAA), aiming to develop SPR-based sensing platforms capable of highly sensitive and selective protein detection (Fig. 11.9a) [156]. The integration of SI-ATRP with molecular imprinting enabled the construction of hydrogel layers approximately 120–130 nm thick on the SPR sensor surfaces. SPR measurements demonstrated that the ConA-imprinted hydrogel layer chips exhibited significantly higher resonance angle shifts and apparent affinity constants toward the target lectin ConA compared to their nonimprinted hydrogel layer, indicating the successful generation of specific recognition sites through molecular imprinting (Fig. 11.9b). This enhanced recognition is attributed to the spatially optimized orientation of GEMA-derived glucose residues within the gel networks,

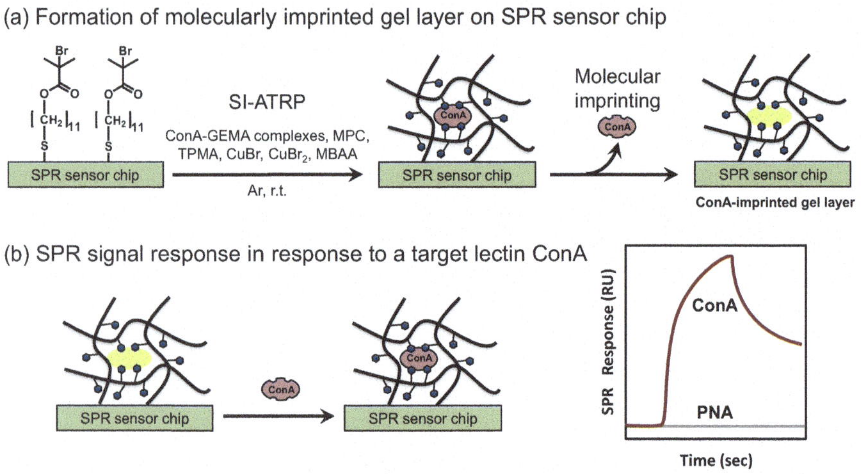


**Fig. 11.9** Adsorption of lectin to molecularly imprinted gel layers fabricated on SPR sensor chip

formed by the molecular imprinting. Moreover, the ConA-imprinted hydrogel layers showed negligible interaction with the non-target lectin peanut agglutinin (PNA), confirming their high selectivity. The incorporation of PMPC as the bioinert backbone component effectively suppressed nonspecific protein adsorption, thereby contributing to the high specificity of the sensor interface. Consequently, employing SI-ATRP in conjunction with molecular imprinting and MPC-based hydrogel networks offers a robust strategy for fabricating SPR sensor devices with exceptional sensitivity and selectivity for target protein detection. The remarkable performance of these molecularly imprinted hydrogel layer SPR sensors highlights their potential as versatile platforms for the development of advanced biosensing technologies across diverse biomedical and analytical applications.

## 11.6 Responsive Films and Hydrogel Surfaces

Surface patterning across nano- to microscale dimensions is pivotal in advancing a wide spectrum of technologies, including electronics, photonics, and biomedical engineering [157, 158]. Responsive polymers have emerged as key candidates for surface patterning due to their chemical versatility, mechanical robustness, and compatibility with diverse fabrication strategies [159]. Conventional approaches such as photolithography and nanoimprint lithography have proven indispensable in achieving high-resolution patterns with controlled geometries [160–163]. In photolithography, patterns are generated by the selective exposure of photoresist films to ultraviolet (UV) light, inducing photochemical reactions like crosslinking or degradation, followed by solvent-based development. While this method enables precise alignment over large areas, its implementation often involves high operational costs and complicated procedures. In contrast, nanoimprint lithography relies on mechanical deformation or UV curing of thermoplastic or liquid polymer films under pressure using pre-patterned molds [158, 164]. Despite its high throughput and resolution, nanoimprint lithography requires mold fabrication via conventional lithographic techniques, and the application of substantial pressure poses a risk of substrate or mold damage. Recent studies have explored the incorporation of azo-polymers with photo-responsive behavior into nanoimprint lithography to enable directionally controlled nanostructures [165]. However, there remains a need for simplified, low-cost micropatterning methodologies that can produce versatile polymeric patterns without extensive infrastructure, especially for the development of affordable, disposable sensing devices and diagnostic chips where ultra-high resolution is not mandatory.

Stimuli-responsive polymers, which undergo significant changes in structure or properties upon external stimuli, offer promising alternatives for adaptive surface design [166–168]. General stimuli-responsive volume changes have traditionally been observed in hydrogels or organogels, wherein polymer networks expand or contract through solvent uptake or release. However, achieving analogous responses in dry films exposed to air has proven challenging. Considering air as a pseudo-solvent, structural changes in photo-responsive polymers could be induced via photochemical modulation of their internal free volume. Recently, a solvent-free strategy for achieving UV-induced volume changes in dry polymer films was proposed using poly(dimethylsiloxane) (PDMS), a polymer known for its intrinsic large free volume, as a base matrix [169]. A novel class of photo-responsive copolymers was developed by incorporating cinnamoyl-functionalized methacrylate monomers capable of photodimerization into a PDMS backbone. 2-Cinnamoyloxyethyl methacrylate (CEMA) as a monomer with a photodimerizable group was copolymerized with PDMS macromonomers to yield PCEMA-*g*-PDMS copolymers. Upon UV irradiation (270 nm), cinnamoyl moieties undergo [2 + 2] cycloaddition, leading to localized crosslinking in exposed regions. Micropatterned surfaces were obtained by exposing PCEMA-*g*-PDMS films to UV through photomasks (Fig. 11.10). Scanning probe microscopy (SPM) visualized distinct topographical changes: unexposed regions remained flat, while UV-exposed regions exhibited pattern depths of ~5.7 nm. Various mask geometries produced corresponding micropatterns without post-exposure solvent treatment. X-ray photoelectron spectroscopy (XPS) and contact angle measurements demonstrated that the surface compositions and hydrophilicity remained unchanged after UV exposure, indicating minimal oxidation or degradation. Thus, the observed reduction in film thickness was attributed to a decrease in free volume upon photodimerization. Upon UV exposure through a photomask, SPM observations revealed that thinner films (<200 nm) slightly increased in height, likely due to lateral chain association during crosslinking, while thicker films (>400 nm) underwent significant volume contraction, forming well-defined micropatterns. This means that the important role of free volume of PDMS is in driving UV-induced pattern formation. Thus, the PCEMA-*g*-PDMS films exhibit controllable film thickness changes in dry environments, enabling cost-effective micropatterning without solvents or etching processes.

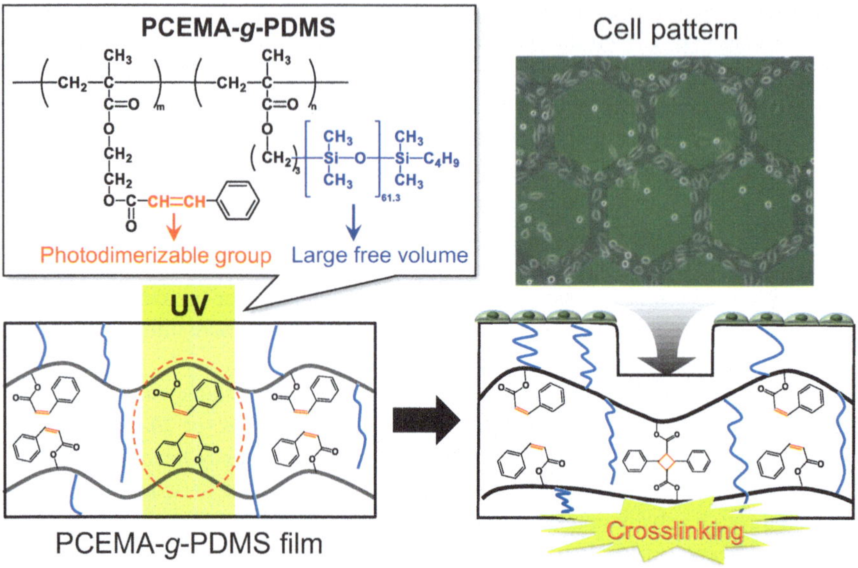


**Fig. 11.10** Micropatterning of the PCEMA-g-PDMS film upon UV exposure through a photomask and cell adhesion on the micropatterned surface

Cells dynamically regulate their behavior in response to surrounding cells and their spatial distribution. For example, mesenchymal stem cells exhibit region-dependent differentiation within cellular aggregates [170, 171], while cardiomyocytes demonstrate enhanced contractile activity when aligned in anisotropic, striped patterns [172]. Consequently, the development of cell culture platforms capable of directing cell localization by modulating adhesion properties is essential for advancing tissue engineering and deciphering the mechanistic underpinnings of cell function [173]. Cell adhesion is governed not only by biochemical factors, such as ligand-receptor interactions and immobilized adhesion proteins [174–177], but also by physical properties of the extracellular environment, including matrix stiffness and surface topography [178–181]. In particular, elastic modulus plays a pivotal role in the differentiation of stem cells to neuronal, muscular, or osteogenic phenotypes [181], underscoring its significance in regenerative medicine. In native tissues, stiffness varies widely, from the compliant nature of brain or adipose tissue to the rigidity of bone and cartilage [182]. Although soft polymeric hydrogels have been widely employed to mimic tissue-like mechanical environments, conventional rigid materials (e.g., ceramics, metals) lack tunability in modulus without altering chemical composition or geometry. This presents a barrier to systematic mechanobiological studies targeting the MPa-range stiffness characteristic of hard tissues. Moreover, even small modulus differences in the kPa range can profoundly affect cellular behavior [183], highlighting the need for high-resolution platforms that mimic physiologically relevant stiffness gradients. Micropatterned substrates have emerged as promising tools for guiding cell adhesion and organization in vitro

[184–187]. Techniques such as photolithography and nanoimprint lithography enable nanoscale control of surface morphology but often involve multistep procedures, costly instrumentation, or the need for pre-fabricated molds. As such, there is a growing demand for simplified and scalable patterning methods capable of generating micro- and nanoscale features in a cost-effective and disposable format for biological applications. A robust strategy was proposed for cell patterning based solely on UV-induced modulation of elastic modulus in photocrosslinkable PCEMA-*g*-PDMS films [188]. When mouse fibroblast L929 cells were cultured on the micropatterned PCEMA-*g*-PDMS films, they exhibited selective adhesion to unexposed regions of the patterned substrates after a 48-hour incubation (Fig. 11.10). Interestingly, similar patterning behavior was observed in the hepatocellular carcinoma cell line HepG2, highlighting the versatility of the substrate across different cell types. Notably, HepG2 cells tended to form multicellular spheroids, in contrast to the flatter morphology of L929, potentially reflecting differences in intrinsic cell adhesion and aggregation behavior, which is a phenomenon influenced by the local microenvironment. Remarkably, cell patterning was achieved without the use of poly(ethylene glycol) (PEG) or other hydrophilic, bioinert coatings, which are typically required to inhibit non-specific cell adhesion. Instead, cells exhibited preferential attachment, spreading, and proliferation on unexposed regions, implying that UV-induced modifications altered cell–material interactions. These findings affirm the biocompatibility of PCEMA-*g*-PDMS films and their potential to guide cell behavior via controlled micropatterning. Protein adsorption measurements, XPS analysis, and contact angle measurements detected no change in the chemical properties of the film surface after UV exposure. On the other hand, the modulus mapping via SPM demonstrated that UV-exposed regions displayed significantly elevated surface modulus compared to unexposed regions, a consequence of increased crosslinking density via photodimerization of cinnamoyl groups. Such changes in physical properties are known to influence mechanosensitive cell responses. Unlike conventional cell culture substrates with uniform moduli, the micropatterned PCEMA-*g*-PDMS films replicate the heterogeneous stiffness of biological tissues, promoting cell localization in regions with more favorable stiffness or softness. Cell patterning techniques are instrumental in understanding biological processes such as intercellular communication and adhesion. Photo-responsive polymer films with controllable elastic modulus offer a biologically favorable and nontoxic approach to spatially directing cell adhesion without chemical surface modification. Thus, the photo-responsive polymer films platform offers a facile, solvent-free approach for creating micropatterned cell culture substrates with tunable mechanical properties. By harnessing UV-induced modulation of surface modulus, the photo-responsive polymer films enable spatiotemporal control of cell adhesion, proliferation, and organization. It holds promise for applications in tissue engineering, mechanobiology, and cellular patterning without the need for complex chemical modifications or post-processing steps.

Since the seminal work by Engler et al., substantial efforts have been made to explore how physical properties, particularly substrate stiffness, govern cellular functions [178–183]. Consequently, numerous studies have aimed to control cell behavior through engineered scaffolds with tunable physical properties, as described above. However, most of these scaffolds possess fixed mechanical characteristics, limiting their ability to mimic the dynamic in vivo environment. Because cells in living systems are continually exposed to time-dependent changes in their surroundings, there is a growing need for scaffolds capable of spatiotemporal modulation of both chemical and physical properties. Such dynamic materials are essential not only for understanding the influence of microenvironmental changes on cellular responses but also for precisely regulating cell behavior. A well-known example includes the work by Okano et al., who fabricated cell sheets by culturing cells on a surface coated with the temperature-responsive PNIPAAm. This polymer exhibits a hydrophilic–hydrophobic switch near its LCST, facilitating temperature-triggered detachment of intact cell layers [11, 189–192]. Beyond temperature-responsiveness, considerable research has been devoted to multi-responsive hydrogels incorporating photo-reactive groups. For instance, photocrosslinked and photodegradable hydrogels have been used to modulate scaffold stiffness in response to light, enabling dynamic regulation of cell behaviors such as spreading, differentiation, and migration [38, 40–42, 193, 194]. Recently, dual-stimuli-responsive hydrogel surfaces, which undergo drastic changes in physical properties such as elastic modulus and chemical properties such as hydrophilicity/hydrophobicity in response to multiple stimuli such as light and temperature, were designed to achieve spatiotemporal regulation of cell behavior through orthogonal control of stiffness and hydrophilicity (Fig. 11.11) [195]. The hydrogels were synthesized via copolymerization of 7-methacryloyloxycoumarin (MAC) as a monomer featuring a photodimerizable group and oligo(ethylene glycol) methacrylate (OEGMA) as a temperature-responsive component exhibiting LCST behavior. Upon UV exposure, the aqueous solution of the resulting copolymer, P(MAC-*co*-OEGMA), underwent a sol–gel transition due to photodimerization of coumarin groups, which facilitated the formation of a physically crosslinked network. The stiffness of the resulting hydrogel could be finely tuned by modulating the duration of UV exposure, enabling precise control over the elastic modulus. Furthermore, P(MAC-*co*-OEGMA) exhibited temperature-responsive changes in the solubility: it remained soluble in aqueous media below 33 °C and became insoluble above 33 °C, indicative of LCST-type behavior. As the temperature increased above the LCST, the hydrogel surface displayed a transition from hydrophilic to hydrophobic characteristics. Photopatterning through a photomask enabled spatial control of crosslink density, producing micropatterned hydrogel surfaces composed of regions with distinct elastic moduli. Cells preferentially adhered to and spread on stiffer regions of the patterned hydrogel surface, demonstrating the influence of mechanical properties on cell behavior. Additionally, temperature-responsive changes in surface hydrophilicity significantly impacted cell morphology: cells maintained a rounded morphology on hydrophilic hydrogel surfaces at temperatures below the LCST, while adopting a spread morphology on hydrophobic surfaces at temperatures above the LCST. Notably, when cells initially cultured at 37 °C on the hydrophobic surface were subsequently exposed to a temperature reduction to 25 °C, they detached from the hydrogel surface, reflecting a reversible control over cell adhesion mediated by

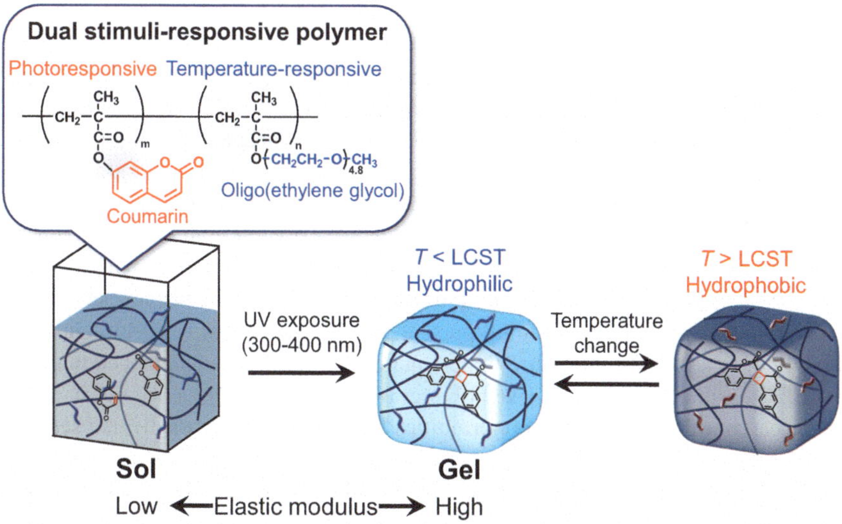


**Fig. 11.11** Sol–gel phase transition of photo/temperature-responsive polymer in aqueous media and their responsive change in the elastic modulus and hydrophilicity/hydrophobicity

temperature-responsiveness. Thus, P(MAC-*co*-OEGMA) hydrogel surfaces offer dual and independent control over mechanical stiffness and surface hydrophilicity/hydrophobicity in response to light and temperature stimuli, respectively. Due to their dynamic and reversible responsiveness, these hydrogel surfaces serve as a promising platform for the spatiotemporal regulation of cell behaviors such as adhesion, spreading, and detachment. Moreover, the exploration of cell responses to such dual stimuli-responsive hydrogel surfaces contributes fundamental insights into the interplay between physicochemical scaffold properties and cellular behavior, paving the way for innovative applications in regenerative medicine, tissue engineering, and high-throughput drug screening.

## 11.7 Conclusion

Stimuli-responsive polymer interfaces are specialized material surfaces and interfaces designed to alter their properties in response to various external stimuli, including temperature, pH, light, and mechanical forces. The polymer interfaces can exist as particles, self-assemblies, layers, or films and have garnered significant attention in the fields of materials science and biomedical engineering due to their versatility and adaptability. Their unique ability to respond dynamically to environmental changes makes them ideal for applications in drug delivery, biosensing, cell regulation, and smart materials, where precise control over material behavior is critical. The design and engineering of stimuli-responsive polymers involve selecting specific monomers and crosslinking strategies to achieve desired responsiveness. Stimuli-responsive polymer interfaces are engineered to undergo significant changes in their structures and properties in response to specific external stimuli. The design and application of stimuli-responsive polymer interfaces face several challenges that need to be addressed for the advancement of the field. One primary challenge is the complexity of achieving precise control over the responsiveness of these polymers to external stimuli such as temperature, pH, and light. Despite their promising capabilities, the field faces several challenges, including ensuring biocompatibility, biodegradability, and scalability for industrial applications. The complexities involved in controlling the precise responsiveness of their polymer interfaces to various stimuli present an ongoing area of research, with significant implications for the future of personalized medicine and advanced material science. Overall, stimuli-responsive polymer interfaces represent a frontier in materials research, offering exciting opportunities for innovation across diverse sectors. Continued interdisciplinary research is essential to overcome current limitations and fully realize the potential of these responsive systems in real-world applications. As new fields such as nanotechnology and biotechnology continue to expand, there are exciting opportunities for the application of stimuli-responsive polymer interfaces. Future work should aim to explore new strategies for designing stimuli-responsive polymer interfaces and innovative applications, such as targeted drug delivery systems, biosensing, cell regulation, and smart coatings, while ensuring that the well-designed polymers can perform effectively in real-world scenarios.

**Acknowledgments** This study was partly supported in part by JSPS KAKENHI (Grant No. JP25K01838) from the Japan Society for the Promotion of Science (JSPS) and the MEXT-Supported Program for Private University Research Branding Project.

## References

1. Stuart MAC, Huck WTS, Genzer J, Muller M, Ober C, Stamm M, Sukhorukov GB, Szleifer I, Tsukruk VV, Urban M, Winnik F, Zauscher S, Luzinov I, Minko S (2010) Emerging applications of stimuli-responsive polymer materials. Nat Mater 9:101–113
2. Theato P, Sumerlin BS, O'Reilly RK, Epps TH III (2013) Stimuli responsive themed issue. Chem Soc Rev 42:7045–7486
3. Boase NRB, Gillies ER, Goh R, Kieltyka RE, Matson JB, Meng F, Sanyal A, Sedláček O (2024) Stimuli-responsive polymers at the interface with biology. Biomacromolecules 25:5417–5436
4. Wei M, Gao Y, Li X, Serpe MJ (2017) Stimuli-responsive polymers and their applications. Polym Chem 8:127–143
5. Singh J, Nayak P (2023) pH-responsive polymers for drug delivery: trends and opportunities. J Polym Sci 61:2828–2850
6. Molina M, Asadian-Birjand M, Balach J, Bergueiro J, Miceliac E, Calderon M (2015) Stimuli-responsive nanogel composites and their application in nanomedicine. Chem Soc Rev 44:6161–6186
7. Palivan CG, Goers R, Najer A, Zhang XY, Car A, Meier W (2016) Bioinspired polymer vesicles and membranes for biological and medical applications. Chem Soc Rev 45:377–411

8. Cabral H, Miyata K, Osada K, Kataoka K (2018) Block copolymer micelles in nanomedicine applications. Chem Rev 118:6844–6892
9. Karg M, Pich A, Hellweg T, Hoare T, Lyon LA, Crassous JJ, Suzuki D, Gumerov RA, Schneider S, Potemkin II, Richtering W (2019) Nanogels and microgels: from model colloids to applications, recent developments, and future trends. Langmuir 35:6231–6255
10. Beach MA, Nayanathara U, Gao YT, Zhang CH, Xiong YJ, Wang YF, Such GK (2024) Polymeric nanoparticles for drug delivery. Chem Rev 124:5505–5616
11. Matsuda N, Shimizu T, Yamato M, Okano T (2007) Tissue engineering based on cell sheet technology. Adv Mater 19:3089–3099
12. Arif M, Raza H, Akhter T (2024) Classification, synthesis, characterization, and applications of metal nanoparticle-containing hybrid microgels: a comprehensive review. RSC Adv 14:24604–24630
13. Guragain S, Bastakoti BP, Malgras V, Nakashima K, Yamauchi Y (2015) Multi-stimuli-responsive polymeric materials. Chem Eur J 21:13164–13174
14. Zhuang J, Gordon MR, Ventura J, Lia L, Thayumanavan S (2013) Multi-stimuli responsive macromolecules and their assemblies. Chem Soc Rev 42:7421–7435
15. Bae YH, Park K (2013) Stimuli-responsive polymers for drug delivery. J Control Release 172:121–127
16. Wei DS, Sun Y, Zhu H, Fu QR (2023) Stimuli-responsive polymer-based nanosystems for cancer theranostics. ACS Nano 17:23223–23261
17. Kocak G, Tuncer C, Bütün V (2017) pH-responsive polymers. Polym Chem 8:144–176
18. Sponchioni M, Palmiero UC, Moscatelli D (2019) Thermo-responsive polymers: applications of smart materials in drug delivery and tissue engineering. Mater Sci Eng C 102:589–605
19. Bertrand O, Gohy JF (2017) Photo-responsive polymers: synthesis and applications. Polym Chem 8:52–73
20. Moad G (2017) RAFT polymerization to form stimuli-responsive polymers. Polym Chem 8:177–219
21. Madineh H, Mansourinia F, Zarrintaj P, Poostchi M, Gnatowski P, Kucinska-Lipka J, Ghaffari M, Hasanin MS, Chapi S, Yazdi MK, Ashrafizadeh M, Bączek T, Saeb MR, Wang G (2025) Stimuli-responsive delivery systems using carbohydrate polymers: a review. Int J Biolog Macromol. in press. https://doi.org/10.1016/j.ijbiomac.2025.142648
22. Bratek-Skicki A (2021) Towards a new class of stimuli-responsive polymer-based materials –recent advances and challenges. Appl Surf Sci Adv 4:100068
23. Rao NV, Ko H, Lee J, Park JH (2018) Recent progress and advances in stimuli-responsive polymers for cancer therapy. Front Bioeng Biotechnol 6:110
24. Yuan Z, Ding J, Zhang Y, Huang B, Song Z, Meng X, Ma X, Gong X, Huang Z, Ma S, Xiang S, Xu W (2022) Components, mechanisms and applications of stimuli-responsive polymer gels. Eur Polym J 177:111473
25. Karimi M, Ghasemi A, Zangabad PS, Rahighi R, Basri SMM, Mirshekari H, Amiri M, Pishabad ZS, Aslani A, Bozorgomid M, Ghosh D, Beyzavi A, Vaseghi A, Aref AR, Haghani L, Bahramia S, Hamblin MR (2016) Smart micro/nanoparticles in stimulus-responsive drug/gene delivery systems. Chem Soc Rev 45:1457–1501
26. Rabiee N, Hajebi S, Bagherzadeh M, Ahmadi S, Rabiee M, Roghani-Mamaqani H, Tahriri M, Tayebi L, Hamblin MR (2019) Stimulus-responsive polymeric nanogels as smart drug delivery systems. Acta Biomater 92:1–18
27. Amalvy JI, Wanless EJ, Li V Y, Michailidou SPA (2004) Synthesis and caracterization of novel pH-responsive microgels based on tertiary amine methacrylates. Langmuir 20:8992–8999
28. Emileh A, Vasheghani-Farahani E, Imani M (2007) Preparation and characterization of pH-sensitive microgels of poly((2-dimethylamino) ethyl methacrylate). Macromol Symp 255:1–7
29. Tanga L, Wanga L, Yanga X, Fenga Y, Lia Y, Fenga W (2021) Poly(N-isopropylacrylamide)-based smart hydrogels: design, properties and applications. Prog Mater Sci 115:100702
30. Pelton RH, Chibante P (1986) Preparation of aqueous latices with N-Isopropylacrylamide. Colloids Surf 20:247–256

31. Guan Y, Zhang Y (2011) PNIPAM microgels for biomedical applications: from dispersed particles to 3D assemblies. Soft Matter 7:6375–6384
32. Cortez-Lemus NA, Licea-Claverie A (2016) Poly(N-vinylcaprolactam), a comprehensive review on a thermoresponsive polymer becoming popular. Prog Polym Sci 53:1–51
33. Sasaoka M, Kawamura A, Miyata T (2022) Core–shell microgels having zwitterionic hydrogel core and temperature-responsive shell prepared via inverse miniemulsion RAFT polymerization. Polym Chem 13:3489–3497
34. Shu T, Hu L, Hunter H, Balasuriya N, Fang C, Zhang Q, Serpe MJ (2022) Multi-responsive micro/nanogels for optical sensing. Adv Phys: X 7:2043185
35. Jiang Y, Chen J, Deng C, Suuronen EJ, Zhong Z (2014) Click hydrogels, microgels and nanogels: emerging platforms for drug delivery and tissue engineering. Biomaterials 35:4969–4985
36. Hao S, Tian C, Bai Y, Wu L, Hao L, Kuang Y, Yang S, Mao H, Gu Z (2023) Photo-crosslinkable hyaluronic acid microgels with reactive oxygen species scavenging capacity for mesenchymal stem cell encapsulation. Int J Biol Macromol 243:124971. https://doi.org/10.1016/j.ijbiomac.2023.124971
37. Hang C, Zou Y, Zhong Y, Zhong Z, Meng F (2017) NIR and UV-responsive degradable hyaluronic acid nanogels for CD44-targeted and remotely triggered intracellular doxorubicin delivery. Colloids Surf B Biointerfaces 158:547–555
38. Ruskowitz ER, DeForest CA (2018) Photoresponsive biomaterials for targeted drug delivery and 4D cell culture. Nat Rev Mater 3:17087
39. Tang S, Richardson BM, Anseth KS (2020) Dynamic covalent hydrogels as biomaterials to mimic the viscoelasticity of soft tissues. Prog Mater Sci 120:100738
40. Yang C, Tibbitt MW, Basta L, Anseth KS (2014) Mechanical memory and dosing influence stem cell fate. Nat Mater 13:645–652
41. Nelson BR, Kirkpatrick BE, Miksch CE, Davidson MD, Skillin NP, Hach GK, Khang A, Hummel SN, Fairbanks BD, Burdick JA, Bowman CN, Anseth KS (2023) Photoinduced dithiolane crosslinking for multiresponsive dynamic hydrogels. Adv Mater 36:2211209
42. Yavitt FM, Kirkpatrick BE, Blatchley MR, Speckl KF, Mohagheghian E, Moldovan R, Wang N, Dempsey PJ, Anseth KS (2023) In situ modulation of intestinal organoid epithelial curvature through photoinduced viscoelasticity directs crypt morphogenesis. Sci Adv 9:eadd5668
43. Miyata T, Namera T, Liu Y, Kawamura A, Yamaoka T (2022) Photoresponsive behaviour of zwitterionic polymer particles with photodimerizable groups on their surfaces. J Mater Chem B 10:2637–2648
44. Nayak S, Lyon LA (2005) Soft nanotechnology with soft nanoparticles. Angew Chem Int Ed 44:7686–7708
45. Ulijn RV, Bibi N, Jayawarna V, Thornton PD, Todd SJ, Mart RJ, Smith AM, Gough JE (2007) Bioresponsive hydrogels. Mater Today 10:40–48
46. Haupt K, Rangel PXM, Bui BTS (2020) Molecularly imprinted polymers: antibody mimics for bioimaging and therapy. Chem Rev 120:9554–9582
47. Miyata T, Uragami T (2002) Biological stimuli-responsive hydrogels. In: Dumitriu S (ed) Polymeric biomaterials. Marcel Dekker, Inc, New York, pp 959–974
48. Miyata T (2010) Biomolecule-Responsive Hydrogels. In: Ottenbrite RM, Park K, Okano T (eds) Biomedical applications of hydrogels handbook. Springer, Berlin, pp 65–86
49. Miyata T (2013) Biomolecule-sensitive Hydrogels. In: Alvarez-Lorenzo C, Concheiro A (eds) RSC smart materials no.3, smart materials for drug delivery: Volume 2. RSC Publishing, London, pp 261–289
50. Miyata T (2015) Target Biomolecule-responsive Hydrogels. In: Schneider HJ (ed) RSC smart materials no.14, Chemoresponsive materials: stimulation by chemical and biological signals. RSC Publishing, London, pp 408–443
51. Miyata T (2022) Target Molecule-responsive Hydrogels. In: Schneider HJ (ed) Chemoresponsive materials: smart materials for chemical and biological stimulation: edition 2, Smart materials series. RSC, London, pp 223–261

52. Ullah B, Khan SR, Ali S, Jamil S (2022) Synthesis, parameters, properties and applications of responsive molecularly imprinted microgels. Rev Chem Eng 38:703–720
53. Zheng J, Zhu C, Xu X, Wang X, Fu J (2023) Supramolecular assemblies of multifunctional microgels for biomedical applications. J Mater Chem B 11:6265–6289
54. Kumar N, Singh S, Sharma P, Kumar B, Kumar A (2024) Single-, dual-, and multi-stimuli-responsive nanogels for biomedical applications. Gels 10:61
55. Cho H, Jeon S, Yang J, Baek SY, Kim D (2020) Hydrogel nanoparticle as a functional coating layer in biosensing, tissue engineering, and drug delivery. Coatings 10:663
56. Pan G, Guo Q, Cao C, Yang H, Li B (2013) Thermo-responsive molecularly imprinted nanogels for specific recognition and controlled release of proteins. Soft Matter 9:3840–3850
57. Peppas NA, Bures CD (2008) Glucose-Responsive Hydrogels. In: Wnek G, Bowlin G (eds) Encyclopedia of biomaterials and biomedical engineering, 2nd edn. CRC Press, Florida, p 11
58. Peters JT, Wechsler ME, Peppas NA (2021) Advanced biomedical hydrogels: molecular architecture and its impact on medical applications. Regen Biomater 8:1–21
59. Zhao L, Xiao C, Gai G, Ding J (2016) Glucose-sensitive polymer nanoparticles for self-regulated drug delivery. Chem Commun 52:7633–7652
60. Sharifzadeh G, Hosseinkhani H (2017) Biomolecule-responsive hydrogels in medicine. Adv Healthc Mater 6:1700801
61. Ishihara K, Kobayashi M, Ishimaru N, Shinohara I (1984) Glucose induced permeation control of insulin through a complex membrane consisting of immobilized glucose oxidase and a poly(amine). Polym J 16:625–631
62. Kost J, Horbett TA, Ratner BD, Singh M (1985) Glucose-sensitive membranes containing glucose oxidase: activity, swelling, and permeability studies. J Biomed Mater Res 19:1117–1133
63. Podual K, Doyle FJ, Peppas NA (2000) Glucose-sensitivity of glucose oxidase-containing cationic copolymer hydrogels having poly(ethylene glycol) grafts. J Control Release 67:9–17
64. Zhao L, Wang L, Zhang Y, Xiao S, Bi F, Zhao J, Gai G, Ding J (2017) Glucose oxidase-based glucose-sensitive drug delivery for diabetes treatment. Polymers 9:255
65. Volpatti LR, Matranga MA, Cortinas AB, Delcassian D, Daniel KB, Langer R, Anderson DG (2019) Glucose-responsive nanoparticles for rapid and extended self-regulated insulin delivery. ACS Nano 14:488–497
66. Kataoka K, Miyazaki H, Bunya M, Okano T, Sakurai Y (1998) Totally synthetic polymer gels responding to external glucose concentration: their preparation and application to on−off regulation of insulin release. J Am Chem Soc 120:12694–12695
67. Matsumoto A, Ishii T, Nishida J, Matsumoto H, Kataoka K, Miyahara Y (2012) A synthetic approach toward a self-regulated insulin delivery system. Angew Chem Int Ed 51:2124–2128
68. Matsumoto A, Tanaka M, Matsumoto H, Ochi K, Morooka Y, Kuwata H, Yamada H, Shirakawa I, Miyazawa T, Ishii H, Kataoka K, Ogawa Y, Miyahara Y, Suganami T (2017) Synthetic "smart gel" provides glucose-responsive insulin delivery in diabetic mice. Sci Adv 3:eaaq0723
69. Ma R, Shi L (2014) Phenylboronic acid-based glucose-responsive polymeric nanoparticles: synthesis and applications in drug delivery. Polym Chem 5:1503–1518
70. Blagojevic L, Kamaly N (2025) Nanogels: a chemically versatile drug delivery platform. Nano Today 61:102645
71. Hoare T, Pelton R (2008) Charge-switching, amphoteric glucose-responsive microgels with physiological swelling activity. Biomacromolecules 9:733–740
72. Xing S, Guan Y, Zhang Y (2011) Kinetics of glucose-induced swelling of p(NIPAM-AAPBA) microgels. Macromolecules 44:4479–4486
73. Guo Q, Zhang X (2019) Synthesized of glucose-responsive nanogels labeled with fluorescence molecule based on phenylboronic acid by RAFT polymerization. J Biomater Sci Polym Ed 30:815–831
74. Miyata T (2010) Preparation of smart soft materials using molecular complexes. Polym J 42:277–289

75. Nakamae K, Miyata T, Jikihara A, Hoffman AS (1994) Formation of poly(glucosyloxyethyl methacrylate)-concanavalin A complex and its glucose-sensitivity. J Biomater Sci Polym Ed 6:79–90
76. Miyata T, Jikihara A, Nakamae K, Hoffman AS (1996) Preparation of poly(glucosyloxyethyl methacrylate)-concanavalin A complex hydrogel and its glucose-sensitivity. Macromol Chem Phys 197:1135–1146
77. Miyata T, Jikihara A, Nakamae K, Hoffman AS (2004) Preparation of reversibly glucose-responsive hydrogels by covalent immobilization of lectin in polymer networks having pendant glucose. J Biomater Sci Polym Ed 15:1085–1098
78. Miyata T, Asami N, Uragami T (1999) A reversibly antigen-responsive hydrogel. Nature 399:766–769
79. Miyata T, Asami N, Uragami T (1999) Preparation of an antigen-sensitive hydrogel using antigen-antibody bindings. Macromolecules 32:2082–2084
80. Miyata T, Jige M, Nakaminami T, Uragami T (2006) Tumor marker-tesponsive behavior of gels prepared by biomolecular imprinting. PNAS 103:1190–1193
81. Kawamura A, Kiguchi T, Nishihata T, Uragami T, Miyata T (2014) Target molecule-responsive hydrogels designed by molecular imprinting using bisphenol A as a template. Chem Commun 50:11101–11103
82. Kawamura A, Hata Y, Miyata T, Uragami T (2012) Synthesis of glucose-responsive bioconjugated gel particles using surfactant-free emulsion polymerization. Colloids Surf B: Biointerfaces 99:74–81
83. Nayak S, Lyon LA (2004) Ligand-functionalized core/shell microgels with permselective shells. Angew Chem Int Ed 43:6706–6709
84. Kim J, Nayak S, Lyon LA (2005) Bioresponsive hydrogel microlenses. J Am Chem Soc 127:9588–9592
85. Kawamura A, Katoh T, Uragami T, Miyata T (2015) Design of molecule-responsive organic-inorganic hybrid nanoparticles bearing cyclodextrin as ligands. Polym J 47:206–211
86. Farooqi ZH, Khan SR, Begum R, Ijaz A (2016) Review on synthesis, properties, characterization, and applications of responsive microgels fabricated with gold nanostructures. Rev Chem Eng 32:49–69
87. Lu Y, Proch S, Schrinner M, Drechsler M, Kempe R, Ballauff M (2009) Thermosensitive core-shell microgel as a "nanoreactor" for catalytic active metal nanoparticles. J Mater Chem 19:3955–3961
88. Siddiq M, Bakhat K, Ajmal M (2020) Stimuli responsive microgel containing silver nanoparticles with tunable optical and catalytic properties. Pure Appl Chem 92:445–459
89. Schafer D, Fink F, Kleinschmidt D, Keisers K, Thomas F, Hoffmann A, Pich A, Herres-Pawlis S (2020) Enhanced catalytic activity of copper complexes in microgels for aerobic oxidation of benzyl alcohols. Chem Commun 56:5601–5604
90. Kakar MU, Khan K, Akram M, Sami R, Khojah E, Iqbal I, Helal M, Hakeem A, Deng Y, Dai R (2021) Synthesis of bimetallic nanoparticles loaded on to PNIPAM hybrid microgel and their catalytic activity. Sci Rep 11:14759
91. Resmini M, Klavin F, Carboni D (2010) Microgels and nanogels with catalytic activity. In: Haupt K (ed) Topics in current chemistry, molecular imprinting. Springer, New York, pp 307–342
92. Daniel M, Astruc D (2004) Gold nanoparticles: assembly, supramolecular chemistry, quantum-size-related properties, and applications toward biology, catalysis, and nanotechnology. Chem Rev 104:293–346
93. Yeh YC, Creran B, Rotello VM (2011) Gold nanoparticles: preparation, properties, and applications in bionanotechnology. Nanoscale 23:1–7
94. Jeong H, Choi E, Ellis E, Lee T (2019) Recent advances in gold nanoparticles for biomedical applications: from hybrid structures. J Mater Chem B 7:3480–3496
95. Suzuki D, Kawaguchi H (2005) Modification of gold nanoparticle composite nanostructures using thermosensitive core-shell particles as a template. Langmuir 21:8175–8179

96. Lian X, Jin J, Tian J, Zhao H (2010) Thermoresponsive nanohydrogels cross-linked by gold nanoparticles. ACS Appl Mater Interfaces 2:2261–2268
97. Liang S, Jasinski J, Hammond GB, Xu B (2015) Supported gold nanoparticle-catalyzed hydration of alkynes under basic conditions. Org Lett 17:162–165
98. Della Pina C, Falletta E, Rossi M (2012) Update on selective oxidation using gold. Chem Soc Rev 41:350–369
99. Corma A, Garci H (2008) Supported gold nanoparticles as catalysts for organic reactions. Chem Soc Rev 37:2096–2126
100. Zhao P, Feng X, Huang D, Yang G, Astruc D (2015) Basic concepts and recent advances in nitrophenol reduction by gold- and other transition metal nanoparticles. Coord Chem Rev 287:114–136
101. Wu S, Lei L, Xia Y, Oliver S, Chen X, Boyer C, Nie Z, Shi S (2021) PNIPAM-immobilized gold-nanoparticles with colorimetric temperature-sensing and reusable temperature-switchable catalysis properties. Polym Chem 12:6903–6913
102. Liu XY, Cheng F, Liu Y, Liua HJ, Chen Y (2010) Preparation and characterization of novel thermoresponsive gold nanoparticles and their responsive catalysis properties. J Mater Chem 20:360–368
103. Liu G, Wang D, Zhou F, Liu W (2015) Electrostatic self-assembly of Au nanoparticles onto thermosensitive magnetic core-shell microgels for thermally tunable and magnetically recyclable catalysis. Small 11:2807–2816
104. Pongsanon P, Oota Y, Kawamura A, Kawasaki H, Miyata T (2023) Controllable catalytic activity of temperature-responsive polymer hybrid microgels designed using a gold nanoparticle monomer with polymerizable groups. Macromolecules 56:9853–9865
105. Pongsanon P, Kawamura A, Kawasaki H, Miyata T (2024) Effect of gold nanoparticle size on regulated catalytic activity of temperature-responsive polymer–gold nanoparticle hybrid microgels. Gels 10:357
106. Karimi M, Ghasemi A, Zangabad PS, Rahighi R, Basri SMM, Mirshekari H, Amiri M, Pishabad ZS, Aslani A, Bozorgomid M, Ghosh D, Beyzavi A, Vaseghi A, Aref AR, Haghani L, Bahrami S, Hamblin MR (2016) Smart micro/nanoparticles in stimulus-responsive drug/gene delivery systems. Chem Soc Rev 45:1457–1501
107. Bordat A, Boissenot T, Nicolas J, Tsapis N (2019) Thermoresponsive polymer nanocarriers for biomedical applications. Adv Drug Deliv Rev 138:167–192
108. Ghosh B, Biswas S (2021) Polymeric micelles in cancer therapy: state of the art. J Control Release 332:127–147
109. Zheng YT, Oz Y, Gu YM, Ahamad N, Shariati K, Chevalier J, Kapur D, Annabi N (2024) Rational design of polymeric micelles for targeted therapeutic delivery. Nano Today 55:28
110. Zhu YQ, Yang B, Chen S, Du JZ (2017) Polymer vesicles: mechanism, preparation, application, and responsive behavior. Prog Polym Sci 64:1–22
111. Rideau E, Dimova R, Schwille P, Wurm FR, Landfester K (2018) Liposomes and polymersomes: a comparative review towards cell mimicking. Chem Soc Rev 47:40
112. Kauscher U, Holme MN, Björnmalm M, Stevens MM (2019) Physical stimuli-responsive vesicles in drug delivery: beyond liposomes and polymersomes. Adv Drug Deli Rev 138:259–275
113. Jiang C, Xu G, Gao J (2022) Stimuli-responsive macromolecular self-assembly. Sustainability 14:11738
114. Mu R, Zhu D, Abdulma-lik S, Wijekoon S, Wei G, Kumbar SG (2024) Stimuli-responsive peptide assemblies: design, self-assembly, and biomedical applications. Bioactive Materials 35:181–207
115. Kataoka K, Kown GS, Yamaoka M, Okano T, Sakurai T (1993) Block copolymer micelles as vehicles for drug delivery. J Control Release 24(1–3):119–132
116. Kataoka K, Harada A, Nagasaki Y (2001) Block copolymer micelles for drug delivery: design, characterization and biological significance. Adv Drug Deliv Rev 47(1):113–131

117. Bae Y, Fukushima S, Harada A, Kataoka K (2003) Design of environment-sensitive supramolecular assemblies for intracellular drug delivery: polymeric micelles that are responsive to intracellular pH change. Angew Chem Int Ed 42(38):4640–4643
118. Discher DE, Eisenberg A (2002) Polymer vesicles. Science 297(5583):967–973
119. Martin F, MacDonald R (1974) Liposomes can mimic virus membranes. Nature 252(5479):161–163
120. Yatvin MB, Weinstein JN, Dennis WH, Blumenthal R (1978) Design of liposomes for enhanced local release of drugs by hyperthermia. Science 202(4374):1290–1293
121. Weinstein JN, Magin RL, Cysyk LR, Zaharko DS (1980) Treatment of solid l1210 murine tumors with local hyperthermia and temperature-sensitive liposomes containing methotrexate. Cancer Res 40(5):1388–1395
122. Kato T, Uchida J, Ichikawa T, Soberats B (2018) Functional liquid-crystalline polymers and supramolecular liquid crystals. Polym J 50(1):149–166
123. Kato T, Uchida J, Ichikawa T, Sakamoto T (2018) Functional liquid crystals towards the next generation of materials. Angew Chem Int Ed 57(16):4355–4371
124. Finkelmann H, Rehage G (1980) Investigations on liquid crystalline polysiloxanes, 2. optical properties of cholesteric phases and influence of the flexible spacer on the mobility of the mesogenic groups. Macromol Chem Rapid Commun 1(12):31–34
125. Finkelmann H, Kock HJ, Rehage G (1981) Investigations on liquid crystalline polysiloxanes, 3. Liquid crystalline elastomers—a new type of liquid crystalline material. Macromol Chem Rapid Commun 2(4):317–322
126. Ohm C, Brehmer M, Zentel R (2010) Liquid crystalline elastomers as actuators and sensors. Adv Mater 22(31):3366–3387
127. Inui K, Miyata T, Uragami T (1996) Permeation and separation of organic liquid mixtures through liquid-crystalline polymer networks. Angew Macromol Chem 240:241–250
128. Inui K, Okazaki K, Miyata T, Uragami T (1998) Effect of mesogenic groups on characteristics of permeation and separation for benzene/cyclohexane mixtures of side-chain liquid-crystalline polymer membranes. J Membr Sci 143(1–2):93–104
129. Inoue Y, Atsumi Y, Kawamura A, Miyata T (2019) Thermoresponsive liquid crystalline molymer membranes that undergo phase transition at body temperature. J Membr Sci 588:117213
130. Inoue Y, Takada K, Kawamura A, Miyata T (2022) Amphiphilic liquid crystalline polymer micelles that exhibit a phase transition at body temperature. ACS Appl Mater Interfaces 14:31513–31524
131. Inoue Y, Hirano Y, Kawamura A, Miyata T (2023) Reversible regulation of drug release from chiral liquid crystalline polymer micelles without leakage. Macromolecules 56:8298–8307
132. Otsuka H, Nagasaki Y, Kataoka K (2003) PEGylated nanoparticles for biological and pharmaceutical applications. Adv Drug Deliv Rev 55(3):403–419
133. Homola J (2008) Surface plasmon resonance sensors for detection of chemical and biological species. Chem Rev 108:462–493
134. Nguyen HH, Park J, Kang S, Kim M (2015) Surface plasmon resonance: a versatile technique for biosensor applications. Sensors 15:10481–10510
135. Singh P (2016) SPR biosensors: historical perspectives and current challenges. Sens Actuators B: Chem 229:110–130
136. Chen C, Wang JS (2020) Optical biosensors: an exhaustive and comprehensive review. Analyst 145:1605–1628
137. Philip A, Kumar AR (2022) The performance enhancement of surface plasmon resonance optical sensors using nanomaterials: a review. Coord Chem Rev 458:30
138. Wang Q, Ren ZH, Zhao WM, Wang L, Yan X, Zhu AS, Qiu FM, Zhang KK (2022) Research advances on surface plasmon resonance biosensors. Nanoscale 14:564–591
139. Das S, Devireddy R, Gartia MR (2023) Surface plasmon resonance (SPR) sensor for cancer biomarker detection. Biosensors-Basel 13:33
140. Uniyal A, Srivastava G, Pal A, Taya S, Muduli A (2023) Recent advances in optical biosensors for sensing applications: a review. Plasmonics 18:735–750

141. Matyjaszewski JX (2001) Atom transfer radical polymerization. Chem Rev 101:2921–2990
142. Satoh K, Kamigaito M (2009) Stereospecific living radical polymerization: dual control of chain length and tacticity for precision polymer synthesis. Chem Rev 109:5120–5156
143. Matyjaszewski K (2012) Atom transfer radical polymerization (ATRP): current status and future perspectives. Macromolecules 45:4015–4039
144. Kuriu Y, Ishikawa M, Kawamura A, Uragami T, Miyata T (2012) SPR signals of three–dimensionally antibody–immobilized gel layers formed on sensor chips by atom transfer radical polymerization. Chem Lett 41:1660–1662
145. Mosbach K (1994) Molecular imprinting, *trends Biochem.* Sci 19:9–14
146. Wulff G (1995) Molecular imprinting in cross–linked materials with the aid of molecular templates - a way towards artificial antibodies. Angew Chem Int Ed Engl 34:1812–1832
147. Alexander C, Davidson L, Hayes W (2003) Imprinting polymers: artificial molecular recognition materials with applications in synthesis and catalysis. Tetrahedron 59:2025–2057
148. Ahmad OS, Bedwell TS, Esen C, Garcia-Cruz A, Piletsky SA (2019) Molecularly imprinted polymers in electrochemical and optical sensors. Trends Biotechnol 37:294–309
149. Byrne ME, Park K, Peppas NA (2002) Molecular imprinting within hydrogels. Adv Drug Deliv Rev 54:149–161
150. Byrne ME, Salian V (2008) Molecular imprinting within hydrogels II: progress and analysis of the field. Int J Pharm 364:188–212
151. T. Takeuchi, T, Hishiya (2008) Molecular imprinting of proteins emerging as a tool for protein recognition, Org Biomol Chem, 6, 2459–2467
152. Lv Y, Tan T, Svec F (2013) Molecular imprinting of proteins in polymers attached to the surface of nanomaterials for selective recognition of biomacromolecules. Biotechol Adv 31:1172–1186
153. Akgönüllü S, Kiliç S, Esen C, Denizli A (2023) Molecularly imprinted polymer-based sensors for protein detection. Polymers 15:19
154. Kuriu Y, Kawamura A, Uragami T, Miyata T (2014) Formation of molecularly imprinted hydrogel thin layers with lectin recognition sites on SPR sensor chips by atom transfer radical polymerization. Chem Lett 43:825–827
155. Naraprawatphong R, Kawanaka G, Hayashi M, Kawamura A, Miyata T (2016) Development of protein-recognition SPR devices by combination of SI-ATRP with biomolecular imprinting using protein ligands. J Mol Imprint 4:21–30
156. Naraprawatphong R, Kawamura A, Miyata T (2018) Preparation of molecularly imprinted hydrogel layer SPR sensor chips with lectin-recognition sites via SI-ATRP. Polym J 50:261–269
157. MacBeath G, Schreiber SL (2000) Printing proteins as microarrays for high-throughput function determination. Science 289:1760–1763
158. del Campo A, Arzt E (2008) Fabrication approaches for generating complex micro - and nanopatterns on polymeric surfaces. Chem Rev 108:911–945
159. Nie Z, Kumacheva E (2008) Patterning surfaces with functional polymers. Nat Mater 7:277–290
160. Campbell M, Sharp DN, Harrison MT, Denning RG, Turberfield AJ (2000) Fabrication of photonic crystals for the visible spectrum by holographic lithography. Nature 404:53–56
161. Moon JH, Ford J, Yang S (2006) Fabricating three-dimensional polymeric photonic structures by multi-beam interference lithography. Polym Adv Technol 17:83–93
162. del Barrio J, Sanchez-Somolinos C (2019) Light to shape the future: from photolithography to 4D printing. Adv Optical Mater 7:1900598
163. Terfort A, Zharnikov M (2021) Electron-irradiation promoted exchange reaction as a tool for surface engineering and chemical lithography. Adv Mater Interfaces 8:2100148
164. Guo LJ (2007) Nanoimprint lithography: methods and material requirements. Adv Mater 19:495–513
165. Yang B, Cai F, Huang S, Yu H (2020) Athermal and soft multi-nanopatterning of azopolymers: phototunable mechanical properties. Angew Chem Int Ed 59:4035–4042

166. Stuart MAC, Huck WTS, Genzer J, Müller M, Ober C, Stamm M, Sukhorukov GB, Szleifer I, Tsukruk VV, Urban M, Winnik F, Zauscher S, Luzinov I, Minko S (2010) Emerging applications of stimuli-responsive polymer materials. Nat Mater 9:101–113
167. Roy D, Cambre JN, Sumerlin BS (2010) Future perspectives and recent advances in stimuli-responsive materials. Prog Polym Sci 35:278–301
168. Vázquez-González M, Willner I (2020) Stimuli-responsive biomolecule-based hydrogels and their applications. Angew Chem Int Ed 59:15342–15377
169. Noguchi T, Akioka N, Kojima Y, Kawamura A, Miyata T (2022) Photoresponsive polymer films with directly micropatternable surface based on the change in free volume by photo-crosslinking. Adv Mater Interfaces 9:2101965
170. Ruiz SA, Chen CS (2008) Emergence of patterned stem cell differentiation within multicellular structures. Stem Cells 26:2921–2927
171. Kilian KA, Bugarija B, Lahn BT, Mrksich M (2010) Geometric cues for directing the differentiation of mesenchymal stem cells. Proc Natl Acad Sci USA 107:4872–4877
172. Takahashi H, Nakayama M, Shimizu T, Yamato M, Okano T (2011) Anisotropic cell sheets for constructing three-dimensional tissue with well-organized cell orientation. Biomaterials 32:8830–8838
173. D'Arcangelo E, McGuigan AP (2015) Micropatterning strategies to engineer controlled cell and tissue architecture in vitro. BioTechniques 58:13–23
174. Palecek SP, Loftus JC, Ginsberg MH, Lauffenburger DA, Horwitz AF (1997) Integrin-ligand binding properties govern cell migration speed through cell-substratum adhesiveness. Nature 385:537–540
175. Miyamoto S, Teramoto H, Coso OA, Gutkind JS, Burbelo PD, Akiuama SK, Yamada KM (1995) Integrin function: molecular hierarchies of cytoskeletal and signaling molecules. J Cell Biol 131:791–805
176. Wang N, Butler JP, Ingber DE (1993) Mechanotransduction across the cell surface and through the cytoskeleton. Science 260:1124–1127
177. Miyamoto S, Akiyama SK, Yamada KM (1995) Synergistic roles for receptor occupancy and aggregation in integrin transmembrane function. Science 267:883–885
178. Bettinger CJ, Langer R, Borenstein JT (2009) Engineering substrate topography at the micro- and nanoscale to control cell function. Angew Chem Int Ed 48:5406–5415
179. Janson IA, Putnam AJ (2015) Extracellular matrix elasticity and topography: material-based cues that affect cell function via conserved mechanisms. J Biomed Mater Res 103:1246–1258
180. von der Mark K, Park J, Bauer S, Schmuki P (2010) Nanoscale engineering of biomimetic surfaces: cues from the extracellular matrix. Cell Tissue Res 339:131–153
181. Engler AJ, Sen S, Sweeney HL, Discher DE (2006) Matrix elasticity directs stem cell lineage specification. Cell 126:677–689
182. Handorf AM, Zhou Y, Halanski MA, Li WJ (2015) Tissue stiffness dictates development, homeostasis, and disease progression. Organogenesis 11:1–15
183. Engler AJ, Carag-Krieger C, Johnson CP, Raab M, Tang HY, Speicher DW, Sanger JW, Sanger JM, Discher DE (2008) Embryonic cardiomyocytes beat best on a matrix with heart-like elasticity: scar-like rigidity inhibits beating. J Cell Sci 121:3794–3802
184. Chen C, Bang S, Cho Y, Lee S, Lee I, Zhang S, Noh I (2016) Research trends in biomimetic medical materials for tissue engineering: 3D bioprinting, surface modification, nano/micro-technology and clinical aspects in tissue engineering of cartilage and bone. Biomater Res 20:10
185. Khademhosseini A, Langer R, Borenstein J, Vacanti JP (2006) Microscale technologies for tissue engineering and biology. Proc Natl Acad Sci USA 103:2480–2487
186. Kim SY, Kang JH, Seo WS, Lee SW, Oh NS, Cho HK, Lee MH (2015) Effect of topographical control by a micro-molding process on the activity of human mesenchymal stem cells on alumina ceramics. Biomater Res 19:23

187. Yang W, Yu H, Li G, Wang B, Wang Y, Liu L (2016) Regulation of breast cancer cell behaviors by the physical microenvironment constructed via projection microstereolithography. Biomater Sci 4:863–870
188. Noguchi T, Higashino M, Kodama N, Kawamura A, Miyata T (2023) Cell patterning on photocrosslinkable polymer films with micropatternable surfaces. Responsive Mater 1:e20230007
189. Yamada N, Okano T, Sakai H, Karikusa F, Sawasaki Y, Sakurai Y (1990) Thermo-responsive polymeric surfaces; control of attachment and detachment of cultured cells. Macromol Chem Rapid Commun 11:571–576
190. Okano T, Yamada N, Sakai H, Sakurai Y (1993) A novel recovery system for cultured cells using plasma-treated polystyrene dishes grafted with poly(N-isopropylacrylamide). J Biomed Mater Res 27:1243–1251
191. Okano T, Yamada N, Okuhara M, Sakai H, Sakurai Y (1995) Mechanism of cell detachment from temperature-modulated, hydrophilic-hydrophobic polymer surfaces. Biomaterials 16:297–303
192. Takahashi H, Nakayama M, Itoga K, Yamato M, Okano T (2011) Micropatterned thermoresponsive polymer brush surfaces for fabricating cell sheets with well-controlled orientational structures. Biomacromolecules 12:1414–1418
193. Zhang K, Feng Q, Fang Z, Gu L, Bian L (2021) Structurally dynamic hydrogels for biomedical applications: pursuing a fine balance between macroscopic stability and microscopic dynamics. Chem Rev 121:11149–11193
194. Kosaka T, Yamaguchi S, Izuta S, Yamahira S, Shibasaki Y, Tateno H, Okamoto A (2022) Biorthogonal photoreactive surfaces for single-cell analysis of intercellular communications. J Am Chem Soc 144:17980–17988
195. Okihara M, Matsuda A, Kawamura A, Miyata T (2024) Design of dual stimuli-responsive gels whose physical and chemical properties vary in response to light and temperature and cell behavior on their surfaces. Polym J 56:193–204

# Chapter 12
# Roles of Cholesterol in Controlling Characteristics of Catanionic Vesicles

**Filip Mravec, Chi-cheng Chiu, and Chien-Hsiang Chang**

**Abstract** Ion pair amphiphile, prepared from mixed cationic/anionic surfactants, has a pseudodouble-chain structure and can be used as a raw material to form a vesicular structure, or catanionic vesicle, in aqueous phase. It has been demonstrated that catanionic vesicles could be fabricated from the ion pair amphiphile hexadecyltrimethylammonium-dodecylsulfate (HTMA:DS) by using a forced formation process with the addition of double-chain cationic surfactants, dialkyldimethylammonium bromide (DXDAB), and the physical stability of the HTMA:DS/DXDAB catanionic vesicles could be further improved in the presence of cholesterol. Through the analysis of bilayer fluidity and solvation, as well as the data integration of different analysis methods, combined with molecular dynamics simulations, the complex effects of the added cholesterol on the molecular arrangement and interaction in the bilayer structures of the HTMA:DS/DXDAB catanionic vesicles were elucidated.

**Keywords** Bilayer fluidity · Catanionic bilayer · Catanionic vesicle · Ion pair amphiphile · Vesicular structure

## 12.1 Introduction

When a single-chain cationic surfactant and a single-chain anionic surfactant are mixed in the aqueous phase, a surface active material with a structure similar to a lipid molecule with two hydrophobic hydrocarbon chains, namely catanionic surfactant [1, 2], will be formed through the strong interaction between the opposite

---

F. Mravec
Materials Research Centre, Faculty of Chemistry, Brno University of Technology, Brno, Czech Republic

C.-c. Chiu · C.-H. Chang (✉)
Department of Chemical Engineering, National Cheng Kung University, Tainan, Taiwan
e-mail: changch@mail.ncku.edu.tw



M. Hashizume, Y. Imura (eds.), *Colloids and Interfacial Dynamics*, Physical Chemistry in Action, https://doi.org/10.1007/978-981-96-6728-4_12

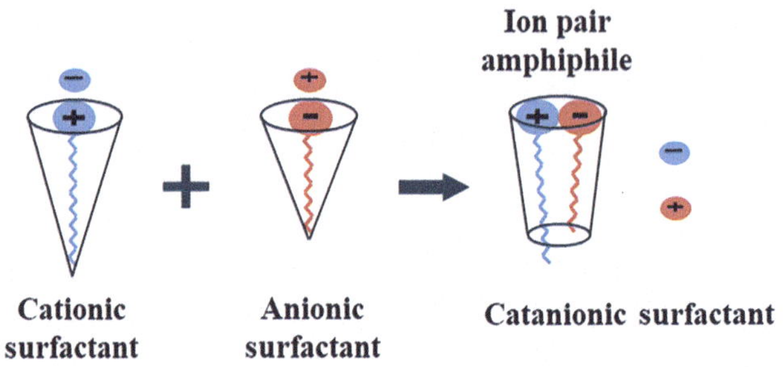


**Fig. 12.1** Catanionic surfactant is formed of mixed cationic/anionic surfactants through the interaction of oppositely charged headgroups. After removing the counter ions from the catanionic surfactant, it is called an ion pair amphiphile

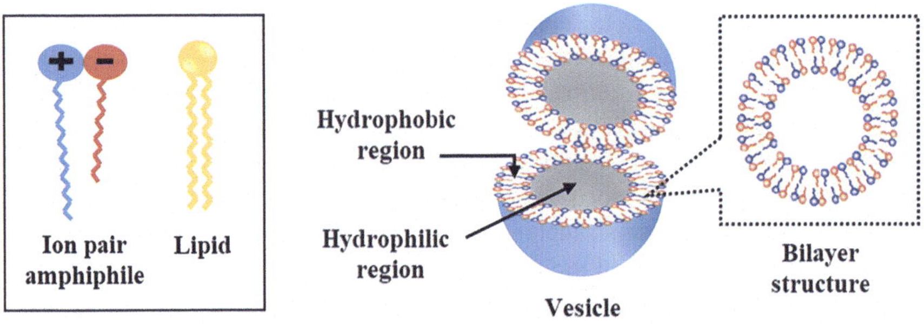


**Fig. 12.2** Ion pair amphiphile has a hydrophobic double-chain structure similar to that of lipid and may aggregate in the aqueous phase to form a hollow sphere with bilayer structure, or catanionic vesicle

charges of the two surfactant head groups. If the hydrocarbon chains are long enough, the catanionic surfactant will be highly hydrophobic and its solubility in the aqueous phase will be very low, thus precipitating from the aqueous phase to form an ion pair amphiphile without the associated counter ions [3–6] (Fig. 12.1).

A vesicle (hollow sphere formed of bilayer structure) formed of lipid molecules with two hydrophobic hydrocarbon chains, that is, a liposome, is traditionally used as a delivery carrier for drugs or cosmetics. The catanionic surfactant or ion pair amphiphile possesses a hydrophobic double-chain structure similar to that of a lipid, and under appropriate conditions, may aggregate in aqueous phase to form a hollow sphere with bilayer structure (Fig. 12.2), or catanionic vesicle, for drug delivery applications [7]. For example, when the cationic surfactant dodecyltrimethylammonium bromide and the anionic surfactant sodium dodecylsulfate were mixed with an appropriate proportion in aqueous phase, a vesicular structure could be spontaneously formed [8]).

Since it is easy to design the pseudodouble-chain structure of catanionic surfactant or ion pair amphiphile with low cost, controlling their vesicle formation behavior has attracted much attention, and their applications have been expanded from drug delivery to nanomaterial preparation, analysis technology, and other fields.

## 12.2 Fabrication of Catanionic Vesicles

### *12.2.1 Spontaneous Formation and Forced Formation*

The fabrication processes for catanionic vesicles can be classified into two categories according to the raw materials [4]. One type is directly formed of mixed cationic/anionic surfactants, or catanionic surfactants, and the other is formed of ion pair amphiphiles, prepared from mixed cationic/anionic surfactants. Most studies on the formation of catanionic vesicles focus on the spontaneous formation of vesicular structures from catanionic surfactants in aqueous phase or in the presence of co-solvents. However, not all mixed cationic/anionic surfactant systems can spontaneously form catanionic vesicles, and the spontaneously formed catanionic vesicles are sensitive to the external environment. In addition, the concentration range of spontaneously formed catanionic vesicles is usually limited, and the charge character of the vesicular structures cannot be really controlled [9, 10], which may limit the applications of spontaneously formed catanionic vesicles.

In order to overcome the problem of limited concentrations of spontaneously formed catanionic vesicles and control the charge character of the catanionic vesicles, co-solvents have been added to prepare catanionic vesicles with specific charge characters by using ion pair amphiphiles as raw materials [11]. However, the use of co-solvents also limits the applications of the catanionic vesicles.

### *12.2.2 Charged Additives and Bilayer Structure Stabilizers*

Ion pair amphiphiles have been used as raw materials to form catanionic vesicles in aqueous phase through a forced formation process in the absence of co-solvents to avoid the problems associated with the spontaneous and semi-spontaneous formation processes of catanionic vesicles. However, the catanionic vesicle fabricated through the forced formation process was not in a thermodynamically stable state, and its physical stability was usually poor. By appropriately adding charged additives (e.g., double-chain cationic surfactants or double-chain anionic surfactants) and membrane stabilizers (e.g., cholesterol), catanionic vesicles with specific charge character can be successfully prepared from ion pair amphiphiles [12–14].

Positively charged catanionic vesicles have been successfully fabricated by using pseudodouble-chain ion pair amphiphiles as raw materials with double-chain cationic surfactants and cholesterol as additives through an appropriate forced formation process [14]. It was found that the toxicity of the catanionic vesicles is, in general, acceptable at the concentrations used in DNA delivery applications [15]. The size stability of the catanionic vesicles was not only related to the electrical repulsion between the vesicles, but also influenced by the molecular interaction/packing in the bilayer structures of the catanionic vesicles. That is, the molar fraction and hydrocarbon chain length of the added double-chain cationic surfactant and cholesterol content played important roles in the physical stability of the catanionic vesicles [16].

A process for preparing new ion pair amphiphiles with longer hydrocarbon chains has been developed to facilitate characteristics control of the catanionic vesicles. By using a new ion pair amphiphile/double-chain anionic surfactant system, in the presence of cholesterol, highly stable and negatively charged catanionic vesicles could be prepared. It is also confirmed that adding cholesterol to the catanionic vesicles could affect the molecular interaction/arrangement in the vesicular bilayer structures, modulate the vesicular bilayer fluidity, and thereby achieve the purpose of stabilizing the catanionic vesicles.

### *12.2.3 Fluidity of Vesicular Bilayer Structure*

The bilayer fluidity plays a very important role in the preparation and application of the vesicular structures, because it significantly affects the permeability, fusion or aggregation, protein binding, and drug encapsulation/release of the vesicular bilayer structures. These properties are closely related to the stability and applications of the vesicles [17, 18]. It has been suggested that the physical stability of catanionic vesicles is related not only to the electrostatic repulsion between the vesicles but also to the fluidity of the vesicular bilayer structures. Adding cholesterol to the catanionic vesicles could affect the molecular interaction/arrangement in the vesicular bilayers and the fluidity of the vesicular bilayer structures, thus improving the physical stability of the catanionic vesicles. Therefore, fabricating catanionic vesicles from ion pair amphiphiles through a forced formation process, molar fraction and hydrocarbon chain length of charged additives and content of the membrane stabilizer cholesterol in the vesicular bilayer structures could affect the physical stability of the catanionic vesicles [16].

There are many methods that can be used to evaluate the fluidity of the vesicular bilayer structures, such as the fluorescence polarization method and infrared spectroscopy [19]. The fluorescence polarization method can analyze the fluidities of different regions in the vesicular bilayer structures or between the head groups according to the hydrophobicity of the fluorescent molecules. For example, 1,6-diphenyl-1,3,5-hexatriene (DPH), a fluorescent probe with high hydrophobicity, is commonly adopted to evaluate the ordering of the hydrophobic regions of bilayer structures [20, 21]. Probes with higher hydrophilicity will be localized at polar headgroups and therefore can be used to obtain information from the interface [22]. The measurement principle of the fluorescence polarization method is that when light passes through a specific polarization filter, the polarized light excites a selected group of fluorophores according to the orientation of their absorption transition moment relative to the plane of the excitation beam (photoselection). The rotation of excited molecules during their lifetimes causes the inconsistency of the plane of polarization of the emitted light with respect to the plane of polarization of the excitation beam. The degree of orderliness of the molecular arrangement in the vesicular bilayer structures determines the degree of rotation of the added fluorescent molecules, and the smaller the fluidity (the higher the order) of the bilayer structures, the slower the rotation of the fluorescent molecules. The higher the fluidity (the lower the order) of the bilayer structures, the faster the rotation of the fluorescent molecules, resulting in more significant depolarization. By detecting the divergent light signal intensities parallel and perpendicular to the original excitation light, respectively, the fluorescence polarization value can be obtained. The larger the fluorescence polarization value, the higher the order of the bilayer structure.

Fourier transform infrared (FT-IR) spectroscopy can also be used to evaluate the orderliness of the molecular packing in the bilayer structures. With infrared spectroscopy, the wavenumber of the $\nu$-$CH_2$ characteristic peak can be analyzed to understand the hydrocarbon chain arrangement in the bilayer structure [23, 24]. The absorption wavenumbers of $\nu$-$CH_2$ peak located around 2920 $cm^{-1}$ for the antisymmetric stretching mode and around 2850 $cm^{-1}$ for the symmetrical stretching mode are often used for analysis. When the order of the vesicular bilayer structures is higher, the $\nu$-$CH_2$ absorption wavenumber of the molecular hydrocarbon chains will be lower. When the fluidity of the bilayer structures is higher, that is, when the order of the molecular packing is lower, the $\nu$-$CH_2$ absorption wavenumber of the molecular hydrocarbon chains will be higher [25, 26].

## 12.3 Effects of Added Cholesterol on Physical Properties of Catanionic Vesicles

By using an ion pair amphiphile, hexadecyltrimethylammonium-dodecylsulfate (HTMA:DS), which is prepared from cationic surfactant hexadecyltrimethylammonium bromide (HTMAB) and anionic surfactant sodium dodecylsulfate (SDS), as the raw material and double-chain cationic surfactants (dialkyldimethylammonium bromide, DXDAB) as additives, positively charged catanionic vesicles with an initial average diameter of about 100 nm and zeta potentials of about 50–60 mV could be fabricated through a forced formation process [12]. However, the average diameter of the vesicular structures was gradually increased with time (Fig. 12.3). It was noted that for the HTMA:DS/DXDAB catanionic vesicles with different compositions, similar zeta potentials were found, but a significant difference in the vesicle stability was observed. This suggested that the physical stability of the catanionic vesicles depended not only on the charge character, but also on the molecular interaction and/or packing in the vesicular bilayer structures.

In order to further improve the physical stability of the HTMA:DS/DXDAB catanionic vesicles, cholesterol, which is known to be able to modulate the bilayer

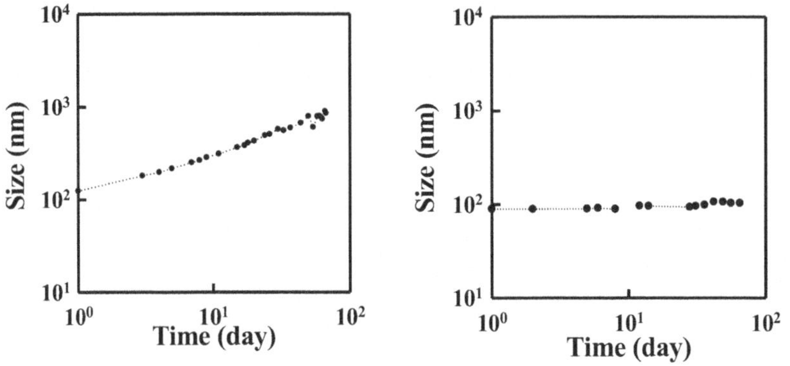


**Fig. 12.3** Time-dependent average diameters of the catanionic vesicles fabricated from HTMA:DS/ditetradecyldimethylammonium bromide (DTDAB) (molar ratio = 5/5) (left) without cholesterol and (right) with 43 mol% cholesterol. (Data obtained from [14])

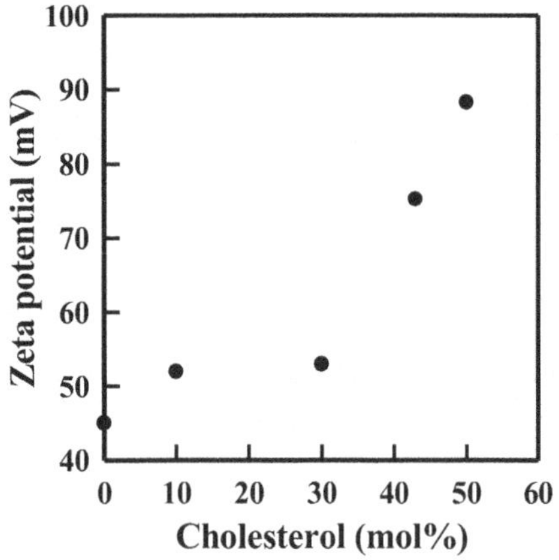


**Fig. 12.4** Zeta potentials of the catanionic vesicles fabricated from HTMA:DS/DTDAB (molar ratio = 5/5) with various contents of cholesterol. (Data obtained from [14])

fluidity of liposomes, has been adopted as an additive [16]. With the addition of cholesterol, the physical stability of the HTMA:DS/ditetradecyldimethylammonium bromide (DTDAB) catanionic vesicles was significantly improved as judged from the time-dependent vesicle size analysis (Fig. 12.3). In the presence of cholesterol, enhanced physical stability was generally detected for the catanionic vesicles of HTMA:DS/DXDAB, for example, HTMA:DS/dihexadecyldimethylammonium bromide (DHDAB) and HTMA:DS/dioctadecyldimethylammonium bromide (DODAB), with different compositions. It is noted that the zeta potential of the catanionic vesicles was increased in the presence of cholesterol, which would be helpful in improving the vesicle stability by providing enhanced repulsive interactions between the vesicles (Fig. 12.4) [14]. With the presence of cholesterol in the bilayer structures, the distance between the charged headgroups would be increased, and thus the counterion binding tendency would be decreased, or the dissociation tendency of the counter ions would be increased, leading to a higher value of zeta potential.

An IR analysis can provide information on how the added cholesterol affects the molecular packing in the bilayer structures of the catanionic vesicles. If the molecules in the vesicular bilayer structures are less closely packed or the bilayer structures become disordered, the wavenumber of the antisymmetric stretching peak of $CH_2$ groups in the hydrocarbon chains will shift to a higher wavenumber. In general, with the addition of cholesterol, the wavenumbers were shifted to higher wavenumbers, which implies the molecular packing in the bilayer structures of the catanionic vesicles became disordered, or the vesicular bilayers became flexible.

A molecular dynamics simulation study has demonstrated how the added cholesterol affects the molecular packing in the vesicular bilayer structures [27]. The added cholesterol would induce disordering effects in the front segment and terminal segment of the hydrocarbon chains of the ion pair amphiphile, and induce a condensing effect in the middle segment of the hydrocarbon chains of the ion pair amphiphile. Apparently, the added cholesterol induced a more pronounced disordering effect than the condensing effect for the catanionic vesicles, and the bilayer structures became disordered or flexible.

The incorporation of cholesterol into the vesicular bilayer structures of the HTMA:DS/DXDAB system not only adjusted the molecular packing but also increased the zeta potential, resulting in improved vesicle stability. That is, the vesicle surface charge and bilayer fluidity can be modulated by added cholesterol for various applications.

## 12.4 Advanced Analysis of the Added Cholesterol Effect on the Characteristics of Catanionic Vesicles

As mentioned in the previous section, the addition of cholesterol affects the zeta potential, physical stability, and fluidity of the catanionic bilayer structures. The description of the physical properties can be supplemented by studying the solvation of the bilayer structure and its solidification under different conditions, particularly when the temperature, pH, or ionic strength changes.

A previously unused measurement in the research area of vesicles formed from ion pair amphiphiles is the determination of the change in bilayer solvation using the fluorescent probe Laurdan (1-[6-(dimethylamino)naphthalen-2-yl]dodecan-1-one). Proper solvation of the bilayer structure is essential for maintaining the bilayer structure's stability. Insufficient or excessive hydration can lead to bilayer destabilization and disruption. Laurdan exhibits a high Stokes shift based on changing the polarity of its microenvironment [28–30]. In a non-polar environment, it emits in the blue region; as the polarity of its surroundings increases, the emission moves to higher wavelengths. If, owing to its aliphatic chain, it is incorporated into the bilayer structure, the maximum emission is around 440 nm in the case of the gel phase and around 490 nm if the bilayer structure is in the liquid phase state. The extent of this shift can then be expressed by a quantity called generalized polarization (GP), which compares the intensities at 440 and 490 nm [31].

$$GP = \frac{I_{440} - I_{490}}{I_{440} + I_{490}} \tag{12.1}$$

The theoretical range of generalized polarization is then from +1 to −1. Increased water penetration into the bilayer structure, as indicated by the higher GP values, can lead to increased bilayer permeability, allowing for the passage of ions and small molecules. Changes in bilayer polarity and order can affect bilayer fluidity. A more polar environment can disrupt the molecular packing, leading to increased bilayer fluidity. The degree of bilayer structure solvation can affect the vesicle stability. Excessive water penetration can destabilize the bilayer structures, leading to vesicle fusion or rupture [32, 33].

Another advanced analysis concept is the interpretation of combined data from different methods, such as dynamic light scattering and fluorescence measurements. For example, the bilayer fluidity determined using the diphenylhexatriene probe varies continuously over a large range of temperatures for different cholesterol contents in the vesicular mixture. Such data can then be used to create a contour graph, where the real representation of cholesterol is on the x-axis and the temperature on the y-axis. The color scale then shows the change in the studied physicochemical quantity; for example, the fluidity of the bilayer structures is expressed as the lipid order parameter (LOP) derived from the time-resolved anisotropy of the DPH probe. The LOP is then defined as the square root of the ratio of the limiting anisotropy ($r_\infty$) to the fundamental anisotropy ($r_0$):

$$LOP = \sqrt[2]{\frac{r_\infty}{r_0}}, \tag{12.2}$$

where $r_\infty$ can be approximated from the steady-state anisotropy ($r_S$) of the DPH probe as [34]:

$$r_\infty = \frac{9}{8} r_S - \frac{1}{20} \tag{12.3}$$

Obtaining such a map is time-consuming; however, it is not easy or intuitive to interpret such data in terms of describing the behavior of the sample. However, such a map can be supplemented with additional information, such as the limit of the polydispersity index (PDI) from dynamic light scattering measurements, which will help define the area in the graph of temperature vs. relative cholesterol content where the polydispersity does not exceed a selected value, typically 0.3 [35].

The PDI can express both the width of the size distribution and its heterogeneity. Therefore, it is possible to add to such advanced graphs the boundary of the area where the size distribution changes from a unimodal to a multimodal character.

With respect to the accuracy of the dynamic light scattering data, it is possible to distinguish areas where the size distribution is represented by one broad peak from areas where the size distribution contains several separate peaks, which may indicate the disintegration of vesicles or, conversely, the formation of larger aggregates, possibly even precipitates, thus indicating unstable regions of temperature or relative cholesterol content.

If a detailed analysis of the temperature dependence of some physicochemical properties of the bilayer structure is performed, one can find other potential boundaries in addition to the previously mentioned PDI values and modality for a more accurate description of the system's behavior. At low contents of cholesterol, the dependences are rather sigmoidal in nature, and with increasing cholesterol content, there is a monotonous change in quantity with temperature change. If one interpolates the sigmoid dependences with, for example, the Boltzmann curve [36], one can find approximate values of the temperature at which the quantity changes and from which temperature the quantity is already constant (in the selected measurement range), as can be seen in Fig. 12.5. If these intersection points for a given relative cholesterol content are plotted in a contour plot, one can define the changes in the studied quantity more precisely than color scaling, which allows one to define new boundaries.

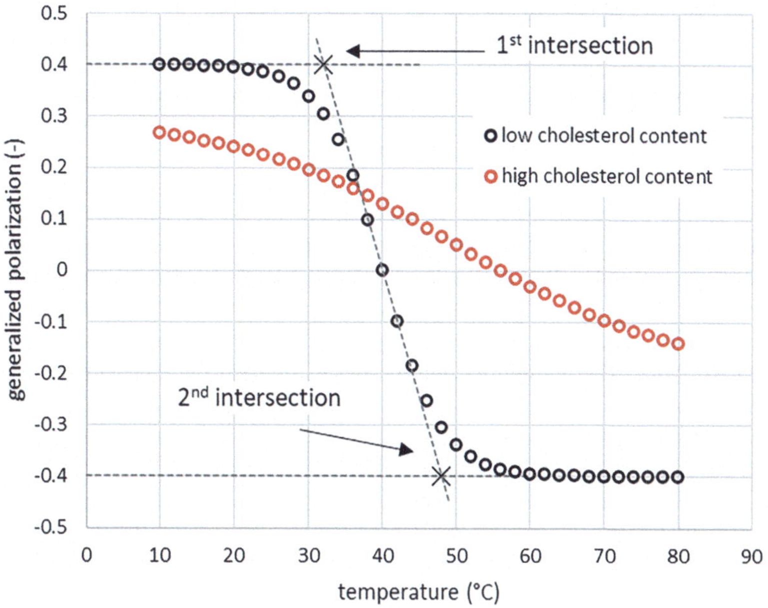


**Fig. 12.5** Model demonstration of the variation of the generalized polarization value with decreasing temperature for low (black points) and high (red) cholesterol contents in the vesicular system. In the course of low cholesterol representation, the determination of the first and second points of intersection of the maximum and minimum, respectively, with the tangent passing through the inflection point from the Boltzmann sigmoidal model, is indicated

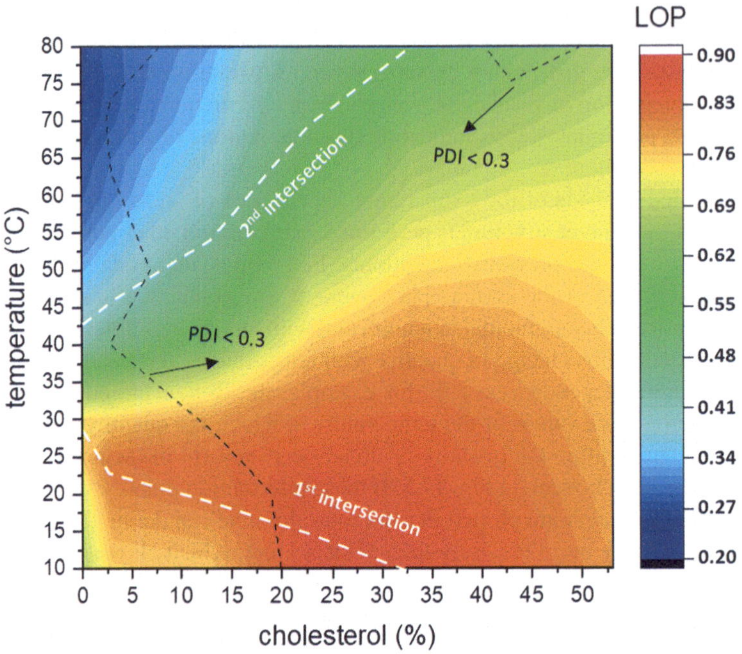


**Fig. 12.6** Map of LOP for given temperatures and cholesterol concentrations (color fill with appropriate scale) of samples. The map contains area delineation of the PDI value = 0.3 (black dashed line with black arrows), as well as by the relevant parameters of the first intersection and second intersection (white dash-dot lines), as obtained from the fitting of the temperature dependences of the LOP with the Boltzmann sigmoid model

As shown in Fig. 12.6, the LOP value varied continuously with increasing temperature for different cholesterol contents. By supplementing this map with the definition of the area with a PDI of less than 0.3 and together with the area where the corresponding LOP value changes, one will then obtain a useful overview of the bilayer state behavior of samples with different cholesterol contents over a wide temperature range. The white lines delimiting the area of LOP change end for samples with a cholesterol content of 33 mol% because for samples with a higher cholesterol content, the values determined using the Boltzmann model lie outside the actual measured values and would thus only be approximations.

Another possible evaluation is a comparison of the generalized polarization values of the DPH probe anisotropy at selected temperatures for different cholesterol contents (Fig. 12.7). These values indicate different aspects of the stability and physicochemical properties of the vesicular bilayer structures. In this type of graph, it is possible to compare systems with different compositions and temperatures. For example, in the case of liposomes, by moving the point of intersection of both axes, the ordered phase area can be demarcated from the disordered or partially ordered phase area. These graphs can thus serve to visually categorize the bilayer phases [37]. The Cartesian plot below shows the reinterpreted data for 20 °C, 40 °C, and

**Fig. 12.7** Cartesian plot of the development of generalized polarization and corresponding steady-state anisotropy values for samples with different cholesterol contents at selected temperatures of 20 °C (blue), 40 °C (green), and 60 °C (red). The labels of the individual points refer to the total cholesterol content in the sample and its color at a given temperature

60 °C at different cholesterol contents. In general, the lower the anisotropy value, the more fluid the bilayer structure is. Similarly, for GP values, the lower the value, the less compact the bilayer structure, thus enabling deeper penetration of water molecules into the bilayer structure and subsequent relaxation of the Laurdan probe. As can be seen, at a low temperature (20 °C), the values of GP and anisotropy are clustered for different cholesterol contents in the region of high values of both GP and anisotropy (blue dots). From these data, it follows that the presence of cholesterol between 3 and 43 mol% shows a similar anisotropy value, which is higher than the values for 0 and 53 mol%. This means that within a certain range of concentrations, the presence of cholesterol contributes to the stiffening of the bilayer structure. A high content of cholesterol then tends to lead to slight destabilization. GP values would have a similar course when they increased to a cholesterol content of 23 mol%, after which further cholesterol additions resulted in a decrease in compactness. At 40 °C (green dots) and 60 °C (red dots), both anisotropy and GP values increase by up to 43 mol% with increasing cholesterol content. The sample containing 53 mol% cholesterol showed a decrease in anisotropy values compared to the sample containing 43 mol% cholesterol, but the GP value was higher.

In a study that also included the use of a derivative of the DPH probe, the trimethylammonium-DPH (TMA-DPH), a catanionic system with 43 mol% cholesterol was found to be stable over time, exhibiting a unimodal size distribution and stable zeta potential [36]. To evaluate catanionic vesicles with different cholesterol contents and temperatures, a PDI value of 0.3 was used as a threshold to define a stable region with a single, well-defined peak in the size distribution. By investigating the bilayer rigidity and solvation using probes with different localizations. The Boltzmann or sigmoid model was found to be suitable for describing changes in parameters with temperature for systems with up to 33 mol% cholesterol. However, at higher cholesterol contents, the accuracy of the model decreased. Increasing the temperature led to a significant increase in the fluidity of the bilayer as well as its solvation, but did not cause significant changes in the vesicle size distribution. These findings provide a foundation for further research on the behavior of these vesicular systems, including their potential applications in areas such as solubilization, stability under different conditions, and interactions with biopolymers.

## 12.5 Molecular Simulations of the Added Cholesterol Effects on the Characteristics of Catanionic Vesicles

Molecular dynamics (MD) simulations are invaluable tools for gaining insights into the molecular effects of cholesterol on the structure and dynamics of ion pair amphiphile bilayers [38–43]. MD simulations offer detailed insights into the molecular interactions, structural organization, and dynamic changes within bilayer structures that are often challenging to capture through experimental approaches alone. By examining how cholesterol modulates key properties, such as alkyl chain packing and mechanical stability, MD simulations complement experimental findings and contribute to a deeper understanding of the behavior and functionality of catanionic vesicles. To enable meaningful comparisons with experimental data, MD simulations must be designed to closely replicate the conditions used in characterizing catanionic vesicles. The effect of cholesterol on ion pair amphiphile bilayer structures can thus be characterized by simulating ion pair amphiphile bilayer compositions with varying cholesterol molar fractions, ranging from cholesterol-free to highly cholesterol-rich membranes (up to $X_{CHOL} = 0.5$).

Figure 12.8 illustrates representative MD configurations of HTMA:DS bilayers with cholesterol molar fractions ranging from 0 to 50 mol%. The addition of cholesterol enhances packing order in the hydrocarbon tails and reduces the tilt of alkyl chains in ion pair amphiphile bilayers with longer alkyl tails. These effects are reflected in the molecular alignment of hydrocarbon tails, where cholesterol acts as a spacer that influences both the packing density and alkyl ordering within the vesicular bilayers [39, 40, 42]. Using MD trajectories, the structural and mechanical properties of the bilayers can be obtained to further reveal key aspects of cholesterol's roles in shaping the physical characteristics of catanionic vesicles.

The addition of cholesterol to HTMA:DS bilayers significantly influences their structural properties, as demonstrated through MD simulations. These simulations reveal how cholesterol modulates the bilayer's molecular area, alkyl chain tilt, gauche fraction, and deuterium order parameter ($|S_{CD}|$), providing valuable insights

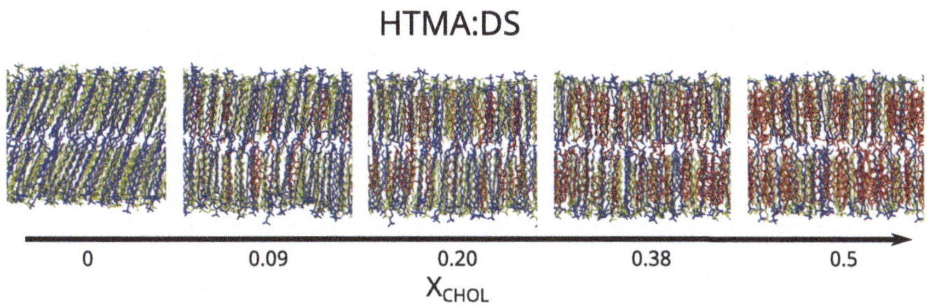


**Fig. 12.8** Representative configurations from MD simulation of ion pair amphiphile bilayers with varying cholesterol contents ($X_{CHOL}$), illustrating changes in bilayer structure and alkyl chain organization under the effects of cholesterol [40, 42, 44]. Figures generated with the VMD package [45]

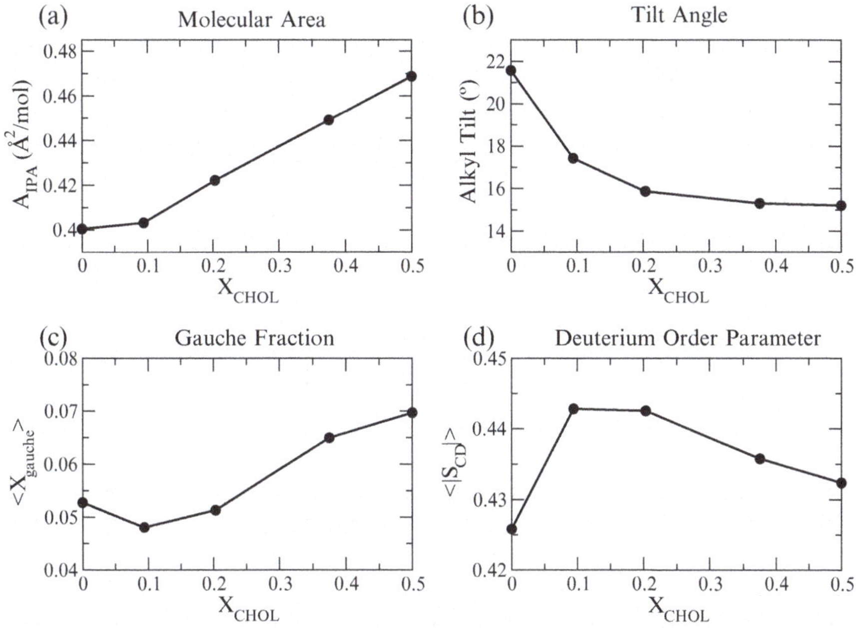


**Fig. 12.9** Effects of varying cholesterol content on the structural properties, including (**a**) molecular area per ion pair amphiphile (IPA) molecule, (**b**) alkyl chain tilt angle relative to the bilayer normal, (**c**) the averaged gauche fraction of alkyl chains, and (**d**) the averaged deuterium order parameter of alkyl chains for HTMA:DS bilayers [40, 42, 44]

into the molecular-level changes that are often difficult to capture experimentally. As shown in Fig. 12.9, cholesterol has a clear effect on the bilayer structure, leading to changes in packing density, hydrocarbon chain orientation, and overall bilayer stability.

As illustrated in Fig. 12.9a, the molecular area per HTMA:DS molecule increases with increasing cholesterol content. This trend is indicative of cholesterol's disordering effect on the bilayer, especially within the solid-like gel phase bilayer structures. In systems with a lower cholesterol content, the ion pair amphiphile molecules are more tightly packed, leading to a smaller molecular area per molecule. As cholesterol is incorporated, its rigid steroid ring structure introduces disruptions in the hydrocarbon chain packing, increasing the effective area for each ion pair amphiphile molecule. For the HTMA:DS bilayer, the insertion of cholesterol into the hydrophobic region disrupts the well-ordered packing of the alkyl chains. Such an increase in molecular area is consistent with FT-IR results, which showed a reduction in alkyl chain ordering as cholesterol content was increased.

Cholesterol with a rigid sterol ring also influences the orientation of alkyl chains within the bilayer, affecting their tilt relative to the bilayer normal [39, 40, 42, 44]. The tilt angle of the alkyl chains relative to the bilayer normal decreases with increasing cholesterol content, as seen in Fig. 12.9b. In the absence of cholesterol, the alkyl chains are oriented at an angle of around 22°. As the cholesterol concentration increases, the tilt angle progressively decreases, indicating that cholesterol forces the alkyl chains to adopt a more upright orientation relative to the bilayer normal. This aligns with experimental findings from fluorescence spectroscopy, which showed that as cholesterol was increased, the fluidity of the bilayer structures was decreased.

The tilt angle converges above $X_{CHOL} = 0.3$, suggesting the saturation of cholesterol in the hydrophobic region [40, 44]. Note that, as cholesterol acts as a structural "spacer" that orders the surrounding alkyl chains, the reduction in alkyl tilt also affects the mechanical stability of the bilayer as discussed later.

The disordering effects of cholesterol indicate that the alignment of alkyl chains is disrupted by inserted cholesterol, which can be further confirmed by the gauche fraction analyses of alkyl chains. The gauche fraction of alkyl chains measures the conformations of the alkyl tails, providing insights into the chain ordering within the ion pair amphiphile bilayer. From MD trajectories, a gauche conformer can be defined for the four connected carbon atoms with the dihedral angle ranging from −120° to 120° and a trans conformer otherwise [40, 42]. A higher gauche fraction reflected more disordering conformations. As shown in Fig. 12.9c, the average gauche fraction along the alkyl chain increases with cholesterol incorporation, indicating a preference for gauche conformations and more disordered chains, potentially introducing more conformational flexibility. The slight increase in the gauche fraction is consistent with FT-IR data, which revealed the slightly increased chain flexibility as cholesterol content was increased.

Another common parameter to quantify the ordering of alkyl chains in vesicular bilayers is the deuterium order parameter, $|S_{CD}|$, with higher values indicating better alignment between alkyl chains. Experimentally, the $|S_{CD}|$ order parameter can be obtained from quadrupolar splitting in solid-state deuterium NMR spectra via analyzing the C–D bond vectors in the lipid alkyl chain with respect to the direction of the external magnetic field [46]. In MD simulations, $S_{CD}$ can be calculated based on the orientation of the carbon–hydrogen (C–H) bond along the alkyl chain [47, 48]:

$$S_{CD} = \frac{1}{2}\left\langle 3\cos^2(\theta_{CH}) - 1\right\rangle \tag{12.4}$$

where $\theta_{CH}$ is the angle between the carbon–hydrogen bond and the bilayer normal, and the bracket denotes the ensemble average. Note that the $|S_{CD}|$ can also be affected by alkyl tail, where a larger tilt angle reduces the $|S_{CD}|$ value.

As shown in Fig. 12.9d, the deuterium order parameter ($|S_{CD}|$) first increases and then decreases with increasing cholesterol content. This trend reflects the complex effects of cholesterol on the bilayer's structural organization, particularly the alignment of alkyl chains relative to the bilayer normal. At lower cholesterol contents (up to $X_{CHOL} \approx 0.2$), the addition of cholesterol leads to a decrease in the alkyl chain tilt angle (as shown in Fig. 12.9b), causing the chains to adopt a more upright orientation relative to the bilayer normal. This alignment improves the ordering of the hydrocarbon tails, resulting in an increase in $|S_{CD}|$. As cholesterol content further increases, the alignment effect of cholesterol on alkyl chains converges. While the bilayer remains relatively ordered in the hydrophobic core due to the cholesterol's rigid structure, the excess cholesterol introduces more disorder and flexibility in the alkyl chains. This disordering effect outweighs the initial ordering influence, resulting in a decrease in the deuterium order parameter. The change in the $|S_{CD}|$ at high cholesterol contents reflects this balance between order and disorder and is consistent with the DPH anisotropy result.

The mechanical properties of catanionic bilayers, including their resistance to deformation and bending, are crucial for understanding their stability and functionality in various applications. Cholesterol is known to significantly influence these mechanical characteristics by modulating parameters, such as the area expansion modulus ($K_A$), tilt modulus ($\chi$), and effective bending modulus ($K_c^{eff}$). Figure 12.10 illustrates how these mechanical properties are generally enhanced by increasing cholesterol content, suggesting an improvement in the internal stability of the HTMA:DS bilayer.

The area expansion modulus, $K_A$, measures the resistance of the bilayer to changes in its surface area upon lateral stretching and can be evaluated from MD trajectories via the fluctuation of lateral membrane area [49, 50]. Higher values of $K_A$ indicate greater mechanical rigidity. Figure 12.10a shows that the area expansion modulus ($K_A$) increases with increasing cholesterol content. The area expansion modulus quantifies the bilayer's resistance to lateral stretching and is an indicator of its mechanical rigidity. This enhancement is most pronounced at higher cholesterol contents, particularly beyond $X_{CHOL}$ of 0.2. This enhancement is attributed to cholesterol's rigid sterol structure, which constrains the neighboring alkyl chains and thereby increases bilayer mechanical strength. This strengthening effect aligns with the observed reduction in molecular flexibility (as seen in the gauche fraction and tilt angle), where cholesterol's alignment effect leads to less lateral movement of the ion pair amphiphile molecules and thus increased rigidity.

The tilt modulus, $\chi$, quantifies the energy required to change the tilting orientation of the alkyl chain relative to the bilayer normal and can be evaluated from the tilt angle distribution obtained from the MD trajectory [51, 52]. As shown in Fig. 12.10b, the tilt modulus increases with cholesterol content, reflecting the strengthening effects of cholesterol on the bilayer mechanical properties. At lower

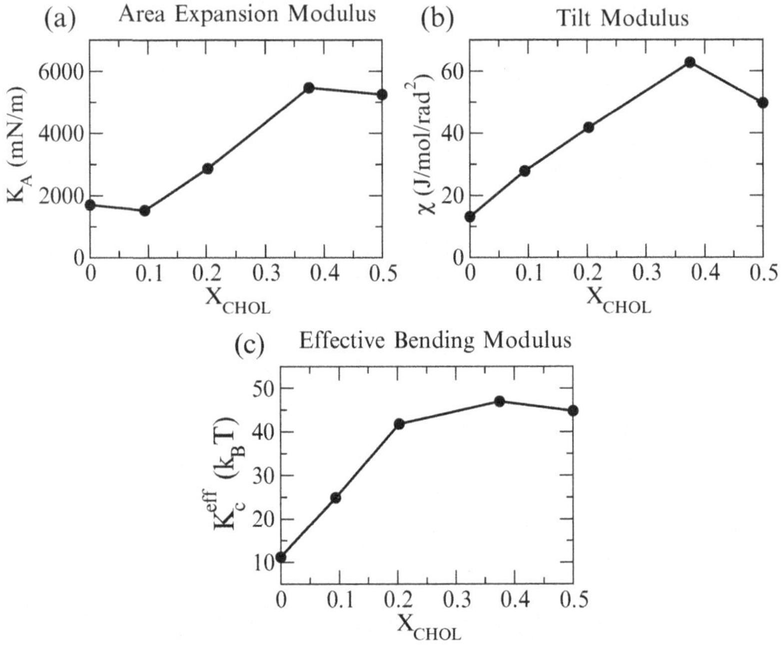


**Fig. 12.10** Influence of cholesterol content on the mechanical properties, including (**a**) area expansion modulus ($K_A$), (**b**) tilt modulus ($\chi$), and (**c**) effective bending modulus ( $K_c^{eff}$ ), of HTMA:DS bilayers [40, 42, 44]

cholesterol contents, the bilayer exhibits more flexibility, allowing the alkyl chains to tilt more easily. However, as cholesterol content rises, the alkyl chains are constrained in their tilt orientation by the rigid sterol ring of cholesterol, leading to an increase in the tilt modulus. This increase is consistent with the decrease in tilt angle observed in Fig. 12.9b, where the alignment of alkyl chains becomes more pronounced as cholesterol reduces the flexibility of the bilayer.

The effective bending modulus, $K_c^{eff}$ , reflects the bilayer's resistance to bending deformations. $K_c^{eff}$ can be evaluated by using the re-weighting method from MD simulations [51, 53]. Figure 12.10c shows that the effective bending modulus increases with increasing cholesterol content, reflecting enhanced resistance to bending deformations. As cholesterol is incorporated into the bilayer, the alkyl chains become more aligned, making the bilayer stiffer and less prone to bending. This higher bending modulus is consistent with the enhanced stability of the vesicle, as a more rigid bilayer is less likely to deform under mechanical stress. However, as cholesterol content continues to increase beyond $X_{CHOL} = 0.3$, the bending modulus plateaus, suggesting that the bilayer reaches a point of saturation where further cholesterol addition leads to minimal additional stiffening.

As the cholesterol content reaches $X_{CHOL} = 0.5$, the mechanical moduli, including the area expansion modulus, tilt modulus, and effective bending modulus, slightly decrease. This slight reduction in mechanical rigidity at high cholesterol contents indicates that cholesterol disrupts the tight packing of alkyl chains and introduces more fluidity into the bilayer. The resulting bilayer flexibility is consistent with the increased bilayer fluidity revealed by fluorescence analyses, which showed that the bilayer became more fluid as cholesterol content was increased beyond a certain point.

## 12.6 Conclusion

The addition of cholesterol to ion pair amphiphile, for example, HTMA:DS, bilayers has a complex, concentration-dependent effect on both structural and mechanical properties of the bilayer structures. Cholesterol increases the molecular area, reflecting its disordering effect on alkyl chain packing, while simultaneously reducing alkyl chain tilt and promoting chain alignment, which leads to enhanced mechanical stability. As cholesterol content rises, the alkyl chains in the ion pair amphiphile bilayer exhibit increased gauche fractions. This correlates well with experimental results from infrared spectroscopy, which showed a decrease in chain ordering with increased cholesterol content, and fluorescence analyses, which indicated a reduction in bilayer fluidity at moderate cholesterol levels. Additionally, cholesterol's impact on vesicle stability aligns with its mechanical strengthening effect, as the bilayer becomes more resistant to deformation, as shown by the increase in area expansion modulus, tilt modulus, and effective bending modulus. However, at high cholesterol contents (above $X_{CHOL} = 0.3$), the bilayer exhibits a slight decrease in mechanical strength and an increase in fluidity, consistent with the experimental fluorescence data showing increased bilayer flexibility. These results suggest that cholesterol plays a dual role in modulating ion pair amphiphile bilayer properties—enhancing mechanical stability through improved chain alignment and reducing packing density and fluidity at higher contents, which affects the bilayer's overall dynamic and stability. The balance between these effects is crucial for optimizing the mechanical and structural properties of catanionic vesicle structures for applications in drug delivery and other vesicular systems.

**Acknowledgments** This work was partly supported by the National Science and Technology Council of Taiwan under the Grants MOST 108-2923-E-006-006-MY3 and NSTC 113-2221-E-006-018-MY3.

## References

1. Jönsson B, Jokela P, Khan A, Lindman B, Sadaghiani A (1991) Catanionic surfactants: phase behavior and microemulsions. Langmuir 7:889–895
2. Filipovic-Vincekovic N, Bujan M, Dragcevic D, Nekic N (1995) Phase behavior in mixtures of cationic and anionic surfactants in aqueous solution. Colloid Polym Sci 273:182–188

3. Blanzat M, Perez E, Rico-Lattes I, Lattes A (1999) Synthesis and anti-HIV activity of catanionic analogs of galactosylceramide. New J Chem 23:1063–1065
4. Tondre C, Caillet C (2001) Properties of the amphiphilic films in mixed cationic/anionic vesicles: a comprehensive view from a literature analysis. Adv Colloid Interf Sci 93:115–134
5. Chien C-L, Yeh S-J, Yang Y-M, Chang C-H, Maa J-R (2002) Formation and encapsulation of catanionic vesicles. J Chin Colloid Interf Soc 24:31–45
6. Marques EF, Regev O, Khan A, Lindman B (2003) Self-organization of double-chained and pseudodouble-chained surfactants: counterion and geometry effects. Adv Colloid Interf Sci 100–102:83–104
7. Dhawan VV, Nagarsenker MS (2017) Catanionic systems in nanotherapeutics–biophysical aspects and novel trends in drug delivery applications. J Control Release 266:331–345
8. Herrington KL, Kaler EW, Miller DD, Zasadzinski ZA, Chiruvolu S (1993) Phase behavior of aqueous mixtures of dodecyltrimethylammonium bromide (DTAB) and sodium dodecyl sulfate (SDS). J Phys Chem 97:13792–13802
9. Fischer A, Hebrant M, Tondre C (2002) Glucose encapsulation in catanionic vesicles and kinetic study of the entrapment/release processes in the sodium dodecyl benzene sulfonate/cetyltrimethylammonium tosylate/water system. J Colloid Interface Sci 248:163–168
10. Antunes FE, Marques EF, Gomes R, Thuresson K, Lindman B, Miguel MG (2004) Network formation of catanionic vesicles and oppositely charged polyelectrolytes. Effect of polymer charge density and hydrophobic modification. Langmuir 20:4647–4656
11. Kuo A-T, Chang C-H (2016) Recent strategies in the developments of catanionic vesicles. J Oleo Sci 65:377–384
12. Wu C-J, Kuo A-T, Lee C-H, Yang Y-M, Chang C-H (2014) Fabrication of positively charged catanionic vesicles from ion pair amphiphile with double-chained cationic surfactant. Colloid Polym Sci 292:589–597
13. Lee C-H, Yang Y-M, Chang C-H (2014) Enhancing physical stability of positively charged catanionic vesicles in the presence of calcium chloride via cholesterol-induced fluidic bilayer characteristic. Colloid Polym Sci 292:2519–2527
14. Kuo A-T, Tu C-L, Yang Y-M, Chang C-H (2018) Enhanced physical stability of positively charged catanionic vesicles: role of cholesterol-adjusted molecular packing. J Taiwan Inst Chem Eng 92:29–35
15. Lee C-H, Yang Y-M, Leu K-L, Lin H-Y, Liang C-H, Chang C-H (2015) Exploring physical stability characteristics of positively charged catanionic vesicle/DNA complexes. Colloid Polym Sci 293:2239–2247
16. Kuo A-T, Tu C-L, Yang Y-M, Chang C-H (2018) Enhanced physical stability of mixed ion pair amphiphile/double-chained cationic surfactant vesicles in the presence of cholesterol. J Oleo Sci 67:727–735
17. New RRC (1990) Preparation of liposomes. In: New RRC (ed) Liposomes: a practical approach. Oxford University Press, New York, pp 33–104
18. Crommelin DJA, Schreier H (1994) Liposomes. In: Kreuter J (ed) Colloidal drug delivery systems. Marcel Dekker, Inc., New York, pp 73–190
19. Fonseca F, Pénicaud C, Tymczyszyn EE, Gómez-Zavaglia A, Passot S (2019) Factors influencing the membrane fluidity and the impact on production of lactic acid bacteria starters. Appli Microbiol Biotechnol 103:6867–6883
20. Merino-Montero S, Montero MT, Hernandez-Borrell J (2006) Effects of lactose permease of *Escherichia coli* on the anisotropy and electrostatic surface potential of liposomes. Biophys Chem 119:101–105
21. Lopes S, Neves C, Eaton P, Gameiro P (2011) Cardiolipin, a key component to mimic the *E. coli* bacterial membrane in model system membranes. Biophys J 100:626a
22. Massey JB (2001) Interfacial properties of phosphatidylcholine bilayers containing vitamin E derivatives. Chem Phys Lipids 109:157–174

23. Kodati VR, El-Jastimi R, Lafleur M (1994) Contribution of the intermolecular coupling and librotorsional mobility in the methylene stretching modes in the infrared-spectra of acyl chains. J Phys Chem 98:12191–12197
24. Chen C, Tripp CP (2008) An infrared spectroscopic based method to measure membrane permeance in liposomes. Biochim Biophys Acta 1778:2266–2272
25. Lewis RNAH, McElhaney RN (1998) The structure and organization of phospholipid bilayers as revealed by infrared spectroscopy. Chem Phys Lipids 96:9–21
26. Wolfangel P, Muller K (2003) Chain order in lipid bilayers: FTIR and solid state NMR studies on bilayer membranes from 1,2-dimyristoyl-sn-glycero-3-phosphoglucose. J Phys Chem 107:9918–9928
27. Kuo A-T, Chang C-H (2014) Cholesterol-induced condensing and disordering effects on a rigid catanionic bilayer: a molecular dynamics study. Langmuir 30:55–62
28. Jay AG, Hamilton JA (2017) Disorder amidst membrane order: standardizing Laurdan generalized polarization and membrane fluidity terms. J Fluoresc 27:243–249
29. Amaro M, Reina F, Hof M, Eggeling C, Sezgin E (2017) Laurdan and di-4-ANEPPDHQ probe different properties of the membrane. J Phys D Appl Phys 501340048:134004
30. Malacrida L, Gratton E (2018) LAURDAN fluorescence and phasor plots reveal the effects of a $H_2O_2$ bolus in NIH-3T3 fibroblast membranes dynamics and hydration. Free Radic Biol Med 128:144–156
31. Sezgin E, Sadowski T, Simons K (2014) Measuring lipid packing of model and cellular membranes with environment sensitive probes. Langmuir 38:8160–8166
32. Sanchez SA, Tricerri MA, Gratton E (2012) Laurdan generalized polarization fluctuations measures membrane packing micro-heterogeneity in vivo. Proceedings of the National Academy of Sciences 109:7314–7319
33. Schönfeldová T, Piller P, Kovacik F, Pabst G, Okur HI, Roke S (2021) Lipid melting transitions involve structural redistribution of interfacial water. J Phys Chem B 125:2457–12465
34. Jähnig F (1979) Structural order of lipids and proteins in membranes: evaluation of fluorescence anisotropy data. Proc Natl Acad Sci 76:6361–6365
35. Danaei M, Dehghankhold M, Ataei S, Hasanzadeh Davarani F, Javanmard R, Dokhani A, Khorasani S, Mozafari MR (2018) Impact of particle size and polydispersity index on the clinical applications of lipidic nanocarrier systems. Pharmaceutics 10:57
36. Havlíková M, Szabová J, Mravcová L, Venerová T, Chang C-H, Pekař M, Jugl A, Mravec F (2021) Cholesterol effect on membrane properties of cationic ion pair amphiphile vesicles at different temperatures. Langmuir 37:2436–2444
37. Suga K, Umakoshi H (2013) Detection of nanosized ordered domains in DOPC/DPPC and DOPC/Ch binary lipid mixture systems of large unilamellar vesicles using a TEMPO quenching method. Langmuir 29:4830–4838
38. Chen C-h, Tian C-a, Chiu C-c (2017) The effects of alkyl chain combinations on the structural and mechanical properties of biomimetic ion pair amphiphile bilayers. Bioengineering 4:84
39. Cheng C-Y, Lai Y-F, Hsieh Y-L, Wu C-H, Chiu C-c, Yang Y-M (2022) Divergent effects of cholesterol on the structure and fluidity of liposome and catanionic vesicle membranes. FEBS Lett 596:1827–1838
40. Huang F-y, Chiu C-c (2017) Interplay between alkyl chain asymmetry and cholesterol addition in the rigid ion pair amphiphile bilayer systems. J Chem Phys 146:035102
41. Tian C-a, Chiu C-c (2018) Importance of hydrophilic groups on modulating the structural, mechanical, and interfacial properties of bilayers: a comparative molecular dynamics study of phosphatidylcholine and ion pair amphiphile membranes. Int J Mol Sci 19(2018):1552
42. Tien W-J, Chen K-y, Huang F-y, Chiu C-c (2021) Effects of cholesterol on water permittivity of biomimetic ion pair amphiphile bilayers: interplay between membrane bending and molecular packing. Int J Mol Sci 20:3252
43. Wang P-Y, Chiu C-c (2024) Molecular mechanism of decoupling between the fluorescence anisotropy thermograms of diphenylhexatriene and the thermotropic phase transitions of biomimetic ion pair amphiphile bilayers. J Mol Liq 395:123959

44. Lai Y-F (2020) Combined effects of ethanol co-solvent and cholesterol additive on the structural, mechanical and phase properties of ion pair amphiphile bilayers: a molecular dynamics study. M.S. Thesis National Cheng Kung University
45. Humphrey W, Dalke A, Schulten K (1996) VMD: visual molecular dynamics. J Mol Graph 14:33–38
46. Vermeer LS, Groot BL, Reat V, Milon A, Czaplicki J (2007) Acyl chain order parameter profiles in phospholipid bilayers: computation from molecular dynamics simulations and comparison with 2H NMR experiments. Eur Biophys J 36:919–931
47. Egberts E, Marrink SJ, Berendsen HJ (1994) Molecular dynamics simulation of a phospholipid membrane. Eur Biophys J 22:423–436
48. Douliez JP, Leonard A, Dufourc EJ (1995) Restatement of order parameters in biomembranes: calculation of C-C bond order parameters from C-D quadrupolar splittings. Biophys J 68:1727–1739
49. Feller SE, Pastor RW (1999) Constant surface tension simulations of lipid bilayers: the sensitivity of surface areas and compressibilities. J Chem Phys 111:1281–1287
50. Shinoda W, Shinoda K, Baba T, Mikami M (2005) Molecular dynamics study of bipolar tetraether lipid membranes. Biophys J 89:3195–3202
51. Khelashvili G, Harries D (2013) How cholesterol tilt modulates the mechanical properties of saturated and unsaturated lipid membranes. J Phys Chem B 117:2411–2421
52. Khelashvili G, Pabst G, Harries D (2010) Cholesterol orientation and tilt modulus in DMPC bilayers. J Phys Chem B114:7524–7534
53. Khelashvili G, Kollmitzer B, Heftberger P, Pabst G, Harries D (2013) Calculating the bending modulus for multicomponent lipid membranes in different thermodynamic phases. J Chem Theory Comput 9:3866–3871

# Chapter 13
# Stability and Structural Analysis of Foams Formed by Surfactants Using Small-Angle Neutron Scattering

**Shiho Yada and Tomokazu Yoshimura**

**Abstract** Small-angle neutron scattering (SANS) is an effective technique for studying varying foam structures. Recently, we succeeded in quantitatively obtaining nano- and micro-level structural information on foams formed by surfactants, including film thickness and average radius of bubbles, by combining various measurement methods centered on SANS. This chapter discusses the structural analysis of foam formed by surfactants, focusing on recent investigations conducted by the authors.

**Keywords** Bubble · Foam · Small-angle neutron scattering · Foam film · Specific surface area · Bubble radius

## 13.1 Introduction

Foams are composed of a combination of interfaces and foam films on the nanometer scale, plateau borders on the micrometer scale, and bubbles on the millimeter scale [1] (Fig. 13.1). Foams, defined as dispersions of gases in liquids, have numerous daily and industrial applications (Fig. 13.2), exemplified by the production of beverages (beer [2] and champagne [3–5]) and detergents/cosmetics. Consequently, numerous foam characterization methods have been developed, such as modified Ross–Miles method (Fig. 13.3a) [6], which evaluates foaming ability and foam stability, shaking, beating [7], rotation [8], stirring [9–11], and air injection [12]. Foam stability can be assessed by photographing a three-dimensional foam [13, 14] or its two-dimensional cross-sections to reconstruct a three-dimensional image [15],

S. Yada
Department of Industrial Chemistry, Tokyo University of Science, Tokyo, Japan
e-mail: yada@rs.tus.ac.jp

T. Yoshimura (✉)
Department of Chemistry, Faculty of Science, Nara Women's University, Nara, Japan
e-mail: yoshimura@cc.nara-wu.ac.jp



M. Hashizume, Y. Imura (eds.), *Colloids and Interfacial Dynamics*, Physical Chemistry in Action, https://doi.org/10.1007/978-981-96-6728-4_13

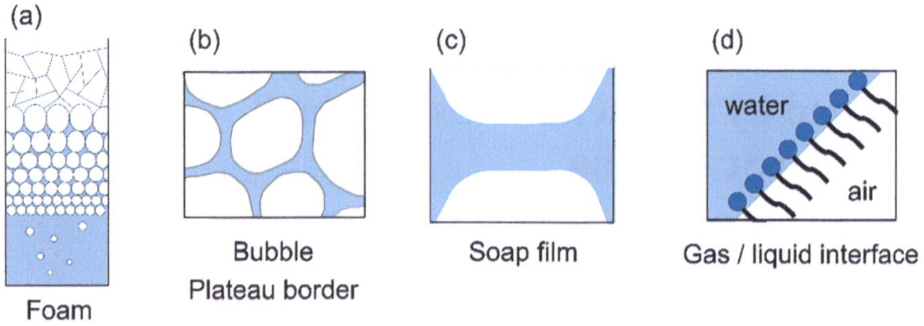


**Fig. 13.1** Structure of foams at different length scales. (**a**) Structure of entire foam: general scale $10^{-2}$–1 m. (**b**) Bubbles that make up the foam: $10^{-4}$–$10^{-2}$ m; plateau border: $10^{-6}$–$10^{-4}$ m. (**c**) Soap film: $10^{-8}$–$10^{-6}$ m. (**d**) Gas/liquid interface containing amphiphilic molecules: $10^{-10}$–$10^{-8}$ m

**Fig. 13.2** Applications of foams. (**a**) Detergent foams are stabilized by low molecular weight synthetic surfactants. (**b**) Beer foams are stabilized by a protective film of proteins. (**c**) Caffe latte is stabilized by protein, such as casein

whereas light interference, Fourier-transform infrared spectroscopy (Fig. 13.3b) [16], Raman scattering, and X-ray diffraction measurements can be used to examine the foam film thickness. In addition, the drainage state of foams can be probed by determining their moisture content distribution [17] and electric conductivity (Fig. 13.3c) [18, 19]. However, despite the abundance of foam characterization techniques, quantitative information such as foam film thickness and foam specific surface area remains limited.

Small-angle neutron scattering (SANS) is an effective method for investigating the dynamic structure of foams. Recently, research using SANS to analyze the nano-micro-level structure of foams has been advancing.

## 13.2 Structural Analysis of Foam Using SANS

In recent years, SANS has been used to directly investigate the nano- and micro-structures of foams. For example, Axelos and Boué [20] used SANS to characterize foams stabilized with sodium dodecyl sulfate (SDS) [21, 22],

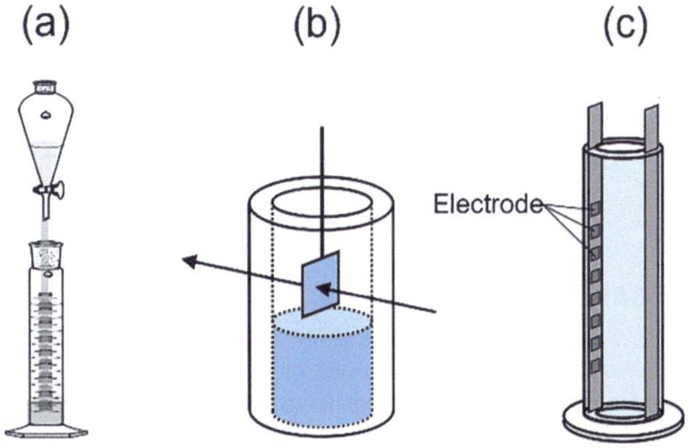


**Fig. 13.3** Evaluation of bubbles and foams. (**a**) Foaming power and foam stability using the Ross–Miles method, (**b**) measurement of the thickness of monolayer using Fourier transform infrared spectroscopy, and (**c**) measurement of water content inside the foam using the electrical conductivity method

cetyltrimethylammonium bromide and pectin [23], and pectin and napin [24]. Micheau et al. used SANS to examine foam stabilized by nonaoxyethylene oleylether carboxylic acid and investigated the effects of pH and salt addition on foam film properties [25]. Hurcom et al. investigated the internal structures of foams stabilized by triblock copolymers comprising hydrophilic polyoxyethylene and hydrophobic polyoxypropylene chains [26], whereas Zhang et al. studied the internal structures of stable foams formed in SDS–salt mixtures [27]. Finally, Mansour et al. used SANS to investigate the structure of foams formed in the presence of sodium alkyl sulfate and alkyltrimethylammonium bromide [28]. Therefore, the application of SANS to foams formed by surfactants enables researchers to obtain structural information on foams at the nano- and micro-levels. However, further studies on the structure of foams using SANS are currently required.

Recently, the authors established a method for evaluating the structure of foams using SANS [29–32]. This section discusses the structural analysis of foams formed by (1) hydroxy group-containing amino acid surfactants [29, 30] (Fig. 13.4a) and (2) homogeneous polyoxyethylene-type nonionic surfactants with multibranched chains [31, 32] (Fig. 13.4b).

## 13.3 SANS Measurement of Foam

SANS measurements were conducted using the small-angle neutron scattering instrument iMATERIA installed at beamline BL20 of the Materials and Life Science Facility (MLF) at the Japan Proton Accelerator Research Complex (J-PARC) in Tokai-mura, Naka-gun, Ibaraki Prefecture, Japan. An image of the sample cell used for measurements is shown in Fig. 13.5. Surfactant $D_2O$ solution was poured into

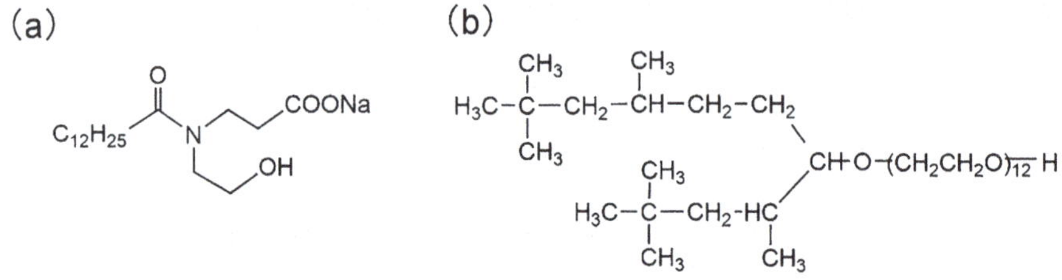


$C_{12}$-EtOH-*β*Ala

*b*$C_7$-*b*$C_9$$EO_{12}$

**Fig. 13.4** Molecular structures and abbreviations of various surfactants: (**a**) hydroxy group-containing amino acid surfactant and (**b**) homogeneous polyoxyethylene (EO)-type nonionic surfactants with multibranched chains

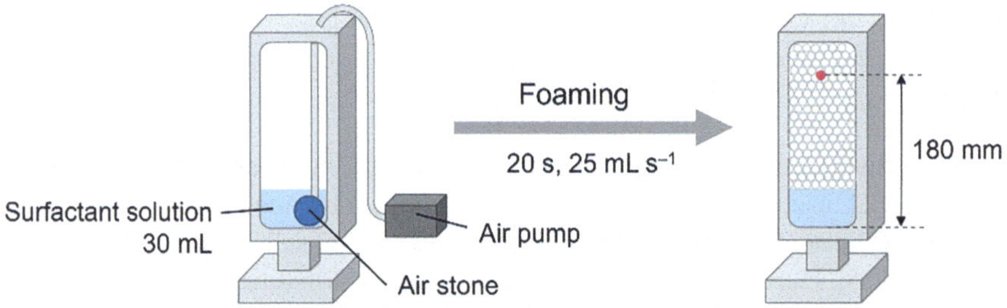


**Fig. 13.5** Schematic illustration of the sample cell used for small-angle neutron scattering (SANS) measurements. (Adapted with permission from Ref. 29. Copyright 2020 American Chemical Society)

the sample cell, and air was introduced through the glass filter at a constant flow rate using an electric air pump to generate foams. After the sample cell was filled with foam, the air pump was stopped, and SANS measurements were conducted by irradiating the foam at a specified height in the sample cell with neutrons for a given period. The obtained data were converted into SANS scattering profiles for the foams every minute immediately after the foams were formed by one-dimensionalizing the data every minute. Please refer to our papers [29, 32] for detailed information on the SANS measurements.

## 13.4 Microstructural Characterization of Foam Formed Using a Hydroxy Group-Containing Amino Acid Surfactant

*N*-acyl amino acid-type surfactants, featuring amide nitrogen-bonded alkyl chains, offer advantages, such as low irritation and high biodegradability, making them widely used in detergents and cosmetics. Recently, we developed a novel surfactant of this type, namely, *N*-dodecanoyl-*N*-(2-hydroxyethyl)-*β*-alanine ($C_{12}$–EtOH–*β*Ala, Fig. 13.4a), by substituting the methyl group of *N*-dodecanoyl-*N*-methyl-*β*--alanine (a popular shampoo component; $C_{12}$–Me–βAla) with a hydroxy group. In alkaline solutions, $C_{12}$–EtOH–*β*Ala surfactant achieved a lower surface tension at the critical micelle concentration (CMC) and a lower occupied area per molecule than $C_{12}$–Me–βAla surfactant [30]. This behavior was ascribed to hydrogen bonding between the hydroxy groups of $C_{12}$–EtOH–βAla and carboxylate ions or water, facilitating dense adsorption and efficient orientation of this surfactant at the air–water interface. Therefore, $C_{12}$–EtOH–βAla-stabilized foams are expected to exhibit excellent stability because of the formation of strong films at the air–water interface. In this study, we investigated the microstructure and stability of $C_{12}$–EtOH–βAla-stabilized foam using SANS and assessed the effect of the hydroxy group introduction on foam properties.

Figure 13.6 shows the SANS profiles of $C_{12}$–EtOH–*β*Ala- and hydroxy group-free surfactants $C_{12}$–Me–*β*Ala-stabilized foams within 0–15 min after foaming. For $C_{12}$–EtOH–*β*Ala, a breakpoint appeared at $q \approx 0.23\ \text{nm}^{-1}$ with increasing time, and the scattering intensity at nonbreakpoint $q$ values decreased, whereas the intensity near the breakpoint remained constant. For $C_{12}$–Me–*β*Ala, the scattering intensity decreased with time at all $q$ values, becoming identical to that of the background after 10 min, indicating foam breakage at a height of 180 mm from the bottom after 10 min.

The scattering intensities $I(q)$ of the SANS profiles were obtained by summing the coherent and incoherent scattering intensities. The profile obtained for $C_{12}$–EtOH-βAla demonstrated that coherent scattering was dominant at $q < 3\ \text{nm}^{-1}$ and flat (i.e., incoherent) scattering was dominant at $q > 3\ \text{nm}^{-1}$. The former region was considered divisible into low- and high-$q$ subregions. At low $q$ ($q < 0.4\ \text{nm}^{-1}$), the scattering intensity depended on $q^{-2}$ ($0.07 < q < 0.23\ \text{nm}^{-1}$) and $q^{-4}$ ($0.23 < q < 0.4\ \text{nm}^{-1}$) and the cross-over value was $q_c = 0.23\ \text{nm}^{-1}$. This reflected the presence of plate-like structures and definite interfaces, providing information on the foam film structure. At high $q$ ($0.4 < q < 3\ \text{nm}^{-1}$), a broad peak coinciding with

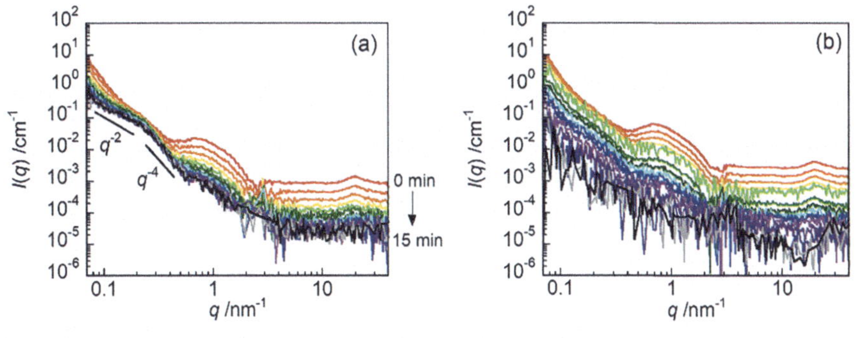


**Fig. 13.6** SANS profiles of (a) $C_{12}$–EtOH–*β*Ala- and (b) $C_{12}$–Me–*β*Ala-stabilized foams at 100 mmol dm$^{-3}$ in $D_2O$ obtained at a height of 180 mm from the cell bottom. (Adapted with permission from Ref. 29. Copyright 2020 American Chemical Society)

that of the micelle solution-scattering profile was observed at approximately 0.7 $nm^{-1}$, indicating that the foam film contained micelles identical to those found in the bulk solution. Analyses of these SANS profiles can provide detailed information regarding the thickness of the foam film and micelles in the film.

The film thickness $d = 2\pi/q_c$ was determined from the cross-over wavenumber $q_c$ [20, 23–25]. Foam films have been reported to reflect incident beams, and data such as reflectivity can be obtained for dry foam, which is characterized by well-pronounced drainage and a very small number of foam films [20]. Figure 13.7 shows the time course of $d$ for foams stabilized by $C_{12}$–EtOH–$\beta$Ala and $C_{12}$–Me–βAla, revealing that $C_{12}$–EtOH–$\beta$Ala formed a stable foam film with $d = 26 \pm 2$ nm within 2–15 min after foaming, whereas $C_{12}$–Me–βAla formed a foam film of the same thickness (25 ± 2 nm) within 6–11 min. Therefore, although the introduction of the hydroxy group increased the foam stability, it did not substantially affect the foam film thickness. The foam film contained micelles and was stabilized by an electric double layer surrounding the micelles and the film.

## 13.5 Structural Analysis of Foam Obtained Using Homogeneous Polyoxyethylene-Type Nonionic Surfactants with Multibranched Chains

Recently, we developed a novel homogeneous polyoxyethylene (EO) alkyl ether-type nonionic surfactant with multibranched chains ($bC_7$–$bC_9EO_{12}$, where $bC_7$ and $bC_9$ represent branched chains with total carbon numbers of seven and nine, respectively, and $EO_{12}$ represents the dodecaoxyethylene chain (Fig. 13.4b) [31]. The surface tension at the CMC for the multibranched chain-type $bC_7$–$bC_9EO_{12}$ decreased considerably to 28.8 mN $m^{-1}$ despite having a large hydrophilic group of $EO_{12}$ [31]. In this study, we investigated the microstructure of a novel homogeneous $bC_7$–$bC_9EO_{12}$-stabilized foam using SANS.

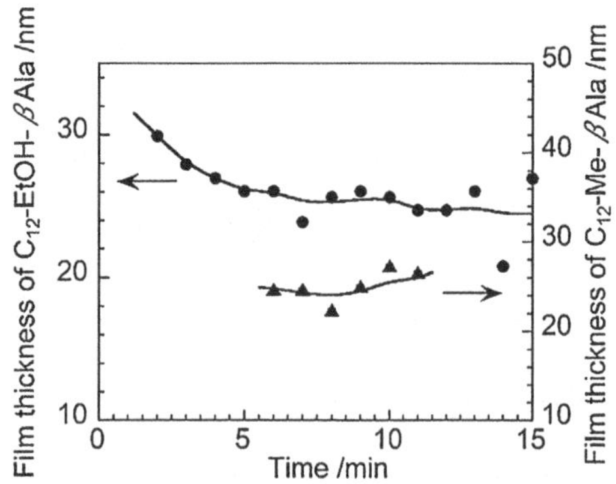


**Fig. 13.7** Change in foam film thickness ($d$) with time for $C_{12}$–EtOH–$\beta$Ala (●) and $C_{12}$–Me–$\beta$Ala (▲). (Adapted with permission from Ref. 29. Copyright 2020 American Chemical Society)

Figure 13.8a shows the SANS profile of $bC_7$–$bC_9EO_{12}$-stabilized foams at 0–1 min after foaming and for the solution that multiplied $I(q)$ by 0.09, where 0.09 is the liquid fraction $\varphi$ in the foams [33]. The shoulder of the foam coincided with that of the bulk solution; that is, the films between the bubbles contained micelles identical to those found in the bulk solution. Figure 13.8b shows the SANS profile obtained by subtracting the $I(q)$ of the solution, which was multiplied by 0.09, from that of the foam. In the region of $q = 0.1–1\ nm^{-1}$, the scattering intensity exhibited a $q^{-4}$ dependence, which indicated the presence of definite interfaces. Therefore, only the scattering of the films between bubbles could be extracted by subtracting the intensity of the solution scattering from that of the foam scattering, revealing that the films between the bubbles containing the plateau border contained micelles identical to those observed in the bulk solution. In the low $q$ region ($q < 0.2\ nm^{-1}$), the SANS profiles were dominated by the scattering of the air–water interfaces following a $q^{-4}$ decay based on Porod's law [20, 24, 34], as expressed below:

$$I(q) = 2\pi(\Delta b)^2 \frac{S}{V} q^{-4} \tag{13.1}$$

where $\Delta b$ is the scattering contrast between $D_2O$ and air, $V$ is the system volume, and $S$ is the interfacial area contained in $V$.

The $I(q)$ depends on the scattering length densities of air and $D_2O$ (0 and $6.39 \times 10^{-4}\ nm^{-2}$, respectively) and specific surface area $S/V$ (amount of the air–water interface per unit volume), with $S = 4\pi NR^2$ and $V = N(4\pi R^3)/\{3(1-\varphi)\}$, resulting in $S/V = 3(1-\varphi)/R$ [33]. Here, $N$ is the number of bubbles and $R$ is the average bubble radius. The average bubble radius $R$ is calculated using the liquid fraction φ, and the specific surface area $S/V$ is obtained using Eq. (13.1). Figure 13.9 shows the time course of the average radii of the bubbles formed by $bC_7$–$bC_9EO_{12}$. The

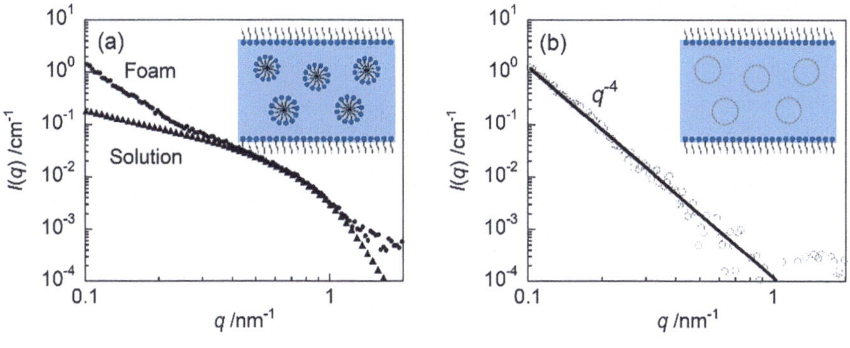


**Fig. 13.8** (**a**) SANS profile of the foam at 0–1 min after foaming and of solution multiplied $I(q)$ by liquid fraction $\varphi$ and (**b**) SANS profile obtained by subtracting $I(q)$ of solution, which is multiplied by $\varphi$, from that of foam for $bC_7$–$bC_9EO_{12}$ in $D_2O$ at 10 mmol dm$^{-3}$. (Reprinted in part with permission from Ref. 32. Copyright 2023 American Chemical Society)

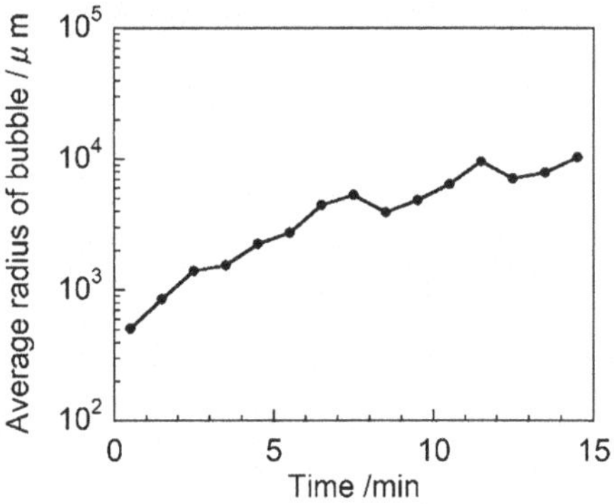


**Fig. 13.9** Variation in average radius of bubble with time for $bC_7$–$bC_9EO_{12}$. (Reprinted in part with permission from Ref. 32. Copyright 2023 American Chemical Society)

average radius of the bubbles immediately after foaming was in the order of hundreds of micrometers and increased with time to thousands of micrometers.

## 13.6 Conclusion

This chapter outlined the structural analysis of foams formed by surfactants using SANS. The structure of foams exhibits substantial variations depending on the molecular structure of the surfactant. Data obtained from the SANS profile of the bubbles is essential in industrial fields, such as cleaning, food, and cosmetics. We hope that research into the structural analysis of foams using surfactants with diverse molecular structures will advance and facilitate the characteristics of these systems.

**Acknowledgments** Research on homogeneous EO-type nonionic surfactants with multibranched chains was conducted in collaboration with Nissan Chemical Corporation. Research on hydroxy-containing amino acid-type surfactants was conducted in collaboration with NOF Corporation and Kracie Ltd. We are grateful to them. We also thank Prof. Satoshi Koizumi of Ibaraki University for their guidance and advice regarding the SANS experiments.

## References

1. Cantat I, Cohen-Addad S, Elias F, Graner F, Höhler R, Pitois O, Rouyer F, Saint-Jalmes A (2013) Foams: structure and dynamics. OUP, Oxford
2. Leike A (2002) Demonstration of the exponential decay law using beer froth. Eur J Phys 23(1):21
3. Liger-Belair G, Jeandet P (2003) Capillary-driven flower-shaped structures around bubbles collapsing in a bubble raft at the surface of a liquid of low viscosity. Langmuir 19(14):5771–5779
4. Liger-Belair G, Voisin C, Jeandet P (2005) Modeling nonclassical heterogeneous bubble nucleation from cellulose fibers: application to bubbling in carbonated beverages. J Phys Chem B 109(30):14573–14580

5. Liger-Belair G, Beaumont F, Jeandet P, Polidori G (2007) Flow patterns of bubble nucleation sites (called fliers) freely floating in champagne glasses. Langmuir 23(22):10976–10983
6. Ross J, Miles GD (1941) An apparatus for comparison of foaming properties of soaps and detergents. Oil & Soap 18(5):99–102
7. Cox MF (1989) Effect of alkyl carbon chain length and ethylene oxide content on the performance of linear alcohol ether sulfates. J Am Oil Chem Soc 66(11):1637–1646
8. Beh H, James K (1977) Foam stabilities of some bubble bath formulations. Cosmetics and Toiletries 92:21
9. Hart JR, DeGeorge MT (1980) The lathering potential of surfactants—a simplified approach to measurement. J Soc Cosmet Chem 31:223
10. Weeks LE, Harris JC, Brown EL (1954) A dynamic foam test for evaluation of hand dishwashing compositions. J Am Oil Chem Soc 31(6):254–257
11. Mittal KL, Shah DO (2013) Surfactants in solution. Springer
12. Ross S, Nishioka G (1980) Foams. Kirk-Othmer Encyclopedia of Chemical Technology 11:127–145
13. Sánchez-Vioque R, Bagger CL, Rabiller C, Guéguen J (2001) Foaming properties of acylated rapeseed (Brassica napus L.) hydrolysates. J Colloid Interface Sci 244(2):386–393
14. Sarker DK, Bertrand D, Chtioui Y, Popineau Y (1998) Characterisation of foam properties using image analysis. J Texture Stud 29(1):15–42
15. Monnereau C, Vignes-Adler M (1998) Optical tomography of real three-dimensional foams. J Colloid Interface Sci 202(1):45–53
16. Tamura T, Kaneko Y, Nikaido M (1997) Stability factors of foam film in contrast to fluctuation induced by humidity reduction. J Colloid Interface Sci 190(1):61–70
17. Małysa K (1992) Wet foams: formation, properties and mechanism of stability. Adv Colloid Interf Sci 40:37–83
18. Feitosa K, Marze S, Saint-Jalmes A, Durian DJ (2005) Electrical conductivity of dispersions: from dry foams to dilute suspensions. J Phys Condens Matter 17(41):6301
19. Weaire D, Phelan R (1996) The physics of foam. J Phys Condens Matter 8(47):9519
20. Axelos MAV, Boué F (2003) Foams as viewed by small-angle neutron scattering. Langmuir 19(17):6598–6604
21. Etrillard J, Axelos MAV, Cantat I, Artzner F, Renault A, Weiss T, Delannay R, Boué F (2005) In situ investigations on organic foam films using neutron and synchrotron radiation. Langmuir 21(6):2229–2234
22. Terriac E, Emile J, Axelos MAV, Grillo I, Meneau F, Boué F (2007) Characterization of bamboo foam films by neutron and X-ray experiments. Colloids Surf A Physicochem Eng Asp 309(1):112–116
23. Ropers MH, Novales B, Boué F, Axelos MAV (2008) Polysaccharide/surfactant complexes at the air−water interface—effect of the charge density on interfacial and foaming Behaviors. Langmuir 24(22):12849–12857
24. Schmidt I, Novales B, Boué F, Axelos MAV (2010) Foaming properties of protein/pectin electrostatic complexes and foam structure at nanoscale. J Colloid Interface Sci 345(2):316–324
25. Micheau C, Bauduin P, Diat O, Faure S (2013) Specific salt and pH effects on foam film of a pH sensitive surfactant. Langmuir 29(27):8472–8481
26. Hurcom J, Paul A, Heenan RK, Davies A, Woodman N, Schweins R, Griffiths PC (2014) The interfacial structure of polymeric surfactant stabilised air-in-water foams. Soft Matter 10(17):3003–3008
27. Zhang L, Mikhailovskaya A, Yazhgur P, Muller F, Cousin F, Langevin D, Wang N, Salonen A (2015) Precipitating sodium dodecyl sulfate to create ultrastable and stimulable foams. Angew Chem 127(33):9669–9672
28. Mansour OT, Cattoz B, Beaube M, Montagnon M, Heenan RK, Schweins R, Appavou M-S, Griffiths PC (2017) Assembly of small molecule surfactants at highly dynamic air–water interfaces. Soft Matter 13(46):8807–8815

29. Yada S, Shimosegawa H, Fujita H, Yamada M, Matsue Y, Yoshimura T (2020) Microstructural characterization of foam formed by a hydroxy group-containing amino acid surfactant using small-angle neutron scattering. Langmuir 36(27):7808–7813
30. Yada S, Wakizaka M, Shimosegawa H, Fujita H, Yamada M, Matsue Y, Yoshimura T (2021) Unique interfacial adsorption behavior of a hydroxy group-containing amino acid surfactant. Colloids Surf A Physicochem Eng Asp 611:125757
31. Yada S, Yoshioka Y, Ohno M, Koda T, Yoshimura T (2022) Adsorption and aggregation properties of homogeneous polyoxyethylene alkyl ether- and ester-type nonionic surfactants with multi-branched double chains. Colloids Surf A Physicochem Eng Asp 648:129247
32. Yada S, Kuroda M, Ohno M, Koda T, Yoshimura T (2023) Stability and structural analysis using small-angle neutron scattering for foam of homogeneous polyoxyethylene-type nonionic surfactants with multibranched chains. Langmuir 39(43):15355–15361
33. Ferreira J, Mikhailovskaya A, Chenneviere A, Restagno F, Cousin F, Muller F, Degrouard J, Salonen A, Marques EF (2017) Interplay between bulk self-assembly, interfacial and foaming properties in a catanionic surfactant mixture of varying composition. Soft Matter 13(39):7197–7206
34. Mikhailovskaya A, Zhang L, Cousin F, Boué F, Yazhgur P, Muller F, Gay C, Salonen A (2017) Probing foam with neutrons. Adv Colloid Interf Sci 247:444–453

# Chapter 14
# Surfactant-Induced Elution of Plasticizers from Medical Devices

**Takehisa Hanawa**

**Abstract** Plastic has become an integral part of our daily lives. This is because plastic has excellent characteristics as a material for various everyday items that surround us and is a valuable material in all industrial fields.

In particular, polyvinyl chloride (PVC) is a chemically stable material that is flexible and durable and is used in various medical equipment, including infusion bags and infusion tubes, in medical settings. Phthalate esters are added to PVC products used in medical equipment as plasticizers. Diethylhexyl phthalate (DEHP), in particular, penetrates between the PVC molecules, giving the resin flexibility, and is often added to PVC, polyvinylidene chloride, cellulose esters, rubber, etc., at a rate of around 40%. DEHP is added to infusion tubes and bags at 30–40%.

However, due to additives in injection solutions, DEHP can leach from PVC infusion tubes and bags. This chapter will introduce a series of studies on DEHP leaching from PVC medical devices, particularly PVC infusion tubes.

Injectable drugs containing immunosuppressants such as Cyclosporin and Tacrolimus contain surfactants such as Tween80 and HCO60. However, the amount of DEHP eluted increased with the concentration of these surfactants, the contact time with PVC, and rising temperatures. In addition, the amount of DEHP eluted increased around the critical micelle concentration (CMC) of Tween80 and HCO60. On the other hand, it was found that the amount of DEHP eluted could be controlled by varying the type and amount of sugar added. PVC infusion tubes and bags containing plasticizers that replace DEHP are now being used, but they are not widely used because they are expensive. When developing new pharmaceuticals, it is probably necessary to continue research on the elution of plasticizers and the adsorption and sorption of pharmaceutical molecules.

**Keywords** PVC · DEHP · Leaching · Surfactant · Tween 80 · HCO60 · Plasticizer

---

T. Hanawa (✉)
Faculty of Pharmaceutical Sciences, Tokyo University of Science, Tokyo, Japan
e-mail: t-hanawa@rs.tus.ac.jp



M. Hashizume, Y. Imura (eds.), *Colloids and Interfacial Dynamics*, Physical Chemistry in Action, https://doi.org/10.1007/978-981-96-6728-4_14

## 14.1 Introduction

The industrialization of plastics in Japan began in 1914 with the production of phenolic resin from coal, followed by the production of polyvinyl chloride in 1949. It was not until around 1958 that mass production began in earnest. However, along with metals and other materials with a history of thousands of years, plastics have become inseparable from our daily lives. This is because plastic has excellent characteristics as a material for various daily necessities that surround us and is a valuable material in all industrial fields.

PVC is a chemically stable material with excellent flexibility and durability and is used in infusion bags, infusion tubes, and various other medical devices in the medical field.

Phthalates are added as plasticizers to PVC products used for medical devices. In particular, diethylhexyl phthalate (DEHP) is added between PVC molecules to give flexibility to the resin. It is often added to PVC, polyvinylidene chloride, cellulose esters, and rubber at about 40%, with infusion tubes and bags containing 30–40% DEHP. DEHP is an endocrine disruptor.

DEHP has been discussed as a candidate endocrine disrupter, but currently, as a general toxicant with mainly testicular toxicity [1, 2], a tolerable daily intake (TDI) of 40–140 μg/kg/day has been established. Its use in toys for infants, cooking gloves, and packaging containers for food made of oil and fat is prohibited in Japan (Food Sanitation Law, 2003, as amended).

It has been known since relatively early on that DEHP leaches from PVC infusion bags. The Ministry of Health, Labour and Welfare (MHLW) has stated in "Safety Information on Drugs and Medical Devices No. 182, June 25," 2025 [3] that although there have been no reports of direct health hazards due to the use of PVC medical devices, it is possible to reduce the amount of DEHP exposure if possible. However, it is necessary to consider reducing the amount of DEHP exposure if possible, and medical institutions were notified that they should consider shifting to and using alternative products that do not elute DEHP.

In the United States, the ATSDR has set the risk level of DEHP as "regulatory." On January 21, 2022, the Agency for Toxic Substances and Disease Registry (ATSDR) of the US Department of Health and Human Services (DHHS) issued an alert to medical institutions to consider shifting the use of DEHP The Agency for Toxic Substances and Disease Registry (ATSDR) has established a minimal risk level (MRL) for the ingestion and inhalation of diethylhexyl phthalate (DEHP). This is an update of the toxicological profile published by the ATSDR and includes the MRLs established at this time. The proposed update was to be published in 2019. The MRLs set by the ATSDR are as follows. However, the ATSDR states that it was unable to set MRLs for acute (14 days or less) and chronic (>1 year) for the inhalation route and chronic (>1 year) for the oral route due to insufficient data.

Intermediate period (15–364 days) by inhalation route MRL = 0.002 ppm.
Acute by oral route (14 days or less) MRL = 0.03 mg/kg/day.
Intermediate period by oral route (15–364 days) MRL = 0.001 mg/kg/day.

ATSDR has established MRLs to identify chemicals that may be of concern at hazardous waste sites; MRLs do not directly regulate but may be used by regulators as a reference when assessing risk (endcrin2022_February.pdf).

Thus, various restrictions are in place for products made of PVC. Therefore, this chapter presents a series of studies on DEHP elution from medical devices made of PVC, especially PVC infusion tubing of micrometers, which results in the fact that the films are robust enough to be treated easily by hand.

## 14.2 The Leaching of Diethylhexyl Phthalate from the Administration Set into the Intravenous Cyclosporine Solutions

The elution of DEHP from PVC infusion tubing was found during a study of the decrease in Cyclosporine (CyA) content in PVC infusion tubing. Figure 14.1 shows a liquid chromatogram of a sample obtained by Nakajima et al. [4–6].

The internal standard was observed about 3 min after sample injection, and the CyA-derived peak was observed at about 5 min. However, as the collection time progressed, a small peak appeared at 13.5 min and gradually increased in intensity. This peak was not observed when CyA injection and saline solution (PS) were injected into the HPLC alone, and we inferred that it was derived from the western part of the infusion bag. When DEHP, a plasticizer, was injected into the HPLC, a peak was observed at 13.5 min, the same time as the sample dropped from the infusion bag, and it became clear that the peak was derived from DEHP.

Since there were cases in which DEHP was not eluted depending on the drug added, Kawano et al. focused on the surfactant added as a solubilizing agent and examined the elution behavior of DEHP when infusion solution was dripped from PVC bags and PVC tubes in detail. Polyoxyethylene hardened castor oil (Cremophor EL) is a hydrophilic nonionic surfactant made by the additional polymerization of hydrogenated castor oil with ethylene oxide, which has a considerable molecular weight, is highly safe, and causes almost no skin irritation. Figure 14.2 shows the

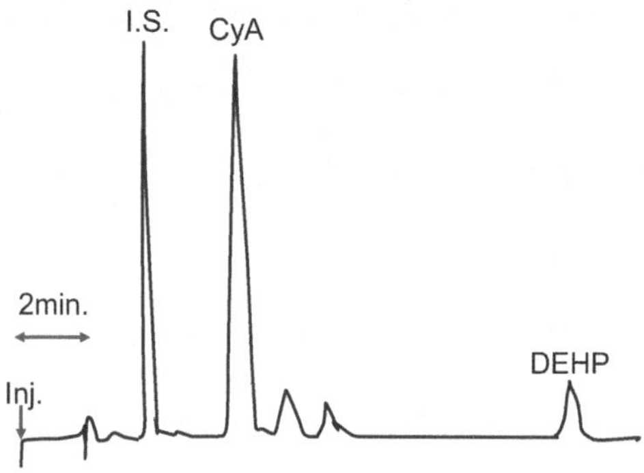


**Fig. 14.1** HPLC chromatogram of Cyclosporine remaining after the passage through the PVC administration set. (Cited and modified from [4])

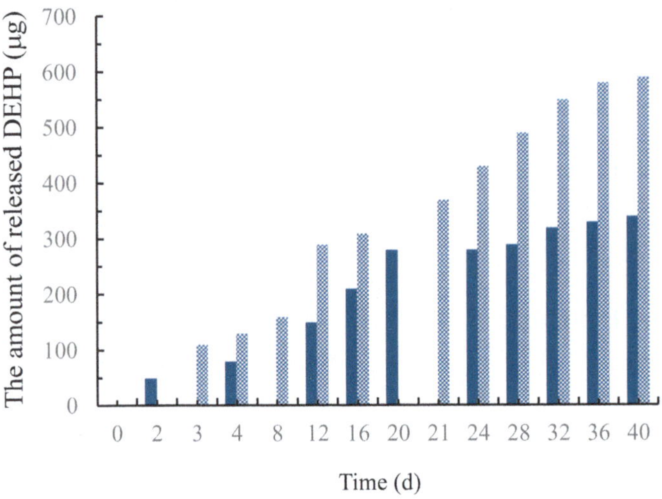


**Fig. 14.2** The amount of DEHP dissolved from the administration set into the aqueous solution containing polyoxyethylated castor oil at 30 °C. The concentration of polyoxyethylene castor oil: 10 mg/mL (blue and white checkered bars), 5 mg/mL (blue solid bars). (Cited and modified from [5])

elution behavior of DEHP when shredded PVC tubes were immersed in solutions containing various concentrations of polyoxyethylene hardened castor oil (Cremophor EL).

However, equilibrium was not reached after 40 days, suggesting that the diffusion of DEHP through the PVC tubing was very slow.

The cumulative amount of DEHP eluted from shredded PVC tubing after 40 days in various concentrations of Cremophor EL solution increased linearly with increasing Cremophor EL concentration (Fig. 14.3).

Based on these phenomena, the elution behavior of CyA injected solution (0.2, 0.5, 1.0 μg/mL) at various concentrations (0.29, 0.58, 1.16 mL/min) in PVC bags and tubes was similar to that in actual use, and the higher the CyA concentration and the slower the drop rate. The higher the CyA concentration and the slower the drop rate, the higher the elution concentration. This is presumably due to the contact time and frequency of Cremophor EL with the PVC wall, but further investigation is needed [7].

## 14.3 Elution of DEHP with Fat Emulsion

Kawano et al. studied the elution behavior of DEHP in alprostadil injection as a fat emulsion. Alprostadil injection is a fat emulsion containing lecithin, oleic acid, glycerol, and soybean oil as additives [8].

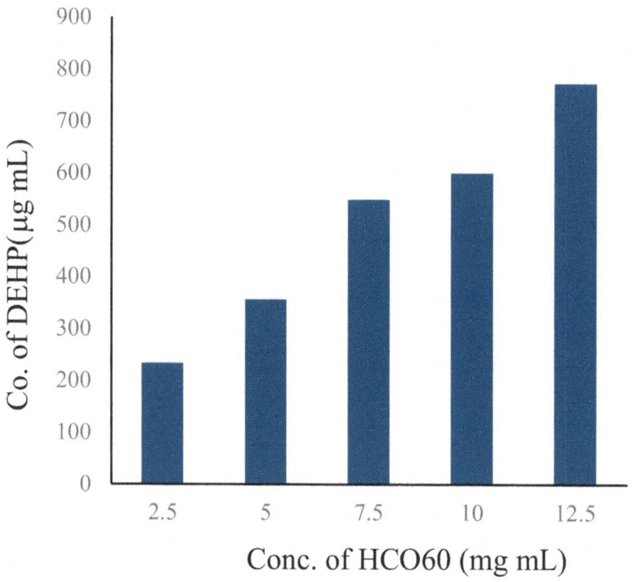


**Fig. 14.3** The relation between the amount of DEHP dissolved from the administration set into the aqueous solutions at 30 °C after 40 days and the concentration of polyoxyethylated castor oil. The weight of the chip of the administration set was about 0.15 g. (Cited and modified from [5])

**Table 14.1** Effect of the lecithin on the dissolved amount of DEHP from the administration set into the distilled water for injection at 30 °C

| Lecithin (mg/mL) | Dilution rate | DEHP (μg/mL) |
|---|---|---|
| 0.36 | × 1/50 | 7.2 |
| 0.90 | × 1/20 | 11.2 |
| 1.80 | × 1/10 | 19.7 |

Cited and modified from [8]

The behavior of DEHP was examined by focusing on each component of the alprostadil injection.

The amount of DEHP eluted into lecithin emulsion (18.0 mg/mL), prepared at the same concentration as alprostadil injection and diluted 10, 20, and 50 times with purified water, was measured (Table 14.1). The amount of DEHP eluted increased with increasing lecithin concentration (Table 14.1), suggesting that the emulsified lecithin emulsion was emulsified as DEHP droplets.

The amount of DEHP eluted from the lecithin emulsion was examined by adding soybean oil at various concentrations (20, 50, and 100 mg/mL) and diluting the emulsion with purified water.

In a recent report, Huang et al. examined the interaction of PVC infusion bags with lipid nanoparticle (LNP) formulations used to encapsulate small interfering RNA (siRNA) and deliver it to in vivo targets for gene silencing via intravenous administration [9].

The plasticizer diethylhexyl phthalate (DEHP) is commonly found in IV bag materials and can leach into the formulation; Huang et al. examined the extent of DEHP leaching in infusion bags containing LNP saline over a 24-h administration period. They reported that the amount of DEHP in LNP saline increased with increasing contact time with the infusion bag material and exceeded the safety limit set by the FDA (3 mg/day or 30 ppm for the proposed dose) after 8 h. The amount of DEHP in the bagged LNP saline solution was also found to increase as the concentration of LNP in the infusion bag increased, suggesting that the elution of DEHP was primarily due to the presence of the lipophilic LNP formulation, they stated. On the other hand, they also examined the effect of eluted DEHP on the encapsulation efficiency of LNPs but found no effect. Based on these results, they conclude that LNP formulations should be administered using DEHP-free infusion bags.

## 14.4 Elution of DEHP from PVC Tubing by Enteral Nutrition

Tanaka et al. investigated DEHP elution from PVC tubing, focusing on the emulsion contained in enteral nutrition (EL) [10].

As a result, they reported that the amount of DEHP elution increased as the environmental temperature of EL increased.

They stated that the mechanism of DEHP elution is caused by the repetition of (a) dissolution of DEHP on the surface of PVC tubes into EL, (b) dispersion of dissolved DEHP into EL solution, and (c) diffusion of DEHP into PVC tubes.

In another experimental system, Tanaka et al. also examined the amount of DEHP elution using a gastrostomy connection tube made of DEHP-added PVC, assuming that various enteral nutrients are administered via gastrostomy to neonates. The results were similar to those described above: DEHP elution increased as the fat content in the EL increased. In contrast, to reduce the amount of DEHP elution, it is desirable to speed up the injection rate when using nutrients with high-fat content. However, the authors note that speeding up the infusion rate of nutritional supplements with high-fat content may cause side effects such as diarrhea, so care should be taken. On the other hand, when using a gastrostomy connection tube made of DEHP-added PVC to administer common enteral nutritional products, the amount of DEHP eluted can easily exceed the tolerable daily intake. DEHP-free nutritional routes are recommended due to the risk of adverse health outcomes with long-term nutrient administration and in children.

In addition, a recent study by Ekici et al. reported a case study examining phthalates (PAEs) in Turkish pediatric enteral nutrition (PEF) samples and their potential impact on consumers [11].

Ekici et al. found that among various types and forms of PEF samples, the so-called type 5 sample showed the highest total PAE concentration (71.52 μg/kg).

This sample was a high-calorie polymeric (PC) formulation, followed by an oligomeric (OC) formulation, Type 1, which had the highest PAEs (60.33 μg/kg). Furthermore, the oligomeric (OC) formulation had a PAE value of 55.05 μg/kg, while the monomeric (MC) formulation, Type 1, had a PAE value of 55.05 μg/kg. However, they stated that no statistically significant differences were found. The ranking of the PEF samples based on total PAEs was as follows: MC (54.94 μg/kg) > OC (54.05 μg/kg) > OC (54.05 μg/kg) > PC (53.32 μg/kg). DEHP, DINP, DNOP, and diethyl phthalate (DEP) were the most abundant PAEs, with the carcinogenic risk associated with DEHP determined to be at an acceptable carcinogenic risk level (Grade: B) in all age and gender groups, with the carcinogenic risk level determined to be at an acceptable carcinogenic risk level. The risk was determined to be an acceptable cancer risk level (Grade: B) for all age and sex groups. Although this is below the tolerable daily intake established by the European Food Safety Authority, continued monitoring and regulation are essential to reduce the risk of PAE exposure, according to the report.

## 14.5 Elution of DEHP by Interaction of Nonionic Surfactants and PVC

### *14.5.1 Elution Behavior of DEHP in Formulations Containing Tween 80*

Lastet®, an antineoplastic drug, contains etoposide as API, Tween 80, PEG400, and ethanol. To examine the effect of these components on DEHP elution, Muramatsu et al. examined the elution behavior of DEHP in (a) a solution containing only Tween 80, (b) a solution containing Tween 80, PEG300, and ethanol, and (c) Lastet® injection solution [12].

The DEHP concentration remained at approximately 1.5 μg/mL after 30 min of the start of the test. As in the CyA experimental system, the amount of DEHP eluted increased by increasing Tween 80 concentration and slower drop acceleration (Fig. 14.4).

The same experiment was conducted with a Neo M.V.-9 injection, used for vitamin B1 supplementation during total parenteral nutrition, as an injection containing Tween 80. In this injection, the amount of DEHP eluted 1 h after the start of the test remained at 0.3–0.4 μg/mL.

In addition to the Neo M.V.-9 injection, the amount of DEHP eluted into a mixed solution of a vitamin K formulation with lecithin added as an emulsifier commonly used in actual medical practice was not significantly different between the two systems. This was presumably because the amount of Tween 80 added in the formulation was higher than that of lecithin, and the amount of lecithin did not reach a concentration that would affect the elution behavior of DEHP with Tween 80.

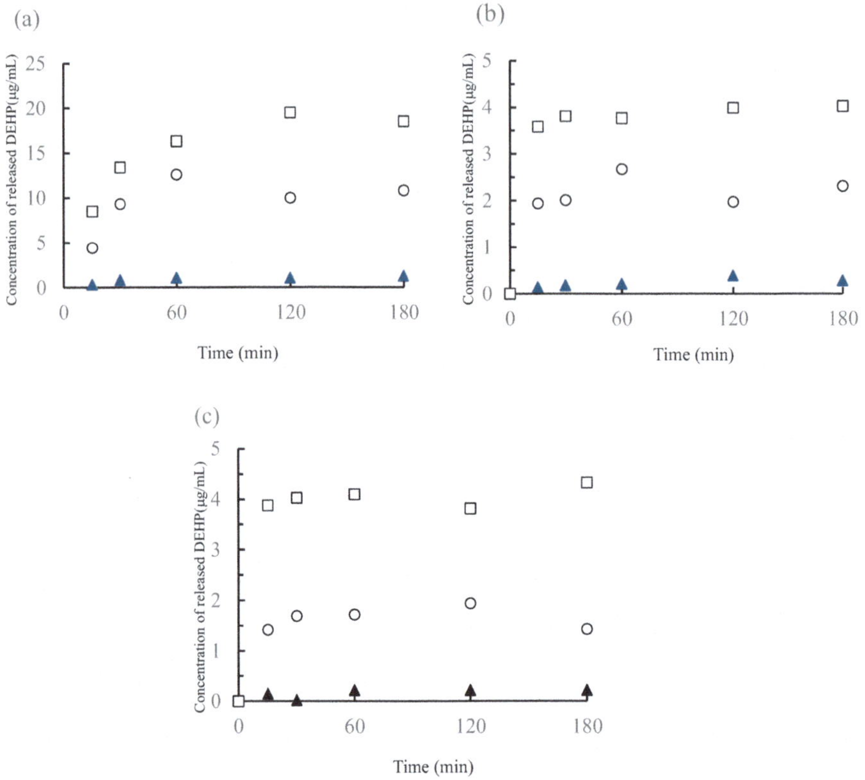


**Fig. 14.4** Effect of the addition of Tween 80 on the dissolution behavior of DEHP into saline at various drip rates. (**a**) The drip rate was 5 mL/h. (**b**) The drip rate was 60 mL/h. (**c**) The drip rate was 80 mL/h. The concentration of Tween 80 are ▲; 0.1 mg/mL, ○; 1 mg/mL, □; 2 mg/mL. (Cited and modified from [12])

### *14.5.2 Elution of DEHP by Formulations Containing HCO60*

Muramatsu et al. also investigated the release behavior of DEHP from PVC tubes in the presence of surfactants such as polysorbate 80 and polyoxyethylene castor oil (Tween 80 and HCO 60) in various solutions. The results showed that the concentration of released DEHP increased with increasing concentrations of Tween 80. The concentration of released DEHP decreased as the drop rate increased. A comparison of the release behavior of DEHP surfactant alone with that observed in the solution containing the surfactant and that observed in the commercial injectable formulation showed that the two were compatible. Based on these results, he stated that it is possible to predict the dissolved concentration of DEHP from the concentration of the surfactant and the rate of titration (Fig. 14.5).

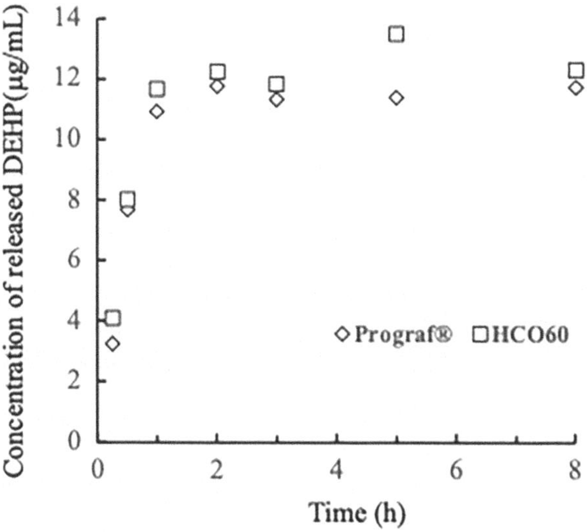


**Fig. 14.5** Effect of Prograf® and HCO60 on the release behavior of DEHP from the PVC tubing into saline. □; Prograf®. ♦; HCO60. (Cited and modified from [12])

## 14.6 Investigation of the DEHP Elution Mechanisms

### *14.6.1 Influence of Different Types of Infusion Solutions on DEHP Elution Behavior*

The release behavior of DEHP from PVC tubing was examined in the presence of various polysorbate 80 (Tween 80) solutions, including saline solution (PS) and distilled water for injection (DWI) [13].

The results showed that the cumulative amount of DEHP released after 5 h was in the order of PS, DWI50%, and traubenzucker, glucose (TZ). Comparing the amount of DEHP released with the critical micelle concentration (CMC) of the various solutions, the lower the CMC of the solutions, the more DEHP was released from the PVC tubes. When the concentration of Tween 80 was kept constant at 1 mg/mL, the cumulative amount of DEHP released at a flow rate of 90 mL/h was more significant than at 60 mL/h. These results suggest that the release of DEHP from PVC tubing is closely related to the interaction of Tween 80 and DEHP, including micelle formation, micelle impingement on the PVC tubing surface, and diffusion characteristics of DEHP and Tween 80 in the liquid medium (Fig. 14.6).

On the other hand, the cumulative elution of the system with 50% glucose solution as solvent and Tween 80 concentration of 2.0 mg/mL after 5 h of testing was about 100 μg, about 1/8 of PS and DWI.

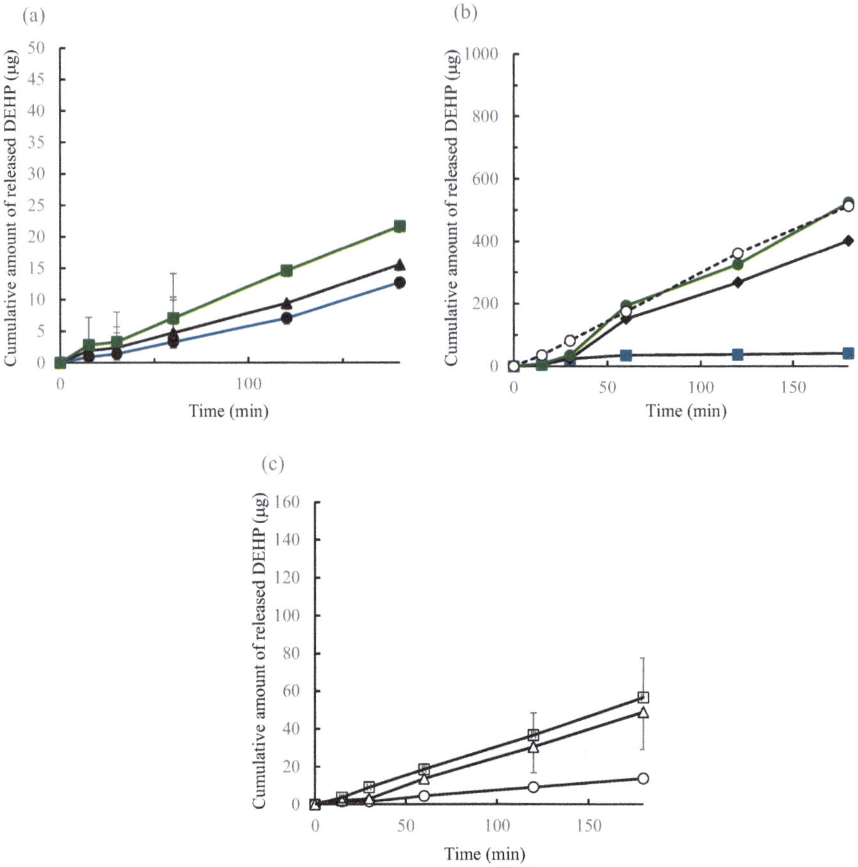


**Fig. 14.6** Release profiles of DEHP from PVC tubing. (**a**) The concentrations of Tween 80 in PS were ●, 0.01 mg/mL; ▲, 0.02 mg/mL; ■, 0.03 mg/mL. (**b**) The concentrations of Tween 80 in PS were ■, 0.03 mg/mL; ◆, 1.0 mg/mL; ●, 2.0 mg/mL; ○, 2.0 mg/mL of Tween 80 in DWI. (**c**) The concentrations of Tween 80 in 50% TZ were ○, 0.04 mg/mL; △, 0.4 mg/mL; □, 2.0 mg/mL. The data are expressed as mean ± SD (n = 3). (Cited and modified from [13])

### *14.6.2 CMC of Tween 80 in PS, DWI, and 50% Glucose*

Since Tween 80 is known to form micelles at concentrations above CMC, we investigated the relationship between the micelle formation of Tween 80 and the dissolution behavior of DEHP in PS, DWI, and 50% glucose. Therefore, we investigated the relationship between the micelle formation of Tween 80 and the elution behavior of DEHP.

First, a surface tension meter measured the CMC of Tween 80 in PS, DWI, and 30.50% glucose solution (Fig. 14.7). The results showed that the CMC in PS was lower than that in DWI. It is known that the CMC of amphiphilic surfactants

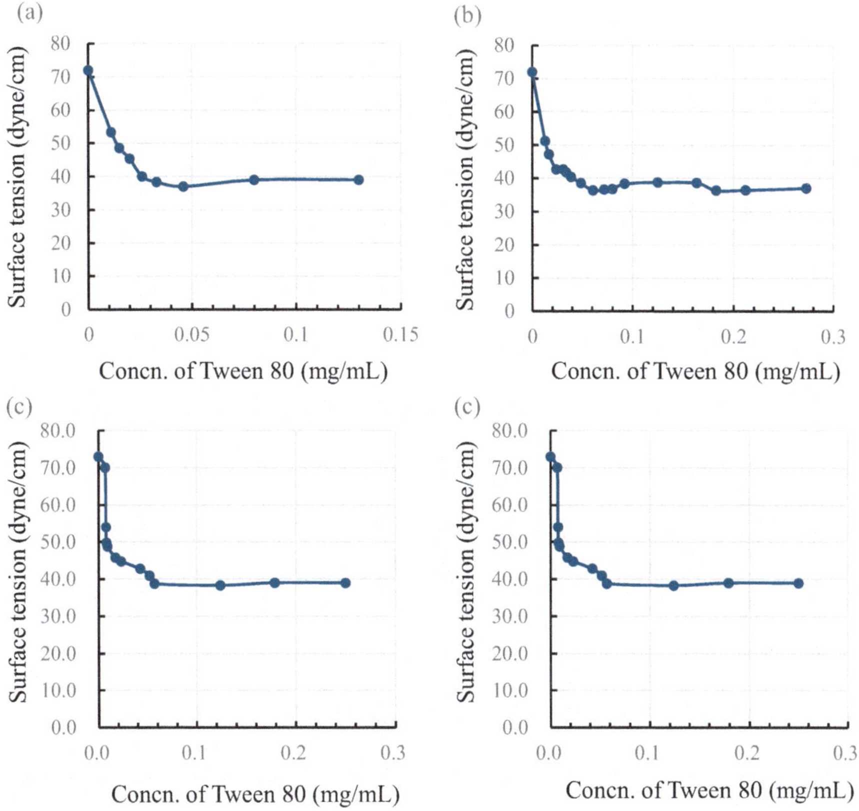


**Fig. 14.7** Surface tension of Tween 80 in various solutions. (**a**) PS, (**b**) DWI, (**c**) 30% TZ, (**d**) 50% TZ. The data are expressed as mean ± SD (n = 3). (Cited and modified from [13])

decreases with the addition of electrolytes, and in this experimental system, NaCl in PS may have affected the formation of Tween 80.

The CMC values in 30% and 50% glucose were about 3.5 and 10 times higher than those in PS, respectively. It is speculated that as the concentration of glucose increases, the number of water molecules surrounding glucose molecules also increases, and the number of water molecules surrounding Tween 80 relatively decreases, resulting in relatively lower mobility of Tween 80 molecules for micelle formation, but a more detailed investigation is needed to determine this. However, a more thorough study is required.

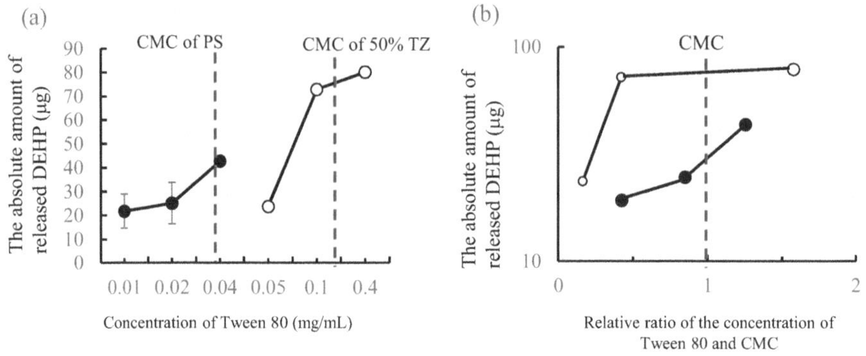


**Fig. 14.8** Release behavior of DEHP from PVC tubing into PS or 50% TZ below or above their CMCs. (**a**) Effect of Tween 80 on the release of DEHP from PVC tubing. (**b**) Relationship between relative concentration of Tween 80 and the amount of DEHP released. ●; PS, ○; 50% TZ; dotted lines show the CMC of each solution. The data are expressed as mean ± SD (n = 3). (Cited and modified from [13])

### *14.6.3 Elution Behavior of DEHP Before and After CMC of Tween 80*

To examine the elution behavior of DEHP before and after CMC of Tween 80, Tween 80 was dissolved in PS and 50% glucose, and the ratio of the concentration of Tween 80 to the CMC value of Tween 80 in each solution was plotted as a relative value. The amount of DEHP eluted at each concentration (Fig. 14.8). In PS, there was no increase in DEHP elution at concentrations lower than CMC, but an increase was observed at concentrations above CMC, suggesting that micelle formation of Tween 80 also contributes to the enhancement of DEHP elution.

Tanaka et al. eluted DEHP from PVC tubes when HCO60 was used as a surfactant, using the drop rate (*S*, mL/h), inner diameter (*R*, cm), length (*H*, cm), and concentration of HCO60 ($C_H$, mg/mL) to determine the amount of DEHP eluted in the total drop time (D_total, μg) (Eq. 14.1) [14].

$$\mathrm{D\_total} = 4.09RH\sqrt{}\left(C_{\mathrm{H}}\right)\left(T - 0.755R^2H/S\right) \qquad (14.1)$$

### *14.6.4 Study of DEHP Release Behavior with Different DEHP Content in PVC*

The release behavior of diethylhexyl phthalate (DEHP) from polyvinyl chloride (PVC) tubing, a component of the intravenous administration set, was investigated using an aqueous solution of polysorbate 80 (Tween 80). To determine the effect of

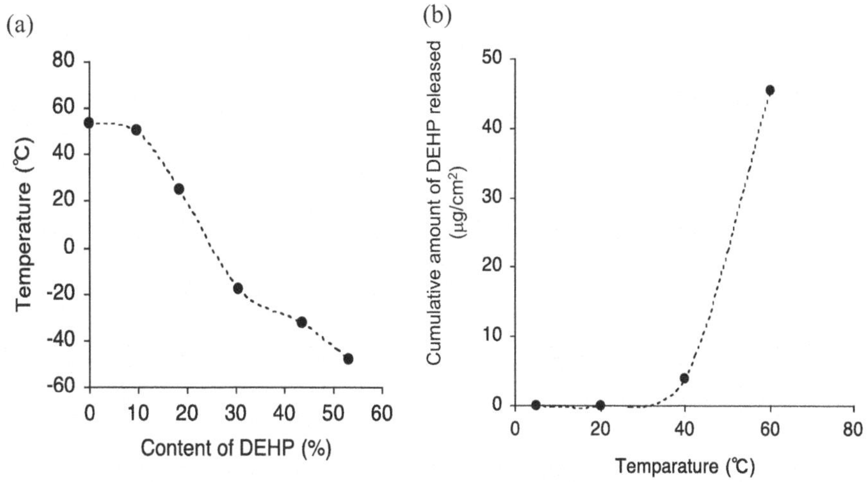


**Fig. 14.9** Effect of the amount of contained DEHP on the *Tg* value and the release behavior of DEHP. (**a**) Effect of the amount of DEHP contained in PVC films on their *Tg*s. (**b**) Release of DEHP from PVC film containing 15.2% DEHP at various temperatures. The data are expressed as mean ± SD (n = 3). (Cited and modified from [15])

temperature on the release behavior of DEHP, PVC films containing varying amounts of DEHP were mounted on cylindrical shafts and rotated at 5 and 40 °C [15].

The films were then placed on the shafts and rotated at 5 and 40 °C, respectively, to increase the amount of DEHP released. The cumulative amount of DEHP released increased with increasing temperature, with diffusion coefficients [$D \times 10^{-10}$ cm$^2$ min$^{-1}$] of 9.1 and 156.0 at 5 and 40 °C, respectively. The glass transition temperature (Tg) of the PVC film measured by differential scanning calorimeter (DSC) decreased with increasing DEHP in the PVC film. The DEHP release occurred at temperatures above Tg. These results indicate that the release of DEHP from PVC tubing is closely related to the state of the PVC and the diffusion of DEHP throughout the PVC (Fig. 14.9).

### 14.6.5 Effect of Sugar Addition on DEHP Elution Behavior

When Hanawa et al. examined the elution behavior of DEHP from PVC tubing with HCO60, they observed a decrease in the cumulative amount of DEHP eluted with increasing glucose concentration (Fig. 14.10) [16]. Furthermore, it was found that the elution behavior of DEHP differed depending on the type of sugar added (Fig. 14.11). Although the details are still under investigation, it is assumed that the difference in hydration characteristics of the surrounding water molecules concerning the added sugar affected the mobility of the HCO60 molecules.

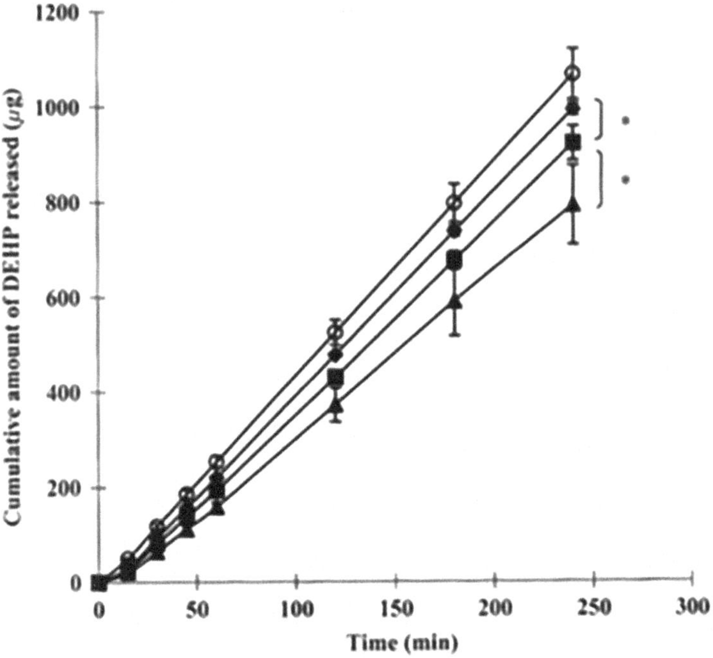


**Fig. 14.10** Effect of TZ concentration on the release of DEHP from PVC tubing. ○; DWI, ◆; 10% TZ, ●; 30% TZ, ■; 50% TZ, ▲; 70% TZ. The concentration of HCO60 was 10 mg/mL. The data are expressed as mean ± SD (n = 3). (Cited and modified from [16])

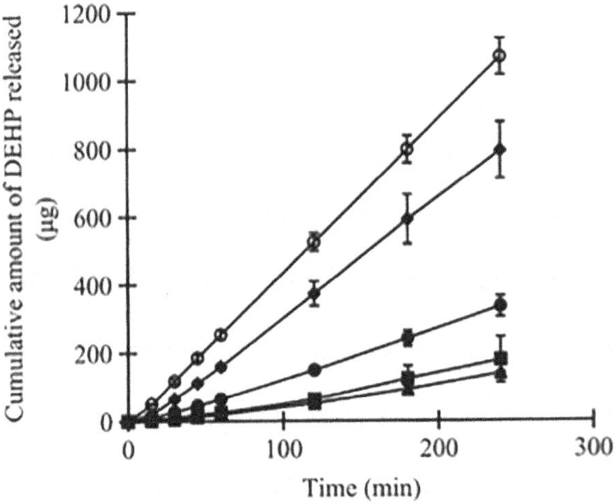


**Fig. 14.11** Release of DEHP from PVC tubing into the various sugar solutions. ○; DWI, ◆; 10% ribose, ■; 10% fructose, ▲; 10% TZ. HCO60 concentration was 10 mg/mL. The data are expressed as mean ± SD (n = 3). (Cited and modified from [16])

## 14.7 Conclusion

In this chapter, we introduced a case in which the surfactant used as a pharmaceutical additive was related to the elution of DEHP, which had been discussed as a candidate endocrine disruptor, from PVC tubes and PVC infusion bags. In recent years, many products have been using triethylhexyl trimellitate (TOTM) as a plasticizer to replace DEHP, and concerns about the elution of TOTM are being resolved. On the other hand, there is a concern that poorly water-soluble pharmaceuticals will dissolve in DEHP and TOTM, reducing the amount of the drug that should be administered.

The cases introduced in this chapter are just a tiny part of the research carried out in the past and up to the present. Still, it is necessary to continue research to ensure that drug therapy is carried out correctly, using the cases introduced here in the formulation design of future pharmaceuticals.

## References


1. Poon R, Lecavalier P, Mueller R, Valli VE, Procter BG, Chu I (1997) Subchronic oral toxicity of di-n-octyl phthalate and di(2-ethylhexyl) phthalate in the rat. Food Chem Toxicol 35:225–239
2. Yifan Hong Y, Zhou LS, Wei Y, Long C, Yan F, Huan W, Wang J, Wu Y, Shengde W, Wei G (2021) Exposure to DEHP induces testis toxicity and injury through the ROS/mTOR/NLRP3 signaling pathway in immature rats. Ecotoxicol Environ Saf 227:112889
3. https://www.mhlw.go.jp/houdou/2002/10/h1031-1a.html#12
4. Nakajima S, Kawano K, Nakazawa K, Terada K (1988) the loss of cyclosporine from solutions into the administration set. Jpn J Hosp Pharm 14:335–338
5. Kawano K, Nakazawa K, Terada K, Nakajima S (1992) The leaching of diethylhexyl phthalate from the administration set into the intravenous cyclosporine solutions. Jpn J Hosp Pharm 18:454–457
6. Kawano K, Terada K, Nakajima S (1992) The mechanism of the leaching of plasticizer from a polyvinyl chloride membrane into the solution. Jpn J Hosp Pharm 18:658–660
7. Kawano K, Terada K, Nakajima S (1992) The effect of the drip condition (drip rate, concentration) on the leaching of diethylhexyl phthalate from the administration set into the intravenous cyclosporine solutions. Jpn J Hosp Pharm 18:658–660
8. Kawano K, Matsunaga A, Nakajima S (1993) The leaching of plasticizer from a polyvinyl chloride container or a parenteral infusion set into the intravenous alprostadil solution. Jpn J Hosp Pharm 19:29–33
9. Huang Z, Fish WP, Sweeney J (2023) Leaching rate of diethylhexyl phthalate (DEHP) from PVC containers with IV administrated lipid nanoparticle formulations. J Drug Deliv Sci Tech 80:104160
10. Tanaka M, Kawano K, Hanawa T, Suzuki M, Nakajima S (2002) Leaching mechanisms of DEHP from PVC tube during the administration of enteral nutrition. J Pharm Sci Technol Jpn 62:146–152
11. Ekici M, Isci G, Bicer NC (2023) Enteral nutrition formulas: analysis of phthalate esters and deterministic assessment of health risks. J Food Compos Anal 126:105848
12. Muramatsu E, Hanawa T, Suzuki M, Tanaka M, Kawanoi K, Nakajima S (2000) Investigation of the effect of the coexistence of surfactant on the release behavior of diethylhexyl phthalate from the polyvinyl chloride tubing. Jpn J Hosp Pharm 26:471–477

13. Hanawa T, Muramatsu E, Asakawa K, Suzuki M, Tanaka M, Kawano K, Seki T, Juni K, Nakajima S (2000) Investigation of the release behavior of diethylhexyl phthalate from the polyvinyl-chloride tubing for intravenous administration. Int J Pharm 210:109–115
14. Tanaka M, Kawano K, Hanawa T, Suzuki M, Nakajima S (2001) Dissolution of DEHP from PVC administration tube. Estimation of DEHP dissolution based on HCO60 concentration and drip conditions. Jpn J Hosp Pharm 27:132–136
15. Hanawa T, Endoh N, Suzuki M, Terada K, Ohkuma M, Sakuta K, Tanaka M, Kawano K, Nakajima S, Oguchi T (2005) Release behavior of DEHP from PVC tubing used for intravenous administration and the plasticized membrane. Int J Pharmaceut 297:30–37
16. Hanawa T, Endoh N, Kazuno F, Suzuki M, Kobayashi D, Tanaka M, Kawano K, Morimoto Y, Nakajima S, Oguchi T (2003) Investigation of the release behavior of DEHP from PVC tubing for intravenous administration based on HCO60. Int J Pharm 267:141–149

# Chapter 15
# Solubility Dynamics and Hydrophobic Interactions of Acidic Analgesics and Basic Anesthetics

**Satoru Goto , Yuya Kurosawa, Chihiro Tsunoda, Kanji Hasegawa, Tomohiro Tsuchida , Kazushi Komatsu, and Hiroshi Terada**

**Abstract** This section reviews the solubility and permeability of various substances, including medicines, food additives, and pollutants, which are often hydrophobic and can cross biological membranes. The membrane permeability, influenced by factors such as lipid rafts, is generally modeled using the partition coefficient (log $P$). The solubility of active pharmaceutical ingredients (APIs), particularly acidic analgesics, is significantly influenced by basic and hydrophobic compounds. For instance, lidocaine enhances the solubility of indomethacin and other analgesics, while more hydrophobic compounds generally reduce it. The complexity of solubility behavior is highlighted, emphasizing the need to investigate solute–solvent interactions further. Dissolving APIs in their solid form involves supersaturation, which can be managed through excipient strategies to enhance bioavailability. The "spring" effect facilitates rapid dissolution followed by recrystallization, while the "parachute" effect prolongs supersaturation by preventing recrystallization. Controlling these effects is crucial for maintaining the therapeutic concentrations of APIs. Lidocaine and diltiazem affect the solubility of α-aryl-propionic and α-aryl-acetic acid-type analgesics through enthalpy–entropy compensation, whereas oxybuprocaine and propranolol do not influence oxicam-type ones. Additionally, it

Hiroshi Terada died before publication of this work was completed.

S. Goto (✉) · H. Terada (Deceased)
Faculty of Pharmaceutical Sciences, Tokyo University of Science, Katsushika, Tokyo, Japan

Center for Drug Delivery Research, Tokyo University of Science, Noda, Chiba, Japan

Division of Colloid and Interface Science (DCIS), Research Institute of Science and Technology (RIST), Tokyo University of Science, Noda, Chiba, Japan
e-mail: s.510@rs.tus.ac.jp

Y. Kurosawa · C. Tsunoda · K. Hasegawa · T. Tsuchida
Faculty of Pharmaceutical Sciences, Tokyo University of Science, Noda, Chiba, Japan

K. Komatsu
Graduate School of Integrated Arts and Sciences, Kochi University, Kochi, Japan


M. Hashizume, Y. Imura (eds.), *Colloids and Interfacial Dynamics*, Physical Chemistry in Action, https://doi.org/10.1007/978-981-96-6728-4_15

reports the protective effects of lidocaine and dibucaine against lipid peroxidation in liposomal membranes, suggesting the need for further research into drug–membrane interactions. This study highlights the importance of exploring the interaction between the structural characteristics of the drug solution and the membranous phase separation to optimize therapeutic efficacy.

**Keywords** Hydrophobic hydration · Partition coefficient · van't Hoff plot · Spring and parachute approach (SPA) · Hansen Solubility Parameter (HSP)

## 15.1 Introduction

We are exposed daily to therapeutic medicines, food flavors, beverages, supplements, cosmetics, fragrances, pesticides, and pollutants containing tiny chemicals, mid-sized organic compounds, macromolecules, minerals, and gases. When these molecules or aggregated particles are dispersed in an aqueous solution and exhibit sufficient hydrophobicity, they can cross the lipid barriers of the digestive tract, rectum, vagina, nasal cavity, lung alveoli, and skin, except for actively transported nutrients. Although washing with water prevents chemical intrusion, maintaining a sanitary environment is not always feasible. Therefore, the hydrophobic hydration and transmembrane uptake of hydrophobic chemicals represent significant challenges.

Hildebrand and Scott highlighted the critical role of solubility in the history of chemistry, noting its intriguing nature that even ancient philosophers had contemplated [1]. Their textbook explains how nineteenth-century thermodynamics defined ideal solutions according to Raoult's law (such as the benzene and toluene mixture) and classified non-ideal binary solutions into those exhibiting positive deviations (e.g., carbon disulfide and acetone) or negative deviations (e.g., chloroform and acetone). These deviations arise from the presence or absence of solute solvation by the solvent. In protic solutions, cations or positively charged solute portions interact with the heteroatoms of the solvent, while anions or negatively charged solute portions bind to the hydrogen atoms of the solvent, leading to an exothermic dispersion.

Claussen and von Stackelberg examined aqueous solutions of gas hydrates, suggesting that solutes integrate into the tetrahedral ice-like structure of water [2]. This phenomenon—called hydrophobic hydration—occurs when the solute, incapable of hydrogen bonding with water, disrupts the hydrogen-bond network of water [3]. The hydrophobic interloping materials perturb the extensively self-associated structure of water, maintained by hydrogen bonding [4]. The extent of this disruption depends significantly on the size of the hydrophobic molecule. The solubility decreases if the solvation of the protic solvent is insufficient to accommodate the size of the solute, categorizing the compounds as hydrophobic.

The biomembrane, composed of a phospholipid bilayer, acts as a lipid barrier between organisms and their environment. This bilayer self-organizes using amphiprotic lipids with twin hydrophobic aliphatic tails, each 16–24 methylene units long. For a substance to enter an organism, it must permeate the environmental fluid to the lipid phase of the biomembrane and be absorbed into the internal matrices. The distribution equilibrium between the membranous and aqueous phases is empirically similar to that between 1-octanol and aqueous phases [5–9]. Consequently, medicinal chemistry and environmental health science evaluate the hydrophobicity of the active pharmaceutical ingredients (APIs) or the hazardous/toxic substances using the partition coefficient log $P$ (or log $K_{OW}$) between 1-octanol and aqueous phases.

According to Lipinski's rule of five, which hypothesizes drug-likeness, the transmembrane ability should not exceed a log $P$ of 5 [10, 11]. This implies that the drug concentration in the 1-octanol phase should be less than $10^5$ times higher than that in the aqueous phase to facilitate appropriate drug transport across the membranes. Drugs with higher log $P$ values tend to accumulate in the membrane, hindering their delivery to the target tissues. Conversely, higher log $P$ values make the hazardous substances more persistent and resistant to excretion. These observations are based on the conventional concept of biological membranes, in which phospholipids are randomly and equally distributed to form Singer–Nicolson's fluid mosaic patterns.

Currently, membranes are considered heterogeneous, with distributions of microdomains called lipid rafts [12–14]. Lipid rafts contain three- to fivefold more cholesterol than bulk membranes and are enriched in sphingolipids and glycosphingolipids containing saturated acyl chains. Saturated lipids are ordered, tightly packed, and thermodynamically stabilized in the ordered liquid phase ($L_o$), which raises the phase transition temperature to a quasi-crystalline state. However, lipid rafts are immiscible in the continuous membrane region, which sustains the appropriate fluidity in the disordered liquid phase ($L_d$). Lipid rafts can be preserved in heterogeneous membranes because the $L_o$ phases induce transient phase separation from the continuous $L_d$ phase. Therefore, lipid rafts have been proposed to influence membrane fluidity and protein trafficking.

Although the membranous transport of solutes could not escape the influence of lipid rafts, the precise effects of lipid rafts on nutrition (which are determined by selective transport systems), APIs, and hazardous substances have yet to be fully elucidated. Under biological conditions, neither thermodynamically ideal solutions nor equilibrated traditional phospholipid bilayers exist. The hydrophobic hydration of the drug in aqueous solutions and the dispersed phase of biomembrane lipid rafts are self-organized as Prigogine's dissipative structures along a non-equilibrated process [3]. Therefore, the membrane permeability of the hydrophobic drugs may be less dependent on their 1-octanol/water partition coefficients, log $P$, as measured in the equilibrium experiments.

## 15.2 Dissolution Profiles of α-Aryl-Propionic Acid-Type and α-Aryl-Acetic Acid-Type Analgesics in the Absence and Presence of Basic Drugs

Many APIs have organic bases/cations, and their aqueous solubility is crucial for regulating their absorption, efficiency, and assessment. The basic anesthetic lidocaine (LDC) enhanced the dissolution rate of the α-aryl-acetic acid-type analgesic indomethacin (INM) in a concentration-dependent manner [15, 16]. In contrast, tetracaine (TTC) and dibucaine (DBC), which are more hydrophobic anesthetics than LDC, reduce the apparent solubility of INM [17]. The basic antihypertensive diltiazem (DTZ), with similar hydrophobicity to TTC, resulted in a reduction of INM solubility to the same extent as TTC [16].

LDC also enhanced the solubility of other α-aryl-propionic acid-type analgesics, such as ketoprofen (KTP) and naproxen (NPX), which are less hydrophobic than INM and ibuprofen (IBP) [18]. It was hypothesized that the basic LDC dissolved the acidic analgesics through their electrical attraction. However, LDC attenuated the dissolution rate of an α-aryl-propionic acid-type analgesic, ibuprofen (IBP), which has a similar hydrophobicity to INM [19]. The IBP and LDC equimolar mixture separated into the waterless and aqueous solution phases, with the aqueous solution phase clouding at 310 K. This suggests that LDC was trapped by an IBP-rich droplet, which was distributed as a dispersed phase, similar to an ionic liquid [20]. The causal relationship of basic/cationic ingredient solubility induced by acidic/anionic analgesics has been inconsistent. An intermolecular interaction between IBP and LDC did not occur in a dimethyl sulfoxide-$d_6$ solvent, excluding water or protic solvents, and the intermolecular interaction between the acidic and basic components was not sustained under electron-spray ionization mass spectrometry (ESI-MS) [20] (Fig. 15.1).

DTZ attenuated the solubilities of IBP, KTP, NPX, aspirin (ASP), INM, and diclofenac (DCF) at temperatures of 277, 283, 298, and 313 K [24]. The solubilities of these neat analgesics decreased with increasing temperatures, indicating an endothermic dissolving process. These APIs dissolve in aqueous solutions via hydrophobic hydration. Adding DTZ further reduced their solubilities, with the effect being approximately half for KTP and NPX, one-third for ASP, and approximately one-tenth for INM and DCF. The decrease in the IBP solubility with the addition of DTZ was reduced at high temperatures, shifting the dissolution process to exothermic. In contrast, LDC enhanced or reduced the solubility, depending on the analgesic and temperature. Hansch–Fujita's analysis using the hydrophobicity, electronic, and steric parameters of analgesics did not provide efficient multiple regression lines, suggesting that the structural causality of solubility changes due to DTZ was not established.

The slopes and intercepts of the van't Hoff plot correspond to the enthalpic and entropic changes during the dissolution process [24]. The intercept-slope diagram shows a linear correlation between analgesics, DTZ, and LDC. Although the plots for these APIs in their mixtures were scattered in the upper-right and lower-left

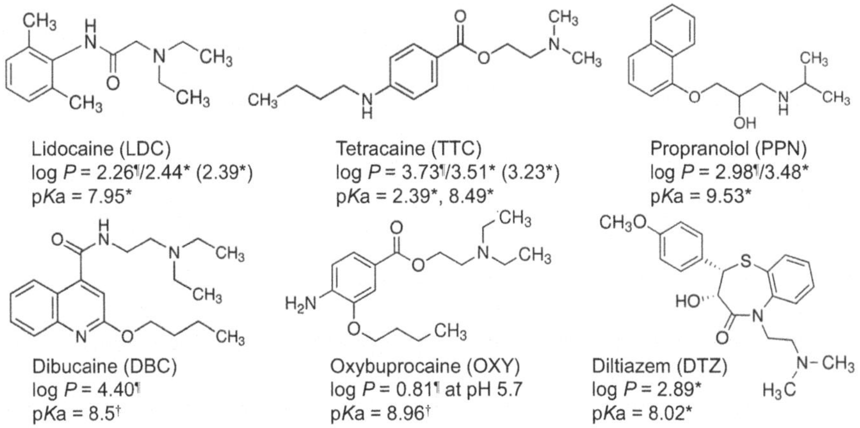


**Fig. 15.1** Chemical structures of the local anesthetics and related basic drugs. *log $P$ and p$K_a$ values were retrieved from Avdeef's Handbook [21–23] and the log $P$ values in parenthesis express the experimental liposome-water partition coefficients. ¶Hansch et al. Handbook [6–9] †Database

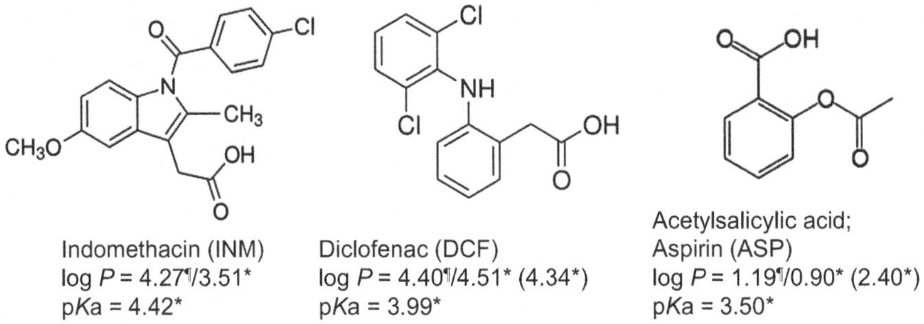


**Fig. 15.2** Chemical structures of the α-aryl-acetic acid type analgesics and aspirin

regions, all plots were adjusted to the correlation line. Their solubility is determined by the balance between the dissolution enthalpy and entropy, with the solute–solvent interaction energy playing a pivotal role. Thus, the entropy–enthalpy compensation governs the solubility of both neat and mixed ingredients (Fig. 15.2).

These results indicate the independence of solubility from the formation of a generally imagined ionic complex or the direct electrical attraction between acidic and basic components [24]. The hydrotropic effects (to improve/impair the solubility) of LDC, DTZ, carnosine (a peptide supplement), and aspartame (an artificial sweetener) on the INM solubility were consistent, regardless of the polymorphic γ- and α-forms of INM [25]. It was certified that the hydrotropic ingredients had no direct interaction with INM. The solubilities of these ingredients are determined by the balance of their solute–solvent interactions (enthalpic contributions) and concentrations (entropic contributions), maintaining a metastable state of the solution

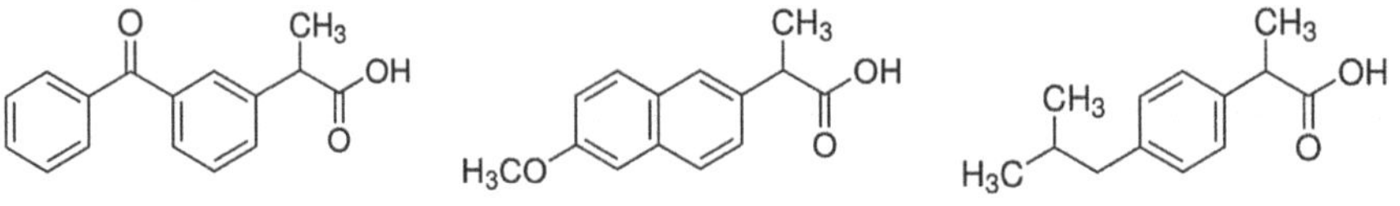

Ketoprofen (KTP)
log *P* = 3.12¶/3.16*
p*K*a = 3.98*

Naproxen (NPX)
log *P* = 3.34¶/3.24*
p*K*a = 4.18*

Ibuprofen (IBP)
log *P* = 3.50¶/4.13* (3.87*)
p*K*a = 4.45*

**Fig. 15.3** Chemical structures of the α-aryl-propionic acid type analgesics

system. This steady-state condition indicates an intermediate and transient energy state, with no ingredients isolated as the most stable or quasi-stable states.

Basic antacids, such as cimetidine (CIM) and famotidine (FAM), have equivalent hydrophobicity at neutral pH. The solubility of CIM is high (60 mM in 100 mM phosphate/NaOH buffer, pH 6.5, at 298 K), whereas FAM exhibits lower solubility (10 mM in the buffer). These basic APIs enhanced the dissolution curve of INM, whereas the solubility of the acidic analgesic drug DCF was amplified by CIM and delayed by FAM [26–28]. The planar amines in 2-diaminomethylimino-1,3-thiazolyl groups of FAM were suggested to interact with the carboxylic acid moiety of DCF, forming an ion-paired complex beyond Bjerrum's threshold distance [26].

Enthalpy–entropy compensation could not explain this solute–solute complexation or the formation of supramolecules. While the carboxyl group can take a proximal approach to planar nitrogen-containing heterocycles or guanidyl groups, a pair of acids and bases that do not cause precipitation but enhance their aqueous solubility would not form a tightly bound complex or supramolecule. The reason for the insufficient interaction between INM and FAM remains unclear, but no evidence of DCF specificity compared to INM was verified. The relationship between DCF and INM for these antacid-related compounds is under further investigation (Fig. 15.3).

## 15.3 Spring and Parachute Effect on the Supersaturation of Piroxicam

The dissolution curves of the drugs from stable crystals were used to determine the saturation profiles. Schefter and Higuchi demonstrated that the concentration profile of theophylline (TPH) initially increased along an incremental curve, but subsequently decreased exponentially to a stable level [29]. This behavior indicated a supersaturation phase during the dissolution process of its anhydrous crystal. Such crystals dissolve rapidly, reaching high concentrations; subsequently, the solute deposits onto more stable hydrate crystals. The temporarily elevated concentration gradually equilibrates with the stable monohydrate crystal saturation level.

Since the stable crystal seeds in solution are promptly deposited, their existence accelerates equilibration, reaching saturation in the liquid phase. In contrast, the solid dispersions of microcrystals/micro-amorphous (suspended or emulsified particles in solid) or the solid solution (molecular dispersion in solid) of APIs lack stable crystal seeds. These systems interfere with deposition, achieving higher dissolution rates than their crystalline forms. The dispersed ingredients have difficulty in escaping from solutions owing to deposition. This allows solid dispersions and solutions to extend the duration of supersaturation.

APIs require appropriate concentrations in the blood to ensure therapeutic efficacy and avoid adverse effects. Therefore, regulating the supersaturation concentration and its persistence period is essential for maintaining a suitable concentration of APIs. This requirement has driven the exploration of various materials, such as hydrophilic polymers, as supporting agents for solid dispersions and solutions. Schefter and Higuchi presented the classic example of a pentanol-solvated form of succinyl sulfathiazole, which exhibited supersaturation at twice the peak height of its anhydrous form [29].

Fujita et al. observed supersaturation of the oxicam-type anesthetic piroxicam (PRX), characterized by a 1,2-benzothiazine-1,1-dioxide scaffold containing a vinyllogue carboxylic acid [30, 31]. In contrast, this dissolving time course was not observed in meloxicam (MLX) with a 5-methyl-2-thiazolyl moiety instead of a 2-pyridinyl moiety present in PRX. PRX forms stable monohydrate crystals (PRX-MH) and quasi-stable anhydrous crystals (PRX-AH). The supersaturation of PRX-AH can reach three times the concentration equilibrated during PRX-MH recrystallization.

Adding a local anesthetic, oxybuprocaine (OXY), to the solution maintained the plateau concentration of PRX-AH supersaturation for 150 h but reduced the efficiency of PRX-MH saturation. The more hydrophobic DBC resulted in a lower plateau concentration of PRX-AH supersaturation compared to OXY. In contrast, for LDC and TTC, the concentration peaks of the PRX concentration were greater than that of the OXY solution but had a shorter supersaturation duration (5 h). The antiarrhythmic and anti-migraine drug propranolol (PPN) demonstrated a more potent effect on PRX-AH supersaturation [30, 31] (Fig. 15.4).

The most stable API crystals (e.g., hydrate crystals) equilibrate at particular concentrations. In contrast, thermodynamically quasi-stable crystals, such as the anhydrous crystals of TPH or PRX and the pentanol-solvated form of succinyl sulfathiazole, dissolve rapidly and generate supersaturation levels exceeding the equilibrium concentration of the most stable crystals. This phenomenon, where the API solids accelerate dissolution, is referred to as the spring effect. While this effect is attributed to the formation of transiently unstable solid states (e.g., anhydrous crystals or unstable solvates), its entirely rationalized structural requirements for compounds such as PRX and TPH remain unidentified compared to structurally related APIs such as meloxicam and caffeine. Nevertheless, co-crystals and amorphous APIs with specific additives can enhance this effect.

Conversely, interference with deposition by solid dispersions and the solid solutions prolongs the supersaturation phase of the solution; a phenomenon termed the parachute effect. By combining accelerated dissolution with prolonged supersaturation, the concentration of APIs can be maintained at therapeutically effective levels for an extended period. For medication administered to the gastrointestinal tract or

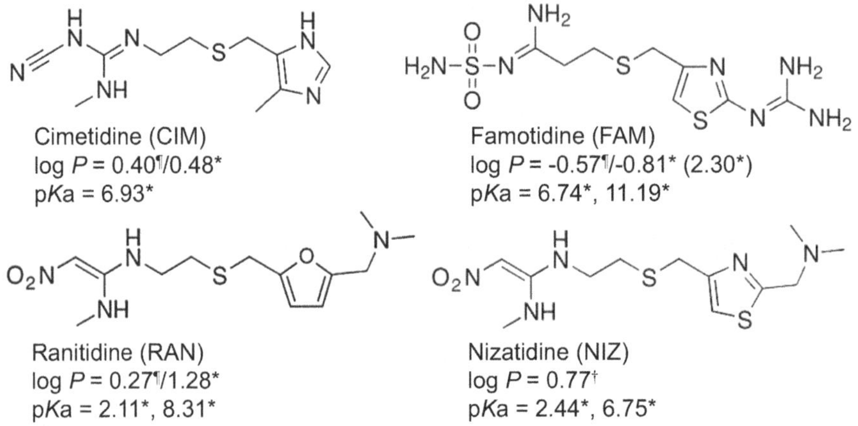


**Fig. 15.4** Chemical structures of the H2-receptor antagonistic antacids

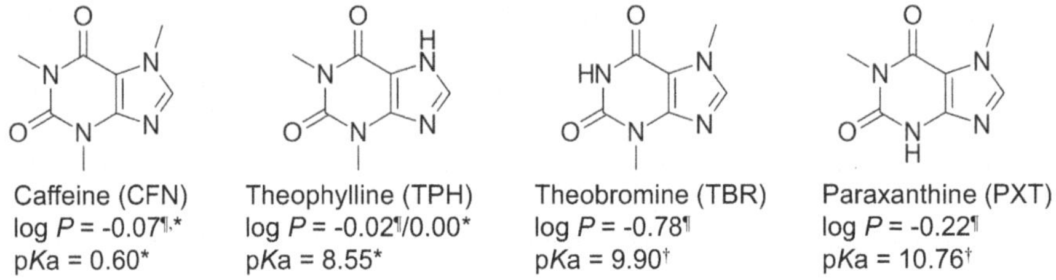


**Fig. 15.5** Chemical structures of the di- or trimethylated xanthine derivatives

topically, a combination of spring and parachute approaches (SPAs) is valuable for maintaining a therapeutically appropriate concentration of APIs in the blood. However, preventing deposition to achieve the parachute effect inherently limits the rapid dissolution associated with the spring effect. In particular, co-crystals or co-amorphous APIs with additives could confer a robust parachute effect, but this is accompanied by gradual dissolution and deposition (Fig. 15.5).

2-Hydroxypropyl-β-cyclodextrin (HP-β-CD) enhanced the supersaturation of PRX, INM, and DCF [32, 33]. The complexation of APIs with HP-β-CD in the solid phase facilitates API release while inhibiting the solute's reprecipitation into the solid phase. These can be emulated using the kinetics of consecutive opposite reactions. This indicates that the supersaturation peak is transient and unsustainable in a steady state. Cyclodextrin inclusion complexes, solid dispersions, and solid solutions can modify the dissolution profiles of solutes to induce rapid dissolution, supersaturation, or delayed deposition. Previous studies reported that a supersaturation of the HP-β-CD complex supports SPA technology [32, 33]. However, this supersaturation effect alone is insufficient to generate the parachute effect, which can be better characterized by the "metastable zone" observed in the time evolution profile. Implementing rapid dissolution, supersaturation, and precipitation delay simultaneously is essential for achieving the combined SPA: i.e., spring (rapid dissolution up to supersaturation) and parachute (sustaining the steady state).

The spring and subsequent parachute effects require certain strategies to sustain supersaturation, with a time-limiting system delaying the start of recrystallization. OXY and PPN derived the key to expanding PRX supersaturation without delaying rapid dissolution. Their parachute effects persisted for a longer period than those of the LDC and TTC. Interestingly, PRX solubility was not influenced by OXY, PPN, LDC, or TTC, with enthalpy–entropy compensation, indicating that this mechanism does not apply. Introducing the most stable PRX-MH during the parachute effect terminated supersaturation. The regulation of PRX supersaturation via the SPAs is governed by the relative proportions of PRX-AH and PRX-MH crystals, highlighting the importance of balancing these crystalline states.

Ostwald ripening was observed in the inhomogeneous structure of dispersed-phase particles over time, where small particles dissolve and redeposit onto larger particles. This phenomenon maintains supersaturation if, and only if, quasi-stable crystals are present in greater quantities than stable crystals. Under this steady-state condition, solute molecules preferentially deposit onto quasi-stable crystals rather than stable ones, sustaining the parachute effect. Conversely, gradually growing stable crystals disrupt the predominance of the quasi-stable crystal growth, terminating the parachute effect. Fujita stated that the SPA processes are phenomenological processes regulated by the steady state between the deposition of anhydrous and monohydrate crystals [30, 31] (Fig. 15.6).

According to van't Hoff plots, the parachute generators did not thermodynamically influence PRX solubility. Neither the dissolving enthalpy nor the entropy of PRX depended on the presence or absence of parachute generators, indicating that the changes in the solubility of PRX during the parachute process were not driven by thermodynamic inevitability. Evidently, the parachute effect was attributed to interfacial rearrangements between the solution and crystals acting as depositing reservoirs. This led to the conclusion that the parachute generators regulated the balance between exposing quasi-stable crystals and concealing stable crystals, thereby extending the steady state [30, 31].

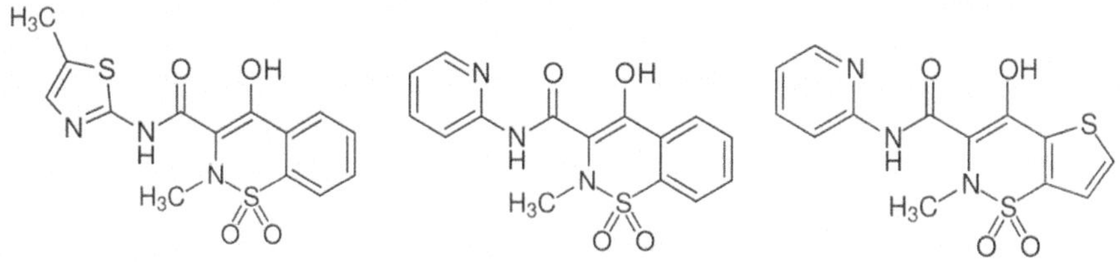

| Meloxicam (MLX) | Piroxicam (PRX) | Tenoxicam (TNX) |
| --- | --- | --- |
| log *P* = 3.01¶/3.43* | log *P* = 3.06¶/1.98* | log *P* = 3.5† |
| p*K*a = 3.43* | p*K*a = 2.33*, 5.07* | p*K*a = 2.21†, 0.26† |

**Fig. 15.6** Chemical structures of the piroxicam-type analgesics

## 15.4 Hydrophobic Hydration of Acidic Analgesics

LDC and DTZ either increased or decreased the solubility of the α-aryl-propionic acid-type analgesics (IBP, KTP, and NPX), α-aryl-acetic acid-type analgesics (INM and DCF), and benzoic acid-type analgesic (ASP), depending on the temperature, exhibiting enthalpy–entropy compensation. In contrast, OXY and PPN did not influence the solubility of the oxicam-type anesthetic PRX. Given that aliphatic amines, such as LDC, DTZ, OXY, and PPN, lack distinguishable physicochemical properties in solution (they have no naked planar amines capable of forming Bjerrum's ion pairs), this suggests that the thermodynamic properties of carboxylic acids differ fundamentally from those of oxicam derivatives.

Experimental observations on the solubility of acidic analgesics in the absence or presence of basic drugs require further clarification. Mixing the neat powders of IBP and LDC resulted in a reduction of IBP solubility by LDC, depending on the LDC molarity [18–20]. However, when solo or mixed crystals of IBP and LDC were preliminarily melted to eliminate structural differences (i.e., rendered amorphous), LDC enhanced the solubility of IBP [24]. On the van't Hoff plot, the negative slope of the IBP solubility became less steep in the presence of LDC, indicating the attenuation of the exothermic dissolving enthalpy of IBP. In the case of crystals, the proportion of IBP and LDC crystal amounts (as determined by their surface areas) likely regulates solute release from crystals and deposition onto crystals, similar to the behavior observed in PRX supersaturation.

The mechanical factors influencing API solubility are the solvent affinity for the solute, solution pH (affecting acid dissociation of the electrolyte solute), ionic strength (Debye shielding distance), solvent viscosity (hydrodynamic resistance to gravitational sedimentation and Brownian motion), surface tension (solute adsorption onto the interface), and ability of the solute to be released from the solid phase and redeposit onto it. Solid dispersion, solid solution, and co-crystal/co-amorphous APIs, in which APIs are blended with effective ingredients or polymers, enhance the release of APIs and reduce their redeposition onto solids.

In the experiments with the mixed powder, IBP solubility decreased in the phosphate buffers depending on the LDC concentration. However, IBP solubility remained unchanged in Good's buffer, which contains organic components instead of oxo acids [20]. The solubility of IBP, KTP, DCF, INM, and MLX increased with rising concentration of the phosphate anion [20]. The resulting solubility–concentration diagrams followed Higuchi–Connors' Bs-type curve [29]. The Bs-type curve consists of a saturating curve at low phosphate concentration and an attenuated precipitation curve at high phosphate concentration. The Bs-type curve suggests that phosphate anions form a water–phosphate hydrogen-bonding network cage, similar to cyclodextrins, to form an inclusion complex with acidic analgesics.

Oxo acids are anions that enhance the dielectric interactions among pure protic solvent molecules. Assuming that the hydrophobic APIs dissolve via hydrophobic hydration, the phosphate concentration would reinforce the hydrogen-bonding cage formed by the solvent molecules encapsulating the APIs. This effect continues until the phosphate anions saturate the solvent-accessible area of the API molecule. However, when phosphate concentration exceeds this saturation point, the continuous phase of the solvent may become condensed, leading to the exclusion of the dispersed solute phase from the solution. This explains the characteristic Bs-type phase solubility diagram observed.

In contrast, high-molarity salts and short-chain alcohols induce solute dehydration. Methanol disrupts both the hydrophilic and hydrophobic hydration of organic solutes. As a result, the acid-dissociative moiety of the acidic ingredient forms fewer hydrogen bonds. This induces the carboxylic acid COOH groups to form an intermolecular association between the two acid ingredient molecules. Consequently, the acidity of the –COOH group decreases, leading to an increase in the apparent acid-dissociation constant ($pK_a$).

## 15.5 Hydrophobic Hydration and Hansen Solubility Parameter

Kitaura and Morokuma's energy decomposition analysis (KM-EDA) describes intermolecular interactions as a combination of four key components: electrostatic interaction energy (ES), exchange repulsion energy (EX), polarization energy, and charge-transfer energy (CT), alongside a highly ordered coupling term (MIX) [34]. In the updated fragment molecular orbital (FMO) approach, ES, EX, CT, and the London dispersion energy (DI) are incorporated as inter-fragment interaction energies [35].

The development of molecular mechanics showcases a reverse progression. Allinger's MM2 introduced empirical force fields consisting of bond stretching, valence angle bending, bond torsion modes, and non-bonding atom pair Coulomb interaction modeled by Lenard–Jones' 12–6 potential, aiming to reproduce the infrared spectra [36, 37]. Later advancements, such as Kollman's AMBER [38–40] and Karplus's CHARMM [41, 42], evaluated the cross terms to adapt these models for biomacromolecules. Subsequently, Class II force field programs employed higher-order functions to resolve multibody problems and enhance the accuracy of the cross terms [43, 44]. These refinements suggest that the higher-order coupling term should not be disregarded, even if those terms are of derivative nature.

Hildebrand and Scott investigated the solubility parameter $\delta$ within solution thermodynamics, defining it as the square root of the cohesive energy density ($[J/m^3]^{1/2} = Pa^{1/2}$) [45–47]. According to their model, the solution's free energy change is equivalent to the volume term (product of volume fractions of solute and solvent and the total mixture volume) multiplied by the square of the difference between the solubility parameters of the solute and solvent. Hansen proposed that $\delta^2$ could be expressed as the linear combination of the squares of $\delta_D$, $\delta_P$, and $\delta_H$, which correspond to the London dispersion forces, the polar cohesive forces, and the hydrogen bonding interaction [47]. Assuming that the Hansen solubility parameters (HSPs) are explained by the KM-EDA energy compositions without quantitative verification, the London dispersion interaction and hydrogen bonding are assigned to the DI and CT terms, respectively. The polar cohesive interaction is broadly attributed to the ES and EX terms [34]. The London dispersion interaction refers to hydrophobic interactions in an organic solvent, whereas the polar cohesive interaction and hydrogen bonding represent the hydrophilic interactions in an aqueous or protic solution.

Based on the concept that materials dissolve more readily in materials with similar properties, Hansen's approach models a material within a sphere in three-dimensional space, where the coordinates represent its HSPs [47]. These spheres group materials with similar affinities, and the proximity of the solute and solvent spheres determines their solubility. Hansen's approximation favors Raoult's ideal binary solutions exhibiting positive deviation, suggesting its applicability in predicting the phase separation between aqueous and hydrophobic materials. In contrast, for solutions with negative deviations from Raoult's ideal model, the strong affinity between the binary components increases the vaporization pressure.

In aqueous solutions, intermolecular interactions between the solute and solvent are categorized as hydration and hydrophobic hydration. This categorization highlights a limitation in Hansen's theory, which does not distinguish between polar cohesive interaction mechanisms. In this framework, the polar cohesive parameter $\delta_H$ is associated with the quantum ES and EX terms. The electrostatic interaction ES aligns with Keesom's orientation force in the classical valence bond theory [47]. Keesom's orientation attraction is observed between permanent dipoles. Meanwhile, the EX-term is related to Debye's induction force between the permanent and induced dipoles. Although Debye's induction force is significantly weaker than Keesom's orientation force, its presence is notable. However, these distinctions are not the focus of current challenges in the application of Hansen's theory.

Hydrophobic hydration, characterized by the absence of direct interactions between the solute and solvent, cannot be predicted by HSPs. Many APIs are hydrophobic; therefore, their dissolution mechanisms involve the formation of hydrophobic hydration rather than polar and cohesive interactions. Hydrogen-bonded cages for hydrophobic hydration consist of protic solvent molecules. A single tetrahedral water unit can accommodate single or twin atomic gases, exemplified by a methane hydrate comprising one methane molecule enclosed by four water molecules arranged in a tetrahedral configuration. These cages are geometrically constructed as a combination of tetrahedral units of water molecules. This effect corresponds to the configuration of the water molecules, as evaluated by statistical thermodynamics based on molecular dynamics simulations. This suggests that the entropy of dissolution describes the thermodynamic contribution to the formation of the water cage, but not the enthalpy of dissolution corresponding to Keesom's orientation and Debye's induction forces. This idea remains speculative and needs further experimental validation to confirm its accuracy.

The authors recommend replacing Hansen's three-dimensional spheres with hyper-dimensional convex hulls (Delaunay hyper-dimensional polygons) for a more accurate application of Hansen's solubility approach to all polar and cohesive solutes. This recommendation is based on two key reasons. First, Hansen's spheres in three-dimensional space often include void spaces that do not contain material. As a result, the proximity between the spheres was quantified based on inaccurate representative points (center of gravity). For example, if the material plots in the HSP space were dispersed to form a bending spindle (often observed), the center of gravity would be situated in a void space, resulting in an inaccurate representation of the material's properties. Second, the model's uneven distribution of the points further complicates the calculation of distances. The authors propose using various procedures to estimate the distance or path length in non-hierarchical cluster analysis instead of the gravitational centers of the spheres.

The primary issue arises from the limitations in the dimensionality of the HSP space, comprising the $\delta_D$, $\delta_P$, and $\delta_H$ parameters. The authors speculate that these parameters were chosen based on the object's visual distinction and computational convenience at the time. Today, hyperdimensional objects have become available for recognizing and utilizing fourth-dimensional (or further hyperdimensional) descriptors, explaining the hydrophobic hydration contribution of the solute. This could be characterized by the entropic density multiplied by the temperature. Despite these possibilities, concrete methods and candidates for incorporating such complex factors are still lacking, requiring further interdisciplinary studies that extend beyond traditional statistical physics or thermodynamics.

The authors examined the behavior of α-aryl-propionic acid-type (IBP, KTP, and NPX) and α-aryl-acetic acid-type analgesics (INM and DCF), in both monomeric and dimeric forms. In diluted solutions, these analgesics exist as monomers, but at higher concentrations, they form dimers through hydrogen bonding. The dimeric form dissolves via hydrophobic hydration because it lacks polar substituents. The partition coefficients of these acidic analgesics were observed to vary significantly, with lower and higher values for individual references. Assuming that the sample drug is hydrophobic, with a log $P$ value of 3–4, its aqueous concentration is about $10^3$–$10^4$ times greater than in octanol. Early experiments with low instrumental sensitivity yielded higher log $P$ values, whereas improved sensitivity in modern experiments produced lower values. Presumably, the aqueous concentration was relatively higher under the old experimental conditions than in modern experiments. At lower concentrations, monomeric analgesics have lower partition coefficients, while at higher concentrations, the dimeric forms, stabilized by intermolecular hydrogen bonds, exhibit higher partition coefficients. Interestingly, this effect was not observed with the oxicam-type analgesics PRX or MLX, which do not form intermolecular hydrogen bonds. This could explain the lack of variation in the partition coefficients of oxicam-type analgesics.

## 15.6 Solution Structures of Acidic Solutes

The hydrophobic hydration of dimeric carboxylic acids involves the transition between monomeric and dimeric molecular states, resulting in various dissolution states, which allows the rearrangement of the dissolution entropy. However, the dissolution enthalpy is impacted by hydrogen bonding interactions between the –COOH group and water molecules or intermolecular bonding between the –COOH groups. This leads to an enthalpy–entropy compensation relationship of analgesics in the presence or absence of aliphatic amine drugs.

FAM contains a heterocyclic planar nitrogen atom, while aliphatic amines and ammoniums contain positive poles in the dipole or the positive charges, with hydrogen atoms or alkyl groups surrounding them. In addition, carboxylic acids expose oxygen atoms with negative charges. The positive and negative charges on organic compounds create asymmetry, with the wrapped positive charge and an exposed negative charge. This prevents the carboxyl group from approaching the aliphatic amine beyond the Bjerrum threshold.

In contrast, the heterocyclic planar nitrogen atom allows the carboxyl group to approach, as there are no obstacles along the normal axis perpendicular to the heterocyclic plane. This interaction delays the DCF dissolution rate in the presence of FAM [26]. However, no such effect was observed for CIM. Heterocyclic and planar moieties of FAM consist of 2-guanidyl-1,3-thiazole, whereas those of CIM contain 2-methyl-1*H*-imidazole and guanidyl groups. This issue remains unresolved. In other experiments, DCF and INM delayed the FAM dissolution [28]. While imidazole and arginine enhanced the solubilities of DCF and INM, histidine showed no effect under the same conditions [28]. The non-charged bases, TPH and caffeine, contain heterocyclic nitrogen atoms but enhance the solubility of DCF [48, 49]. Preliminary experiments demonstrated that ranitidine (RAN) and nizatidine (NIZ) had a lower reduction of DCF solubility. The approach of DCF to the planar nitrogen atom might require specific electronic density orbital characteristics, which warrants further investigation into their structure–property relationship.

The oxicam-type analgesics PRX and MLX form tautomeric intramolecular hydrogen bonds between the vinylogous carboxylic oxygen and its hydroxyl hydrogen, as well as between the hydroxyl oxygen and amido hydrogen. These intramolecular bonds are more stable compared to intermolecular hydrogen bonding. However, their structural diversity is more limited than that of carboxylic acid-type analgesics, which can form dimers, polymers, and various rearrangements of water's hydrogen-bonding cages during hydrophobic hydration. This could explain why enthalpy–entropy compensation is not observed for oxicams, even after the addition of basic drugs.

Pain control is an essential therapeutic category, but the reliance solely on opioids and steroids is limited due to their severe adverse effects. Non-narcotic analgesics, including those administered for daily use, are among the most widely used polypharmaceutical medicines. These analgesics offer a range of pharmacological efficacies and dosage forms. However, individual analgesics have a particular affinity for combined APIs, vitamins, supplements, or dietary flavors. Therefore, careful consideration must be given to the types of analgesics administered, such as carboxylic acids and oxicams. However, these issues are not generally recognized. Carboxylic acid-based analgesics, including DCF, INM, felbinac, IBP, NPX, KTP, loxoprofen, and flurbiprofen, exhibit structural variation in solution. Depending on their concentration, these compounds can exist as monomers, dimers, or polymeric aggregations, leading to diverse thermodynamic states. Dimers and polymeric aggregates exhibit increased hydrophobicity due to intermolecular hydrogen bonding, influencing their bio-membranous permeability. Consequently, these changes can affect the bioavailability of the administered medicine, impacting both medication compliance and therapeutic effectiveness. A decline in these parameters may affect a patient's quality of life.

Pharmaceutical additives, such as polymers, quasi-stable co-crystals/co-amorphous materials, solubilizers, and cyclodextrins, can enhance the solubility and dissolution rates of poorly soluble APIs [32, 33, 50–55]. These additives also influence the partition coefficients and membrane permeability of APIs.

Horizumi et al. showed that local anesthetics LDC and DBC inhibited lipid peroxidation in egg yolk phosphatidylcholine liposomes induced by Fenton's reaction despite exhibiting no radical scavenging activity for 2,2-diphenyl-1-picrylhydrazyl (DPPH) [56–61]. The hydroxyl radical generated in the reaction attacked the polyunsaturated acyl chains in the liposomal membrane, forming thiobarbiturate reactive substances (TBARSs). The radical scavenger Trolox® (6-methoxy-2,5,7,8-tetramethylchromane-2-carboxylic acid) inhibited TBARS production, whereas LDC and DBC stabilized membrane perturbation to prevent a radical attack. Cholesterol, inserted between the phospholipids of the lipid bilayer membrane, stabilizes membranous structures and supports the formation of the lipid raft phase, as described in Sect. 15.1. LDC and DBC compete with cholesterol to inhibit TBARS production, probably through a similar mechanism. Their assay reported the membranous protective effects of the drugs.

The hydrophobic hydration of drugs in aqueous solutions and their interaction with the dispersed phase of biomembrane lipid rafts represent self-organizing behaviors consistent with Prigogine's concept of dissipative structures under non-equilibrated processes [3]. This section underscores the importance of investigating the interaction between the drug's solution-phase structure and the phase separation dynamics within biomembranes.

## References


1. Hildebrand JH, Scott RL (1924) The solubility of nonelectrolytes: monograph series, American Chemical Society No. 17. Reinhold Publishing Co, New York
2. Moelwyn-Hughes EA (1961) Physical chemistry, 2nd edn. Pergamon Press, Oxford
3. Kondepudi D, Prigogine I (2015) Modern thermodynamics: from heat engines to dissipative structures. John Wiley & Sons, New York. https://doi.org/10.1002/9781118698723
4. Yaminsky VV, Vogler EA (2001) Hydrophobic hydration. Curr Opin Colloid Interface Sci 6:342–349. https://doi.org/10.1016/S1359-0294(01)00104-2
5. Rekker RF (1977) The hydrophobic fragmental constant, its derivation and application a means of characterizing membrane systems. Elsevier, Amsterdam
6. Hansch C, Leo A (1979) Substituent constants for correlation analysis in chemistry and biology. John Wiley & Sons, New York
7. Hansch C, Leo A, Taft RW (1991) A survey of Hammet substituent constants and resonance and field parameters. Chem Rev 91:165–195. https://doi.org/10.1021/cr00002a004

8. Hansch C, Leo AL, Hoekman D (1995) Exploring QSAR, hydrophobic, electronic, and steric constants. American Chemical Society, Wasghington, DC
9. Hansch C, Leo AL, Hoekman D (1995) Exploring QSAR, fundamentals and applications in chemistry and biology. American Chemical Society, Wasghington, DC
10. Lipinski CA, Lombardo F, Dominy BW, Feeney PJ (1997) Experimental and computational approaches to estimate solubility and permeability in drug discovery and development settings. Adv Drug Deliv Rev 23:3. https://doi.org/10.1016/S0169-409X(96)00423-1
11. Lipinski CA (2004) Lead- and drug-like compounds: the rule-of-five revolution. Drug Discov Today Technol 1:337. https://doi.org/10.1016/j.ddtec.2004.11.007
12. Thomas S, Preda-Pais A, Casares S, Brumeanu T-D (2004) Analysis of lipid rafts in T cells. Mol Immunol 41(4):399–409. https://doi.org/10.1016/j.molimm.2004.03.022
13. Thomas S, Kumar SR, Brumeanu T-D (2004) Role of lipid rafts in T cells. Arch Immunol Ther Exp 52:215–224
14. Korade Z, Kenworthy AK (2008) Lipid rafts, cholesterol, and the brain. Neuropharmacology 55:1265–1273. https://doi.org/10.1016/j.neuropharm.2008.02.019
15. Tateuchi R, Sagawa N, Shimada Y, Goto S (2015) Enhancement of the 1-octanol/water partition coefficient of the anti-inflammatory indomethacin in the presence of lidocaine and other local anesthetics. J Phys Chem B 119:9868–9873. https://doi.org/10.1021/acs.jpcb.5b03984
16. Shimada Y, Tateuchi R, Chatani H, Goto S (2018) Mechanisms underlying changes in indomethacin solubility with local anesthetics and related basic additives. J Mol Struct 1155:165–170. https://doi.org/10.1016/j.molstruc.2017.10.101
17. Shimada Y, Komaki H, Hirai A, Goto S, Hashimoto Y, Uchiro H, Terada H (2018) Decarboxylation of indomethacin induced by heat treatment. Int J Pharm 545:51–56. https://doi.org/10.1016/j.ijpharm.2018.04.022
18. Hasegawa K, Ogawa S, Chatani H, Kataoka H, Tsuchida T, Goto S (2024) Thermodynamic and kinetic analysis of the melting process of S-ketoprofen and lidocaine mixtures. RSC Pharm 1:536–547. https://doi.org/10.1039/D4PM00039K
19. Kataoka H, Sakaki Y, Komatsu K, Shimada Y, Goto S (2017) The melting process of the peritectic mixture of lidocaine and ibuprofen was interpreted using the site percolation theory model. J Pharm Sci 106:3016–3021. https://doi.org/10.1016/j.xphs.2017.04.010
20. Chatani H, Goto S, Kataoka H, Fujita M, Otsuka Y, Shimada Y, Terada H (2019) Effects of phosphate on drug solubility behavior of mixture ibuprofen and lidocaine. Chem Phys 525:110415. https://doi.org/10.1016/j.chemphys.2019.110415
21. Avdeef A (2003) Absorption and drug development, solubility, permeability, and charge state. John Wiley & Sons, Chichester
22. Avdeef A, Box KJ, Comer JEA, Hibbert C, Tam KY (1998) pH-metric log P 10. Determination of liposomal membrane-water partition coefficients of ionizable drugs. Pharm Res 15:209–215. https://doi.org/10.1023/a:1011954332221
23. Sangster J (1997) Octanol-water partition coefficients: fundamentals and physical chemistry. John Wiley & Sons, Chichester
24. Kinoshita T, Tsunoda C, Goto S, Hasegawa K, Chatani H, Fujita M, Kataoka H, Katahara Y, Shimada Y, Otsuka Y, Komatsu K, Terada H (2022) Enthalpy–entropy compensation in the structure-dependent effect of nonsteroidal anti-inflammatory drugs on the aqueous solubility of diltiazem. Chem Pharm Bull 70:120–129. https://doi.org/10.1248/cpb.c21-00834
25. Hasegawa K, Goto S, Kataoka H, Chatani H, Kinoshita T, Yokoyama H, Tsuchida T (2024) Quantification of crystallinity during indomethacin crystalline transformation from α- to γ-polymorphic forms and of the thermodynamic contribution to dissolution in aqueous buffer and solutions of solubilizer. RSC Adv 14:4129. https://doi.org/10.1039/d3ra08481g
26. Kasai T, Shiono K, Otsuka Y, Shimada Y, Terada H, Komatsu K, Goto S (2020) Molecular recognizable ion-paired complex formation between diclofenac/indomethacin and famotidine/cimetidine regulates their aqueous solubility. Int J Pharm 590:119841. https://doi.org/10.1016/j.ijpharm.2020.119841

27. Tsunoda C, Hasegawa K, Hiroshige R, Kasai T, Yokoyama H, Goto S (2023) Effect of cyclodextrin complex formation on solubility changes of each drug due to intermolecular interactions between acidic NSAIDs and basic H2 blockers. Mol Pharm 20:5032–5042. https://doi.org/10.1021/acs.molpharmaceut.3c00291
28. Tsunoda C, Goto S, Hiroshige R, Kasai T, Okumura T, Yokoyama H (2023) Optimization of the stability constants of the ternary system of diclofenac/famotidine/β-cyclodextrin by nonlinear least-squares method using theoretical equations. Int J Pharm 638:122913. https://doi.org/10.1016/j.ijpharm.2023.122913
29. Higuchi T, Connors KA (1965) Phase-solubility techniques. In: Advances in analytical chemistry and instrumentation. John Wiley & Sons, New York, pp 117–212
30. Fujita M, Goto S, Chatani H, Otsuka Y, Shimada Y, Terada H, Inoo K (2020) The function of oxybuprocaine: a: parachute effect that sustains the supersaturated state of anhydrous piroxicam crystals. RSC Adv 10:1572–1579. https://doi.org/10.1039/c9ra09952b
31. Fujita M, Tsuchida T, Kataoka H, Tsunoda C, Moritake K, Goto S (2024) Spring and parachute approach for piroxicam dissolution; its phenomenological model on the thermodynamics of irreversible processes. Int J Pharm 667 A:124886. https://doi.org/10.1016/j.jphotochem.2024.116142
32. Wada-Hirai A, Shimizu S, Ichii R, Tsunoda C, Hiroshige R, Fujita M, Li Y-P, Shimada Y, Otsuka Y, Goto S (2021) Stabilization of the metastable α-form of indomethacin-induced by the addition of 2-hydroxypropyl-β-cyclodextrin, causing supersaturation (*spring*) and its sustaining deployment (*parachute*). J Pharm Sci 110:3623–3630. https://doi.org/10.1016/j.xphs.2021.07.002
33. Oshite Y, Wada-Hirai A, Ichii R, Kuroda C, Hasegawa K, Hiroshige R, Yokoyama H, Tsuchida T, Goto S (2024) Comparative study on the effects of the inclusion complexes of non-steroidal anti-inflammatory drugs with 2-hydroxypropyl-β-cyclodextrins on dissociation rates and supersaturation. RSC Pharm 1:80–97. https://doi.org/10.1039/d3pm00039g
34. Kitaura K, Morokuma K (1976) A new energy decomposition scheme for molecular interactions within the Hartree-Fock approximation. Int J Quantum Chem 10:325–340. https://doi.org/10.1002/qua.560100211
35. Kitaura K, Ikeo E, Asada T, Nakano T, Uebayasi M (1999) Fragment molecular orbital method: an approximate computational method for large molecules. Chem Phys Lett 313:701–706. https://doi.org/10.1016/S0009-2614(99)00874-X
36. Allinger NL (1977) Conformational analysis. 130. MM2. A hydrocarbon force field utilizing V1 and V2 torsional terms. J Am Chem Soc 99:8127–8134. https://doi.org/10.1021/ja00467a001
37. Allinger NL, Kok RA, Imam MR (1988) Hydrogen bonding in MM2. J Comput Chem 9:591–595. https://doi.org/10.1002/jcc.540090602
38. Weiner PK, Kollman PA (1981) AMBER: assisted model building with energy refinement. A general program for modeling molecules and their interactions. J Comput Chem 2:287–303. https://doi.org/10.1002/jcc.540020311
39. Pearlman DA, Case DA, Caldwell JW, Ross WS, Cheatham TE III, DeBolt S, Ferguson D, Seibel G, Kollman P (1995) AMBER, a package of computer programs for applying molecular mechanics, normal mode analysis, molecular dynamics and free energy calculations to simulate the structural and energetic properties of molecules. Comput Phys Commun 91:1–41. https://doi.org/10.1016/0010-4655(95)00041-D
40. Kollman P, Caldwell JW, Ross WS (2002) AMBER: a program for simulation of biological and organic molecules. In: Encyclopedia of computational chemistry. Wiley, New York. https://doi.org/10.1002/0470845015.cfa005
41. Brooks BR, Bruccoleri RE, Olafson BD, States DJ, Swaminathan S, Karplus M (1983) CHARMM: a program for macromolecular energy, minimization, and dynamics calculations. J Comput Chem 4:187–217. https://doi.org/10.1002/jcc.540040211
42. MacKerell AD Jr, Brooks B, Brooks CL III, Nilsson L, Roux B, Won Y, Karplus M (1998) CHARMM: the energy function and its parameterization with an overview of the program. In:

von Ragué Schleyer P et al (eds) The encyclopedia of computational chemistry, vol 1. John Wiley & Sons, Chichester, pp 271–277
43. Maple JR, Hwang M-J, Stockfisch TP, Dinur U, Waldman M, Ewig CS, Hagler AT (1994) Derivation of class II force fields. I. Methodology and quantum force field for the alkyl functional group and alkane molecules. J Comput Chem 15:162–182. https://doi.org/10.1002/jcc.540150207
44. Hwang MJ, Stockfisch TP, Hagler AT (1994) Derivation of class II force fields. 2. Derivation and characterization of a class II force field, CFF93, for the alkyl functional group and alkane molecules. J Am Chem Soc 116:2515–2525. https://doi.org/10.1021/ja00085a036
45. Grant DJW, Higuchi T (1990) Solubility behavior of organic compounds, techniques of chemistry, vol XXI. John-Wiley & Sons, Chichester
46. Yalkowsky SH (1999) Solubility and solubilization in aqueous media. Oxford University Press, New York
47. Hansen CH (2000) Hansen solubility parameters—a user's handbook. CRC Press, Boca Raton
48. Suenaga S, Kataoka H, Hasegawa K, Koga R, Tsunoda C, Kuwashima W, Tsuchida T, Goto S (2024) How does the powder mixture of ibuprofen and caffeine attenuate the solubility of ibuprofen? Comparative study for the xanthine derivatives to recognize their intermolecular interactions using Fourier-transform infrared (FTIR) spectra, differential scanning calorimetry (DSC), and X-ray powder diffractometry (XRPD). Mol Pharm 21:4524–4540. https://doi.org/10.1021/acs.molpharmaceut.4c00429
49. Koga R, Tsuchida T, Tsunoda C, Kataoka H, Suenaga S, Goto S (2024) Physicochemical properties of diclofenac regulated by its intermolecular interactions with caffeine evaluated via vector space model, nuclear magnetic resonance, and Fourier-transform infrared analyses. Mater Chem Phys 333:130158. https://doi.org/10.1016/j.matchemphys.2024.130158
50. Li Y-P, Goto S, Shimada Y, Komatsu K, Yokoyama Y, Terada H, Makino K (2015) Study of intermolecular interaction of hydroperoxyl-β-cyclodextrin complex through phase diagrams of the fusion entropy: contrast between nifedipine and nicardipine hydrochloride. Phys Chem Biophys 5:5. https://doi.org/10.4172/2161-0398.1000187
51. Li Y-P, Goto S, Shimada Y, Makino K (2016) Phase solution and solution recrystallization equilibrium constants of hydroxypropyl-β-cyclodextrin complexes with nifedipine and nicardipine hydrochloride. J Pharm Sci Technol Jpn 76:267–273. https://doi.org/10.14843/jpstj.76.267
52. Shimizu S, Wada-Hirai A, Li Y-P, Shimada Y, Otsuka Y, Goto S (2020) Relationship between phase solubility diagrams and crystalline structures during dissolution of cimetidine/cyclodextrin complex crystals. J Pharm Sci 109:2206–2212. https://doi.org/10.1016/j.xphs.2020.03.029
53. Otsuka Y, Kuwashima W, Tanaka Y, Yamaki Y, Shimada Y, Goto S (2021) Effects of heat treatment on indomethacin-cimetidine mixture; investigation of drug-drug interaction using singular value decomposition in FTIR spectroscopy. J Pharm Sci 110:1142–1147. https://doi.org/10.1016/j.xphs.2020.09.049
54. Hiroshige R, Goto S, Ichii R, Shimizu S, Wada-Hirai A, Li Y-P, Shimada Y, Otsuka Y, Makino K, Takahashi H (2022) Protective effects of cyclodextrins on edaravone degradation induced by atmospheric oxygen or additive oxidant. J Incl Phenom Macrocycl Chem 102:327–338. https://doi.org/10.1007/s10847-021-01122-1
55. Hiroshige R, Goto S, Tsunoda C, Ichii R, Shimizu S, Otsuka Y, Makino K, Takahashi H, Yokoyama H (2022) Trajectory of the spectral/structural rearrangements for the photo-oxidative reaction of neat ketoprofen and its cyclodextrin complex. J Incl Phenom Macrocycl Chem 102:791–800. https://doi.org/10.1007/s10847-022-01160-3
56. Takara A, Kobayashi K, Watanabe S, Okuyama K, Shimada Y, Goto S (2017) Dibucaine inhibits ketoprofen photodegradation via a mechanism different from that of antioxidants. J Photochem Photobiol A Chem 333:208–212. https://doi.org/10.1016/j.jphotochem.2016.10.026
57. Horizumi Y, Goto S, Takatsuka M, Yokoyama H (2023) Effects of local anesthetics on liposomal membranes determined by their inhibitory activity of lipid peroxidation. Mol Pharm 20:2911–2918. https://doi.org/10.1021/acs.molpharmaceut.2c01053

58. Kurosawa Y, Otsuka Y, Goto S (2022) Increased selectivity of sodium deoxycholate to around Tryptophan213 in bovine serum albumin upon micellization as revealed by singular value decomposition for excitation-emission matrix. Coll Surf B: Biointerface 212:112344. https://doi.org/10.1016/j.colsurfb.2022.112344
59. Kurosawa Y, Goto S, Mitsuya K, Otsuka Y, Yokoyama H (2023) Interaction mode of hydroxypropyl-β-cyclodextrin with vaccine adjuvant components Tween 80 and Triton X-100 revealed by fluorescence increasing-quenching analysis. Phys Chem Chem Phys 25:6203–6213. https://doi.org/10.1039/d3cp00094jK
60. Tsurushima M, Kurosawa Y, Goto S (2023) Propionic acid groups and multiple aromatic rings induce binding of ketoprofen and naproxen to the hydrophobic core of bovine serum albumin. Mol Pharm 20:3549–3558. https://doi.org/10.1021/acs.molpharmaceut.3c00169
61. Minamide M, Tsurushima M, Koga R, Hasegawa K, Kurosawa Y, Tsuchida T, Goto S (2024) Distinguishing the transitions of fluorescence spectra of tryptophan-134 and 213 in BSA induced by bindings of UV filters, oxybenzone-3, and avobenzone. Bull Chem Soc Jpn 97:uoae058. https://doi.org/10.1093/bulcsj/uoae058

# Chapter 16
# Antioxidative Mechanism of Radical Scavenger and Local Anesthetics

**Tomohiro Tsuchida and Satoru Goto**

**Abstract** Photodegradation of ketoprofen (KET) generates radicals. It initiates chain oxidative reactions, leading to severe inflammatory conditions such as photosensitivity. Generally, antioxidants eliminate radicals generated in the body; however, incomplete elimination results in adverse side effects, such as oxidative stress. Previous studies have shown that local anesthetics inhibit KET photodegradation. However, local anesthetics do not exhibit radical-scavenging activity. Investigating the differences in the antioxidant activities of radical scavengers and local anesthetics through comparative analysis is crucial for identifying novel antioxidants that can replace traditional radical scavengers. Based on the characteristics of their antioxidant mechanism, radical scavengers can be classified into those targeting protein oxidation and those acting on lipid peroxidation. Focusing on the effects of lipid peroxidation, the relationship between local anesthetics and radical scavengers has been investigated. Similar to their protective effects against KET photodegradation, lidocaine (LDC) and dibucaine (DBC) inhibited the progression of lipid peroxidation reactions. This activity appears to be related to the structural flexibility of the LDC and DBC. Our findings suggest that LAs may inhibit lipid peroxidation by altering the orderliness of lipid membranes, which will be a new lipid-antioxidant approach.

---

T. Tsuchida
Faculty of Pharmaceutical Sciences, Tokyo University of Science, Katsushika-ku, Tokyo, Japan
e-mail: t.tsuchida@rs.tus.ac.jp

S. Goto (✉)
Faculty of Pharmaceutical Sciences, Tokyo University of Science, Katsushika-ku, Tokyo, Japan

Division of Colloid and Interface Science, Research Institute of Science and Technology, Tokyo University of Science, Noda, Chiba, Japan
e-mail: s.510@rs.tus.ac.jp


M. Hashizume, Y. Imura (eds.), *Colloids and Interfacial Dynamics*, Physical Chemistry in Action, https://doi.org/10.1007/978-981-96-6728-4_16

**Keywords** Antioxidation · Radical scavenger · Oxygen radical absorption capacity (ORAC) assay · Lipid peroxidation · Local anesthetics · 2-Thiobarbituric acid reactive substances (TBARS) assay

## 16.1 Introduction

Ketoprofen (KET) undergoes photodegradation, leading to inflammation known as photosensitivity [1, 2]. Previous studies have proposed various processes for KET photodegradation. It has been suggested that the primary reaction involves an intermolecular electron transfer from the carboxyl group, resulting in decarboxylation. In this process, UV irradiation generates the KET anion, which subsequently reacts with oxygen molecules in the solution to generate reactive oxygen species such as superoxide anions, hydroxyl radicals, and singlet oxygen [3, 4]. A minor reaction was the photoionization of KET in its triplet state. Monti et al. reported a mechanism in which the triplet state of KET ions undergoes decarboxylation from KET molecules, resulting in the formation of KET benzyl radicals. This process is driven by intramolecular charge transfer or charge donation to the molecules in solution [5]. Furthermore, the radicals generated from photodegradation are believed to react with unreacted KET, initiating a radical chain reaction that accelerates its decomposition. Antioxidants have been proposed to efficiently eliminate KET radicals formed during photodegradation [6]. The use of cyclodextrin has been proposed as a method to suppress KET photodegradation. Monti et al. reported that a complex formed through the interaction between cyclodextrin and KET inhibited photodecarboxylation [7]. Hiroshige et al. demonstrated that the reaction mechanism of the photodegradation of KET changes due to the inclusion of cyclodextrin [8]. Takara et al. hypothesized that substances capable of interacting with KET could protect it from photodegradation. They investigated the effects of local anesthetics (LAs) on KET photodegradation [6]. LAs react with nonsteroidal anti-inflammatory drugs (NSAIDs) and alter their physicochemical properties [9]. Dibucaine (DBC), an LA, inhibits the photodegradation of KET through a mechanism different from its radical-scavenging activity. This protective effect is thought to be due to the intermolecular interactions between KET and DBC. The observation that compounds, such as DBC and cyclodextrins, which lack radical-scavenging activity, exhibit antioxidant activity is necessary to examine the relationship between various radical scavengers and these substances. In this chapter, we introduce the classification of various radical scavengers based on their antioxidation characteristics and the relationship between the antioxidation features of LAs, such as lidocaine (LDC) and DBC, and the features of radical scavengers.

## 16.2 Classification of Antioxidants Based on Their Inhibitory Abilities Against Single-Electron Transfer and Hydrogen Atom Transfer Reactions

Several assays have been developed to rapidly predict the antioxidant and radical-scavenging activities of natural compounds and extracts. The stable radical 1,1-diphenyl-2-picrylhydrazyl (DPPH; structure shown in Fig. 16.1a) quantifies the antioxidant (mainly phenolic) content [10–13]. The DPPH assay monitors the decrease in DPPH radicals by antioxidants using UV–Vis absorbance measurements. The 2,2′-azinobis-(2,3-dihydro-3-ethylbenzothiazole)-6-sulfonic acid (ABTS; structure shown in Fig. 16.1b) assay is a stronger radical-scavenging assay than the DPPH assay and is often used to screen complex antioxidant mixtures such as plant extracts, beverages, and biological fluids [14]. Both assays are the most frequently used among the methods used to estimate radical-scavenging activity.

The reaction between free radicals and target molecules such as fatty acids and antioxidants can occur through different mechanisms, including single-electron transfer (SET), hydrogen atom transfer (HAT), or a combination of both (Fig. 16.2) [15]. Liao et al. cited the DPPH assay as an example of a SET reaction and the oxygen radical absorption capacity (ORAC) assay as an example of an HAT reaction

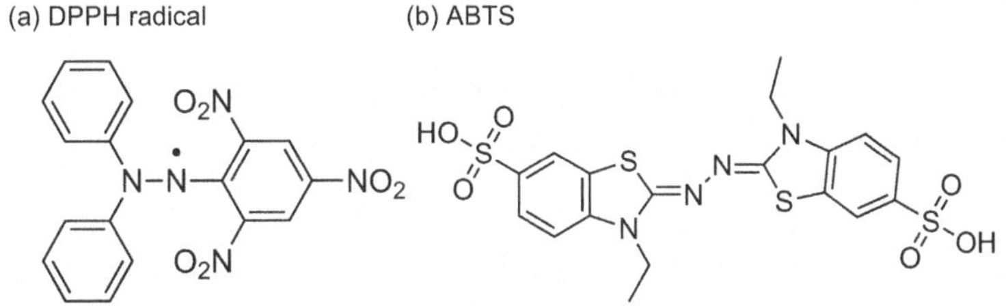


**Fig. 16.1** Structures of DPPH radical (**a**) and ABTS (**b**)

**Fig. 16.2** Free radical-scavenging process in SET and HAT reactions

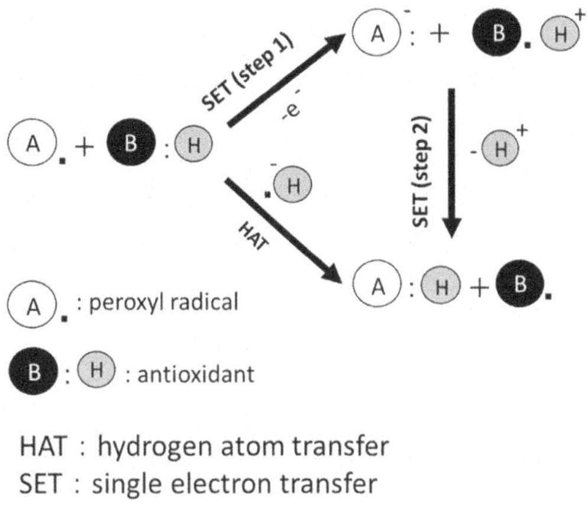

[16]. The ORAC assay is a classic tool for measuring the antioxidant capacity of various biomolecules and is based on the oxidation of a fluorescent probe by peroxyl radicals, e.g., 2,2′-azobis-(2,4-dimethylvaleronitrile) (AAPH) [17]. The antioxidants in the assay prevent the peroxyl radical oxidation of the fluorescent probe until the antioxidant activity of the sample is depleted. The antioxidant capacity of the sample is evaluated using a fluorescence decay curve, typically represented by the area under the curve (AUC) [18]. These assays are widely regarded as universal standards for evaluating the antioxidant capacity of foods and functional foods. Ou et al. developed and validated an improved ORAC assay using fluorescein (FLH) as a fluorescent probe (ORAC-FLH) [19].

Takatsuka et al. classify antioxidant reaction behavior during radical-scavenging activity by applying Voronoi polyhedral division to the results of the DPPH, ABTS, and ORAC-FLH assays [20]. The four model antioxidants used as the basis for classification were TRO, ASC, cysteine (CYS), and 2,6-di-*tert*-butyl-*p*-cresol (BHT), standard antioxidants. In addition, edaravone (EDA), flavonoids (quercetin, QUE, myricetin, and MYR), spices (capsaicin, CAP; curcumin, CUR), and steroids (hydrocortisone, HYD) were compared with model antioxidants.

### 16.2.1 Comparison of DPPH and ABTS Radical-Scavenging Capacities Between EDA and Other Antioxidants

The radical-scavenging activities of the antioxidants against DPPH and ABTS radicals were measured according to the method described by Liao et al. [21]. The absorbance is recorded at 521 nm for the DPPH spectrum and at 744 nm for the ABTS spectrum, both in the presence and absence (control) of antioxidants, to calculate the radical-scavenging activity. Under steady-state conditions, the proportion of the remaining radicals shows a linear correlation with the concentration of each antioxidant up to a specific threshold. Above this concentration, 100% of the radicals were consumed. The anti-radical activity was defined as the amount of antioxidant required to decrease the initial DPPH radical concentration by 50% ($EC_{50}$). The logarithm of the reciprocal of $EC_{50}$ for DPPH and ABTS radical values is shown as $EC_{50}^{\mathrm{DPPH}}$ and $EC_{50}^{\mathrm{ABTS}}$.

Using the DPPH assay, the observed $EC_{50}^{\mathrm{DPPH}}$ of EDA was 4.21 μM. Tokumaru et al. previously reported the concentration at which 50% of DPPH radicals are scavenged by EDA is 4.7 ± 0.3 μM [22], and Kraus et al. reported this concentration as 6.5 ± 1.3 μM [23]. The lower $EC_{50}^{\mathrm{DPPH}}$ of EDA observed in this study, compared to that in previous reports, may be related to differences in solvent composition, which could also compete with antioxidants for radical scavenging. While the flavonoids (QUE and MYR) showed very high activity in the DPPH assay, no significant activity was observed for the steroid (HYD).

The $EC_{50}^{\mathrm{ABTS}}$ of EDA was 5.52 μM, which is lower than the $EC_{50}$ of 24 ± 5 μM reported by Kraus et al. [23]. As with the $EC_{50}^{\mathrm{DPPH}}$, differences in the solvent and pH conditions between the two studies may account for the observed discrepancy. The $EC_{50}^{ABTS}$ of EDA was more potent than those of TRO, ASC, and BHT, although it was similar to that of CYS. The ABTS radical-scavenging rates of QUE, MYR, CAP, and CUR decreased gradually with increasing antioxidant concentration, similar to an exponential function. This behavior likely resulted from a two-step radical-scavenging reaction. Therefore, the $EC_{50}^{ABTS}$ values of QUE, MYR, CAP, and CUR should not be directly compared to those of TRO, ASC, CYS, BHT, and EDA.

### *16.2.2 Comparison of Antioxidant Activity Using the ORAC-FLH Method*

In the ORAC-FLH assay, the ORAC value is defined as the ratio of the AUC of the antioxidant curve for the sample per unit concentration to that of TRO per unit concentration when TRO was added. All the tested antioxidants show an increase in ORAC values with increasing concentrations. This indicates that the decomposition of FLH by AAPH radicals was inhibited. The observed ORAC value of EDA was 5.65 μM of TRO equivalent. Qiang et al. reported an ORAC value for EDA of 0.72 ± 0.03 μM of TRO equivalent, which is considerably smaller than the value reported here [24]. The ORAC value of the flavonoid QUE was overwhelmingly high, and EDA was the second most active compound. However, changes in the antioxidant curve of FLH fluorescence versus time varied significantly depending on the type of antioxidant added. This indicates differences in the changes observed between the antioxidants, such as an extension of the lag time before the onset of fluorescence quenching and a decrease in the quenching rate. Therefore, the ORAC-FLH assay results reflected the distinct reaction mechanisms of each antioxidant.

### *16.2.3 Reinterpretation of the ORAC-FLH Assay Based on Lag Time and Rate Constants*

In the ORAC-FLH assay, the symmetrical decomposition of heated AAPH into two peroxyl radical equivalents occurs only in the presence of reactive oxygen species (ROS), making this step exclusively responsive to antioxidants that can act via the SET mechanism. Thus, the ORAC-FLH assay mixes reactions involving hydrogen-independent SET processes with those involving HAT processes (Fig. 16.3) [25].

The rate of change in the fluorescence intensity over time was defined as the observed quenching rate constant ($k_{obs}$), and the period before the fluorescence intensity began to decrease was defined as the lag time (Fig. 16.4). While the lag time reflects the process of peroxyl radical production from AAPH, the quenching rate constant $k_{obs}$ reflects the reaction between the AAPH peroxyl radicals and FLH. The changes in the lag time and $k_{obs}$ provide important qualitative information

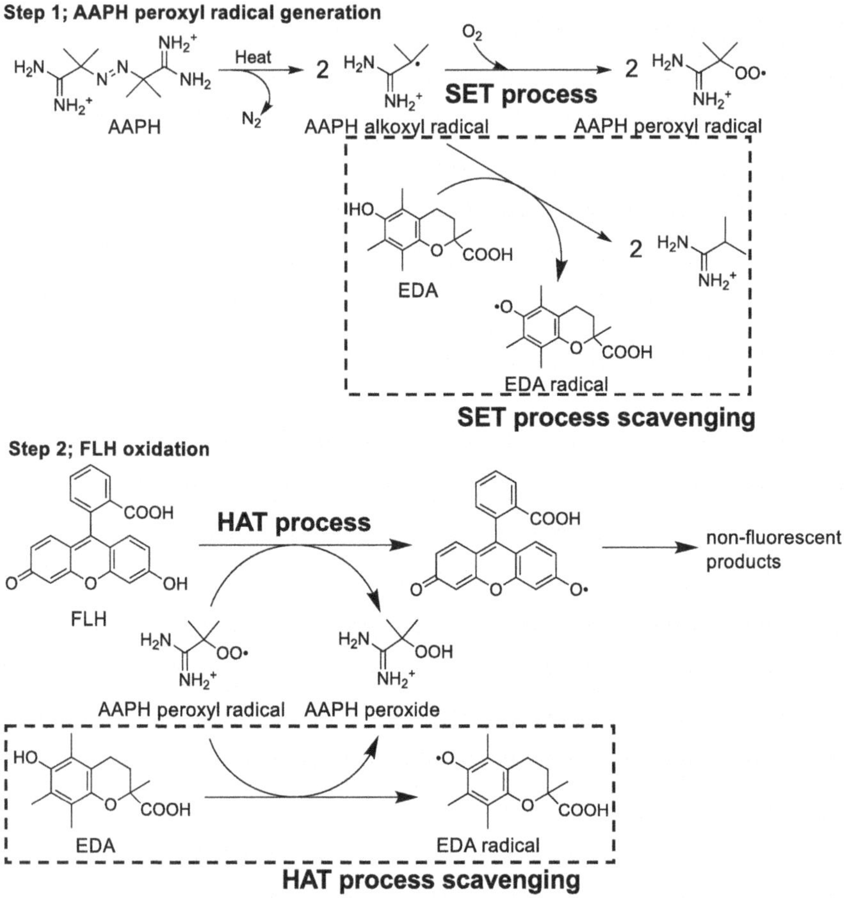


**Fig. 16.3** Process of AAPH radical production and ORAC-FLH assay. Mechanism for prevention of FLH oxidation through SET or HAT processes by EDA

regarding the reaction mechanism of antioxidants. Furthermore, there is no stoichiometric equivalence between the two. Therefore, using the AUC value to compare the magnitude of the antioxidant capacity among compounds is useful for evaluating the overall antioxidant effects of the test substances; however, these results do not reflect the intrinsic characteristics of their antioxidant activity.

The lag time was extended for CYS, and the quenching rate of FLH fluorescence was reduced. Similarly, EDA prolonged the lag time and reduced the extinction rate of FLH fluorescence in a concentration-dependent manner. The quenching rates for TRO and ASC remained constant with respect to the antioxidant concentration, and the lag time was extended. No lag time corresponding to the non-reactive period was observed for BHT, and the quenching rate was significantly reduced even at the lowest BHT concentration tested. The lag time was extended with the flavonoids

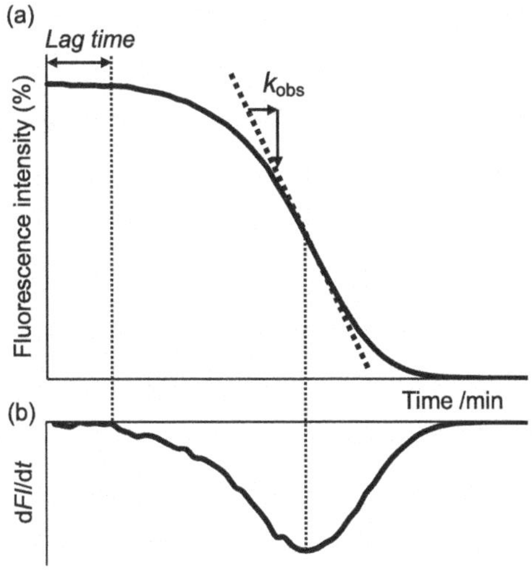


**Fig. 16.4** Illustration of lag time and reaction rate constant in the result of ORAC-FLH assay. Antioxidant curve from the fluorescence of FLH versus time (**a**) and plot of the first derivative of fluorescence intensity over time (**b**)

QUE and MYR, and the quenching rate decreased. The behavior of CAP and CUR was similar to that of the flavonoids.

### 16.2.4 Antioxidant Classification Based on the Character of the Radical-Scavenging Reaction

Takatsuka et al. defined two parameters: lag time per unit, logarithm of the antioxidant concentration, and quenching rate per unit [20]. These parameters were standardized in an orthogonal space with equal variances along both axes based on their averages and standard deviations. Takatsuka then created two-dimensional Voronoi polygons for four representative antioxidants, TRO, ASC, CYS, and BHT, by dividing this orthogonal space by the perpendicular bisectors drawn between each pair of antioxidants.

EDA showed a short extension of the lag time with increasing concentrations, whereas its reactivity with peroxyl radicals remained high. As noted above, the lag time represents the delay from AAPH to peroxyl radical formation, suggesting SET reaction to ROS. It has been suggested that $k_{obs}$ competitively antagonizes the quenching of FLH fluorescence by attacking peroxyl radicals, followed by a HAT reaction with the radicals. Conventional ORAC-FLH analysis is a standard method for comparing these processes using a single numerical value. However, comparing them within a vector space, as shown here, provides a more schematic understanding (Fig. 16.5).

The high reactivity of QUE and CUR is utilized in reactions involving ROS in biological membranes in the body. In contrast, MYR and CAP, which are delivered intracellularly, may retain their activity. A similar situation was observed for CYS,

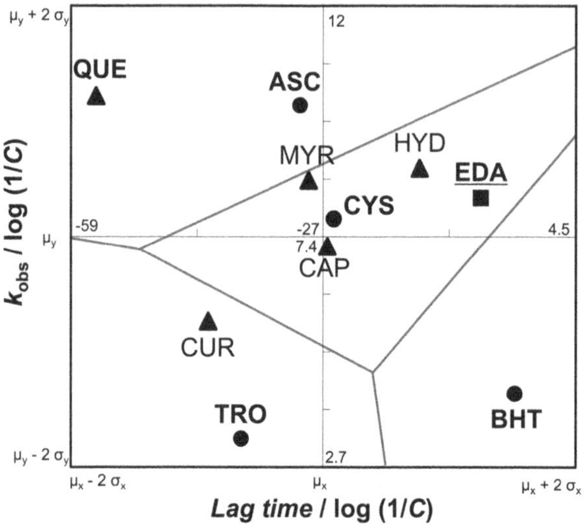


**Fig. 16.5** Classification of antioxidant reactions by Voronoi polygonal division. The higher the value on the vertical axis, the greater the reactivity with peroxyl radicals. The lower the value on the horizontal axis, the longer the lag time. Model antioxidants are represented by circles; EDA by squares; and flavonoids, spices, and steroids by triangles. The population means and the standard deviations are represented by μ and σ, respectively

which exhibited a bimolecular reaction. EDA, MYR, HYD, and CAP are grouped with the model antioxidant, CYS, which, similar to glutathione, suppresses protein oxidation. If EDA has similar effects as CYS, it may also reduce oxidative stress, which might inhibit TDP-43 protein mutations in amyotrophic lateral sclerosis (ALS). Fisetin, similar to MYR, is a flavonoid that demonstrated neuroprotective effects in ALS mice expressing multiple mutant forms of human copper/zinc superoxide dismutase [26]. Therefore, compounds within the CYS group may be potential new candidates for ALS therapy.

TRO is a highly hydrophobic antioxidant that functions effectively in hydrophobic environments such as within biological membranes. CUR belongs to the same group as TRO, and its antioxidant activity is expected to be prominent in hydrophobic environments. A notable characteristic of TRO and CUR in antioxidant reactions is their pronounced effect of extending the lag time. In contrast, BHT had minimal influence on the extension of lag time and the quenching rate.

## 16.3 Inhibitory Effect of LAs and TRO on Liposomal Membrane Peroxidation

Cell membranes, which contain abundant polyunsaturated fatty acids (PUFAs), are highly vulnerable to damage by ROS, including lipid peroxidation [27]. This process, closely tied to aging and cancer development, has garnered significant interest owing to its potential relevance in future pharmacological strategies [28]. Oxidative stress induces partial brain atrophy through lipid peroxidation, which may contribute to depression and schizophrenia [29, 30]. Biological membranes are frequently exposed to environments rich in unsaturated fatty acids, oxygen, and metals that promote this process [31]. Lipid peroxidation is initiated by the removal of hydrogen atoms from unsaturated fatty acids, resulting in the generation of lipid radicals [31]. These radicals, in turn, attack PUFAs, leading to a chain reaction involving oxidative damage. This chain reaction can be interrupted by various free radical scavengers, such as α-tocopherol (α-TOC), a well-established antioxidant in the body that effectively inhibits lipid peroxidation [32]. Studies have shown that lipid membrane fluidity is closely associated with the extent of lipid peroxidation [33]. However, the role of drugs influencing membrane fluidity in reducing lipid peroxidation has not been sufficiently explored.

Over a century ago, Meyers and Overton demonstrated that the hydrophobicity of various volatile chemicals plays a significant role in their anesthetic activity [34, 35]. However, the precise mechanisms underlying this anesthesia remain unclear. Peterson et al. found that general anesthetics, such as isoflurane and chloroform, disrupted lipid rafts and fluid lipid regions composed of sphingolipids and cholesterol [36]. This disruption causes a thorough mixing of ganglioside rafts and bulk lipids. These inhaled anesthetics also disturb lipid rafts, activating phospholipase D2 through their effects on lipid structures. These findings led us to revisit the effect of drugs targeting lipid membranes on biomembranes. Similar mechanisms have been proposed for both general anesthetics and LAs, which may influence the membranes of the central nervous system [37]. Because most LA molecules contain large hydrophobic regions, they can interact with the hydrophobic regions of cell membranes and alter their physical properties. Consequently, channel proteins in the membranes may be indirectly affected by these changes in membrane properties. LDC (Fig. 16.6a), a common LA, has been shown to solubilize lipid membranes at high concentrations, resulting in neurotoxicity [38]. DBC (Fig. 16.6b), another LA, exhibited actions similar to those of LDC. The ability of LAs to directly interact with biomembranes stems from their amphiphilic cationic nature, which enables electrostatic interactions with negatively charged groups on the membrane surface [39, 40]. Additionally, the high hydrophobicity of LAs allows them to penetrate biomembranes, modify their physical properties, and increase membrane

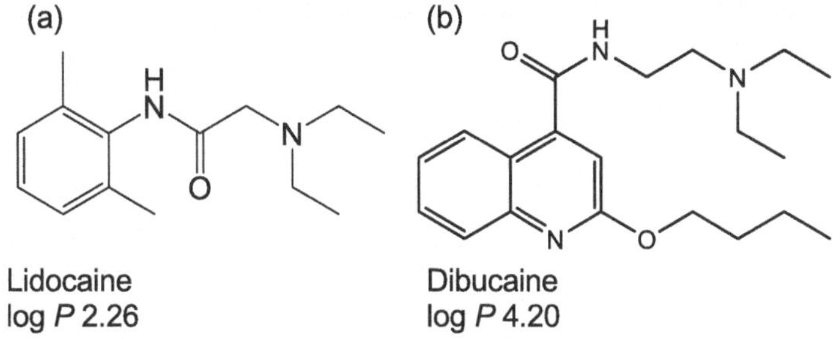


**Fig. 16.6** Structure of lidocaine (**a**) and dibucaine (**b**). log *P* value is cited from [61]

fluidity [41]. By inhibiting chemical transmission in neurons, LAs have been linked to cardiotoxicity associated with their effects on cardiolipin-containing biomembranes as well as their role in disrupting lipid rafts [42–44].

In Sect. 16.2, various antioxidants are classified into protein- and lipid-peroxidation-targeting types. As mentioned earlier, the photodegradation of KET is strongly inhibited by DBC [6]. The inhibitory effect of DBC on KET photodegradation suggests that the structural characteristics of DBC play a crucial role [6]. Similarly, structural characteristics of LAs may influence lipid molecules' physicochemical properties, thereby inhibiting lipid peroxidation. Horizumi investigated the relationship between the antioxidant activity of lipid-targeting antioxidants and LDC and DBC, which do not exhibit radical-scavenging activity. This section introduces the development of a model that mimics the oxidative states of biological membranes using liposomes (membrane protection model) and explains the relationship between the antioxidative effects of LAs and TRO on membrane oxidation [45]. This study includes an important discussion on the mechanism by which substances lacking radical-scavenging activity achieve their antioxidative effects.

### *16.3.1 Construction of a Membrane Protection Model Using EyPC Liposomes*

The intricate structure of biomembranes makes it challenging to investigate lipid raft stability in the presence of drugs. To address this, liposomes, as lipid models of biological membranes, are commonly used to study the interactions between lipid membranes and pharmaceuticals owing to their exceptional ability to mimic basic biological membranes. Among these, egg yolk lecithin (EyPC) liposomes are particularly advantageous models for biomembranes rich in PUFAs such as linoleate and arachidonate [46].

First, Horizumi constructed a model to replicate the oxidation reactions of the cell membrane using EyPC liposomes. The EyPC liposomes were prepared as multilamellar vesicles (MLVs) using the Bangham method [47, 48]. The oxidizing agent employed was Fenton's reagent (FR), composed of $Fe(NH_4)_2(SO_4)_2$ and $H_2O_2$. TRO was added to the system to simulate the action of α-TOC, a natural antioxidant, in vivo. This model, named the "Membrane protection model." Lipid peroxidation levels can be assessed using the 2-thiobarbituric acid reactive substance (TBARS) assay.

Although the TBARS method has been criticized for its lack of specificity, with reports indicating that non-lipid peroxide substances can also react with TBA, because separation analysis techniques such as high-performance liquid chromatography have their limitations [49]. Specifically, although they can separate multiple dyes with an absorbance of 530 nm, fine peaks often fall below the detection limit, making accurate quantification unreliable. So, to quantify the concentration dependence of the reagent, Horizumi opted for UV–Vis spectral measurements, which allowed for total quantification.

Malondialdehyde (MDA) is the standard substance used in the TBARS assays. When the TBARS assay is applied to MDA, an absorption spectrum with a maximum wavelength of 530 nm is obtained. Additionally, absorbance shows a linear relationship with MDA concentration, demonstrating its quantifiability. However, when the TBARS assay was applied to the MLVs subjected to oxidative stress induced by FR addition, a new absorption peak was observed at 455 nm. This peak was also observed when the TBARS assay was performed on the FR solution alone (without MLVs), indicating that it does not originate from peroxidized MLVs but rather reflects the formation of a non-specific product through the TBA reaction.

### *16.3.2 Stronger Peroxide Suppression Effect of DBC Compared to LDC*

In the membrane protection model, TRO inhibited the lipid peroxidation of PUFA in MLVs in a concentration-dependent manner. When LDC was added to the membrane protection model, a reduction in the TBARS spectra was observed (Fig. 16.7). Moreover, in the presence of high concentrations of LDC, the reducing effect of TRO on the TBARS spectrum was attenuated, which weakened the lipid

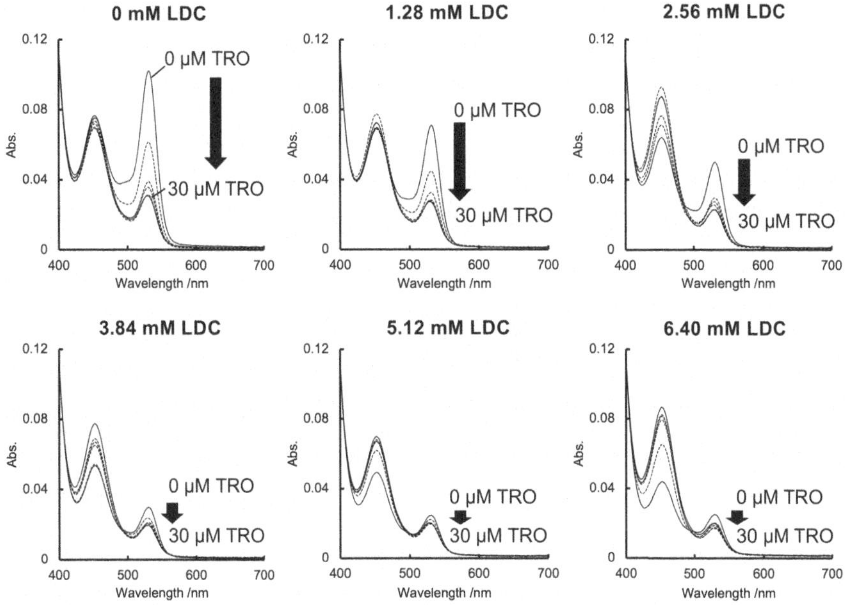


**Fig. 16.7** TBARS spectra of MLV subjected to lipid peroxidation under each concentration of LDC and TRO

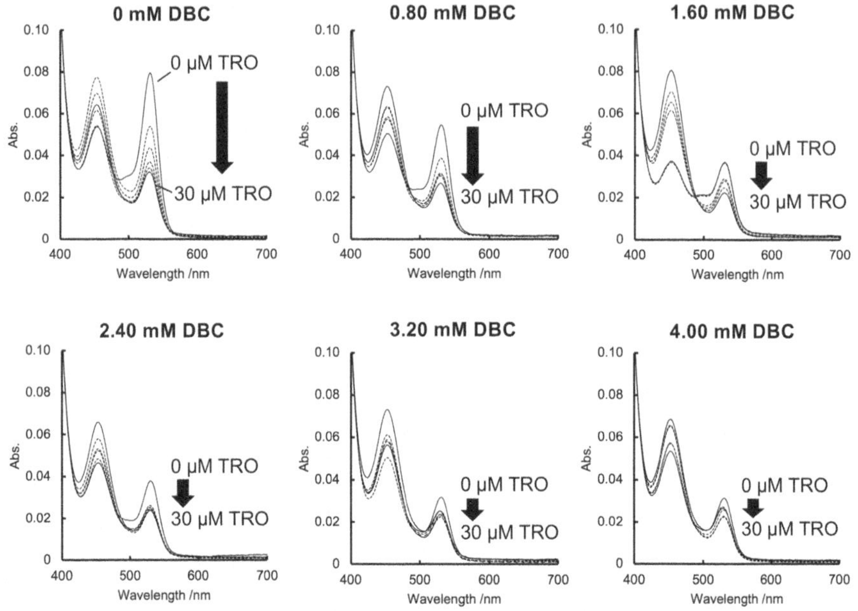


**Fig. 16.8** TBARS spectra of MLV subjected to lipid peroxidation under each concentration of DBC and TRO

peroxidation-inhibitory activity of TRO. Thus, the inhibitory effect of LDC on lipid peroxidation could be attributed to its suppressive effect on TRO inhibition. Similarly, reductions in the TBARS spectrum and attenuation of TRO's lipid peroxidation-inhibitory effect are observed with DBC addition (Fig. 16.8). These findings suggest that both LDC and DBC have lipid peroxidation-inhibitory effects on FR-induced PUFA oxidation in MLVs. However, the nonspecific absorption peak at 455 nm did not exhibit concentration-dependent changes upon treatment with LAs or TRO. If the absorption peak at 455 nm, presumed to originate from nonspecific products, and the peak at 530 nm, derived from TBA products, co-occur with LDC or DBC concentration, it is necessary to analyze the signal at 455 nm simultaneously. Applying singular value decomposition (SVD) to the TBARS spectrum is recommended to verify the presence or absence of co-occurrence.

SVD, a matrix decomposition technique, breaks down the observed spectra into three components: basis vectors (U), singular values (Σ), and eigenvectors ($V^t$), as illustrated in Fig. 16.9. The basis vectors represent the minimal set of fundamental functions required to reconstruct the vector space of the observed spectra and are shared across all spectra. Singular values are the diagonal components that indicate the degree of contribution of each corresponding basis vector to the spectra. Eigenvectors are also referred to as singular vectors and define the direction of variation for each basis vector under specific experimental conditions. Generally, SVD analysis identifies the average spectra as the first component and extracts the differential spectra that account for variations under particular experimental conditions. In other words, SVD can uncover dependencies among vector components related to experimental results and isolate co-occurring features [8, 45, 50–60]. Unlike artificial intelligence and deep learning methods, the basis vector space model in SVD is entirely transparent, without the influence of black-box processes or automated opaque interpretations. This transparency allows for explainable basis vectors and manual visualization of their contributions aligned with each observed spectrum.

When SVD was applied to the TBARS spectra obtained at various concentrations of LDC, DBC, and TRO, the cumulative contribution ratio of the first component was 77.3%, increasing to 89.4% when the second component was included and reaching 94.9% when the third component was included (Fig. 16.10). The basis vector of the first component was found to have decremental signals at 455 and 530 nm, suggesting that it corresponds to the average spectrum, as previously described. The basis vector of the second component exhibits a decremental signal

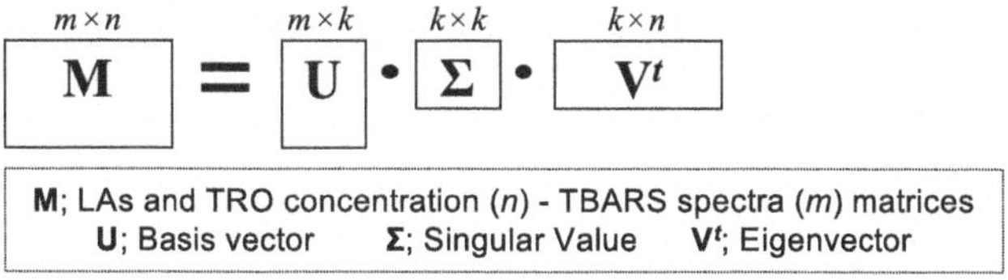


**Fig. 16.9** Scheme of SVD. This method extracts basis vector, singular value, and eigenvector from the "LAs and TRO concentration-TBARS spectra" matrices

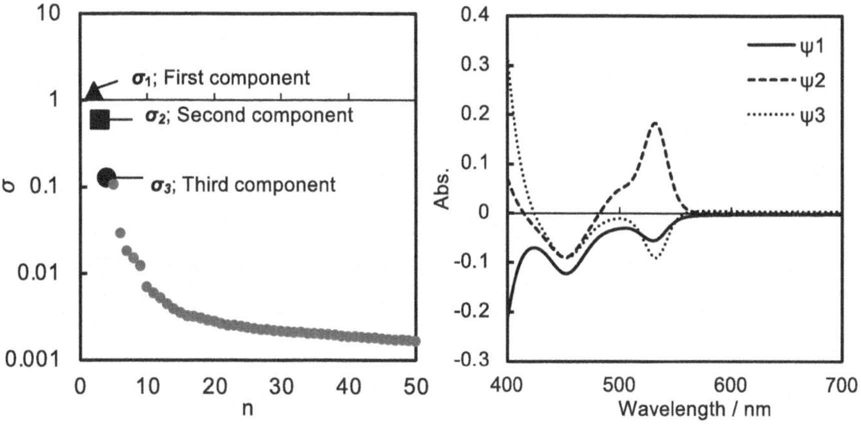


**Fig. 16.10** Top 50 singular values obtained from the SVD results of 300 TBARS spectra (**a**). Cumulative contribution ratio for the first component was 77.3%, which increased to 89.4% and 94.9% when including the second and third components, respectively. Basis vectors up to the third component (**b**)

at 455 nm and an incremental signal at 530 nm. This reflects the dissolution of the nonspecific absorption peak of the 455 nm signal. Thus, the 455 nm absorbance peak was thought not to co-occur. In the third component, decremental signals were observed at 455 and 530 nm. However, they may have been ineffective because their contributions were small. Based on these results, it was concluded that the 455 nm peak in the TBARS spectrum does not need to be observed for the quantification of PUFA, and only the absorbance at the 530 nm peak should be considered.

### *16.3.3 Inhibitory Effects of LDC and DBC on the Peroxide Suppression Activity of TRO*

A three-dimensional plot of LDC concentration, TRO concentration, and absorbance at 530 nm in the TBARS spectrum is shown in Fig. 16.11a. First, the inhibitory effect of LDC on TRO's suppression of MLV peroxidation at various LDC concentrations was calculated as the 50% inhibitory concentration ($pI_{50}^{TRO}$) using curve fitting. Similarly, the inhibitory effect of LDC on the peroxidation-suppression effect of MLV at various TRO concentrations was calculated as the 50% inhibitory concentration ($pI_{50}^{LDC}$). The results of plotting the calculated $pI_{50}$ values at the LDC and TRO concentrations are shown in Fig. 16.12a, b.

The values of $pI_{50}^{TRO}$ tend to decrease with increasing LDC concentrations. In contrast, the values of $pI_{50}^{LDC}$ remained largely unchanged, regardless of the TRO

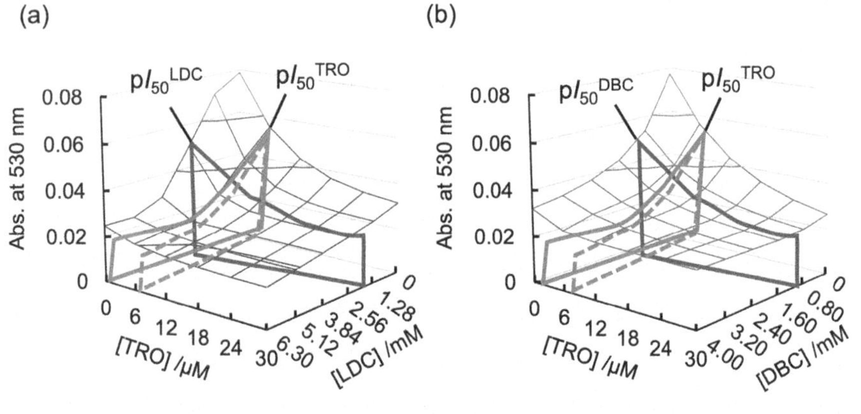


**Fig. 16.11** 3D plots between the absorbance at 530 nm in the TBARS spectra, TRO, and LDC (**a**) or DBC (**b**) concentrations. The solid lines in the graphs represent $pI_{50}^{TRO}$ as a function of LDC or DBC concentrations and $pI_{50}^{LDC}$ or $pI_{50}^{DBC}$ as a function of TRO concentrations. The dashed lines indicate the LA concentration-dependent changes in $pI_{50}^{TRO}$. The $pI_{50}$ values at all concentrations of TRO and LAs are shown in the graphs in Fig. 16.12

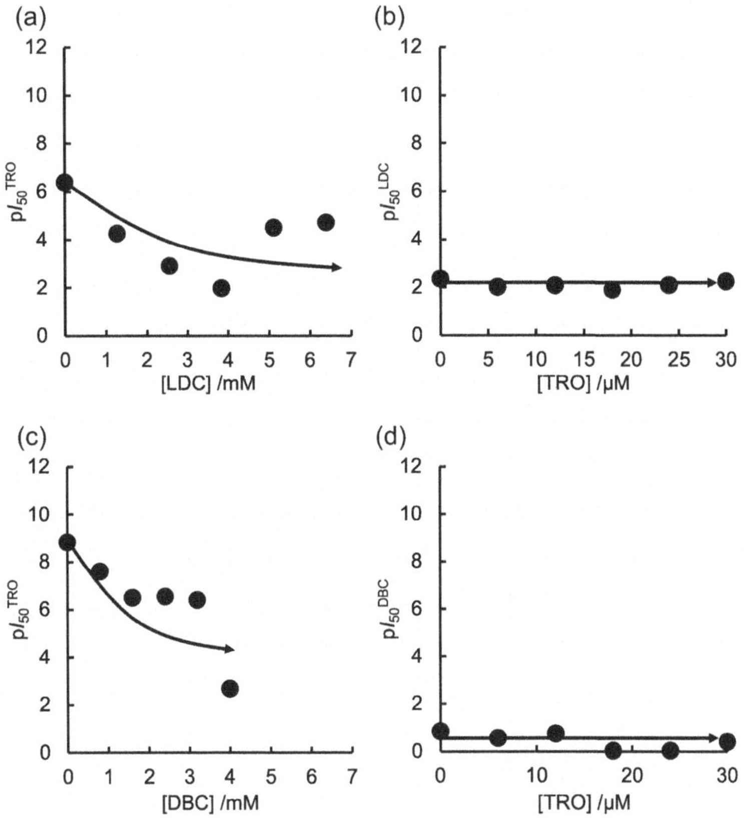


**Fig. 16.12** Plots of $pI_{50}^{TRO}$ versus LDC concentration (**a**), $pI_{50}^{LDC}$ versus TRO concentration (**b**), $pI_{50}^{TRO}$ versus DBC concentration (**c**), and $pI_{50}^{DBC}$ versus TRO concentration (**d**)

concentration. This difference suggests that the peroxidation-suppression effect of LDC on PUFA in the MLV affects the lipid peroxidation-suppression effect of TRO. In contrast, TRO does not influence the inhibitory effect of LDC. Similar results were observed for DBC, where DBC tended to decrease $pI_{50}^{TRO}$ at lower concentrations than LDC (Figs. 16.11b and 16.12c, d).

The hydroxyl radicals produced by FR access the unsaturated carbon chains of PUFA after their distribution into the lipid membrane of the MLV, and achieve the oxidation of PUFA. TRO partitions into the lipid membrane and scavenges hydroxyl radicals generated by FR or lipid radicals generated by oxidation. In contrast, LDC and DBC do not exhibit any radical-scavenging activity [45]. The lipid peroxidation-suppressive effects of LDC and DBC are thought to occur via mechanisms distinct from those of TRO. Thus, Horizumi hypothesized that LDC and DBC inhibit this process, possibly by altering the properties of the lipid membrane (e.g., fluidity), thereby creating an environment that prevents the penetration of hydroxyl radicals.

### 16.3.4 Peroxide Suppression Effects of LDC and DBC on a Membrane Protection Model Using Small Unilamellar Liposomes

Previously, $pI_{50}^{TRO}$ showed a decreasing trend with increasing concentrations of LDC and DBC; however, these changes were not continuous. Because MLV possesses a multilamellar structure, it is challenging to establish stable conditions under which LAs saturate the lipid membrane. Small unilamellar liposomes (SUVs) with a simple unilamellar membrane structure were prepared to address this issue, and similar experiments were conducted [45]. The changes in TBARS spectra observed in the membrane protection model using SUV were similar to those obtained using MLVs. However, the results of the SVD analysis differed, as the 455 nm peak originating from nonspecific products was successfully separated into the third component. Accordingly, $pI_{50}^{TRO}$, $pI_{50}^{LDC}$, and $pI_{50}^{DBC}$ can be calculated by linear regression based on the first and second components. While the $pI_{50}^{LDC}$ and $pI_{50}^{DBC}$ observed with MLV did not correlate with the concentration of TRO, the results using SUV revealed a decrease in these values with increasing TRO concentrations. This discrepancy can be attributed to the structural differences between MLV and SUV. In MLV with multilamellar structures, TRO suppressed PUFA peroxidation in lipid membranes that were not fully penetrated by LAs, creating an environment in which LAs and TRO did not compete. In contrast, the unilamellar structure of the SUV observed only competitive interactions between LAs and TRO.

The $pI_{50}^{LDC}$ and $pI_{50}^{DBC}$ values decreased with increasing TRO concentration, as observed for the MLVs. Notably, the ability of DBC to reduce $pI_{50}^{TRO}$ was approximately twice that of LDC. LDC and DBC did not exhibit radical-scavenging activity against DPPH or galvinoxyl radicals. Furthermore, the radical-scavenging activity of TRO against both radicals was not inhibited in the presence of LDC or DBC. This result is particularly intriguing because it suggests that LDC and DBC exhibit antioxidant activity only in the presence of lipid membranes. Therefore, it can be concluded that both drugs affect the lipid membrane properties of SUVs. The log *P* values of LDC and DBC are 2.26 and 4.20, respectively [61], suggesting that DBC partitions into the lipid membrane to a greater extent than LDC. Consequently, DBC likely significantly influences the membrane properties, even at lower concentrations.

### 16.3.5 Antioxidant Mechanism of LAs Against Lipid Peroxidation

DBC suppressed PUFA peroxidation in both MLVs and SUVs more effectively than LDC and inhibited the lipid peroxidation-suppressing activity of TRO. However, neither LDC nor DBC exhibited radical-scavenging activity, and they only inhibited

TRO's radical-scavenging activity in the presence of lipid membranes. The authors interpreted that their interactions with lipid membranes mediated the lipid oxidation-suppressing and TRO-inhibiting effects of LDC and DBC. This suggests that LDC and DBC partition into the lipid membrane, potentially altering membrane properties, such as fluidity. Indeed, DBC, which demonstrates stronger peroxidation inhibition than LDC, is more hydrophobic and is likely to partition more extensively into the lipid membrane. This may lead to changes in membrane fluidity, which could hinder the access of the hydroxyl radicals generated from FR or TRO to the lipid membrane. High concentrations of LDC and DBC have been reported to dissolve lipid membranes, a process associated with changes in global or local membrane fluidity [48]. Even at lower concentrations, at which LDC and DBC do not dissolve the membranes, they may still influence membrane fluidity. Fluorescence anisotropy measurements were performed using the fluorescent probe DPH to quantify the effects of LAs on membrane fluidity. These experiments revealed the tendency of DBC to enhance membrane fluidity more strongly than LDC. In the "Membrane Protection Model," TRO within the lipid membrane suppresses membrane oxidation caused by ROS (Fig. 16.13a). DBC, due to its high hydrophobicity, localizes deep within the lipid membrane, significantly altering membrane fluidity and thereby inhibiting the penetration of ROS into the membrane interior (Fig. 16.13b). In contrast, LDC, which has lower hydrophobicity compared to DBC, localizes closer to the surface of the lipid membrane and exerts a lesser effect on membrane fluidity (Fig. 16.13c). Moreover, a potential relationship between the solvation enthalpy ($\Delta S_{fus}$) of the compounds and their inhibitory effects on TRO was identified. From a quantitative structure–activity relationship perspective, solvation enthalpy can be interpreted as a measure of the structural instability of a molecule's subdomains. Based on this, highly dynamic regions, such as the alkyl chain of DBC, may oscillate within the lipid membrane, disrupting its order and consequently enhancing its fluidity.

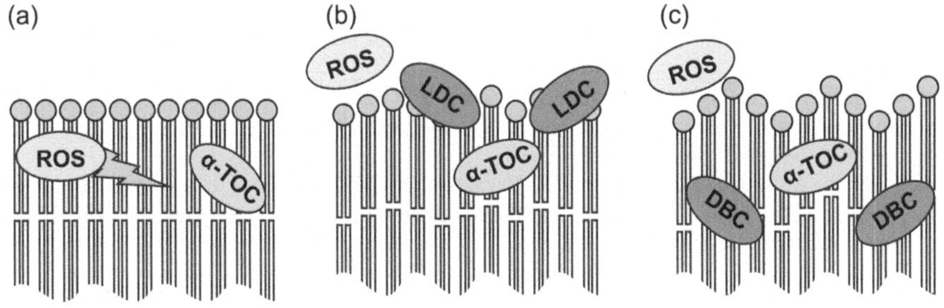


**Fig. 16.13** Scheme of the "Membrane Protection Model" (**a**) and membrane antioxidant effect of LDC (**b**) and DBC (**c**)

## 16.4 Conclusion

Radical scavengers generally suppress chain oxidation reactions based on the photodecomposition of KET. Even LAs, which lack radical-scavenging activity, can suppress photodecomposition. This is an intriguing phenomenon, as it suggests that substances without radical-scavenging activity can achieve antioxidation. Radical scavengers are categorized as those that act on protein oxidation and those that act on lipid peroxidation. Horizumi investigated the relationship between the antioxidant characteristics of radical scavengers and LAs by focusing on lipid peroxidation. He developed a membrane protection model using EyPC liposomes and compared the antioxidant activities of TRO with those of LAs using this model. The antioxidant activities of LAs and TRO antagonized each other. Interestingly, this antagonistic effect was observed only in the presence of lipid membranes, suggesting that LAs may alter properties such as membrane fluidity. The antioxidant response of LAs toward lipid membranes appears to structurally suppress the progression of radical reactions, as seen in their suppressive effect on KET photodecomposition. Moving forward, it will be necessary to examine the relationship between LAs and the group of radical scavengers that act on protein oxidation identified in the classification. These studies highlighted the potential of using non-radical-scavenging substances to suppress oxidation.

## References

1. Sugiura M, Hayakawa R, Xie ZL, Sugiura K, Hiramoto K, Shamoto M (2002) Experimental study on phototoxicity and the photosensitization potential of ketoprofen, suprofen, tiaprofenic acid and benzophenone and the photocross-reactivity in Guinea pigs. Photodermatol Photoimmunol Photomed 18:82–89. https://doi.org/10.1034/j.1600-0781.2002.180205.x
2. Le Coz CJ, Bottlaender A, Scrivener JN, Santinelli F, Cribier BJ, Heid E, Grosshans EM (1998) Photocontact dermatitis from ketoprofen and tiaprofenic acid: cross-reactivity study in 12 consecutive patients. Contact Derm 38:245–252. https://doi.org/10.1111/j.1600-0536.1998.tb05737.x
3. Nakajima A, Tahara M, Yoshimura Y, Nakazawa H (2005) Determination of free radicals generated from light exposed ketoprofen. J Photochem Photobiol A 174:89–97. https://doi.org/10.1016/j.jphotochem.2005.03.015
4. Nakajima A, Tahara M, Yoshimura Y, Nakazawa H (2007) Study of compounds suppressing free radical generation from UV-exposed ketoprofen. Chem Pharm Bull 55:1431–1438. https://doi.org/10.1248/cpb.55.1431
5. Monti S, Sortino S, DeGuidi G, Marconi G (1997) Photochemistry of 2-(3-benzoylphenyl)propionic acid (ketoprofen): part 1. A picosecond and nanosecond time resolved study in aqueous solution. J Chem Soc Faraday Trans 93:2269–2275. https://doi.org/10.1039/a700367f
6. Takara A, Kobayashi K, Watanabe S, Okuyama K, Shimada Y, Goto S (2017) Dibucaine inhibits ketoprofen photodegradation via a mechanism different from that of antioxidants. J Photochem Photobio A Chem 333:208–212. https://doi.org/10.1016/j.jphotochem.2016.10.026
7. Monti S, Sortino S, De Guidi G, Marconi G (1998) Supramolecular photochemistry of 2-(3-benzoylphenyl)propionic acid (ketoprofen). A study in the β-cyclodextrin cavity. New J Chem 22:1013. https://doi.org/10.1039/A709179F

8. Hiroshige R, Goto S, Tsunoda C, Ichii R, Shimizu S, Otsuka Y, Makino K, Takahashi H, Yokoyama H (2022) Trajectory of the spectral/structural rearrangements for photo-oxidative reaction of neat ketoprofen and its cyclodextrin complex. J Incl Phenom Macrocycl Chem 102:791–800. https://doi.org/10.1007/s10847-022-01160-3
9. Tateuchi R, Sagawa N, Shimada Y, Goto S (2015) Enhancement of the 1-octanol/water partition coefficient of the anti-inflammatory indomethacin in the presence of lidocaine and other local anesthetics. J Phys Chem B 119:9868–9873. https://doi.org/10.1021/acs.jpcb.5b03984
10. Huang D, Qu B, Prior RL (2005) The chemistry behind antioxidant capacity assays. J Agric Food Chem 53:1841–1856. https://doi.org/10.1021/jf030723c
11. Dong H, Secundo F, Xue C, Mao X (2017) Whole-cell biocatalytic synthesis of cinnamyl acetate with a novel esterase from the DNA library of *Acinetobacter hemolyticus*. J Agric Food Chem 65:2120–2128. https://doi.org/10.1021/acs.jafc.6b05799
12. Garzón A, Bravo I, Barbero AJ, Albaladejo J (2014) Mechanistic and kinetic study on the reactions of coumaric acids with reactive oxygen species: a DFT approach. J Agric Food Chem 62:9705–9710. https://doi.org/10.1021/jf5011148
13. Zhang X, Zhang M, Dong L, Jia X, Liu L, Ma Y, Huang F, Zhang R (2019) Phytochemical profile, bioactivity, and prebiotic potential of bound phenolics released from rice bran dietary fiber during in vitro gastrointestinal digestion and colonic fermentation. J Agric Food Chem 67:12796–12805. https://doi.org/10.1021/acs.jafc.9b06477
14. Re R, Pellegrini N, Proteggente A, Pannala A, Yang M, Rice-Evans C (1999) Antioxidant activity applying an improved ABTS radical cation decolorization assay. Free Radic Biol Med 26(9–10):1231–1237. https://doi.org/10.1016/S0891-5849(98)00315-3
15. Liang N, Kitts DD (2014) Antioxidant property of coffee components: assessment of methods that define mechanisms of action. Molecules 19:19180–19208. https://doi.org/10.3390/molecules191119180
16. Liao W, Gu L, Zheng Y, Zhu Z, Zhao M, Liangb M, Ren J (2016) Analysis of the quantitative structure–activity relationship of glutathione-derived peptides based on different free radical scavenging systems. Med Chem Commun 7:2083–2093. https://doi.org/10.1039/C6MD00006A
17. Glazer AN (1990) Phycoerythrin fluorescence-based assay for reactive oxygen species. Methods Enzymol 186:161–168. https://doi.org/10.1016/0076-6879(90)86106-6
18. Prior RL, Wu X, Schaich K (2005) Standardized methods for the determination of antioxidant capacity and phenolics in foods and dietary supplements. J Agric Food Chem 18:4290–4302. https://doi.org/10.1021/jf0502698
19. Ou B, Hampsch-Woodill M, Prior RL (2001) Development and validation of an improved oxygen radical absorbance capacity assay using fluorescein as the fluorescent probe. J Agric Food Chem 49:4619–4626. https://doi.org/10.1021/jf010586o
20. Takatsuka M, Goto S, Kobayashi K, Otsuka Y, Shimada Y (2021) Leading individual features of antioxidant systematically classified by the ORAC assay and its single electron transfer and hydrogen atom transfer reactivities; analyzing ALS therapeutic drug Edaravone. BBA Adv 1:100030. https://doi.org/10.1016/j.bbadva.2021.100030
21. Liao W, Ning Z, Ma L, Yin X, Wei Q, Yuan E, Yang J, Ren J (2014) Recrystallization of dihydromyricetin from Ampelopsis grossedentata and its anti-oxidant activity evaluation. Rejuvenation Res 17:422–429. https://doi.org/10.1089/rej.2014.1555
22. Tokumaru O, Shuto Y, Ogata K, Kamibayashi M, Bacal K, Takei H, Yokoi I, Kitano T (2018) Dose-dependency of multiple free radical-scavenging activity of edaravone. J Surg Res 228:147–153. https://doi.org/10.1016/j.jss.2018.03.020
23. Kraus RL, Pasieczny R, Lariosa-Willingham K, Turner MS, Jiang A, Trauger JW (2005) Antioxidant properties of minocycline: neuroprotection in an oxidative stress assay and direct radical-scavenging activity. J Neurochem 94:819–827. https://doi.org/10.1111/j.1471-4159.2005.03219.x
24. Qiang X, Li Y, Yang X, Luo L, Xu R, Zheng Y, Cao Z, Tan Z, Deng Y (2017) DL-3-n-butylphthalide-Edaravone hybrids as novel dual inhibitors of amyloid-β aggregation and

monoamine oxidases with high antioxidant potency for Alzheimer's therapy. Bioorg Med Chem Lett 27:718–722. https://doi.org/10.1016/j.bmcl.2017.01.050
25. Litwinienko G, Ingold KU (2007) Solvent effects on the rates and mechanisms of reaction of phenols with free radicals. Acc Chem Res 40:222–230. https://doi.org/10.1021/ar0682029
26. Wang TH, Wang SY, Wang XD, Jiang HQ, Yang YQ, Wang Y, Cheng JL, Zhang CT, Liang WW, Feng HL (2018) Fisetin exerts antioxidant and neuroprotective effects in multiple mutant hSOD1 models of amyotrophic lateral sclerosis by activating ERK. Neuroscience 379(21):152–166. https://doi.org/10.1016/j.neuroscience.2018.03.008
27. Su LJ, Zhang JH, Gomez H, Murugan R, Hong X, Xu D, Jiang F, Peng ZY (2019) Reactive oxygen species-induced lipid peroxidation in apoptosis, autophagy, and ferroptosis. Oxidative Med Cell Longev 2019:1–13. https://doi.org/10.1155/2019/5080843
28. Niki E (1987) Antioxidants in relation to lipid peroxidation. Chem Phys Lipids 44:227–253. https://doi.org/10.1016/0009-3084(87)90052-1
29. Köhler S, Thomas AJ, Lloyd A, Barber R, Almeida OP, O'Brien JT (2010) White matter hyperintensities, cortisol levels, brain atrophy and continuing cognitive deficits in late-life depression. Br J Psychiatry 196:143–149. https://doi.org/10.1192/bjp.bp.109.071399
30. Romeo Z, Spironelli C (2022) Hearing voices in the head: two meta-analyses on structural correlates of auditory hallucinations in schizophrenia. Neuroimage Clin 36:103241. https://doi.org/10.1016/j.nicl.2022.103241
31. Ernster L, Nordenbrand K (1967) Microsomal lipid peroxidation. Methods Enzymol 10:574–580. https://doi.org/10.1016/0076-6879(67)10099-2
32. Yuen KS, Halliday GM (1997) α-Tocopherol, an inhibitor of epidermal lipid peroxidation, prevents ultraviolet radiation from suppressing the skin immune system. Photochem Photobiol 65:587–592. https://doi.org/10.1111/j.1751-1097.1997.tb08610.x
33. Jovanović AA, Balanć BD, Ota A, Grabnar AP, Djordjević VB, Šavikin KP, Bugarski BM, Nedović VA, Ulrih PN (2018) Comparative effects of cholesterol and $\beta$-sitosterol on the liposome membrane characteristics. Eur J Lipid Sci Technol 120:1800039. https://doi.org/10.1002/ejlt.201800039
34. Urban BW, Bleckwenn M, Barann M (2006) Interactions of anesthetics with their targets: non-specific, specific or both? Pharmacol Ther 111:729–770. https://doi.org/10.1016/j.pharmthera.2005.12.005
35. Lugli AK, Yost CS, Kindler CH (2009) Anaesthetic mechanisms: update on the challenge of unraveling the mystery of anaesthesia. Eur J Anaesthesiol 26:807–820. https://doi.org/10.1097/EJA.0b013e32832d6b0f
36. Petersen EN, Chung H-W, Nayebosadri A, Hansen SB (2016) Kinetic disruption of lipid rafts is a mechanosensor for phospholipase D. Nat Commun 7:13873. https://doi.org/10.1038/ncomms13873
37. Meltzer HY, Fibiger HC (1996) Olanzapine: a new atypical antipsychotic drug. Neuropsychopharmacology 14:83–85. https://doi.org/10.1016/0893-133X(95)00197-L
38. Kanai Y, Katsuki H, Takasaki M (2000) Lidocaine disrupts neuronal membrane of rat sciatic nerve *in vitro*. Anesth Analg 91:944–948. https://doi.org/10.1097/00000539-200010000-00033
39. Matsumori N, Murata M, Tachibana K (1995) Conformational analysis of natural products using long-range carbon-proton coupling constants: three-dimensional structure of okadaic acid in solution. Tetrahedron 51:12229–12238. https://doi.org/10.1016/0040-4020(95)00790-F
40. Bechinger B, Lohner K (2006) Detergent-like actions of linear amphipathic cationic antimicrobial peptides. Biochim Biophys Acta 1758:1529–1539. https://doi.org/10.1016/j.bbamem.2006.07.001
41. Girke S, Mohr K, Schrape S (1989) Comparison between the activities of cationic amphiphilic drugs to affect phospholipid-membranes and to depress cardiac function. Biochem Pharmacol 8:2487–2496. https://doi.org/10.1016/0006-2952(89)90093-2
42. Boiteux C, Vorobyov I, French RJ, French C, Yarov-Yarovoy V, Allen TW (2014) Local anesthetic and antiepileptic drug access and binding to a bacterial voltage-gated sodium channel. Proc Natl Acad Sci U S A 111:13057–13062. https://doi.org/10.1073/pnas.1408710111

43. Grage SL, Culetto A, Ulrich AS, Weinschenk S (2021) Membrane-mediated activity of local anesthetics. Mol Pharmacol 100:502–512. https://doi.org/10.1124/molpharm.121.000252
44. Mildner J, Wnętrzak A, Dynarowicz-Latka P (2019) Cholesterol and cardiolipin importance in local anesthetics–membrane interactions: the Langmuir monolayer study. J Membr Biol 252:31–39. https://doi.org/10.1007/s00232-018-0055-6
45. Horizumi Y, Goto S, Takatsuka M, Yokoyama H (2023) Effects of local anesthetics on liposomal membranes determined by their inhibitory activity of lipid peroxidation. Mol Pharmacol 20:2911–2918. https://doi.org/10.1021/acs.molpharmaceut.2c01053
46. Liu T, Xu H, Han T, Wang J, Yin F, Wang C (2021) Effect of dietary egg yolk lecithin levels on survival, growth, lipid metabolism, and antioxidant capacity of early juvenile green mud *crab Scylla paramamosain*. Aquaculture 540:736706. https://doi.org/10.1016/J.AQUACULTURE.2021.736706
47. Bangham AD, Standish MM, Watkins JC (1965) Diffusion of univalent ions across the lamellae of swollen phospholipids. J Mol Biol 13:238–252. https://doi.org/10.1016/S0022-2836(65)80093-6
48. Deamer D, Bangham AD (1976) Large volume liposomes by an ether vaporization method. Biochim Biophys Acta 443:629–634. https://doi.org/10.1016/0005-2736(76)90483-1
49. Bektasģğlu B, Esin Çelik S, Özyürek M, GüÇlü K, Apak R (2006) Novel hydroxyl radical scavenging antioxidant activity assay for water-soluble antioxidants using a modified CUPRAC method, Biochem Biophys Res Commun 345:1194–1200 https://doi.org/10.1016/j.bbrc.2006.05.038
50. Tsunoda C, Goto S, Hiroshige R, Kasai T, Okumura T, Yokoyama H (2023) Optimization of the stability constants of the ternary system of diclofenac/famotidine/β-cyclodextrin by nonlinear least-squares method using theoretical equations. Int J Pharm 638:122913. https://doi.org/10.1016/j.ijpharm.2023.122913
51. Tsunoda C, Hasegawa K, Hiroshige R, Kasai T, Yokoyama H, Goto S (2023) Effect of cyclodextrin complex formation on solubility changes of each drug due to intermolecular interactions between acidic NSAIDs and basic H2 blockers. Mol Pharm 20:5032–5042. https://doi.org/10.1021/acs.molpharmaceut.3c00291
52. Hasegawa K, Goto S, Kataoka H, Chatani H, Kinoshita T, Yokoyama H, Tsuchida T (2024) Quantification of crystallinity during indomethacin crystalline transformation from α- to γ-polymorphic forms and of the thermodynamic contribution to dissolution in aqueous buffer and solutions of solubilizer. RSC Adv 14:4129. https://doi.org/10.1039/d3ra08481g
53. Otsuka Y, Kuwashima W, Tanaka Y, Yamaki Y, Shimada Y, Goto S (2021) Effects of heat treatment on indomethacin-cimetidine mixture; investigation of drug-drug interaction using singular value decomposition in FTIR spectroscopy. J Pharm Sci 110:1142–1147. https://doi.org/10.1016/j.xphs.2020.09.049
54. Shiratori T, Goto S, Sakaguchi T, Kasai T, Otsuka Y, Higashi K, Makino K, Takahashi H, Komatsu K (2021) Singular value decomposition analysis of the secondary structure features contributing to the circular dichroism spectra of model proteins. Biochem Biophys Rep 28:101153. https://doi.org/10.1016/j.bbrep.2021.101153
55. Hasegawa K, Goto S, Tsunoda C, Kuroda C, Okumura Y, Hiroshige R, Wada-Hirai A, Shimizu S, Yokoyama H, Tsuchida T (2023) Using singular value decomposition to analyze drug/β--cyclodextrin mixtures: insights from X-ray powder diffraction patterns. Phys Chem Chem Phys 25:29266–29282. https://doi.org/10.1039/D3CP02737F
56. Takatsuka M, Goto S, Kobayashi K, Otsuka Y, Shimada Y (2022) Evaluation of pure antioxidative capacity of antioxidants: ESR spectroscopy of stable radicals by DPPH and ABTS assays with singular value decomposition. Food Biosci 48:101714. https://doi.org/10.1016/j.fbio.2022.101714
57. Kurosawa Y, Otsuka Y, Goto S (2022) Increased selectivity of sodium deoxycholate to around Tryptophan213 in bovine serum albumin upon micellization as revealed by singular value decomposition for excitation-emission matrix. Colloids Surf B Biointerface 212:112344. https://doi.org/10.1016/j.colsurfb.2022.112344

58. Kurosawa Y, Goto S, Mitsuya K, Otsuka Y, Yokoyama H (2023) Interaction mode of hydroxypropyl-β-cyclodextrin with vaccine adjuvant components Tween 80 and Triton X-100 revealed by fluorescence increasing-quenching analysis. Phys Chem Chem Phys 25:6203–6213. https://doi.org/10.1039/d3cp00094j
59. Koga R, Tsuchida T, Tsunoda C, Kataoka H, Suenaga S, Goto S (2024) Physicochemical properties of diclofenac regulated by its intermolecular interactions with caffeine evaluated via vector space model, nuclear magnetic resonance, and Fourier-transform infrared analyses. Mater Chem Phys 333:130158. https://doi.org/10.1016/j.matchemphys.2024.130158
60. Minamide M, Tsurushima M, Koga R, Hasegawa K, Kurosawa Y, Tsuchida T, Goto S (2024) Distinguishing the transitions of fluorescence spectra of tryptophan-134 and 213 in BSA induced by bindings of UV filters, oxybenzone-3, and avobenzone. Bull Chem Soc Jpn 97(6):uoae058. https://doi.org/10.1093/bulcsj/uoae058
61. Hansch C, Leo A (1979) Substituent constants for correlation analysis in chemistry and biology. JoHn Wiley & Sons, New York

# Chapter 17
# In Situ Observations of Friction Interfaces

**Kaisei Sato, Hikaru Okubo, Shouhei Kawada, Seiya Watanabe, and Shinya Sasaki**

**Abstract** Tribology is an engineering field that deals with friction, wear, and lubrication. It plays a crucial role in improving the energy efficiency, durability, and reliability of mechanical systems. While friction and wear phenomena are observed as macroscopic events, their fundamental mechanisms are largely governed by atomic and molecular behaviors at the frictional interfaces. Therefore, surface analysis techniques aimed at elucidating these mechanisms are key to advancing tribology technologies. However, frictional interfaces exist in a dynamic, non-equilibrium state, exposed to high temperatures, high pressures, and high shear fields. This has made it challenging to obtain sufficient information using conventional surface analysis methods. In recent years, the application of in situ or operando analysis techniques, which allow for real-time observation of the states at frictional interfaces, has gradually shed light on friction and wear phenomena that were previously considered black boxes. This article introduces in situ analysis techniques for investigating frictional surfaces.

**Keywords** Tribology · Friction · Wear · Tribo-chemical reaction · In situ observation · SFG · Raman spectroscopy · AFM

K. Sato · S. Sasaki (✉)
Department of Mechanical Engineering, Tokyo University of Science, Katsushika, Tokyo, Japan
e-mail: s.sasaki@rs.tus.ac.jp

H. Okubo
Yokohama National University, Yokohama, Japan

S. Kawada
Department of Mechanical Engineering, Kyushu University, Fukuoka, Fukuoka, Japan

S. Watanabe
Department of Chemistry, Graduate School of Science, Osaka University, Toyonaka, Osaka, Japan



M. Hashizume, Y. Imura (eds.), *Colloids and Interfacial Dynamics*, Physical Chemistry in Action, https://doi.org/10.1007/978-981-96-6728-4_17

## 17.1 Introduction

Tribology is an engineering field that deals with friction, wear, and lubrication. It plays a crucial role in improving the energy efficiency, durability, and reliability of mechanical systems. While friction and wear phenomena are observed as macroscopic events, their fundamental mechanisms are largely governed by atomic and molecular behaviors at the frictional interfaces. Therefore, surface analysis techniques aimed at elucidating these mechanisms are key to advancing tribology technologies.

The use of surface analysis techniques in the field of tribology has evolved significantly over time. Starting with optical microscopy, the 1980s saw the widespread adoption of SEM for observing frictional surfaces and cross-sections, and electron probe micro analyzer (EPMA) for elemental analysis. Since then, TEM, SPM, and XRD have been actively utilized for structural observation and analysis, while AES, XPS, SIMS, IR, and Raman spectroscopy have been commonly applied for analyzing chemical composition and states. In recent years, confocal microscopy and optical interferometry have become standard for surface texture measurements, and nanoindentation and atomic force microscopy (AFM) are now widely used for mechanical property evaluation. Additionally, synchrotron radiation-based surface analysis techniques have also advanced, enabling further progress in understanding friction and wear phenomena that were previously regarded as black boxes. These insights have undoubtedly contributed significantly to the development of sliding materials and lubricant additives [1].

A defining characteristic of tribological surfaces is their exposure to mechanical stimulation due to friction. Even if the pre-friction surface originally consists of a crystalline structure with a single element, the frictional interface often features disrupted crystal structures with complex mixtures of opposing surface materials and environmental elements. Furthermore, in most practical applications, the surface is covered by adsorbed molecules or reaction products derived from lubricants. Therefore, obtaining information about the composition and structure of these complex surface layers is critical for tribological surface analysis.

Unlike standard surface analysis, which often involves preparing the surface to match the analysis method, preparing tribological surfaces can result in the loss of critical information (Fig. 17.1). This is a fundamental difference. In particular, when applying surface analysis techniques that require high vacuum to tribological surfaces, maintaining the original state of the surface while maximizing the analytical performance of the equipment becomes a significant challenge.

Moreover, tribological surfaces, which govern tribological characteristics, are in a state of constant dynamic change due to mechanical stimulation. To truly understand the state of frictional interfaces, in situ or operando observations that allow for real-time analysis are essential.

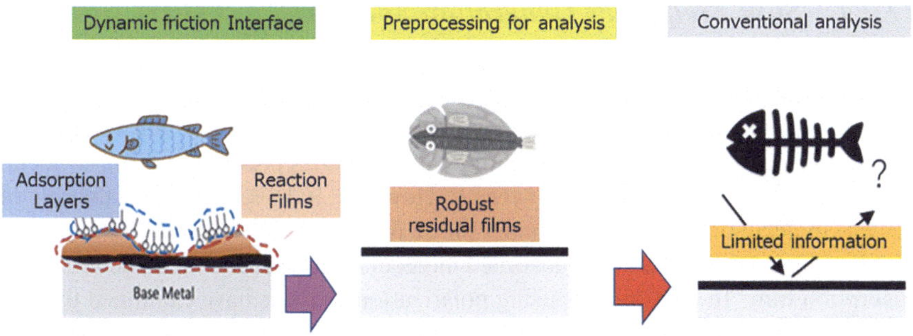


**Fig. 17.1** Importance of in situ observation and analysis of tribo-surfaces

## 17.2 In Situ Observation of Friction Interface

The difficulty in observing tribological surfaces in situ lies in the fact that the phenomena occurring in the microscopic regions between solid surfaces are dynamic in nature, involving phase transitions, chemical reactions, and material damage. Moreover, even if the contact surface itself is not directly observed, analyzing the surface condition immediately after friction requires maintaining the original environmental conditions. For example, in boundary lubrication, observation must be conducted in the presence of lubricants, making it challenging to apply analysis techniques that require a vacuum environment. Together with the constraint of measurement environments, analysis techniques with high sensitivity are often required. For example, to detect molecular adsorption films, which have mere nano-thickness but play an important role in friction reduction, sensitivity on the order of nanometers or sub-nanometers is required. Given the constraints of space, time, and environmental conditions, obtaining necessary information through in situ observation of frictional interfaces requires the establishment of customized observation techniques tailored to specific objectives.

## 17.3 Spectroscopic Analysis for In Situ Observation

In situ observation system integrates a spectroscopic analysis device with a friction apparatus, enabling real-time observation of tribological interfaces under operating conditions.

### 17.3.1 Sum Frequency Generation Spectroscopy (SFG)

When explaining the mechanisms of friction under boundary lubrication, the single-molecule adsorption model proposed by Bowden and Tabor is often referenced. This model depicts an adsorption film of polar molecules preventing metal-to-metal contact as a boundary lubrication layer [2]. However, it does not address the role of solvent molecules. Furthermore, under shear resistance at the frictional interface, questions remain about how the adsorbed molecular groups form the structure of the adsorption film. To date, studies using polarization analysis have examined the orientation of adsorbed molecular films. However, despite being one of the most fundamental issues for discussing boundary lubrication mechanisms, many aspects remain unclear.

Sum frequency generation spectroscopy (SFG) has emerged as a powerful tool for investigating these unresolved questions [3]. By selectively probing interfacial molecular structures with high sensitivity and specificity, SFG can provide critical insights into the arrangement and behavior of adsorption molecules under boundary lubrication conditions. Such findings are crucial for advancing our understanding of friction and wear mechanisms at the molecular level. SFG spectroscopy is a highly sensitive analytical technique for probing interfacial structures, characterized by its ability to provide detailed information about the species and orientation of adsorbed molecules. We successfully integrated a friction mechanism into an SFG spectroscopic system, enabling real-time observation of structural changes in adsorbed molecules induced by friction [4]. A particularly noteworthy finding from our research was the revelation that, under lubrication with stearic acid and *n*-dodecane, structural changes at the interface under shear stress were not observed in the stearic acid adsorbed molecules themselves. Instead, the changes occurred in the *n*-dodecane solvent molecules constrained by the stearic acid.

This result challenged the conventional understanding of boundary lubrication mechanisms, which has historically focused solely on adsorption films composed of polar molecules. The findings provided direct evidence that solvent molecules, which do not participate in forming robust chemisorbed films, play a significant role in friction mechanisms. Moreover, we demonstrated through a series of studies that the shear field induced by friction has a substantial influence on the formation of adsorption films [5]. This work highlights the complex interplay between adsorbed molecules and solvent molecules in determining the lubrication performance under boundary conditions (Fig. 17.2).

### 17.3.2 Raman Spectroscopy

The composition and properties of reaction films formed at frictional interfaces provide critical insights when discussing boundary lubrication mechanisms. Conventionally, surface analysis techniques such as SEM-EDX, XPS, and

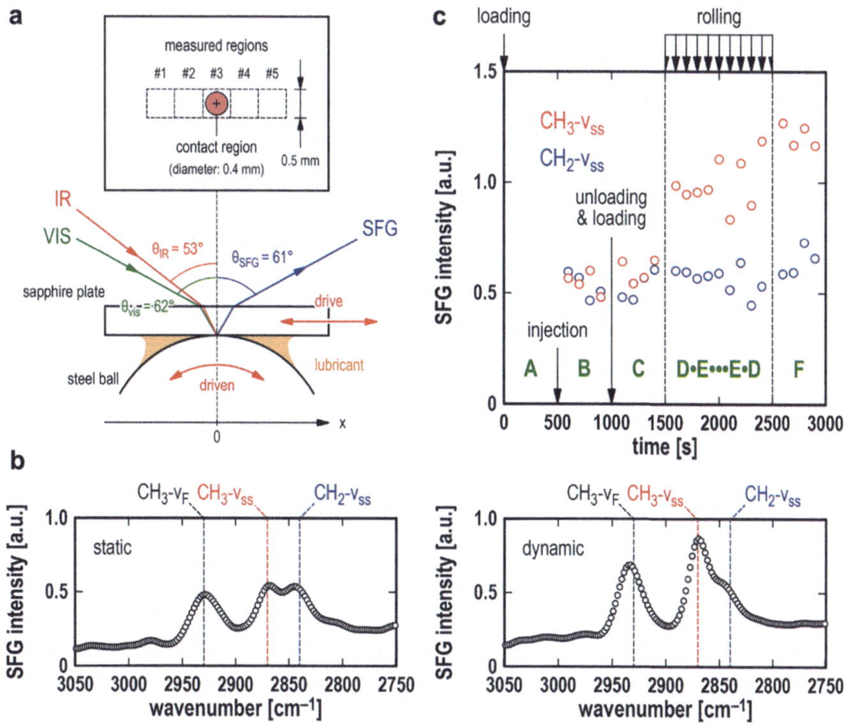


**Fig. 17.2** SFG measurements. (**a**) Schematic of the SFG apparatus using the pure rolling reciprocating rig, which is the same as the unkra thin film interferometry (UTFI) apparatus. (**b**) SFG spectra of SA/HD-$d_{34}$ under static and dynamic conditions. (**c**) Temporal change in the intensity of $CH_2$-$v_{ss}$ and $CH_3$-$v_{ss}$ of SA/HD-$d_{34}$ measured according to the same rig motion protocol as the film thickness measurements [5]

ToF-SIMS are employed after cleaning the test specimens following friction tests. These methods deliver high-resolution, high-sensitivity information on elemental composition and chemical bonding states. However, they do not capture the dynamic behavior of "live" frictional interfaces in real time.

For in situ observation of frictional surfaces in atmospheric conditions, particularly in the presence of lubricants, electron-based techniques cannot be utilized due to immediate absorption. Instead, light (including X-rays) serves as the only viable medium for transmitting information. While SFG spectroscopy is one such technique, infrared spectroscopy has long been utilized as well. Raman spectroscopy, which also provides molecular vibrational spectrum information, complements infrared spectroscopy due to differing sensitivities based on the material being analyzed. We developed an in situ micro-Raman spectroscopy system by integrating a friction mechanism with a micro-Raman spectrometer. This setup enabled simultaneous measurements of structural changes at frictional interfaces and reaction film formation processes alongside frictional behavior. Using this system, they elucidated the abnormal wear mechanisms of diamond-like carbon (DLC) films. The mechanisms can be explained by both the structural changes occurring on the surface of the DLC film and the formation process of lubricant-derived tribofilms, as monitored by the in situ Raman tribometer [6].

The in situ micro-Raman observation method holds great promise for furthering our understanding of the interactions between additives and frictional surfaces under boundary lubrication conditions since Raman spectroscopy enables the simultaneous acquisition of diverse molecular information such as the "concentration," "internal strain," "temperature change," and "crystallinity" of materials through spectral analysis and is used for the tribological in situ analysis of various materials, including inorganic and organic substances, low-molecular-weight compounds, and polymers. By directly observing these interactions and their effects on lubrication performance, this approach is expected to advance both tribological science and the development of more effective lubrication technologies.

## 17.4 Atomic Force Microscopy (AFM)

Atomic force microscopy (AFM) can be broadly divided into two operational modes. The contact mode does not excite the cantilever but detects its static displacement by making the probe contact the surface. This measurement method is based on the basic operating principle described above. However, due to strong interactions with the surface, it has disadvantages, such as probe tip damage and the inability to achieve reproducible high-resolution imaging. The other mode, the dynamic mode, involves vibrating the cantilever near its resonance frequency. The amplitude/frequency changes as the cantilever tip approaches the sample surface in this state. By suppressing contact between the probe and the surface, it is possible to achieve stable, high-resolution imaging. The dynamic mode includes the amplitude modulation (AM) mode, which detects changes in vibration amplitude by forcibly oscillating the cantilever near the resonance frequency, and the frequency modulation (FM) mode, which directly detects resonance frequency shifts.

FM-AFM achieves high resolution by utilizing the resonance of a cantilever with a high Q factor and detecting frequency changes with high sensitivity [7]. It has advanced as a high-resolution observation method in high vacuum; therefore, it has been considered unsuitable for measurements in solutions where the Q factor significantly decreases. In recent years, new methods for detecting minute displacements and techniques for signal noise countermeasures have been developed, making it possible to perform high-sensitivity and high-resolution imaging even in solutions [8]. One characteristic of FM-AFM measurements is cross-sectional imaging of solid–liquid interfaces. As shown in Fig. 17.3, the probe approaches the sample surface vertically, and the frequency change ($\Delta f$) during this process is measured as a force curve. By continuously acquiring force curves in the $x$-direction on the $Z$-$X$ plane, the structure of the solid–liquid interface can be visualized as a cross-sectional image.

For in situ observation of frictional surfaces, contact mode is used during nanofriction testing. AM mode is employed for surface topography measurements. FM

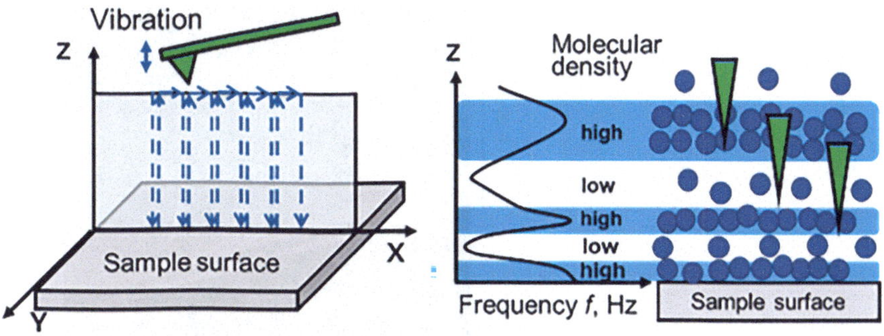


**Fig. 17.3** Schematic diagram of cross-sectional imaging by FM-AFM

mode is utilized for the observation of solid–liquid interfacial structures. Additionally, the shape, material, and spring constant of the probe and cantilever must be selected according to the application.

### *17.4.1 Ionic Liquids on Mica Surfaces*

Ionic liquids, composed entirely of charged ions, are strongly influenced by the potential of solid surfaces. Electrical interactions occur between ionic liquids and solid surfaces, leading to the formation of an electric double layer. In the case of ionic liquids, the formation of an electric double layer at the solid–liquid interface is considered to contribute to friction reduction. For this study, 1-ethyl-3-methylimidazolium dicyanamide was used as the ionic liquid, and the solid–liquid interfacial structure with mica was analyzed. Figure 17.4 shows the cross-sectional imaging obtained by FM-AFM [9]. In the imaging, the brighter areas indicate stronger interaction forces. Near the surface, in addition to the electric double layer, the formation of a diffuse double layer was also confirmed. The thickness of each layer matched the size of the ionic liquid molecules.

Furthermore, when water was added to the ionic liquid under this condition, the layered structure disappeared. This observation aligns with the increase in the friction coefficient upon adding water, supporting the hypothesis that the formation of an electric double layer contributes to friction reduction.

### *17.4.2 DLC Surface Under Lubrication with Fatty Acid Esters*

Diamond-like carbon (DLC) exhibits a low friction coefficient of approximately 0.02–0.04 when lubricated with fatty acid ester oils. To elucidate the mechanism behind this behavior, the solid–liquid interfacial structure was observed using FM-AFM. Figure 17.5 shows cross-sectional imaging before and after friction. The

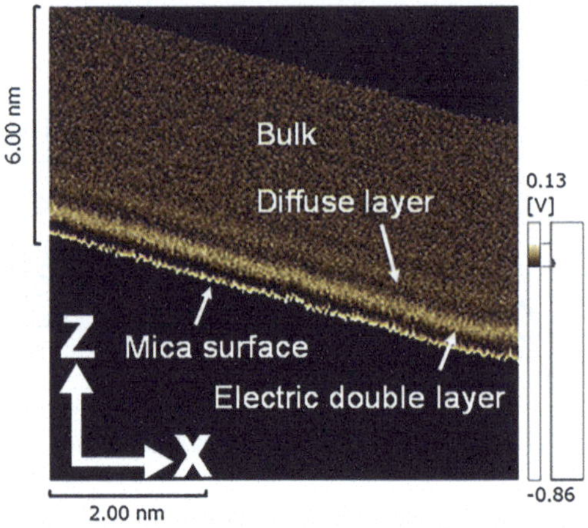


**Fig. 17.4** Cross-sectional imaging of the mica and ionic liquid interface

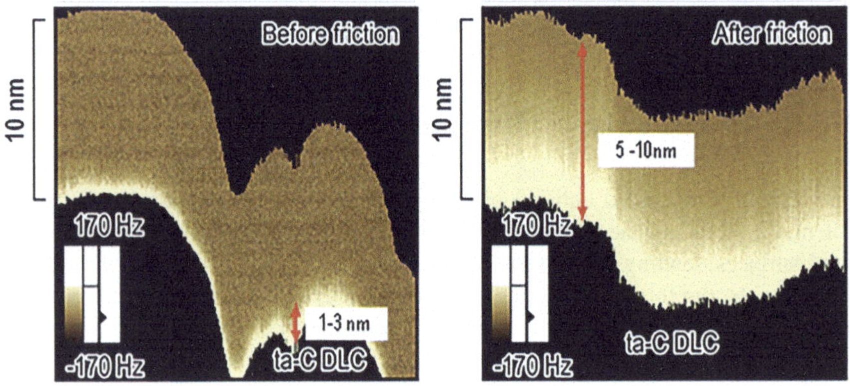


**Fig. 17.5** Cross-sectional imaging of the DLC and fatty acid ester interface

thickness of the bright layer with strong interaction forces was approximately 2 nm before friction, whereas it increased up to about 10 nm, 4–5 times larger, after friction. From these observations, it was inferred that the friction-reducing effect of fatty acid ester oil on DLC is attributed to the formation of a tribofilm on the friction surface [10].

### 17.4.3 CNF in Ester-Added Oil

A molded body composed entirely of cellulose nanofiber (CNF) exhibits an ultra-low friction phenomenon with a friction coefficient below 0.01 in ester-added oil. Figure 17.6 shows cross-sectional imaging of the surface of the 100% CNF molded body in base oil (poly-α-olefin, PAO) and ester-added oil. In PAO, the bright layer

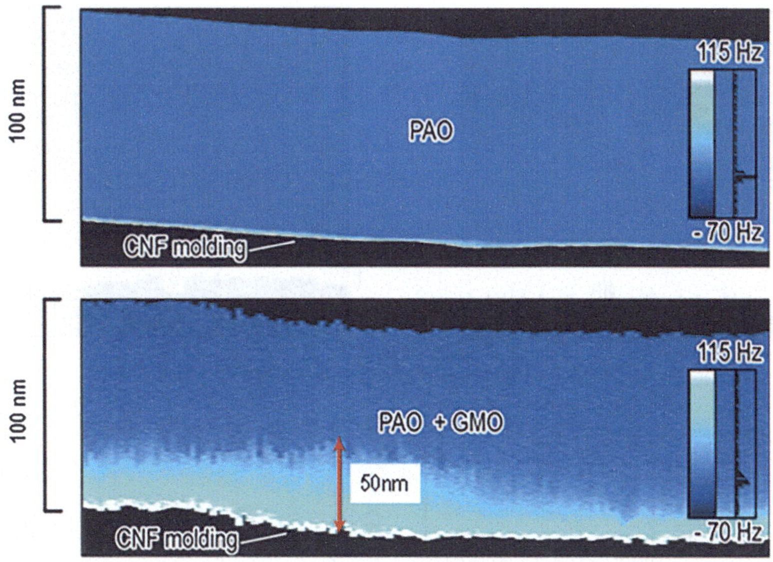


**Fig. 17.6** Cross-sectional imaging of CNF surface in lubricating oil. Top: in PAO; bottom: in ester-added oil

indicating strong interaction forces is scarcely observed. In contrast, in ester-added oil, a repulsive layer approximately 50 nm thick was identified on the CNF molded body surface. From this, it was inferred that the ultra-low friction phenomenon is attributed to the formation of a swollen layer of CNF fibers induced by the ester [11].

### 17.4.4 Soft Contact Lenses

The observation of solid–liquid interfacial structures using FM-AFM is a powerful tool for understanding ultra-low friction phenomena in soft matter. In the case of soft contact lenses, it has been hypothesized that the ultra-low friction observed in aqueous solutions is due to the formation of a soft hydrated layer on the surface. Cross-sectional imaging via FM-AFM has confirmed the existence of the hydrated layer, which was previously difficult to verify. It also revealed the relationship between the disappearance of the hydrated layer due to damage and the increase in friction, as well as the phenomenon of the hydrated layer recovering over time. Figure 17.7 shows cross-sectional images of the *Z*–*X* plane of a contact lens (CL) surface in saline solution obtained before and after friction testing, along with the frequency shift ($\Delta f$) along line segment AB [12)]. In the cross-sectional images of the *Z*–*X* plane, the $\Delta f$ changes caused by the interaction forces between the cantilever and the sample are displayed as brightness variations. Therefore, brighter regions indicate a larger positive $\Delta f$ change, signifying repulsive interaction regions.

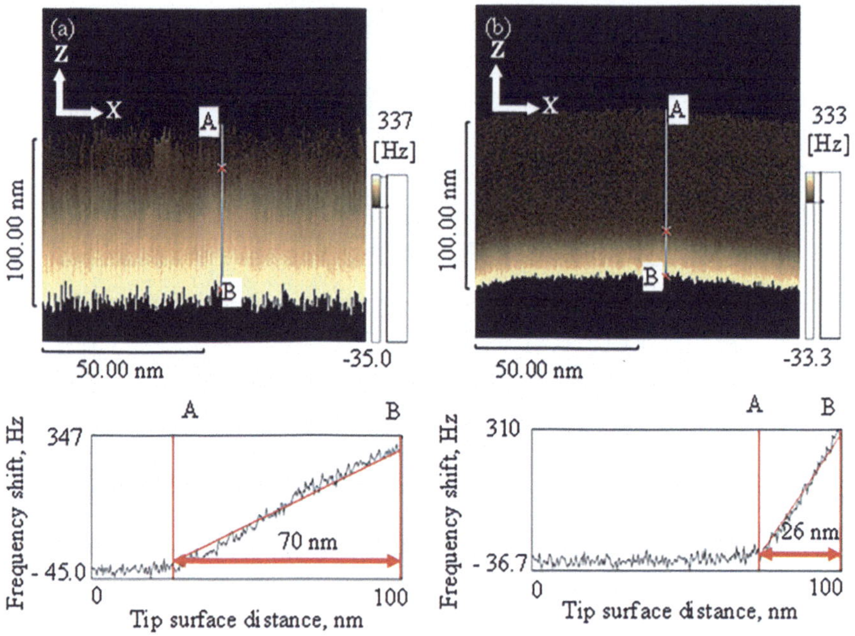


**Fig. 17.7** Z–X cross-section imaging and $\Delta f$ curve of the soft contact lens surface: (**a**) before and (**b**) after the friction test

On the CL surface before the friction test, a broad repulsive interaction region of approximately 70 nm was observed (Fig. 17.7a). In contrast, the CL surface after the friction test showed a narrower repulsive region of about 26 nm (Fig. 17.7b), confirming a reduction in the repulsive region compared to before the test. Additionally, in both pre- and post-friction surfaces, $\Delta f$ changes were observed to increase monotonically as the cantilever approached the interface. Focusing on the slope of the $\Delta f$ changes, it was confirmed that the slope after the friction test was steeper than that before the test.

### *17.4.5 In Situ Observation of Tribofilm Formation*

Conventional AFM devices are equipped with functions to measure various physical information in addition to surface topography. One such function is FFM or LFM (lateral force microscopy), which operates in contact mode to scan the probe and detects the lateral force (friction force) acting between the probe and the surface as the torsion of the cantilever, visualizing it as an image. Carpick et al. regarded this operation as a nano-tribotest capable of single asperity contact control and established it as a measurement technique for in situ observation of tribofilm formation [13]. In AFM-based surface measurements, sharp probes are typically used to

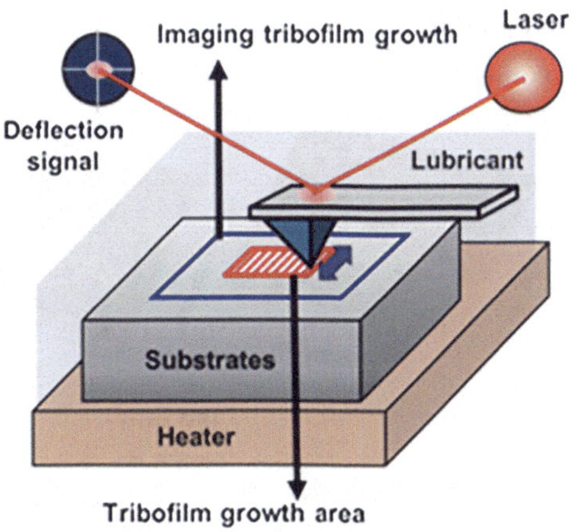


**Fig. 17.8** Schematic overview of in situ observation of tribofilm formation process

achieve high resolution. However, when using AFM as a nano-tribotester, the probe tip's shape and material must be selected according to the friction conditions. Especially for observing tribofilm formation processes, measures such as using colloidal probes or pre-worn probes are necessary to suppress shape changes caused by probe tip wear.

The observation of tribofilm formation processes is performed as illustrated in Fig. 17.8. Within a confined region, the probe is scanned in contact mode under high load. After a set number of scans, the surrounding area is measured in low-load contact mode or dynamic mode, and this process is repeated. Figure 17.9 shows the results of in situ AFM observation of the formation process of reaction films derived from ZDDP (zinc dialkyldithiophosphates). In base oil (PAO4) with added ZDDP (Fig. 17.8a), the reaction film was observed to grow uniformly within the scanned region. On the other hand, in oil with ZDDP and GMO (glycerol monooleate) added, non-uniform growth of the reaction film was observed (Fig. 17.8b), and after 240 min, the film thickness was approximately 20 nm, about half of the thickness observed with ZDDP alone [14].

In situ AFM observation not only captures shape changes but also enables the detection of friction characteristics of the reaction film using LFM mode and the time-dependent changes in adhesion force through force curve measurements [15].

## 17.5 Conclusions

In tribology, performance is evaluated based on phenomena manifested on a macroscopic scale. However, understanding the mechanisms underlying these phenomena requires tracing back to the molecular and atomic levels at the friction interface. In practical friction interfaces, various organic and inorganic substances, including oxygen and water from the atmosphere, as well as components of surfaces and lubricants, undergo continuous adsorption, reactions, and desorption under

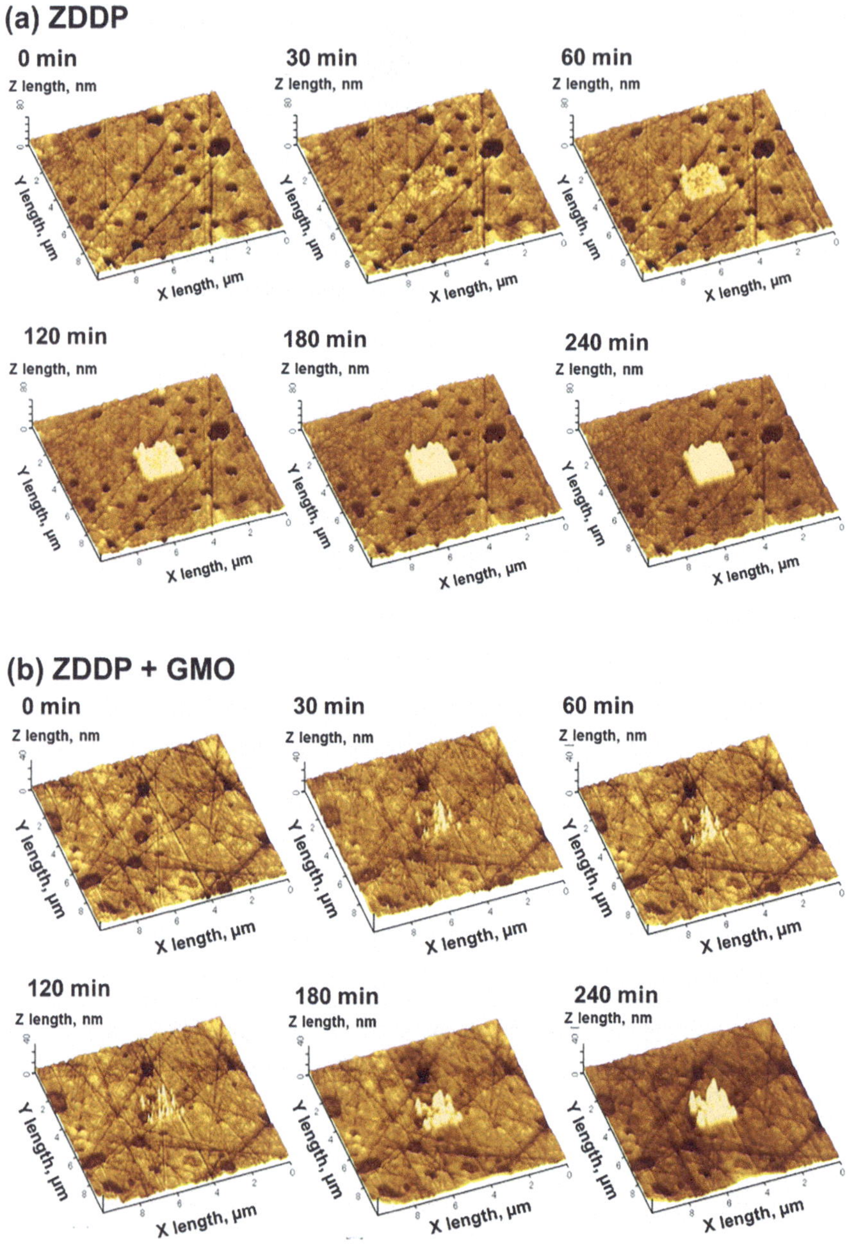


**Fig. 17.9** In situ observation of tribofilm formation process: (**a**) in ZDDP containing oil; (**b**) in ZDDP and GMO combined oil

high-pressure and high-temperature conditions. These processes result in a dynamic and highly complex state of constant change. In addition to the aforementioned techniques, various methods have been developed for in situ observation of friction interfaces. These include the use of optical microscopy and scanning electron microscopy for observing wear morphologies, transmission electron microscopy for studying adhesion and fracture processes between single asperities, as well as neutron reflectometry for structural analysis of adsorption films. The key to unraveling lubrication mechanisms lies in clearly identifying what specific information is needed from in situ observations and building a system that combines appropriate analytical techniques with the friction mechanism.

Recently, the integration of digital technologies has significantly enhanced spatial resolution, detection sensitivity, and the speed of data acquisition and analysis. This advancement is expected to expand the range of analytical techniques applicable to in situ observation of friction surfaces.

## References


1. Nito M et al (2021) Effects of mechanical properties of tribo-film on frictional phenomena under lubrication with CVTF. Trans Soc Autom Eng Jpn 52(3):574–581. (in Japanese)
2. Tadokoro C et al (2016) Polarization observations of adsorption behavior of fatty acids using optical anisotropy of liquid crystal. Tribol Lett 64(2):30–39
3. Miyamae T, Akaike K (2023) Analysis of molecular surface/interfacial layer by sum-frequency generation (SFG) spectroscopy. In: Interfacial phenomena in adhesion and adhesive bonding. Springer Nature, London, pp 291–360
4. Watanabe S et al (2016) Analysis of the interfacial molecular behavior of a lubrication film of n-dodecane containing stearic acid under lubricating conditions by sum frequency generation spectroscopy. Langmuir 32(51):13649–13656
5. Watanabe S et al (2022) Processes of molecular adsorption and ordering enhanced by mechanical stimuli under high contact pressure. Sci Rep 12:3870
6. Okubo H et al (2017) In situ Raman observation of the graphitization process of tetrahedral amorphous carbon diamond-like carbon under boundary lubrication in poly-alpha-olefin with an organic friction modifier. Tribol Online 12(5):229–237
7. Albrecht TR, Grütter P, Horne D, Rugar D (1991) Frequency modulation detection using high-Q cantilevers for enhanced force microscope sensitivity. J Appl Phys 69:668
8. Fukuma T, Kobayashi K, Matsushige K, Yamada H (2005) True molecular resolution in liquid by frequency-modulation atomic force microscopy. Appl Phys Lett 86:193108
9. Kawada S, Kodama E, Sato K, Ogawa S, Watanabe M, Okubo H, Sasaki S (2019) Effect of water on the interfacial structures of room-temperature ionic liquids. Surf Interface Anal 51(1):17
10. Okubo H, Sasaki S (2019) Frequency-modulation atomic force microscopic observation for ultralow frictional solid–liquid interface of diamond-like carbon in an environmentally friendly oil. Tribol Lett 67:3
11. Okubo H et al (2023) Novel environmentally superior tribomaterial with superlow friction: 100% cellulose nanofiber molding. Tribol Lett 71:83
12. Nakajima A, Sato K, Watanabe S, Sasaki S (2022) Proposal of a lubrication mechanism for contact lens surfaces based on hydrated layer observation using FM-AFM. Tribologist 67(8):581–588. (in Japanese )

13. Gosvami NN, Bares JA, Mangolini F, Konicek AR, Yablon DG, Carpick RW (2015) Mechanisms of antiwear tribo-film growth revealed in situ by single-asperity sliding contacts. Science 348:102
14. Sato K, Ouchi H, Okubo H, Watanabe S, Sasaski S (2022) Investigation of the lubrication mechanism by in-situ AFM observation for the formation process of the tribo-films under lubrication with combined solutions of ZDDP and organic friction modifiers. Tribologist 67(5):354–365. (in Japanese)
15. Sato K, Watanabe S, Sasaki S (2022) High friction mechanism of ZDDP tribofilm based on in situ AFM observation of nano-friction and adhesion properties. Tribol Lett 70:94

# Chapter 18
# Friction Dynamics on Human-Skin Surfaces

**Yoshimune Nonomura**

**Abstract** Skin-surface friction phenomena have been extensively studied since the 1950s, because of the nature of human skin as a viscoelastic soft material comprising the epidermis, dermis, and subcutaneous tissue, which induce unique mechanical phenomena. Additionally, as the skin-surface friction phenomenon reflects the biological state of the skin as well as the textures of creams or powders, a biomimetic tactile-sensing system that mimics the natural, smooth motion of humans when they touch an object, constantly changing its movement speed, was developed. This system can evaluate the friction profiles, friction-force velocity dependence, and sliding-out hysteresis. The developed system was used to evaluate the friction characteristics of the human skin, and the observed delay times and oscillatory phenomena reflected skin deformation. However, after applying moisturizers, such as glycerol, Vaseline, and silicone oil, the oscillations either disappeared or an adhesive phenomenon appeared (the strong attachment of the contact probe and skin). Overall, this manuscript presents the history of the studies on the friction phenomena occurring on human-skin surfaces and the friction dynamics observed using tactile-sensing systems.

**Keywords** Human skin · Friction · Sinusoidal motion · Moisturizer · Thickener

## 18.1 Motivation for Studying the Friction Dynamics of the Human Skin

The human skin is the most seen organ of the body. This organ must not be considered mundane and trivial. This organ consists of the epidermis, dermis, and subcutaneous tissue, each of which exhibits significant physical and chemical differences

Y. Nonomura (✉)
Department of Applied Chemistry, Chemical Engineering, and Biochemical Engineering, Academic Assembly, Yamagata University, Yamagata, Japan
e-mail: nonoy@yz.yamagata-u.ac.jp



M. Hashizume, Y. Imura (eds.), *Colloids and Interfacial Dynamics*, Physical Chemistry in Action, https://doi.org/10.1007/978-981-96-6728-4_18

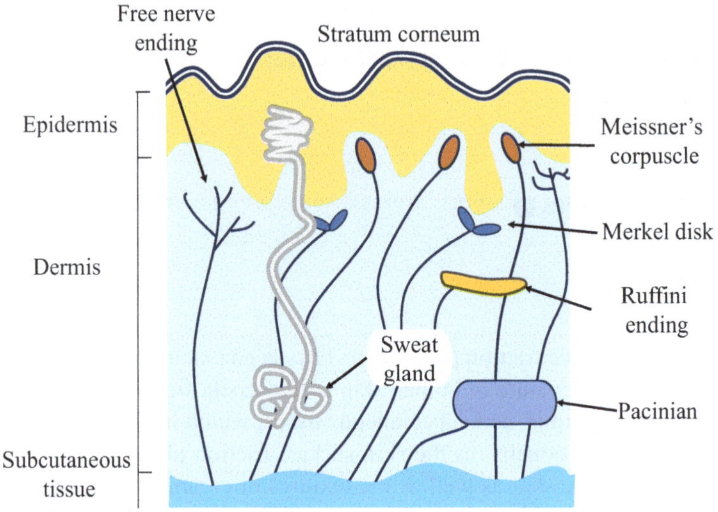


**Fig. 18.1** Structure of the human skin

(Fig. 18.1). Its top surface is covered by the stratum corneum, which is a thin, hard membrane composed of proteins, lipids, and water [1]. The skin protects us by performing numerous functions, such as ultraviolet protection and active-oxygen scavenging, in addition to its barrier and immune functions that prevent water evaporation as well as the entry of chemical substances, bacteria, and microorganisms into the body. Thus, the "No skin, no life" claim of some skin specialists is unsurprising.

The human skin perceives mechanical stimulations via the following four types of tactile receptors located inside the skin: Meissner's corpuscles, Merkel disks, Ruffini endings, and Pacinian [2]. These receptors exhibit different wavenumbers and response characteristics and combine to detect stimuli ranging from a few Hz to several 100 Hz. Additionally, the sensations of warmth, coldness, and pain perceived at the free nerve endings are added to evoke a wide variety of sensations.

Numerous methods have been developed over time to measure the hydration, elasticity, and surface topography of the human skin, as these physical and morphological properties strongly influence skin health as well as beauty [3–5]. Conversely, the frictional properties of the skin have also gained attention over the past decades. The tactile-dimension concept holds that the friction phenomenon occurring on the skin surface, along with the hardness, thermal properties, and surface topography, is among the factors constituting tactile sensations, such as moistness and smoothness. In the cosmetics industry, the textures of the skin and hair, following the application of powders and creams, have been analyzed using friction parameters [6–9]. In the textile industry, the friction between the skin and the fabric has been evaluated to improve the comfort of clothing and bedding. In addition to subjective judgments by touching, the fit of textiles has been quantified by objective physical evaluation methods based on the mechanical and surface properties of fabrics measured by the

Kawabata Evaluation System [10]. Furthermore, the evaluation of friction dynamics of the skin surface can elucidate several interesting phenomena in everyday life. When humans manipulate an object, they generally adjust their grip force in response to friction. This feedback effect occurs when the differences in pre-lift-off strain patterns are used by the nervous system to modulate the grip force [11]. Finger snapping, which is deployed in communication and music, is executed by the fastest motion produced by the human body, a peak angular acceleration of $106°s^{-2}$, and is controlled by skin friction [12].

## 18.2 Traditional Studies on the Friction Dynamics of Human Skin

### *18.2.1 Friction Evaluations of the Human Skin*

To the best of the author's knowledge, the frictional properties of human skin were first reported in the 1950s. Authors have reported that the friction coefficient ($\mu$) of the skin surface, when rubbed with various materials, depended on the surface properties of the materials [13]. In detail, when the skins of the hands, arms, and legs of ten healthy subjects were rubbed with five materials (aluminum, nylon, polyethylene, polyethylene terephthalate, and silicone), the calculated average $\mu$ was $0.46 \pm 0.15$. Notably, the palm produced the highest $\mu$, $0.62 \pm 0.22$. Additionally, silicone and nylon exhibited the highest ($0.61 \pm 0.21$) and lowest ($0.37 \pm 0.22$) $\mu$, respectively [14].

The skin behavior was not represented by a simple law of friction; rather, it was proportional to the power of the vertical force [15]. These phenomena occur because the skin undergoes viscoelastic deformation rather than a purely plastic one. The optical coherence tomography (OCT)-based observation of the interface between the skin and the contacting solid surface revealed an increase in the contact length of the OCT image as a power of the applied normal force [16].

Furthermore, it has been reported that the skin exhibits a more complex mechanical response. When a finger was slid across a sandpaper to identify its surface roughness, the vibrational acceleration increased with the increasing sliding speed, resulting in high-frequency stick-slip phenomena [17]. Moreover, the greater the magnitude of this vibration, the more favorable it was for bump recognition. Moreover, the anisotropy of the finger structure ensured that the friction phenomenon and tactile sensation depended on the hand-movement direction: when the fingers slipped in the distal direction, stick-slip phenomena occurred, reducing the bump-recognition ability [18].

The skin-friction properties may depend on the attributes of the subject. In a classic study, the kinetic-$\mu$ depended on the body part, although it did not differ significantly among age and gender groups [19]. However, recent data have identified age as one of the most important factors that change skin friction: older individuals exhibit lower skin hydration and friction than their younger counterparts; they exhibit lower abilities to perceive object bumps [20]. Furthermore, simulations have attributed this age-related difference to the increase in microscale roughness with aging, which reduces the true contact area between the contact probe and skin [21].

Friction reflects the physiological condition of the skin. Notably, the frictions on the hands, arms, and backs of atopic patients are lower than those of normal controls [22]. This is probably due to the dryness and hardness of the patient's skin. However, exfoliation with adhesive tapes decreases the skin's elastic modulus $E$ and increases $\mu$ [23].

### 18.2.2 Modeling of the Human-Skin Friction

Several mechanical models integrating the skin condition and viscoelasticity have been deployed to simulate skin-friction characteristics, although providing a universal understanding from the findings of these simulations is challenging, as the friction properties depend on the probe material and movement method, as described in Sect. 18.2.1. The frictional force $F$ observed when the skin is rubbed with a spherical probe can be explained by the interfacial component $F_{\text{int}}$ (originating from the energy dissipation to break the skin–probe contact point) and deformation component $F_{\text{def}}$ (generated by the viscoelastic deformation occurring at the front of the probe during sliding) (Fig. 18.2) [24]. $F$ is mathematically defined as follows:

$$F = F_{\text{int}} + F_{\text{def}} \tag{18.1}$$

Furthermore, $F_{\text{def}}$ is expressed in Eq. (18.2) based on the mechanical properties of the skin and the shape of the probe.

$$F_{\text{def}} = \left(\frac{9}{128R}\right)^{2/3} \left(\frac{1-\nu^2}{E}\right)^{1/3} W^{3/4}, \tag{18.2}$$

where $\beta$, $E$, and $\nu$ are the hysteresis loss ratio, elastic modulus, and Poisson's ratio of the skin, respectively, and $R$ is the probe radius. Conversely, $F_{\text{int}}$ is expressed as the product of the shear strength $\tau$ (between the skin and probe) and $A$.

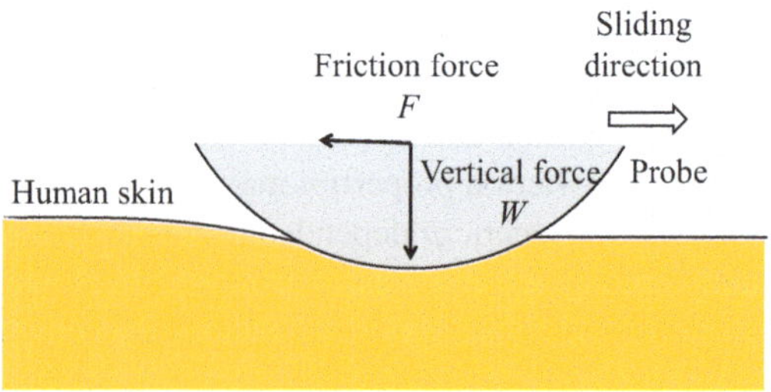


**Fig. 18.2** Contact condition of the probe on the human-skin surface

$$F_{\text{int}} = \tau A. \tag{18.3}$$

Moreover, as $\tau$ depends on the pressure applied to the skin, it can be expressed as follows:

$$\tau = \tau_0 + \alpha p, \tag{18.4}$$

where $\tau_0$ is the intrinsic $\tau$, $\alpha$ is the pressure coefficient, and $p = W/A$.

### *18.2.3 Effects of Water, Moisturizers, and Lubricants*

The frictional properties of the skin change when something is applied to it. This is because the applied substance exhibits adhesive and lubricating properties and also affects the skin condition. It is well known that water, among other substances, exerts significant effects. A study revealed that $\mu$ of the skin exceeded 1.6 when water or surfactant solutions were applied using a motor-driven wheel [25]. This water-induced friction force increase may be attributed to hydration, which softened the skin and increased the skin–solid surface true contact area. A Raman spectroscopy–based study revealed that the water content of the entire stratum corneum increased significantly when water was applied to the forearm skin for as long as 90 min, reaching the granular layer [26]. Finite-element simulations to compare the mechanical properties of skin under dry, natural, and wet conditions revealed that the bulk tissue of the finger softens upon hydration [27]. The findings elucidate why human fingers can securely grasp flat objects under a wide variety of conditions.

Furthermore, the effects of applying so-called moisturizers on friction have been reported. For example, the application of glycerol (or Vaseline) with water to the human skin increased its moisture content and $\mu$ [28]. Although these substances are viscous lubricating emollients that reduce $\mu$ of the skin immediately after application, they increase the coefficient via occlusive effects and hydration.

### *18.2.4 Friction Measurement Under Dynamic Conditions*

A friction-evaluation system that mimics natural and smooth movements was developed. Humans rarely move their hands at a constant speed when they touch objects; instead, they move naturally and smoothly. An observation of the motion with a high-speed camera reveals that the movement is often characterized by acceleration, "gradually speeding up, slowing down, and subsequently stopping" [29, 30]. Here, a friction-evaluation system was developed using a finger model and a scotch-yoke mechanism to convert elliptical motion into sinusoidal motion [31, 32]. Figures 18.3 and 18.4 show the biomimetic tactile-sensing system comprising a finger-model

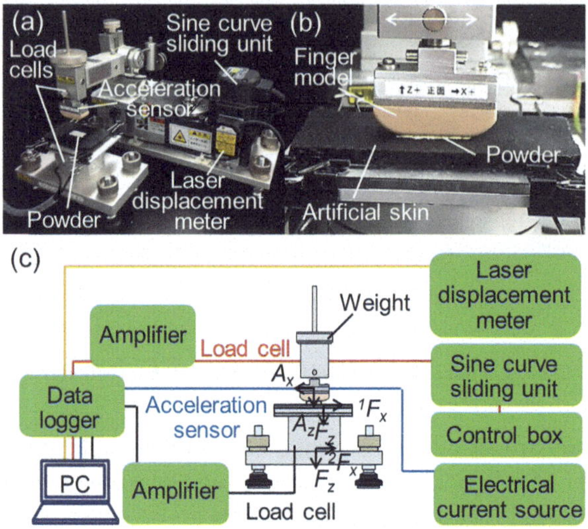


**Fig. 18.3** Biomimetic tactile-sensing system with a finger-model contact probe. (**a**) Overall view, (**b**) contact probe, and (**c**) schematic. (Reprinted with permission from [33])

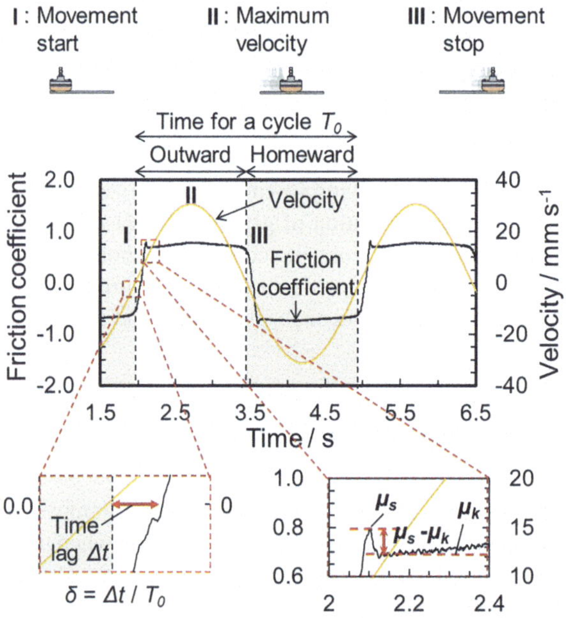


**Fig. 18.4** Typical temporal friction profile of a cosmetic powder. (Reprinted with permission from [33])

contact probe and the friction profile of a cosmetic powder under sinusoidal motion [33]. When the contact probe began to slide, $\mu$ increased, reaching the static $\mu$ ($\mu_s$) and slightly decreasing to the dynamic $\mu$ ($\mu_k$). A delay $\Delta t$ in the response of $\mu$ to the sliding of the contact probe was observed, and this delay time ($\delta$) was expressed as the ratio of the sinusoidal-motion period, $T_0$.

$$\delta = \Delta t \, / \, T_0. \tag{18.5}$$

An evaluation under sinusoidal motion provides information that cannot be obtained from the measurements under constant-velocity conditions. The first is $\delta$. On soft material surfaces, such as the human skin, material deformation occurs during the pre-sliding of the contact probe, and this also affects the tactile sensation. As the sliding velocity often changes under sinusoidal motion, the velocity dependence can be quantitatively analyzed by a single measurement. The velocity dependence of the mechanical stimulus applied to the skin during dynamic friction strongly influences tactile perception.

A model was also proposed for analyzing the pre-sliding and dynamic-friction processes. In a typical friction profile, $\mu$ in the pre-sliding region ($f_{st}$) increases at the start of the movement and can be expressed by Eq. (18.6):

$$f_{st} = \frac{1}{2}\frac{k_s}{W}\frac{V^2}{|\dot{V}|} + \frac{\eta A}{W}V - f_0, \tag{18.6}$$

where $k_s$ and $\eta A$ are the elastic and viscous moduli, respectively, and $V$ and $W$ are the sliding velocity and vertical force, respectively. This equation reveals that $f_{st}$ can be analyzed as a quadratic function of the velocity. Thus, $\mu$ in the dynamic-friction process ($f_d$) is represented by Eq. (18.7):

$$f_d = \frac{C}{W}V + f_c, \tag{18.7}$$

where $C$ is the viscosity coefficient for this process.

## 18.3 Friction Dynamics of the Human Skin

### *18.3.1 Naked Skin*

The biomimetic tactile-sensing system was used to evaluate the friction on the fingers, palms, and forearms of 20 subjects (Fig. 18.5) [34]. The skin surfaces at points 1–6 were rubbed with a contact probe, after which the effects of the changing water content of the stratum corneum at each position, as well as the skin $E$ on the friction parameters, were analyzed to determine the factors governing the friction dynamics.

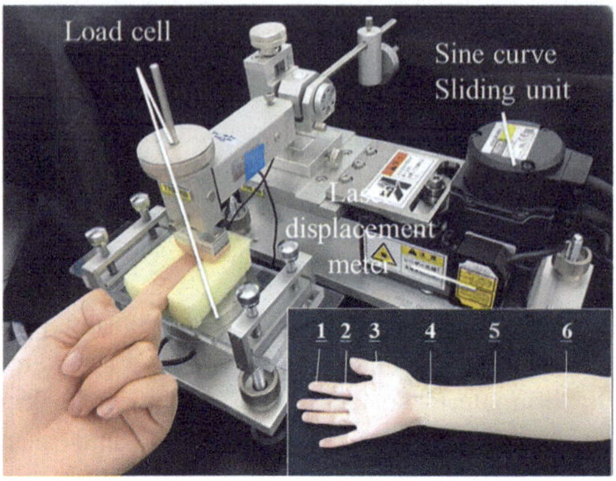


**Fig. 18.5** Biomimetic tactile-sensing system and the evaluation points 1–6 on the skin surface (inset image). (Reprinted from [34] under the Creative Commons Attribution 4.0 International license)

Figure 18.6 shows a typical friction profile when the skin was rubbed with the contact probe. The parameter, $T/T_0$, is the ratio of the round-trip to elapsed times. A stable pattern, wherein the friction profile on the return path was almost identical to that on the forward path, was obtained under all the conditions. However, the stick-slip phenomenon, wherein the friction force fluctuates periodically, can occur depending on the subject and site. Figure 18.6a shows a stable profile without the occurrence of stick-slip phenomena. The fingertip, 1, was rubbed at an angular velocity of 2.1 rad $s^{-1}$ with a vertical load of 0.98 N. The friction coefficient increased with the increase in the sliding velocity and reached 0.97 at $T/T_0 = 0.409$. A delay was observed in the frictional response to the movement of the contact probe. Furthermore, $\delta$ was normalized by dividing it by $T_0$, yielding 0.148. Figure 18.6b shows a typical profile with a stick-slip pattern. The palm, 3, was rubbed with a vertical load of 0.98 N at an angular velocity of 2.1 rad $s^{-1}$, and $\mu$ reached 0.73 at $T/T_0 = 0.373$ and oscillated with a width of approximately 0.5 ($\delta$ was 0.112).

Figure 18.7 shows $\mu$ and $\delta$. These parameters indicate that the long $\delta$ and occasional stick-slip phenomena are characteristics of skin-friction dynamics. Furthermore, $\mu$ at points 1–6 were 0.360 ± 0.085, 0.302 ± 0.061, 0.444 ± 0.112, 0.141 ± 0.048, 0.166 ± 0.053, and 0.233 ± 0.076, respectively. These values were smaller than that of the acrylic resin, S2 (1.511 ± 0.229), exhibiting high surface energy and frictional resistance. However, they were similar to or higher than the coefficients (0.134 ± 0.019) for the polytetrafluoroethylene resin, S3, exhibiting low surface energy and frictional resistance. Additionally, $\mu$ of the finger or palm points (1–3) exceeded those of the forearm points (4–6).

Furthermore, $\delta$ for points 1–6 were 0.138 ± 0.048, 0.119 ± 0.042, 0.122 ± 0.045, 0.202 ± 0.061, 0.188 ± 0.052, and 0.158 ± 0.056, respectively. These values were extremely large compared with those obtained with the soft materials. Additionally, the occurrence rates of stick-slip phenomena were 65%, 93%, 85%, 35%, 70%, and 80% for points 1–6, respectively; however, this phenomenon did not occur for the following materials: S1, S2, and S3.

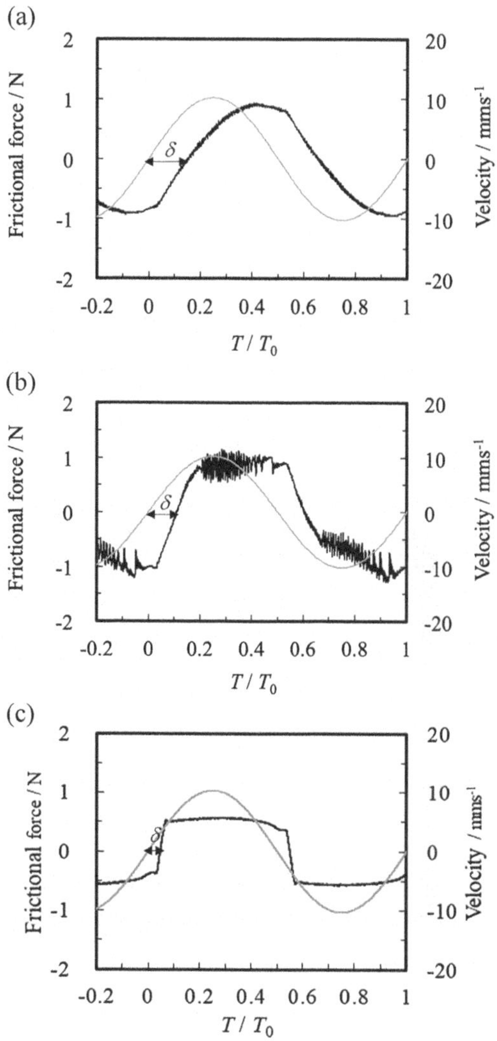


**Fig. 18.6** Temporal profiles of the friction force (black line) and velocity (gray line). (**a**) Stable pattern at 1, (**b**) stick-slip pattern at 3, and (**c**) artificial skin made of a urethane resin. (Reprinted from [34] under the Creative Commons Attribution 4.0 International license)

The effects of the water content of the skin stratum corneum and skin elasticity were examined. The moisture content of 1 (47.1 ± 21.7 (A.U.)) exceeded those at points 2–6 (26–32 (A.U.)). However, these values exceeded those of S1 (6.2 ± 2.1) and S2 and S3 (0 each). This indicated that the skin contains a larger amount of water than any general soft material and that the water content of the 1 was particularly significant. The elastic parameters, Ur/Uf, of 1–6 were 37.0% ± 12.4%, 43.2% ± 14.0%, 48.9% ± 13.1%, 69.8% ± 9.4%, 75.6% ± 7.6%, and 74.1% ± 13.3%, respectively. Regarding the definition of Ur/Uf, Uf and Ur are the skin-elongation and return-of-elongation degrees, respectively. Ur/Uf of S1 was 74.1% ± 13.3%, and those of S2 and S3 were not measured. Furthermore, Ur/Uf of 4–6 was greater

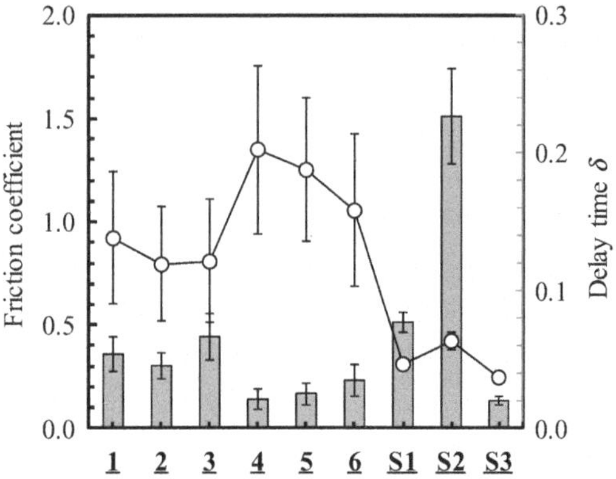


**Fig. 18.7** Friction parameters of skin surfaces 1–6 and materials S1–S3: average $\mu$ (gray bar) and $\delta$ (white circle). (Reprinted from [34])

than that of 1–3: The reversible return of deformation to mechanical stimulation was greater for locations 4–6.

The skin surface exhibited significant $\delta$. Here, $\delta$ of skin surfaces 1–6 were several times longer than those of conventional soft-matter surfaces (S1–S3). The author assumed that the large $\delta$ was caused by the characteristic deformation behavior of the human skin during the sliding process. The author observed that the contact probe significantly deformed the skin surface horizontally at the early stages of sliding because the soft and adhesive skin comprised a layered structure.

The stick-slip phenomenon occurred on the skin surface. As the skin exhibits viscoelasticity, the contact probe deforms and sinks into it. It is thought that stick-slip only occurs on the skin surface, as the contact probe sticks to the skin and is repeatedly peeled off. Finally, $\mu$ and $\delta$ of the finger/palm and forearm differed: $\mu$ on the finger/palm was 0.3–0.4 and larger than that on the forearm (0.1–0.2). Additionally, $\delta$ averaged 0.11–0.14 and 0.16–0.20 for the finger/palm and forearm, respectively; the value for the forearm exceeded that for the finger/palm, probably because the mechanical properties and deformation behaviors depend on the skin-surface location.

### *18.3.2 Effect of Moisturizers and Thickeners*

As mentioned in Sect. 18.2.3, water, moisturizers, and lubricants dramatically change the mechanical properties of the skin. Here, the effects of these liquids on the friction dynamics are presented. First, the friction characteristics of untreated human skin and the human skin moistened with water or a 10 wt% aqueous solution of glycerol were evaluated in 20 subjects using the biomimetic tactile-sensing system to demonstrate the effects of moisturization on nonlinear friction dynamics [35]. Here, the stick pattern, in which the contact probe and skin adhered, in addition to the stable and oscillation patterns, was observed. When the skin was rubbed immediately after the 10 wt% glycerol-aqueous solution treatment, a stick pattern was observed, in which extremely high $F$ values were obtained. The friction coefficient increased with increasing sliding velocity. At a maximum velocity of 10.3 mm $s^{-1}$, $\mu$ was 0.21, increasing during deceleration and reaching a maximum value of 1.67 at a velocity of 2.2 mm $s^{-1}$. The average $\mu$ and $\delta$ were 0.74 and 0.211 s, respectively, which were greater than those of the other friction patterns.

Here, the author proposed an expression mechanism for the three friction patterns. The mechanical energies generated by the friction and deformation modes on the human skin are represented as $E_1$ and $E_2$, respectively. The stable pattern occurred when the adhesion between the contact probe and skin was low, and the skin itself was barely deformed: $E_1 > E_2$, where the friction between the skin and contact probe was the dominant mechanical phenomenon. In the oscillation pattern, the elastic energy was stored by the adhesion between the contact probe and the skin. When the friction force exceeded the adhesion force, the adhesion was broken by the movement of the contact probe. At this time, $E_1$ dominated, and the stored elastic energy was dissipated on the skin surface, resulting in a friction pattern in which oscillation was expressed. Contrarily, a stick pattern was observed at a high adhesion force between the contact probe and skin surface, and the skin itself underwent significant deformation before sliding: $E_1 < E_2$, where the deformation energy of the skin dominated.

Next, the author described the effect of petrolatum or silicone on the friction dynamics. Skin moisturization using petrolatum and silicone oils decreased $\mu$ and $\delta$ [36]. When petrolatum or silicone oil was applied to the human skin, $\Delta\mu$ was $-0.19 \pm 0.16$ and $-0.26 \pm 0.16$, respectively, which were significantly lower than those of the untreated skin. Conversely, their $\Delta\delta$ values were $-0.000 \pm 0.030$ and $-0.030 \pm 0.035$, respectively, and were only significantly decreased in the silicone-oil-treated skin.

The lubricating effects of the oils accounted for these phenomena. This effect eases the breaking of the adhesion at the true $A$ between the skin and the contact probe and decreases the hysteresis and elastic energy stored by skin deformation. Skin moisturization using water or glycerol increased $\mu$ and $\delta$, as it caused swelling and softening of the stratum corneum, thereby enhancing the adhesion at the skin–contact probe true $A$. Nevertheless, improved moisturization was observed, although no corresponding change was observed in the skin's softness. These findings indicate that the lubricating influences of the petrolatum and silicone oils were stronger on the friction characteristics than their moisturizing impacts.

Why were the frictional properties of the oil-treated skin different from those of the water- or glycerol-treated ones? The moisturizing impact varied depending on whether the moisturizer was oil-soluble or water-soluble, as reflected by the conditions and mechanical characteristics of the moisturizer-treated human skin.

Figure 18.8 shows the water- or glycerol-treated human skin as well as the petrolatum- or silicone-oil-treated skin: Water and glycerol infiltrated the stratum corneum as well as the upper epidermis. Contrarily, the petrolatum and silicone oils remained on the skin surface (forming an oil film) or infiltrated the stratum corneum. Numerous studies have demonstrated such persistence on the skin surface as well as penetration of the stratum corneum interior. Egawa et al. used confocal

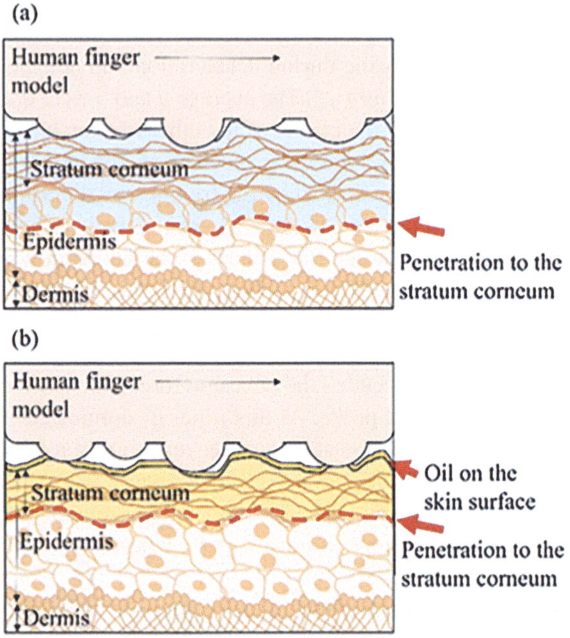


**Fig. 18.8** Surface condition of human skin treated with (**a**) water and a 10 wt% aqueous solution of glycerol and (**b**) petrolatum and silicone oils. (Reprinted from [36] under the Creative Commons Attribution 4.0 International license)

Raman spectroscopy to demonstrate how water molecules penetrate the skin surface and inner stratum corneum when water is applied to the human skin, thereby increasing the thickness of the stratum corneum [37]. Furthermore, they confirmed that glycerol infiltrated the upper epidermis, including the stratum corneum. Zhang and Richters et al. illustrated the incorporation of glycerol inside the stratum corneum using in vivo confocal Raman microspectroscopy [38, 39].

Finally, the author demonstrated the effect of a thickener aqueous solution. When the human skin was treated with this polymer aqueous solution, $\mu$ and $\delta$ were 0.69–0.99 and 0.171–0.179, respectively, immediately after treatment; these values exceeded those for untreated skin [40]. This increase was induced by the swelling and softening of the stratum corneum owing to the penetration of water in the polymer aqueous solution, increasing the skin–contact probe true $A$. Furthermore, $\mu$ of the skin exhibited a significant difference immediately after the treatment with different polymer aqueous solutions. Among the polymers, one with a low salt resistance and low yield stress exhibited the lowest $\mu$ owing to the collapse of the polymer-network structures via shearing and reduced viscosity due to the salt contents of human skin. The gel-cream-treated skin also exhibited a higher $\mu$ than the untreated skin immediately after treatment and after 90 min. This phenomenon was caused by the occlusive effect of the oil content of the gel cream.

## 18.4 Conclusion

The friction phenomena on the human skin surface were evaluated. The extant studies revealed that the mechanical response at skin surfaces follows a power law, rather than the Coulomb–Amontons law observed for general solid surfaces, and that contact with water significantly increases $F$ and amplifies oscillatory phenomena. The temporal profile of the tactile-sensing system under dynamic conditions revealed delayed mechanical response to the movement of the contact probe during sliding, and three types of responses (stable, vibration, and stick patterns) were observed depending on the water content of the stratum corneum. Moreover, a significant increase was observed in the occurrence probability of the stick pattern following the application of water or glycerin. These characteristics were attributable to the nonlinear mechanical response of soft skin.

These findings on the friction dynamics of the human skin will aid the assessment of skin health and enhance the efficacy of skin care. They will also ease the use of topical skin-care products and cosmetics. Moreover, the fact that many physicians and consumers could determine the physical health of a person by simply touching indicates that the skin-surface friction parameters provide a great deal of information.

**Acknowledgments** This study was supported by JSPS KAKENHI (JP22540417, JP16H01661, and JP23K28172), JST A-STEP AS251Z01095K, and the Yamagata University YU-COE (S) program.

## References

1. Sakamoto K, Lochhead RY, Maibach HI, Yamashita Y (2017) Cosmetic science and technology: theoretical principles and applications. Elsevier, Amsterdam
2. Abraira VE, Ginty DD (2013) The sensory neurons of touch. Neuron 79:618–639
3. Cua AB, Wilhelm KP, Maibach HI (1990) Elastic properties of human skin: relation to age, sex, and anatomical region. Arch Dermatol Res 282:283–288
4. Lévêque JL (1999) EEMCO guidance for the assessment of skin topography. J Eur Acad Dermatol Venereol 12:103–114
5. Gidado IM, Qassem M, Triantis IF, Kyriacou PA (2022) Review of advances in the measurement of skin hydration based on sensing of optical and electrical tissue properties. Sensors 22:7151
6. Okamoto S, Nagano H, Yamada Y (2013) Psychophysical dimensions of tactile perception of textures. IEEE Trans Haptics 6:81–93
7. Akanny E, Kohlmann C (2024) Predicting tactile sensory attributes of personal care emulsions based on instrumental characterizations: a review. Int J Cosmet Sci 46:1035–1063
8. Aita Y, Nonomura Y (2016) Friction and surface temperature of wet hair containing water, oil, or oil-in-water emulsion. J Oleo Sci 65:493–498
9. Nonomura Y, Kurita K, Kashimoto A, Hotta H, Suzuki T, Kaneko Y, Iitaka K (2004) The crystal structure and tribology of a novel organic crystalline powder: the mechanism of lubricity. Bull Chem Soc Jpn 75:2305–2308

10. Kawabata S, Niwa M (1991) Objective measurement of fabric mechanical property and quality. Int J Cloth Sci Technol 3:7–18
11. Delhaye BP, Schiltz F, Crevecoeur F, Thonnard JL, Lefèvre P (2024) Fast grip force adaptation to friction relies on localized fingerpad strains. Sci Adv 10:eadh9344
12. Acharya R, Challita EJ, Ilton M, Bhamla MS (2021) The ultrafast snap of a finger is mediated by skin friction. J R Soc Interface 18:20210672
13. Naylor PFD (1955) The skin surface and friction. Br J Dermatol 67:239–248
14. Zhang M, Mak AFT (1999) In vivo friction properties of human skin. Prosthetics Orthot Int 23:135–141
15. Comaish S, Bottoms E (1971) The skin and friction: deviations from Amonton's laws and the effects of hydration and lubrication. Br J Dermatol 184:37–43
16. Liu X, Lu Z, Lewis R, Carré MJ, Matcher SJ (2013) Feasibility of using optical coherence tomography to study the influence of skin structure on finger friction. Tribol Int 63:34–44
17. Zhou X, Mo JL, Li YY, Xu JY, Zhang X, Cai S, Jin ZM (2018) Correlation between tactile perception and tribological and dynamical properties for human finger under different sliding speeds. Tribol Int 123:286–295
18. Zhou X, Mo JL, Li YY, Xiang ZY, Yang D, Masen MA, Jin ZM (2020) Effect of finger sliding direction on tactile perception, friction and dynamics. Tribol Lett 68:85
19. Cua AB, Wilhelm KP, Maibach HI (1990) Frictional properties of human skin: relation to age, sex and anatomical region, stratum corneum hydration and transepidermal water loss. Br J Dermatol 123:473–479
20. Skedung L, El Rawadi C, Arvidsson M, Farcet C, Luengo GS, Breton L, Rutland MW (2018) Mechanisms of tactile sensory deterioration amongst the elderly. Sci Rep 8:5303
21. Diosa JG, Moreno R, Chica EL, Villarraga JA, Tepole AB (2021) Changes in the three dimensional microscale topography of human skin with aging impact its mechanical and tribological behavior. PLoS One 16:e0241533
22. Lodén M, Olsson H, Axéll T, Linde YW (2006) Friction, capacitance and transepidermal water loss (TEWL) in dry atopic and normal skin. Br J Dermatol 126:137–141
23. Chen CY, Yu CA, Hong TF, Chung YL, Li WL (2015) Contact and frictional properties of stratum corneum of human skin. Biosurf Biotribol 1:62–70
24. Adams MJ, Briscoe BJ, Johnson SA (2007) Friction and lubrication of human skin. Tribol Lett 26:239–253
25. Highley DR, Coomey M, Beste MD, Wolfram LJ (1977) Frictional properties of skin. J Invest Dermatol 69:303–305
26. Egawa M, Tagami H (2008) Comparison of the depth profiles of water and water-binding substances in the stratum corneum determined in vivo by Raman spectroscopy between the cheek and volar forearm skin: effects of age, seasonal changes and artificial forced hydration. Br J Dermatol 158:251–260
27. Nam S, Kuchenbecker KJ (2021) Optimizing a viscoelastic finite element model to represent the dry, natural, and moist human finger pressing on glass. IEEE Trans Haptics 14:303–309
28. Nacht S, Close JA, Yeung D, Ganset EH (1981) Skin friction coefficient: changes induced by skin hydration and emollient application and correlation with perceived skin feel. J Soc Cosmet Chem 32(2):55–65
29. Nonomura Y, Miura T, Miyashita T, Asao Y, Shirado H, Makino Y, Maeno T (2012) How to identify water from thickener aqueous solutions by touch. J R Soc Interface 9:1216–1223
30. Tanaka Y, Tiest WMB, Kappers AML, Sano A (2014) Contact force and scanning velocity during active roughness perception. PLoS One 9:e93363
31. Aita Y, Asanuma N, Takahashi A, Mayama H, Nonomura Y (2017) Nonlinear friction dynamics on polymer surface under accelerated movement. AIP Adv 7:045005
32. Aita Y, Asanuma N, Mayama H, Nonomura Y (2018) Sliding profile and energy at static friction between polymer surfaces. Chem Lett 47:767–769
33. Tsuchiya R, Kuhara R, Kikegawa K, Nagatani N, Nonomura Y (2020) Tactile and physical properties of cosmetic powders with a *shittori* feel. Bull Chem Soc Jpn 93:399–405

34. Nonomura Y, Ouchi H (2020) Friction dynamics on human skin surfaces. J Oleo Sci 69:461–465
35. Sakata Y, Mayama H, Nonomura Y (2022) Friction dynamics of moisturized human skin under nonlinear motion. Int J Cosmet Sci 44:20–29
36. Sakata Y, Mayama H, Nonomura Y (2024) Friction dynamics of human skin treated with oil under nonlinear motion. J Oleo Sci 73:177–186
37. Egawa M, Hirao T, Takahashi M (2007) In vivo estimation of stratum corneum thickness from water concentration profiles obtained with Raman spectroscopy. Acta Derm Venereol 87:4–8
38. Zhang SL, Caspers PJ, Puppels GJ (2005) In vivo confocal Raman microspectroscopy of the skin: effect of skin care products on molecular concentration depth-profiles. Microsc Microanal 11:790–791
39. Richters RJH, Falcone D, Uzunbajakava NE, Varghese B, Caspers PJ, Puppels GJ, Van Erp PEJ, Van de Kerkhof PCM (2017) Sensitive skin: assessment of the skin barrier using confocal Raman microspectroscopy. Skin Pharmacol Physiol 30:1–12
40. Kikuchi R, Sakata Y, Nonomura Y (2024) Friction dynamics of human skin treated using polymer aqueous solutions. J Oleo Sci 73:1015–1026

# Index

**A**
Accumulation layer, 148
Adsorption, 193–198, 200, 203, 207
Amphiphilic block copolymers, 223
Amphiphilic liquid crystalline polymers, 225
Amphiphilic polymers, 223
Anatase, 143
Antibody, 227, 229
Antioxidation, 314, 330
Atomic Force Microscopy (AFM), 340–345
Atom transfer radical polymerization (ATRP), 214
Attenuated total reflection (ATR), 82
Attenuated Total Reflection Surface-Enhanced Infrared Absorption Spectroscopy (ATR-SEIRAS), 181

**B**
Back-reactions, 177–179, 187
Band alignment, 160
Band bending, 143
Bilayer, 83
Bilayer fluidity, 250–254, 263
Biointerface
Biomembrane, 83
Bound water, 76
Bubble radius, 273
Bubbles, 267, 273, 274
Bulk water, 76

**C**
Capsules, 212
Carrier dynamics, 143
Carrier lifetime, 144
Carrier transfer, 181–184
Catalysts, 128–130
Catalytic activity, 222
Catanionic bilayer, 253, 261
Catanionic vesicles, 248–253, 257, 258, 263
Cell arrays, 51
Cell patterning, 233
Cells, 232
Cellulose nanofiber (CNF), 342
Charge carriers, 175, 176, 180
Cloud point, 92
Cocatalysts, 174–185
Concanavalin A (Con A), 219
Conduction band minimum (CBM), 151
Core–shell structures, 222
Critical Packing Parameter, 92
Cyclodextrin (CD), 221

**D**
Delay time, 147
Denatured state, 88
Depletion layer, 148
Diethylhexyl phthalate (DEHP), 278–286, 288, 289, 291


M. Hashizume, Y. Imura (eds.), *Colloids and Interfacial Dynamics*, Physical Chemistry in Action, https://doi.org/10.1007/978-981-96-6728-4

Diffusion-limited aggregation, 128, 129, 133, 134
Dynamic crosslinks, 220
Dynamics, 60, 61, 63, 65, 67, 70, 72, 73

**E**
Elastic modulus, 232
Electron-hole recombination, 144
Emulsion polymerization, 212

**F**
First hydration shell, 77
FM-AFM, 340
Foam, 267–274
Foam film, 267–269, 271, 272
Formation dynamics, 129, 133–135, 137–139
Free-standing films, 15
Free water, 76
Frequency modulation (FM) mode, 340
Friction, 336, 350
Functional surfaces, 99, 104, 111, 112, 117, 119

**G**
Glass transition, 190, 191, 206, 207
Glucose oxidase (GOx), 219
Glucose-responsive gel particles, 219
2-glucosyloxyethyl methacrylate, 220
Gold nanoparticles (AuNPs), 221
Grafting-from, 222
Grafting-to, 222

**H**
Hansen solubility parameters (HSPs), 303–305
HCO60, 284, 288, 289
Human skin, 349
Hydration
Hydration water, 76
Hydrogel layers, 227
Hydrogels, 227
Hydrogel scaffolds, 48
Hydrogen bond, 60, 65, 68, 70–72, 85
Hydrogen bond network, 77
Hydrogen evolution reaction (HER), 174
Hydrophobic hydration, 86, 90, 294–296, 302–307
Hypermobile water, 79

**I**
Ice-berg, 86, 90
In situ, 336
In situ monitoring, 180–184, 187
Interactions, 98–105, 108–110, 119
Interfaces, 60–63, 69–73
Interfacial gelation, 198
Ionic liquids, 341
Ion pair amphiphile, 248–251, 253, 258, 260, 261, 263

**L**
Lamellar, 83
Leaching, 282
Lectins, 229
Lipid peroxidation, 320–324, 327, 328, 330
Liquid crystalline–isotropic phase transition, 225
Liquid crystalline–isotropic phase transition temperature, 225
Liquid crystalline polymers, 225
Local anesthetics (LAs), 314
Lower critical solution temperature (LCST), 92, 215, 217
Lubrication, 336

**M**
Membrane fusion, 87
Metal nanoparticles, 221
Micelles, 212, 223
Micropatterning, 29, 34, 36
Moisturizer, 353
Molecular binding sites, 228
Molecular dynamics simulations, 85
Molecular imprinting, 220, 228, 229
Molecular loading, 15
Molecular permeation, 16, 22
Molecularly stimuli-responsive polymer particles, 219

**N**
Nanocrystals, 128–131, 133–137, 139
Nanoimprint lithography, 230
Nanospheres, 98–101, 104–112, 114, 115, 117, 119, 120
Nanowires, 128–138

Native state, 88
Neutron quasi-elastic scattering, 85
Neutron reflectivity (NR), 198, 200

**O**
Operando, 336
Organic–inorganic hybrids, 221
Osmolytes, 90
Oxygen evolution reaction (OER), 174
Oxygen radical absorption capacity (ORAC) assay, 315, 316

**P**
Particles, 212
Partition coefficient, 295, 305, 307
PEGylation
Phenylboronic acid (PBA), 219
Phospholipid, 76
Photocatalysis, 142
Photocatalysts, 173–175, 179–182
Photocatalytic water splitting, 172–180, 185
Photochromism, 5–7, 9
Photodimerizable group, 234
Photodimerization, 231
Photolithography, 230
Photo-responsive polymer particles, 216, 218
Photo-responsive polymers, 215
pH-responsive polymer particles, 216
pH-responsive polymers, 215
Plasticizer, 278, 279, 282, 291
Poly(dimethylsiloxane) (PDMS), 231
Poly(ethylene glycol) (PEG), 229
Polyion complexes, 14, 24
Polymer, 189–193, 195–199, 202–204, 207
Polymer thin films, 191, 203
Poly(2-methacryloyloxyethyl phosphorylcholine) (PMPC), 217, 229
Poly(*N*-isopropylacrylamide) (PNIPAAm), 215
Polysaccharides, 14–16, 18, 19, 21, 23, 24
Polyvinyl chloride (PVC), 278–282, 284, 285, 288, 289, 291
Potential gradient, 148
Precipitation polymerization, 212, 217

**Q**
Quartz crystal microbalance (QCM), 192, 195

**R**
Radical scavengers, 314, 321, 330
RAFT polymerization, 217
Raman spectroscopy, 338–340
Reversible addition-fragmentation chain transfer (RAFT), 214
Rutile, 142
Rutile/anatase boundary, 160

**S**
Second hydration shell, 77
Self-assembled structure, 76
Self-assembly, 98–112, 114–120
Silver nanoparticles (AgNPs), 221
Sinusoidal motion, 353
Small-angle neutron scattering (SANS), 268, 269
Soft contact lenses, 343–344
Sol–gel transition, 234
Space-charge region (SCR), 148
Specific surface area, 268, 273
Spring and parachute approaches (SPAs), 300, 301
Stimuli-responsiveness
Stimuli-responsive polymer materials, 212
Stimuli-responsive polymer particles, 215
Stimuli-responsive polymer self-assemblies, 223
Sum frequency generation spectroscopy (SFG), 338
Surface-initiated atom transfer radical polymerization (SI-ATRP), 221, 227, 228
Surface patterning, 230
Surface photovoltage (SPV), 148
Surface plasmon resonance (SPR), 227
Surfactant, 2, 3, 5, 6, 10, 279, 284, 286, 288, 291
Synchrotron radiation, 145
Synergistic effect, 144

**T**
TBARS assay, 322, 323
Temperature-responsive polymer particles, 216
Temperature-responsive polymers, 215
Terahertz spectroscopy
Thermal denaturation, 88
Thickener, 360
3D culture, 28, 29, 42
Time-domain spectroscopy, 81

Time-resolved X-ray photoelectron spectroscopy (TR-XPS), 143
Titanium dioxide ($TiO_2$), 142
Tribofilm, 342
Tribology, 336
Tween 80, 283–288

**U**
Unfreezable (non-freezing) water, 76, 88
Upper critical solution temperature (UCST), 215, 217

**V**
Valence band maximum (VBM), 156
van't Hoff plot, 296, 301, 302
Vesicles, 212, 223
Vesicular structures, 248–251
Viscoelasticity, 190, 191, 195–197, 203, 207

**W**
Water, 60, 61, 65, 67–73
Water interfaces, 60
Water's dynamics, 76
Weakly affected water
Wear, 336

**X**
X-ray absorption fine structure (XAFS), 180
X-ray absorption spectroscopy (XAS), 160
X-ray photoelectron spectroscopy (XPS), 143

The manufacturer's authorised representative in the EU is Springer
Nature Customer Service Centre GmbH, Europaplatz 3, 69115 Heidelberg,
Germany. If you have any concerns regarding our products, please
contact ProductSafety@springernature.com

Printed and bound by CPI Group (UK) Ltd, Croydon, CR0 4YY
01/10/2026
02225923-0016